Biofertilizers for Sustainable Soil Management

The alkaline calcareous nature, high pH, salinity, heavy metals pollution, and low organic matter content of soils in many parts of the world have diminished the soil fertility and made essential nutrients unavailable to crops. To cope with the poor availability of soil nutrients, improve soil health, and feed the fast-growing global population, the farming community is using millions of tons of expensive chemical fertilizers in their fields to maintain an adequate level of nutrients for crop sustainability as well as to ensure food security. In this scenario, the exploitation of biofertilizers has become of paramount importance in the agricultural sector for their potential role in food safety and sustainable crop production. Bearing in mind the key importance of biofertilizers, this book examines the role of biofertilizers in sustainable management of soil and plant health under different conditions of the changing climate. Finally, it provides a platform for scientists and academicians all over the world to promote, share, and discuss various new issues, developments, and limitations in biofertilizers, crops, and beneficial microbes.

Salient Features:

- Mainly focuses on the role of biofertilizers in managing soils for improving crop and vegetable yields as a substitute for chemical fertilizers.
- Highlights the valuable information for the mechanism of action, factors affecting, and limitations of biofertilizers in the wider ecosystem.
- Presents a diversity of techniques used across plant science.
- Designed to cater to the needs of researchers, technologists, policy makers, and undergraduates and postgraduates studying in the fields of organic agriculture, soil microbiology, soil biology, soil fertility, and fertilizers.
- Addresses plant responses to biofertilizers.

Biofertilizers for Sustainable Soil Management

Edited by

Shah Fahad, Shah Saud, Fazli Wahid, and Muhammad Adnan

CRC Press is an imprint of the
Taylor & Francis Group, an informa business

First edition published 2024
by CRC Press
6000 Broken Sound Parkway NW, Suite 300, Boca Raton, FL 33487-2742

and by CRC Press
4 Park Square, Milton Park, Abingdon, Oxon, OX14 4RN

CRC Press is an imprint of Taylor & Francis Group, LLC

Library of Congress Cataloging-in-Publication Data
Names: Fahad, Shah (Assistant professor of agriculture), editor. | Saud, Shah, editor. |
Wahid, Fazli (Soil scientist), editor. | Adnan, Muhammad (Lecturer in agriculture), editor.
Title: Biofertilizers for sustainable soil management / edited by Shah Fahad, Shah Saud,
Fazli Wahid, and Muhammad Adnan.
Description: First edition. | Boca Raton, FL : CRC Press, 2024. |
Includes bibliographical references and index.
Identifiers: LCCN 2023012718 (print) | LCCN 2023012719 (ebook) | ISBN 9781032260419 (hardback) |
ISBN 9781032260433 (paperback) | ISBN 9781003286233 (ebook)
Subjects: LCSH: Biofertilizers. | Sustainable agriculture. | Soil management.
Classification: LCC S654.5 .B567 2024 (print) | LCC S654.5 (ebook) |
DDC 631.8/6–dc23/eng/20230530
LC record available at https://lccn.loc.gov/2023012718
LC ebook record available at https://lccn.loc.gov/2023012719

ISBN: 978-1-032-26041-9 (hbk)
ISBN: 978-1-032-26043-3 (pbk)
ISBN: 978-1-003-28623-3 (ebk)

DOI: 10.1201/9781003286233

Typeset in Times
by Newgen Publishing UK

Contents

Editor Biographies

Shah Fahad works as Assistant Professor in the Department of Agronomy, Abdul Wali Khan University Mardan, Khyber Pakhtunkhwa, Pakistan. He obtained his PhD in Agronomy from Huazhong Agriculture University, China, in 2015. After completing his postdoctoral research in Agronomy at the Huazhong Agriculture University (2015–17), he accepted the position of Assistant Professor at the University of Haripur. He has published over 430 peer-reviewed papers (impact factor 1834.76) with more than 370 research and 60 review articles, on important aspects of climate change, plant physiology and breeding, plant nutrition, plant stress responses and tolerance mechanisms, and exogenous chemical priming-induced abiotic stress tolerance. He has also contributed 80 book chapters to various book editions published by Springer, Wiley-Blackwell, and Elsevier. He has edited 17 book volumes, including this one, published by CRC Press, Springer, and Intech Open. He won the Young Rice International Scientist award and Distinguished Scholar award in 2014 and 2015, respectively. He has won 17 projects from international and national donor agencies. Dr Shah Fahad figured in the Clarivate's 2021 and 2022 lists of Highly Cited Researchers in the field of Plant and Agriculture sciences. His name is included among the top two percent of scientists in a global list compiled by Stanford University, USA. He has worked and is presently continuing to work on a wide range of topics, including climate change, greenhouse gas emissions, abiotic stresses tolerance, roles of phytohormones and their interactions in abiotic stress responses, heavy metals, and the regulation of nutrient transport processes.

Shah Saud received his PhD and Post Doctorate in Turf grasses (Horticulture) from Northeast Agricultural University, Harbin, China. He is currently working as Assistant Professor in the College of Life Science, Linyi University, Linyi, Shandong, 276000, China. Dr. Shah Saud has published over 200 research publications in peer-reviewed journals. He has edited six books and written 40 book chapters on important aspects of plant physiology, plant stress responses, and environmental problems in relation to agricultural plants. According to Scopus®, Dr. Shah Saud's publications have received roughly 8566 citations with an h-index of 44.

Fazli Wahid completed his PhD in the field of soil microbiology and plant nutrition at the University of Agriculture Peshawar and University of Natural Resources and Life Sciences, Vienna, Austria. He also completed his postdoc at Nigde University, Turkey, in plant nutrition. Currently, he is working as Assistant Professor in the Department of Agriculuture, University of Swabi. His research work is mainly focused on the isolation and purification of beneficial microbes, and the production of biofertilizers to improve soil and plant health. He has isolated several bacterial strains for solubilization of P from rock phosphate in high-pH soils. He has published 70 research articles in well-reputed journals with 10 book chapters, while he has two books in progress as editor. He is also working as a potential reviewer for *Chemosphere* and *Pedospher*. He has won one international and two national research projects and organized several international and national conferences and seminars at the University of Swabi.

Muhammad Adnan is a lecturer at the Department of Agriculture at the University of Swabi (UOS), Pakistan. He completed his PhD (soil fertility and microbiology) at the Department of Soil and Environmental Sciences (SES), the University of Agriculture Peshawar, Pakistan, and the Department of Plant, Soil and Microbial Sciences, Michigan State University, USA. He received his MSc and BSc (Hons) in Soil and Environmental Sciences, from the Department of SES at the University of Agriculture, Peshawar, Pakistan.

Contributors

Asad Abbas
School of Horticulture, Anhui Agricultural University, Hefei, China

Zainul Abideen
Institute of Sustainable Halophyte Utilization, University of Karachi, Sindh, Pakistan

Muhamad Adnan
Department of Agriculture, University of Swabi, Pakistan

Syed Muhammad Afzal
Center of Biotechnology and Microbiology, University of Peshawar, KP, Pakistan

Rafiq Ahmad
Department of Microbiology, The University of Haripur, KP, Pakistan

Shakeel Ahmad
Guangxi Colleges and Universities, key laboratory of Crop Cultivation and Tillage, National Demonstration Center for Experimental Plant Science Education, Agricultural College of Guangxi University, Nanning, China

Adeel Ahmad
Institute of Soil and Environmental Sciences, University of Agriculture, Faisalabad, Pakistan

Nazeer Ahmed
Department of Agriculture, University of Swabi, Khyber Pakhtunkhwa, Pakistan

Saeed Ahmed
Agricultural Research Center, Londrina State University, Londrina, Brazil

Waseem Ahmed
Department of Horticulture, The University of Haripur, Khyber Pakhtunkhwa, Pakistan

Muhammad Akmal
Institute of Soil & Environmental Sciences, PMAS-Arid Agriculture University, Rawalpindi, Pakistan

Mukhtar Alam
Department of Agriculture, University of Swabi, Swabi, Pakistan

Izhar Ali
College of Agriculture, Guangxi University, Nanning, China

Mazhar Ali
Department of Environmental Sciences, COMSATS University, Islamabad, Punjab, Pakistan

Abdel Rahman Altawaha
Department of Biological Sciences Al Hussein Bin Talal University, Ma'an, Jordan

Amanullah
Department of Agronomy, the University of Agriculture Peshawar, Khyber Pakhtunkhwa, Pakistan

Muhammad Arif
Department of Agronomy, the University of Agriculture Peshawar, Khyber Pakhtunkhwa, Pakistan

Umair Ashraf
Department of Botany, Division of Science and Technology, University of Education, Lahore, Punjab, Pakistan

Sofia Asif
Department of Food Science and Technology, The University of Haripur, Khyber Pakhtunkhwa, Pakistan

Atique-ur-Rehman
Department of Agronomy, Bahauddin Zakariya University, Multan, Punjab, Pakistan

Muhammad Azeem
Institute of Soil & Environmental Sciences, PMAS-Arid Agriculture University, Rawalpindi, Pakistan

Amir Aziz
Department of Soil and Environmental Sciences, College of Agriculture, University of Sargodha, Sargodha, Pakistan

Muhammad Adnan Bukhari
Department of Agronomy, Faculty of Agriculture and Environment, The Islamia University of Bahawalpur, Bahawalpur, Punjab, Pakistan

Koushik Chakraborty
Department of Crop Physiology & Biochemistry, ICAR-National Rice Research Institute Bidyadharpur, Cuttack, Odisha, India

Ayman EL Sabagh
Department of Agronomy, Faculty of Agriculture, University of Kafrel-Sheikh, Kafrel-Sheikh, Egypt

Shah Fahad
Department of Agronomy, Abdul Wali Khan University Mardan, Khyber Pakhtunkhwa, Pakistan

Bushra Gul
Department of Biosciences, University of Wah, Punjab, Pakistan

Farhana Gul
Department of Botany, Government Girls Degree College Dargai, Malakand, Khyber Pakhtunkhwa, Pakistan

Ali Raza Gurmani
Department of Soil & Climate Sciences, The University of Haripur, Khyber Pakhtunkhwa, Pakistan

Ishfaq Hameed
Department of Botany, University of Chitral, Khyber Pakhtunkhwa, Pakistan

Muhammad Haroon
Department of Agriculture, University of Swabi, Khyber Pakhtunkhwa, Pakistan

Shah Hassan
Department of Agricultural Extension Education and Communication, The University of Agriculture, Peshawar, Pakistan

Muhammad Umair Hassan
Research Center on Ecological Sciences, Jiangxi Agricultural University, Nanchang, China

Muhammad Hayat
State Key Laboratory of Microbial Technology, Institute of Microbial Technology, Shandong University Qingdao, China

Qaiser Hussain
Institute of Soil & Environmental Sciences, PMAS-Arid Agriculture University, Rawalpindi, Pakistan

Muhammad Zahid Ihsan
The Cholistan Institute of Desert Studies, Faculty of Agriculture and Environment Sciences, The Islamia University of Bahawalpur, Bahawalpur, Punjab, Pakistan

Shahzada Sohail Ijaz
Institute of Soil & Environmental Sciences, PMAS-Arid Agriculture University, Rawalpindi, Pakistan

Anas Iqbal
College of Agriculture, Guangxi University, Nanning, China

Kinza Iqbal
Institute of Soil & Environmental Sciences, PMAS-Arid Agriculture University, Rawalpindi, Pakistan

Muhammad Irfan
School of Environmental Science and Engineering, Tianjin University, Tianjin, PR China

Ayesha Jabeen
Department of Soil & Climate Sciences, The University of Haripur, Khyber Pakhtunkhwa, Pakistan

Muhammad Jamil
Soil and Water Testing Laboratory, Sahiwal, Pakistan

Hafiz Muhammad Rashad Javeed
Department of Environmental Sciences, COMSATS University, Islamabad, Punjab, Pakistan

Muhammad Jawad
Department of Soil & Climate Sciences, The University of Haripur, Khyber Pakhtunkhwa, Pakistan

Servat Jehan
Institute of Soil & Environmental Sciences, PMAS-Arid Agriculture University, Rawalpindi, Pakistan

Ligeng Jiang
Department of Botany, University of Chitral, Khyber Pakhtunkhwa, Pakistan

Muhammad Junaid
Graduate School of Life and Environmental Sciences, University of Tsukuba, Tsukuba, Ibaraki, Japan

Mehwish Kanwal
Ministry of National Food Security & Research, Pakistan Tobacco Board, Peshawar, Khyber Pakhtunkhwa, Pakistan

Samina Khalid
Department of Environmental Sciences, COMSATS University Islamabad, Punjab, Pakistan

Rabia Khalid
Institute of Soil & Environmental Sciences, PMAS-Arid Agriculture University, Rawalpindi, Pakistan

Sohail Khan
College of Life Science and Technology, State Key Laboratory for Conservation and Utilization of Subtropical Agro-bioresources, Guangxi University, Nanning, Guangxi, China

Jallat Khan
Department of Chemistry, Khwaja Fareed University of Engineering & Information Technology, Rahim Yar Khan, Punjab, Pakistan

Nazish Huma Khan
Department of Environmental Sciences, University of Swabi, Pakistan

Muhammad Sufyan Khan
Department of Environmental Sciences, University of Swabi, Pakistan

Khalid Saifullah Khan
Institute of Soil & Environmental Sciences, PMAS-Arid Agriculture University, Rawalpindi, Pakistan

Sammina Mahmood
Department of Botany, Division of Science and Technology, University of Education, Lahore, Punjab, Pakistan

Sajid Masood
Institute of Soil & Environmental Sciences, PMAS-Arid Agriculture University, Rawalpindi, Pakistan

Muhammad Mubeen
Department of Environmental Sciences, COMSATS University Islamabad, Punjab, Pakistan

Javed Muhammad
Department of Microbiology, University of Haripur, Khyber Pakhtunkhwa, Pakistan

Sahar Mumtaz
Division of Science and Technology, Department of Botany University of Education, Lahore, Pakistan

Jamal Nasar
Guangxi Colleges and Universities Key Laboratory of Crop Cultivation and Tillage, Agricultural College of Guangxi University, Nanning, China

Muhammad Nawaz
Department of Agricultural Engineering, Khwaja Fareed University of Engineering and Information Technology, Rahim Yar Khan, Punjab, Pakistan

Fahim Nawaz
Department of Agronomy, Muhammad Nawaz Shareef University of Agriculture, Multan, Punjab, Pakistan

Taufiq Nawaz
Department of Food Science and Technology, The University of Agriculture, Peshawar, Pakistan

Rainaz Parvez
Department of Botany, Government Girls Degree College Dargai, Malakand, Khyber Pakhtunkhwa, Pakistan

Muther Mansoor Qaisrani
Department of Bioscience & Technology Khwaja Fareed University of Engineering and Information Technology, Rohm Yar Khan, Punjab, Pakistan

Rafi Qamar
Department of Agronomy, College of Agriculture, University of Sargodha, Sargodha, Punjab, Pakistan

Abdul Qayyum
Department of Agronomy, The University of Haripur, Khyber Pakhtunkhwa, Pakistan

Mazhar Rafique
Department of Soil & Climate Sciences, The University of Haripur, Khyber Pakhtunkhwa, Pakistan

Faiza Rawish
Superior College of Commerce Sat Sira Chowk, Mandi Bahauddin, Punjab, Pakistan

Muhammad Romman
College of Agriculture, Guangxi University, Nanning, China

Muhammad Saeed
Department of Agriculture, University of Swabi, Anbar, Swabi, Khyber Pakhtunkhwa, Pakistan

Muhammad Saeed
Department of Weed Science and Botany, University of Agriculture, Peshawar, Pakistan

Tooba Saeed
National Centre of Excellence in Physical Chemistry, University of Peshawar, Pakistan

Muhammad Hamzah Saleem
MOA Key Laboratory of Crop Ecophysiology and Farming System in the Middle Reaches of the Yangtze River, College of Plant Science and Technology, Huazhong, China

Adeel Sattar
Department of Pharmacology and Toxicology, University of Veterinary and Animal Sciences, Lahore, Pakistan

Shah Saud
College of Life Science, Linyi University, Linyi, Shandong, 276000, China

Muhammad Izhar Shafi
Department of Soil and Environmental Sciences, Faculty of Crop Production Sciences, The University of Agriculture, Peshawar, Pakistan

Adnan Noor Shah
Department of Agricultural Engineering, Khwaja Fareed University of Engineering and Information Technology, Rahim Yar Khan, Punjab, Pakistan

Muhammad Shahab
Department of Agriculture, University of Swabi, Anbar, Swabi, Khyber Pakhtunkhwa, Pakistan

Muhammad Shahzad
Department of Agronomy, University of Poonch Rawalakot, AJK, Pakistan

Fazli Subhan
Department of Agriculture, University of Swabi, Anbar, Swabi, Khyber Pakhtunkhwa, Pakistan

Adnan Hassan Tahir
Department of Clinical Science, Faculty of Veterinary and Animal Sciences, PMAS-Arid Agriculture University, Rawalpindi, Pakistan

Mukkram Ali Tahir
Department of Soil and Environmental Sciences, College of Agriculture, University of Sargodha, Sargodha, Pakistan

Rafi Ullah
Department of Agriculture, University of Swabi, Khyber Pakhtunkhwa, Pakistan

Hidayat Ullah
Department of Agriculture, University of Swabi, Swabi, Pakistan

Zia Ur Rehman
Department of Agriculture, University of Swabi, Swabi, Pakistan

Fazli Wahid
Department of Agriculture, University of Swabi, Anbar, Swabi, Khyber Pakhtunkhwa, Pakistan

Waqar-un-Nisa
Department of Soil & Climate Sciences, The University of Haripur, Khyber Pakhtunkhwa, Pakistan

Humaira Yasmin
Department of Biosciences, COMSATS University Islamabad, Islamabad, Pakistan

Noman Younas
Institute of Soil and Environmental Sciences, University of Agriculture, Faisalabad

Maid Zaman
Department of Entomology, The University of Haripur, Khyber Pakhtunkhwa, Pakistan

Muhammad Zamin
Department of Agriculture, University of Swabi, Anbar, Swabi, Khyber Pakhtunkhwa, Pakistan

Muhammad Shahid Ibni Zamir
Department of Agronomy, University of Agriculture, Faisalabad, Punjab, Pakistan

Muhammad Zia-Ur-Rehman
Institute of Soil and Environmental Sciences, University of Agriculture, Faisalabad, Punjab, Pakistan

Fazli Zuljalal
Department of Environmental Sciences, University of Swabi, Pakistan

Acknowledgements

Words are bounded and knowledge is limited in praising ALLAH, the Instant and Sustaining Source of all Mercy and Kindness, and the Sustainer of the Worlds. My greatest and ultimate gratitude is due to ALLAH (Subhanahu wa Taqadus). I thank ALLAH with all my humility for everything that I can think of. His generous blessing and exaltation succeeded my thoughts and caused my ambitions to thrive and to bare the cherished fruit of my modest efforts in the form of this piece of literature from the blooming spring of blossoming knowledge. May ALLAH forgive my failings and weaknesses, strengthen and enliven my faith in HIM, and endow me with knowledge and wisdom. All praise and respect are for Holy Prophet Muhammad Salle Allah Alleh Wassalam, the greatest educator, the everlasting source of guidance and knowledge for humanity. He taught the principles of morality and eternal values and enabled us to recognize our Creator. I have a deep sense of obligation to my parents, brothers, sisters, and son. Their unconditional love, care, and confidence in my abilities helped me achieve this milestone in my life. For this and much more, I am forever in their debt. It is to them that I dedicate this book. In this arduous time, I also appreciate the patience and serenity of my wife, who brought joy to my life in so many different ways. It is indeed on account of her affections and prayers that I was able to achieve something in my life.

Shah Fahad

1 Endophytic Microbes

Deciphering Their Role and Mechanism to Mitigate Abiotic Stresses in Plants as an Eco-friendly Tool

Sammina Mahmood,[*1] *Adeel Sattar,*[2] *Umair Ashraf,*[1]
Adnan Hassan Tahir,[3] *and Muhammad Zia-Ur-Rehman*[4]
[1] Department of Botany, Division of Science and Technology, University of Education, Lahore, Punjab, Pakistan
[2] Department of Pharmacology and Toxicology, University of Veterinary and Animal Sciences, Lahore, Pakistan
[3] Department of Clinical Science, Faculty of Veterinary and Animal Sciences, PMAS-Arid Agriculture University, Rawalpindi, Pakistan
[4] Institute of Soil and Environmental Sciences, University of Agriculture, Faisalabad, Punjab, Pakistan
* Correspondence: sammina.mahmood@ue.edu.pk

CONTENTS

1.1 INTRODUCTION

Under natural conditions, plants are continuously exposed to a number of biotic and abiotic challenges which pose a threat to plant productivity. Particularly, abiotic stresses like drought, salinity, heavy metal stress etc. are more severe than biotic threats as these change the physiochemical properties of soil and hence the above- and belowground ecology of the plant. The severity of and long exposure to stress induces toxicity at the physiological, biochemical, and molecular levels of the plant (Ventura-Lima et al. 2011; Al-Zahrani et al. 2022; Rajesh et al. 2022; Anam et al. 2021; Deepranjan et al. 2021; Haider et al. 2021; Amjad et al. 2021). Mostly, physiochemical strategies are applied to combat abiotic stresses, but these contribute to ecological toxicity. Several bacterial and fungal strains have been discovered which upon exposure to stress situations undergo structural alterations and secrete certain compounds such as organic acid, humic acid, polysaccharides,

DOI: 10.1201/9781003286233-1

proteins, etc., which in turn interact with the soil conditions and contaminants in an eco-friendly manner. This interaction greatly affects the physical and chemical properties of soil and helps to combat the toxic effects by adopting differential strategies based on the types of stresses imposed on the plant (Bruins et al. 2000; Etesami 2018; Sajjad et al. 2021a,b; Fakhre et al. 2021; Khatun et al. 2021; Ibrar et al. 2021; Bukhari et al. 2021; Haoliang et al. (2022); Sana et al. 2022; Abid et al. 2021). Under drought stress, plants adopt a strategy to produce ACC deaminase, IAA, phosphate solubilization, exopolysaccharide synthesis (EPS), growth on N-free culture, osmolyte accumulation, e.g. HCN, in multiple crop species (Kumar et al. 2016; Mariotti et al. 2021; Niu et al. 2018; Vettori et al. 2010; Zaman et al. 2021; Sajjad et al. 2021a,b; Rehana et al. 2021; Yang et al. 2022; Ahmad et al. 2022; Shah et al. 2022; Muhammad et al. 2022; Wiqar et al. 2022). The bacterial cells and their metabolites promote plant growth under water stress conditions. Plants infected with inoculants were found to contain higher amounts of chlorophyll a, b, total chlorophyll contents, carotenoids, and stress-related phytohormones like ABA, JA, and SA (Heidari et al. 2011; Mariotti et al. 2021). The modified level of enzymes under salinity stress has also been found to have a positive modulating effect on plant growth and development. However, the rate of this effect depends upon the severity of the stress, inoculated bacteria used, growth stage of the plant, and the type and concentration of enzyme produced (Ilangumaran & Smith 2017; Khan et al. 2017; Sun et al. 2020; Farhat et al. 2022; Niaz et al. 2022; Ihsan et al. 2022; Chao et al. 2022; Qin et al. 2022; Xue et al. 2022; Ali et al. 2022; Mehmood et al. 2022; El Sabagh et al. 2022; Ibad et al. 2022). Under a heavy metal stress situation, microbes use strategies to adsorb, physical entrapment on the cell surface, and a reduction of the uptake of metals by plants by altering the cellular architecture. Cellular architecture modulations include reduced uptake of metal ions, chelation with other organic compounds, and efflux pump activation in the case of huge uptake (Mazhar et al. 2020; Wei et al. 2018; Zhang et al. 2019; Deepranjan et al. 2021; Haider et al. 2021; Huang Li et al.2021; Ikram et al. 2021; Jabborova et al. 2021; Khadim et al. 2021a,b; Manzer et al. 2021; Muzammal et al. 2021; Abdul et al. 2021a,b; Ashfaq et al. 2021; Amjad et al. 2021; Atif et al. 2021; Athar et al. 2021). Each microbe has its own unique evolutionary route of emergence, adaptation, and significance under a range of environmental conditions. Local ecological conditions also influence the establishment and development of particular microbial communities in an ecological zone. This ecology dominance not only intensifies the range of microbiota but also diversifies their role in sustainable agriculture (Compant et al. 2019; Lawson et al. 2019; Adnan et al. 2018a,b; Adnan et al. 2019; Akram et al. 2018a,b; Aziz et al. 2017a,b; Chang et al. 2021; Chen et al. 2021; Emre et al. 2021; Habib et al. 2017; Hafiz et al. 2016; Hafiz et al. 2019; Ghulam et al. 2021; Guofu et al. 2021; Hafeez et al. 2021; Khan et al. 2021; Kamaran et al. 2017; Muhmmad et al. 2020; Safi et al. 2021).

In this chapter, we aim to disentangle the comprehensive details of microbiota, their mechanistic relationship with soil ecology and plant growth, and their potential use under the most pronounced abiotic stresses, i.e. drought, salinity, and heavy metal stress situations as plant growth-promoting/modulating tools. In this respect, microbe community architecture and their interactions in the root are also discussed in this chapter.

1.2 ECOLOGY OF THE MICROBIOME

An inherently enriched gene pool of microbes called the microbiome, which includes mycobiomes and bacteria, has been described by the several researchers and reviewers. Their enormous richness has been attributed to the natural actions and habitats of microbes as they rarely occur as a single species in a single environment. Multiple sites, as well as a huge range of experimental areas (soil, plants, animals, sediments, etc.), have shown that great variation exists, even in a small sample, which is attributed to their hypermutability that accelerates their evolution. Their co-occurrence in an environment leads to co-evolution and adaptation, resulting in the establishment of a large variety of relationships. Untapping this relationship has revealed two major levels of interaction: (1)

interactions among the organisms involved, e.g. bacteria–bacteria, bacteria–fungus, bacteria–plant, bacteria–fungus–plant, fungus–fungus, and fungus–plant (van Elsas et al. 2012), and (2) based on the microbial interaction, the type of relationship/mechanism (e.g. epi/endo symbiosis, mutualistic, pathogenic, parasitic, competitive, antagonistic, etc.) that facilitates the cohabitation and coevolution (Faust & Raes 2012).

Each interaction and mechanism has its own unique evolutionary route of emergence, adaptation, and significance under a range of environmental conditions (Sajjad et al. 2019; Saud et al. 2013; Saud et al. 2014; Saud et al. 2017; Saud et al. 2016; Shah et al. 2013; Saud et al. 2020; Saud et al. 2022a,b; Qamar et al. 2017; Hamza et al. 2021; Irfan et al. 2021;Wajid et al. 2017; Yang et al. 2017; Zahida et al. 2017; Depeng et al. 2018). The establishment of new species, called "invaders," clearly shows the development and adaptation pattern in a new environment, that could affect the local microbial community of a particular niche either positively or negatively. This dominance of ecology not only intensifies the range of microbiota but also diversifies their role in sustainable agriculture (Compant et al. 2019; Lawson et al. 2019). The recruitment of plant microbiota, either bacteria or fungi, to the plant host needs extensive physiological, metabolic, and genome reprogramming that shapes the natural evolution of microbiota. The particular level of reprogramming recapitulates the shift between free living and symbiotic lives of an organism under the selection pressure of the host plant (Phelan et al. 2012; Nadal Jimenez et al. 2012; Semchenko et al. 2018; Rodriguez et al. 2009; Genre et al. 2020).

Based on the mode of occurrence and association, plant-associated microbiota are mainly categorized into two classes: rhizospheric (present in soil) and endophytic (present inside the host). However, bacteria (bacteriota) have many more applications and success stories relevant to plant stress mitigation compared to fungi (mycobiota), owing to the size and complexity of genomes. Indeed, the lack of a reference genome, complexity of life cycle and somatic structures, and limited genetic transformation ability hampered its potential research and put mycobiota far behind bacteriota ((Rodriguez et al. 2009; Naranjo-Ortiz & Gabaldón 2020).

1.3 REFERENCE ENVIRONMENT FOR PLANT–MICROBE INTERACTIONS

Plants, being biologically intact organisms, are directly dependent on the soil composition for their growth and development (Hussain et al. 2020; Hafiz et al. 2020 a,b; Shafi et al. 2020; Wahid et al. 2020; Subhan et al. 2020; Zafar-ul-Hye et al. 2020a,b; Zafar et al. 2021; Adnan et al. 2020; Ilyas et al. 2020; Saleem et al. 2020a,b,c; Rehman 2020; Farhat et al. 2020; Wu et al. 2020; Mubeen et al.2020; Farhana 2020; Jan et al. 2019; Wu et al.2019; Ahmad et al,. 2019). Ground and aerial parts are exposed to different types of environments and hence face diverse challenges (Baseer et al. 2019; Hafiz et al. 2018; Tariq et al. 2018; Fahad and Bano 2012; Fahad et al. 2017; Fahad et al. 2013; Fahad et al. 2014a,b). In reference to microbial interactions with plants, these environments are categorized into rhizosphere, endosphere, and phyllosphere. Each environment presents different microbial loads and variants, which further depend upon the ecology of the particular niche. However, by all means, the microbial genome is considered as the second genome of plants, which directly or indirectly determines the plant fitness and development and hence its productivity under multiple biotic and abiotic circumstances (Mendes & Raaijmakers 2015; Haldar & Sengupta 2015; Lakshmanan et al. 2014; Fahad et al. 2016a,b,c,d; Fahad et al. 2015a,b; Fahad et al. 2018a,b; Fahad et al. 2019a,b; Fahad et al. 2020; Fahad et al. 2021a,b,c,d,e,f; Fahad et al. 2022a,b; Hesham & Fahad 2020).

1.3.1 Plant–microbe Interactions in the Rhizosphere

The roots are the anchoring appendages of plants, which exude carbohydrates, amino acids, and organic acids depending upon the type of plant species and the physical and chemical condition of the soil, which determine the biotic and abiotic conditions of the rhizosphere (Iqra et al. 2020; Akbar et al. 2020; Mahar et al. 2020; Noor et al. 2020; Bayram et al. 2020; Amanullah 2018a,b; Amanullah 2017; Amanullah et al.2020; Amanullah et al.2021; Rashid et al.2020; Arif et al.2020; Amir et al.2020; Saman et al. 2020; Muhammad Tahir ul Qamar et al. 2020; Jakirand Allah 2020; Mahmood et al. 2021; Farah et al. 2020; Sadam et al. 2020; Unsar et al. 2020; Fazli et al. 2020). Exudates do not only act as signaling molecules, which are recognized by the microbial community, but also as screening agents for the recruitment of specific microorganisms to specific plant species (Berg et al. 2016; Haldar & Sengupta 2015). In bulk soil, the microbial community presents a great diversity and is influenced by soil type and environmental factors. Microbe colonization and development are strongly and directly influenced by the prevailing soil chemical and physical conditions. Further complexity arises when the plant root type is taken into consideration, as each plant species has specific types of roots, which further dependsupon the prevailingd soil type for its establishment and development. Only specialized communityiesof microbes present in the area close to the root exudate are able to infect, enter, and establish inside the plant cell. Microbial communities further specialize their activation spatially and temporally within the plant organs (Akiyama et al. 2005; Reinhold-Hurek et al. 2015; Ottesen et al. 2013). Dynamicity takes place after infection when microbes impact plants either beneficially or negatively.

1.4 ROLE OF BACTERIA IN SALINITY TOLERANCE

Salinity refers to increased concentrations of salts within soil which negatively affect the morphological and physiological growth and development of plants. Due to increased concentrations of certain salts (Na^+/Cl^-), hypertonic conditions are generated which impede the uptake of water by plant cells. A reduction in water contents results in an osmotic stress condition, and if the situation is prolonged, nutrient imbalance and ion toxicity further exacerbate the situation within the cell. Ionic toxicity results in an imbalance in the cellular homeostasis and starts damaging the cell wall. Ultimately, the cell is collapsed physiologically and biochemically gradually, appearing as wilted phenotypically (Munns 2002). Each plant species has a different salinity stress tolerance capacity depending upon its built-in power to compartmentalize salt ions and produce organic solutes successfully to contribute to the adjustment of osmotic balance (ionic/osmotic homeostasis) within the cell. Plants are therefore categorized as salt-sensitive vs. salt-tolerant species (Ruiz-Lozano et al. 2012; Karakas et al. 2020). As soil salinity changes, the soil physical and chemical properties, and hence the microbial density/activity, are also affected, mostly negatively due to the reduction in organic matter (Canfora et al. 2014; Rath et al. 2019). A reduction in organic matter leads to lowered air spaces, creating situations that are non-ideal for microbial establishment, growth, and development (Canfora et al. 2014; Enamul et al. 2020; Gopakumar et al. 2020; Zia-ur-Rehman 2020; Ayman et al. 2020; Al-Wabel et al. 2020a,b; Senol 2020; Amjad et al. 2020; Ibrar et al. 2020; Sajid et al. 2020; Muhammad et al. 2021; Sidra et al. 2021; Zahir et al. 2021; Sahrish et al. 2022). This presents an undeniable picture that salinity affects plant productivity both quantitatively and qualitatively.

As with plants, a plethora of microorganisms exist with differential responses to multiple levels of salt stress conditions. These are also classified as salt-resistant vs. salt-sensitive microbes, with their survival dependent on differential stress conditions. Plants resistant in nature usually establish an interaction with microbes which can thrive under stressful conditions (Liu et al. 2017; Etesami & Glick 2020; Ventosa et al. 2008). Bacteria which can withstand salinity stress almost up to 25% are described as halotolerant bacteria, which are plant growth-promoting bacteria belonging to *Bacillus*, *Pseudomonas*, *Halomonas*, *Oceanobacillus*, and *Streptomyceas* with very similar mechanisms (Ventosa et al. 2008; Etesami & Glick 2020). The mechanism of action of halobacterium in mitigating salt stress includes protecting the cell wall from damage by synthesizing specific membranes, maintaining ionic homeostasis with the cell by synthesizing compatible organic solutes, pumping salt ions out of the cell, and adjusting the osmotic balance. These strategies help to adapt the cell to salt stress by enhancing the production of exoploysaccharides that limit salt entry into the cell (Ruppel et al. 2013). Overall, cell energy is augmented by adjusting osmotic balance, which relieves the cell from ionic toxicity and nutrient imbalance, detoxification, and modulation of action and function of phytohormones present in each plant type (Abd-Allah et al. 2018; Barnawal et al. 2016; Egamberdieva et al. 2016).

Endophytic bacteria with indol-3-acetic acid (IAA), 1-aminocylopropane-1-carboxylate (ACC) deaminase, phosphate-solubilizing and nitrogen-fixing properties helped to promote ionic and osmotic adjustment in plants (Khan et al. 2019; Jorge et al. 2019; Kang et al. 2019; Khan et al. 2020; Tufail et al. 2021). Endophytic bacterial inoculation (*Bacillus firmus*, *Bacillus pumilus*) was found to have a positive impact on reactive oxygen species (ROS)-scavenging enzymes CAT, POD, SOD, MDA, glutathione reductase, and dehydroascordáte reductase reported in *Solanum tuberosum* (Egamberdieva et al. 2017; Gururani et al. 2013). This modified level of enzymes had a positive modulating effect on the process of photosynthesis by safeguarding the chlorophyll pigment, however the rate of effect depends on the severity of stress, inoculated bacteria used, growth stage of the plant, and the type and concentration of enzyme produced (Ilangumaran & Smith 2017; Khan et al. 2017; Sun et al. 2020). Besides the ROS-scavenging strategy, osmoregulation by synthesis of osmolytes under salt stress also proved to be a promising route to avoid the harmful effects of salinity on photosynthetic apparatuses (Sun et al. 2020). Inoculation of *Arabidopsis thaliana* with *Enterobacter* species up-regulated the salt-responsive genes and increased the proline contents within cells (Hasegawa et al. 2000; Kim et al. 2014), which promoted the ionic homeostasis and stabilization of cellular structure under stress conditions (Maggio et al. 2002). Modulation of ethylene concentration in cells (lethal in higher concentrations by inhibiting germination and seedling growth) was also accomplished with the help of endophytic bacteria species *Enterobacter* P23. It reduces the ethylene toxicity by activating ACC deaminase, which in turn converts the ACC to ammonia and alpha-ketobutyrate, hence reducing ethylene contents in cells (Glick 2014; Kasotia et al. 2016; Sarkar et al. 2018). The modulation of the ionic interaction Na^+/K^+ was also found to be effective. Na^+ toxicity interferes with physiological processes (Basu et al. 2021). *B. subtilis* downregulates the HKT1 (high affinity K^+ transporter) in roots, thus reducing the uptake of Na^+. Intriguingly, HKT1 expression was enhanced in shoot which promotes Na^+ shoot–root recirculation (Liu et al. 2017; Zhang et al. 2008).

1.5 DROUGHT STRESS

T scarcity of water, drought, has a negative impact if imposed during any developmental stage of plant life. The morphological effects include reduced seed germination, negative effects on root and shoot architectural features, reproductive disorders associated with flower functioning, and ultimately a yield lower than the expected rate (Abdelaal et al. 2021; Hu et al. 2020; Ilyas et al. 2021; Khan et al. 2021). All these morphological losses/impairments are associated with disrupted physiology, anatomy, and biochemical functioning of plants. The overall impacts from all areas include cell wall thickening, increased cuticle deposition, variations in vascular and mesophyll cell diameters, and stomatal closing. Reduced hydraulic conductivity also limits nutrient uptake, stomatal conductance, relative water content, and an enhanced transpiration rate, which ultimately impairs the photosynthetic apparatus (Abdelaal et al. 2021; Abdelaal et al. 2018; Hafez et al. 2020; Liu et al. 2005; Reddy et al. 2004; Shao et al. 2008; Ullah et al. 2017). In trying to cope with drought by altering the above-mentioned attributes, plants are exposed to another closely related but lethal stress situation, which is high temperature stress, which ultimately leads to disruption of cellular membranes (Moretti et al. 2010; Rysiak et al. 2021).

Plants use multiple strategies to cope with stress situations which are categorized into three types: escape, avoidance, and tolerance. Drought escape is defined as the adjustment of life cycle such as to avoid confronting a stress situation, which mostly occurs by shortening the life cycle. Avoidance means to continue the life cycle near to the normal rate by managing or delaying the physiological processes. Tolerance describes the built-in power of the plant genome to resist dehydration with minimal loss of yield (Luo et al. 2018; Manavalan et al. 2009). Multiple strategies have been developed through crop breeding programs to address this dynamic stress condition with variable success depending upon the type of plant, environment, and severity of stress. However, in recently, a great deal of attention has been paid to using the microbial community to infect and stimulate natural triggers against stress situations in an eco-friendly manner. The success of this biological application depends upon the survival, establishment, and development of target microbes in a given set of environmental conditions. Several drought-resistant bacteria (*B. subtilis*, *B. altitudinis*, *B. mojavensis*, and *Brevibacillus laterosporus*) have been isolated from desert soil or drought-tolerant plants (Astorga-Eló et al. 2021; Milet et al. 2016). Besides the *Bacillus* species, other species of bacteria from the *Pseudomonas*, *Klebsiella*, and *Azotobacter* families have shown a growth-stimulating effect under drought stress.

All these microbes are accompanied with various properties which include ACC deaminase, IAA, phosphate solubilization, exopolysaccharide synthesis (EPS), growth on N-free culture, and osmolyte accumulation, e.g. HCN, under drought stress conditions in multiple crop species (Kumar et al. 2016; Mariotti et al. 2021; Niu et al. 2018; Sandhya et al. 2010; Sarma & Saikia 2014). The bacterial cells and their metabolites promote plant growth under water stress conditions. Plants infected with inoculants were found to contain higher amounts of chlorophyll a, b, total chlorophyll, carotenoids, and stress-related phytohormones like ABA, JA, and SA (Heidari et al. 2011; Mariotti et al. 2021). A mixture of three PGPR strains (*B. subtilis* SM21, *B. cereus* AR156, and *Serratia* sp. XY21) alleviated drought stress in cucumber by maintaining root growth, stabilizing osmotic potential, reducing plasma lemma peroxidation, enhancing production of SOD and APX in leaves, and ultimately maintaining the net photosynthesis rate without involving the ACC deaminase to lower the ethylene level (Wang et al. 2012). Pepper plants inoculated with B. *licheniformis* K11 enhanced the production of plant small heat shock proteins (sHSP) by upregulating the transcription of Ca-PR10, HSP, Cadhn, and VA genes (Kaushal & Wani 2016; Lim & Kim 2013).

Fungi have been found to be more effective under drought stress than bacteria due to their body structure and properties (Treseder et al. 2018). The main mechanisms involved thickening of the cell wall, osmolyte accumulation, and melanin, which made them better candidates than bacteria under drought stress (Treseder & Lennon 2015). The protective cell wall and filamentous nature of fungi connected and intermingled together allow water and solutes to flow between them. The hyphae grow longer and deeper into the soil profile, allowing the extraction of water from deep sites in the soil. Fungi can tolerate and flourish even more under water stress conditions. Due to this property, they maintain their density under drought stress, continue their decomposition function, and maintain the C and N balance and circulation in the soil (Treseder et al. 2018). Endophytic arbuscular mycorrhizal fungi (AMF) promote plant growth under adverse environmental conditions, particularly drought and salinity. An extensive network of hyphae collect water from distant locations and also provide mineral nutrition absorbed from the soil to the host plant, especially phosphorous and ammonia which it can take up more efficiently than plants (Adamec & Andrejiová 2018; Bizos et al. 2020; Bona et al. 2015; Marulanda et al. 2009; Matias et al. 2009). Phytohormones JA and SA are also produced by AMF (Dreischhoff et al. 2020). Various plants species inoculated with multiple fungal strains in various experiments have produced osmolytes (sugar, protein, proline), ROS-scavenging enzymes SOD, POX, and CAT under drought stress conditions and proved to be promising organisms for a strategy to cope with stress situations (Azad & Kaminskyj 2016; Dastogeer et al. 2018; Liao et al. 2021; Morsy et al. 2020; Sun et al. 2010; Verma et al. 2021).

1.6 HEAVY METAL STRESS

Another environmental contaminant stress that threatens plant growth and development recently is heavy metal which is gaining momentous attention due to its deadly consequences. In an effort to attain more yield from the agriculture sector and comfortable lives, the enforced use of pesticides, herbicides, chemical fertilizers, electrical home appliances, industrialization, mining, vehicles, fuel combustion etc. all release heavy metals in different forms into environment. The most common heavy metals released from these sources are cadmium (Cd), copper (Cu), nickel (Ni), cobalt (Co), arsenic (As), zinc (Zn), lead (Pb), mercury (Hg), and chromium (Cr), which are all persistent and recalcitrant in nature, damaging eco-dynamics both above and below the ground (Anbia & Amirmahmoodi 2016). Long-term exposure to heavy metals induces physiological and biochemical impairments in plant architecture. The overall range of toxicity includes inhibition of seed germination, reduction in chlorophyll content and hence photosynthetic activity, variations in the cellular metabolic profile which result in oxidative stress and a reduction in enzymatic antioxidant activity, and variations in anatomical features of plant foliage. These all ultimately lead toward abnormalities in mitotic activities, which cause reductions in root and shoot growth, wilting, and chlorosis (Amin & Latif 2017; Garg & Singla 2011; Luziatelli et al. 2020; Mallick et al. 2015).

Bacteria promoting growth in plants under arsenic stress belong to the Gram-positive (firmicute and action-bacteria in which *Bacillus* spp. is prominent) and Gram-negative (Proteobacteria where *Pseudomonas* spp. holds the prominent position) phyla. The mechanisms used by bacteria to mitigate arsenic stress include biotransformation via enzymatic oxidation and reduction, bioaccumulation on the surface, arsenite efflux, methylation, and bio-adsorption (Thomas et al. 2004). The arsenite oxidation system is well characterized in bacterial isolates from *Alcaligenes faecalis*, which holds an island of arsenic genes. These genes alone, or in combination, are responsible for the assemblage of arsenic oxidase enzyme and its translocation from cytoplasm to periplasm by following differential mechanisms depending upon the organism (Mallick et al. 2015; Phung et al. 2012). Mercuric ions (Hg^{2+}) and methyl mercury (CH_3Hg^+) are the two bioavailable toxic forms of mercury. The mechanism of detoxification by bacteria includes conversion of Hg^{2+} into the homodimeric form which then ultimately reduces to Hg^O, which being volatile diffuses into the intracellular environment. CH_3Hg^+ is downregulated by proton attack, which breaks the bond and releases Hg^{2+} and organic acids such as methane (Amin & Latif 2017; Begley & Ealick 2004; Silver & Phung 2005; Zhang et al. 2020). Reduced uptake, membrane efflux pump, and enzymatic reduction are the most imperative mechanisms used to mitigate chromium stress by bacteria. These mechanisms are among the characteristics of *Shewanella* sp., *Sphingomonas* spp., and *Variovorax* spp. bacterial species (Bilal et al. 2018; Viti et al. 2014).

Certain heavy metals like cadmium (Cd^{2+}), zinc, cobalt, and nickel are difficult to detoxify by bacterial amendments. The reason behind this complexity is the conversion of divalent cations (reoxidation) into the highly reactive former form which is not volatile in nature like Hg^O (Bruins et al. 2000). Due to the reactive nature of metallic ions and their interaction with cellular organelles it is difficult to predict the exact mechanism of detoxification. However, adsorption on the cell surface, ion exchange, and chelation prevent the metallic ions from gaining entry into the cell. After entry, activation of the efflux pump in plants infected with certain bacterial species like *Bacillus*, *Pseudomonas*, *Enterobacter*, and *Rhizobium* was found to be active as the resistance mechanism (Bruins et al. 2000; Silver & Phung 2005). The upregulation of metallo-proteins in proteomic analysis of a culture infected with *Pseudomonas aeruginosa* further strengthens the role of bacteria in cadmium stress management and maintenance of cellular homeostasis (Izrael-Živković et al. 2018; Zhang et al. 2019). Many other reports highlighted the role of multiple bacterial families in tolerance of lead (Pb) (Chen et al. 2005; Feng et al. 2015; Piazza et al. 2012), copper and its derivatives (Asaf et al. 2018; Hao et al. 2015; Wei et al. 2009), cobalt (Eitinger et al. 2005), zinc (Wang et al. 2020; Webb et al. 2017), and nickel (Mazhar et al. 2020; Nanda et al. 2019).

Plant growth-promoting bacteria are found in diverse ecological habitats, such as the rhizosphere, phyllosphere, root nodules, endophytes, industrial and sewage effluents, etc. and hence have diverse effects under variable stress conditions. Overall, they are promising biological tools to enhance plant growth under metal stress conditions as well as metal detoxification. *Bacillus*,

Pseudomonas, *Sphingomonas*, *Variovorax*, *Enterobacter*, and *Rhizobium* spp. are particularly involved in metal tolerance by adopting various genomic pathways. The induced genetic activation/deactivation of genes results in a variation in the profile of phytohormones which act as a front-line defense in plants (Chen et al. 2005; Mazhar et al. 2020; Wang et al. 2020). Phytohormones help to sustain plant growth by modulating the physiological changes which aid in nutrient uptake under heavy metal stress conditions. These findings are supported by several investigations describing indole acetic acid increases (Hafez et al. 2020) under stress conditions which ultimately result in loosening of the cell wall of root cells. This loosening results in the secretion of root exudate, rich in organic acid and phytochelatins, attracting the microbial community close to the root zone and hence restricting the entry of heavy metal in the symplastic pathway (Bacilio-Jiménez et al. 2003; Etesami 2018; Schutzendubel & Polle 2002). On the other hand, certain bacteria produce ACC deaminase under metal stress conditions which inhibit the production of ethylene, a worse growth repressor in plants under stress condition (Manivasagan et al. 2014). Phosphate-solubilizing bacteria (PSB) modulate the pH of soil and increase the concentration of bioavailable phosphate, which eventually immobilizes certain heavy metals by adsorption and chelation (Wei et al. 2018; Yuan et al. 2017).

1.7 CONCLUSION

Abiotic stresses pose serious threats to global food security due to their deleterious effects on plant physiology, biochemistry, and cellular and molecular biology. Besides the intense debate on the drastic effects of drought and salinity on forums designed to promote plant production strategies, heavy metal toxicity also has gained huge attention due to its deadly consequences not only on plant growth but also on human and animal health. This reinforces the need to adopt an eco-friendly strategy to combat these stresses with minimal or no compromise of crop yield. Microbe-assisted plant growth, being a biological and eco-friendly tool, has changed the global scenario addressing the impacts and strategies to secure plant production under stress situations. Several bacterial (*Bacillus*, *Pseudomonas*, *Rhizobium*, *Acinetobacter*) and fungal genera have been found to exhibit a multi-tolerance ability under different abiotic stress and ecological conditions. Under all types of abiotic circumstances, microbial infestation modifies the profile of ACC deaminase, IAA, phosphate solubilization, exopolysaccharide synthesis (EPS), osmolyte production, and accumulation at various rates depending upon the crop species, microbial strain, ecological condition, growth stage of the plant, and concentration of enzyme produced. Plants infected with inoculants were found to contain higher amounts of chlorophyll a, b, total chlorophyll, carotenoids, and stress-related phytohormones like ABA, JA, and SA. Microbes can also be used in the strategy to adsorb, physical entrapment on the cell surface, and reduction in uptake, particularly with metal toxicity to plants and altering cellular architecture. Cellular architecture modulations include reduced uptake of metal ions, chelation with other organic compounds, and efflux pump activation in the case of huge uptake. Microbe-assisted plant growth has arisen as a promising eco-friendly and biological strategy in sustainable agriculture projects designed under both normal and stressful situations.

REFERENCES

Abd-Allah EF, Alqarawi AA, Hashem A, Radhakrishnan R, Al-Huqail AA, Al-Otibi FON, Malik JA, Alharbi RI, Egamberdieva D (2018): Endophytic bacterium Bacillus subtilis (BERA 71) improves salt tolerance in chickpea plants by regulating the plant defense mechanisms. *J Plant Interact* 13, 37–44.

Abdelaal K, AlKahtani M, Attia K, Hafez Y, Király L, Künstler A (2021): The role of plant growth-promoting bacteria in alleviating the adverse effects of drought on plants. *Biol* 10, 520.

Abdelaal KA, Hafez YM, El-Afry MM, Tantawy DS, Alshaal T (2018): Effect of some osmoregulators on photosynthesis, lipid peroxidation, antioxidative capacity, and productivity of barley (Hordeum vulgare L.) under water deficit stress. *Environ Sci Pollut Res* 25, 30199–30211.

Abdul S, Muhammad AA, Shabir H, Hesham A El E, Sajjad H, Niaz A, Abdul G, RZ Sayyed, Fahad S, Subhan D, Rahul D (2021a): Zinc nutrition and arbuscular mycorrhizal symbiosis effects on maize (Zea mays L.) growth and productivity. *J Saudi Soc Agri Sci* 28(11), 6339–6351 https://doi.org/10.1016/j.sjbs.2021.06.096

Abdul S, Muhammad AA, Subhan D, Niaz A, Fahad S, Rahul D, Mohammad JA, Omaima N, Muhammad Habib ur R, Bernard RG (2021b): Effect of arbuscular mycorrhizal fungi on the physiological functioning of maize under zinc-deficient soils. *Sci Rep* 11, 18468.

Abid M, Khalid N, Qasim A, Saud A, Manzer HS, Chao W, Depeng W, Shah S, Jan B, Subhan D, Rahul D, Hafiz MH, Wajid N, Muhammad M, Farooq S, Fahad S (2021): Exploring the potential of moringa leaf extract as bio stimulant for improving yield and quality of black cumin oil. *Sci Rep* 11, 24217 https://doi.org/10.1038/s41598-021-03617-w

Adamec S, Andrejiová A (2018): Mycorrhiza and stress tolerance of vegetables: A review. *Acta Horticulturae et Regiotecturae* 21, 30–35.

Adnan M, Fahad S, Khan IA, Saeed M, Ihsan MZ, Saud S, Riaz M, Wang D, Wu C (2019). Integration of poultry manure and phosphate solubilizing bacteria improved availability of Ca bound P in calcareous soils. *3 Biotech* 9(10), 368.

Adnan M, Fahad S, Muhammad Z, Shahen S, Ishaq AM, Subhan D, Zafar-ul-Hye M, Martin LB, Raja MMN, Beena S, Saud S, Imran A, Zhen Y, Martin B, Jiri H, Rahul D (2020): Coupling phosphate-solubilizing bacteria with phosphorus supplements improve maize phosphorus acquisition and growth under lime induced salinity stress. *Plants* 9(7), 900. doi: 10.3390/plants9070900

Adnan M, Shah Z, Sharif M, Rahman H (2018b). Liming induces carbon dioxide (CO_2) emission in PSB inoculated alkaline soil supplemented with different phosphorus sources. *Environ Sci Poll Res* 25(10), 9501–9509.

Adnan M, Zahir S, Fahad S, Arif M, Mukhtar A, Imtiaz AK, Ishaq AM, Abdul B, Hidayat U, Muhammad A, Inayat-Ur R, Saud S, Muhammad ZI, Yousaf J, Amanullah, Hafiz MH, Wajid N (2018a): Phosphate-solubilizing bacteria nullify the antagonistic effect of soil calcification on bioavailability of phosphorus in alkaline soils. *Sci Rep* 8, 4339 https://doi.org/10.1038/s41598-018-22653-7

Ahmad N, Hussain S, Ali MA, Minhas A, Waheed W, Danish S, Fahad S, Ghafoor U, Baig, KS, Sultan H, Muhammad IH, Mohammad JA, Theodore DM (2022): Correlation of soil characteristics and citrus leaf nutrients contents in current scenario of Layyah district. *Hortic* 8, 61 https://doi.org/10.3390/horticulturae8010061

Ahmad S, Kamran M, Ding R, Meng X, Wang H, Ahmad I, Fahad S, Han Q (2019): Exogenous melatonin confers drought stress by promoting plant growth, photosynthetic capacity and antioxidant defense system of maize seedlings. *PeerJ* 7, e7793 http://doi.org/10.7717/peerj.7793

Akbar H, Timothy JK, Jagadish T, Golam M, Apurbo KC, Muhammad F, Rajan B, Fahad S, Hasanuzzaman M (2020): Agricultural land degradation: processes and problems undermining future food security. In: Fahad S, Hasanuzzaman M, Alam M, Ullah H, Saeed M, Khan AK, Adnan M (Eds.), *Environment, Climate, Plant and Vegetation Growth.* Springer Publ Ltd, Springer Nature Switzerland AG. Part of Springer Nature. pp. 17–62. https://doi.org/10.1007/978-3-030-49732-3

Akiyama K, Matsuzaki K-i, Hayashi H (2005): Plant sesquiterpenes induce hyphal branching in arbuscular mycorrhizal fungi. *Nature* 435, 824–827.

Akram R, Turan V, Hammad HM, Ahmad S, Hussain S, Hasnain A, Maqbool MM, Rehmani MIA, Rasool A, Masood N, Mahmood F, Mubeen M, Sultana SR, Fahad S, Amanet K, Saleem M, Abbas Y, Akhtar HM, Waseem F, Murtaza R, Amin A, Zahoor SA, ul Din MS, Nasim W (2018a): Fate of organic and inorganic pollutants in paddy soils. In: Hashmi, MZ and Varma, A (Eds.), *Environmental Pollution of Paddy Soils, Soil Biology*. Springer International Publishing AG, Gewerbestrasse 11, Cham, CH-6330, Switzerland, pp. 197–214.

Akram R, Turan V, Wahid A, Ijaz M, Shahid MA, Kaleem S, Hafeez A, Maqbool MM, Chaudhary HJ, Munis, MFH, Mubeen M, Sadiq N, Murtaza R, Kazmi DH, Ali S, Khan N, Sultana SR, Fahad S, Amin A, Nasim W (2018b): Paddy land pollutants and their role in climate change. In: Hashmi, MZ and Varma, A (Eds.), *Environmental Pollution of Paddy Soils, Soil Biology*. Springer International Publishing AG, Gewerbestrasse 11, Cham, CH-6330, Switzerland, pp. 113–124.

Ali S, Hameed G, Muhammad A, Depeng W, Fahad S (2022): Comparative genetic evaluation of maize inbred lines at seedling and maturity stages under drought stress. *J Plant Growth Regul* 42, 989–1005 https://doi.org/10.1007/s00344-022-10608-2

Al-Zahrani HS, Alharby HF and Fahad S (2022): Antioxidative defense system, hormones, and metabolite accumulation in different plant parts of two contrasting rice cultivars as influenced by plant growth regulators under heat stress. *Front Plant Sci* 13, 911846. doi: 10.3389/fpls.2022.911846

Amanullah, Fahad S (Eds.) (2017): Rice – technology and production. *IntechOpen Croatia* 2017 http://dx.doi.org/10.5772/64480

Amanullah, Fahad S (Eds.) (2018a): Corn – production and human health in changing climate. *IntechOpen United Kingdom* 2018. http://dx.doi.org/10.5772/intechopen.74074

Amanullah, Fahad S (Eds.) (2018b): Nitrogen in agriculture – updates. *IntechOpen Croatia* 2018. http://dx.doi.org/10.5772/65846

Amanullah, Muhammad I, Haider N, Shah K, Manzoor A, Asim M, Saif U, Izhar A, Fahad S, Adnan M et al. (2021): Integrated foliar nutrients application improve wheat (Triticum aestivum L.) productivity under calcareous soils in drylands. *Commun Soil Sci Plant Anal* 52(21), 2748–2766 https://Doi.Org/10.1080/00103624.2021.1956521

Amanullah, Shah K, Imran, Hamdan AK, Muhammad A, Abdel RA, Muhammad A, Fahad S, Azizullah S, Brajendra P (2020): Effects of climate change on irrigation water quality. In: Fahad S, Hasanuzzaman M, Alam M, Ullah H,Saeed M, Khan AK, Adnan M (Eds.), *Environment, Climate, Plant and Vegetation Growth*. Springer Publ Ltd, Springer Nature Switzerland AG. Part of Springer Nature. pp. 123–132. https://doi.org/10.1007/978-3-030-49732-3

Amir M, Muhammad A, Allah B, Sevgi Ç, Haroon ZK, Muhammad A, Emre A (2020): Bio fortification under climate change: the fight between quality and quantity. In: Fahad S, Hasanuzzaman M, Alam M, Ullah H, Saeed M, Khan AK, Adnan M (Eds.), *Environment, Climate, Plant and Vegetation Growth*. Springer Publ Ltd, Springer Nature Switzerland AG. Part of Springer Nature. pp. 173–228. https://doi.org/10.1007/978-3-030-49732-3

Amin A, Latif Z (2017): Screening of mercury-resistant and indole-3-acetic acid producing bacterial-consortium for growth promotion of Cicer arietinum L. *J Basic Microbiol* 57, 204–217.

Amjad I, Muhammad H, Farooq S, Anwar H (2020): Role of plant bioactives in sustainable agriculture. In: Fahad S, Hasanuzzaman M, Alam M, Ullah H, Saeed M, Khan AK, Adnan M (Eds.), *Environment, Climate, Plant and Vegetation Growth*. Springer Publ Ltd, Springer Nature Switzerland AG. Part of Springer Nature. pp. 591–606. https://doi.org/10.1007/978-3-030-49732-3

Amjad SF, Mansoora N, Din IU, Khalid IR, Jatoi GH, Murtaza G, Yaseen S, Naz M, Danish S, Fahad S et al. (2021): Application of zinc fertilizer and mycorrhizal inoculation on physio-biochemical parameters of wheat grown under water-stressed environment. *Sustainability* 13, 11007 https://doi.org/10.3390/su131911007

Anam I, Huma G, Ali H, Muhammad K, Muhammad R, Aasma P, Muhammad SC, Noman W, Sana F, Sobia A, Fahad S (2021): Ameliorative mechanisms of turmeric-extracted curcumin on arsenic (As)-induced biochemical alterations, oxidative damage, and impaired organ functions in rats. *Environ Sci Pollut Res* 28(46), 66313–66326 https://doi.org/10.1007/s11356-021-15695-4

Anbia M, Amirmahmoodi S (2016): Removal of Hg (II) and Mn (II) from aqueous solution using nanoporous carbon impregnated with surfactants. *Arabian J Chem* 9, S319–S325.

Arif M, Talha J, Muhammad R, Fahad S, Muhammad A, Amanullah, Kawsar A, Ishaq AM, Bushra K, Fahd R (2020): Biochar: a remedy for climate change. In: Fahad S, Hasanuzzaman M, Alam M, Ullah H,Saeed M, Khan AK, Adnan M (Eds.), *Environment, Climate, Plant and Vegetation Growth*. Springer Publ Ltd, Springer Nature Switzerland AG. Part of Springer Nature. pp. 151–172. https://doi.org/10.1007/978-3-030-49732-3

Asaf S, Khan AL, Khan MA, Al-Harrasi A, Lee I-J (2018): Complete genome sequencing and analysis of endophytic Sphingomonas sp. LK11 and its potential in plant growth. *3 Biotech* 8, 1–14.

Ashfaq AR, Uzma Y, Niaz A, Muhammad AA, Fahad S, Haider S, Tayebeh Z, Subhan D, Süleyman T, Hesham AelE, Pramila T, Jamal MA, Sulaiman AA, Rahul D (2021): Toxicity of cadmium and nickel in the context of applied activated carbon biochar for improvement in soil fertility. *Saudi Society Agricultural Sci* 29(2), 743–750 https://doi.org/10.1016/j.sjbs.2021.09.035

Astorga-Eló M, Gonzalez S, Acuña JJ, Sadowsky MJ, Jorquera MA (2021): Rhizobacteria from "flowering desert" events contribute to the mitigation of water scarcity stress during tomato seedling germination and growth. *Sci Rep* 11, 1–12.

Athar M, Masood IA, Sana S, Ahmed M, Xiukang W, Sajid F, Sher AK, Habib A, Faran M, Zafar H, Farhana G, Fahad S (2021): Bio-diesel production of sunflower through sulphur management in a semi-arid subtropical environment. *Environ Sci Pollut Res* 29(9), 13268–13278 https://doi.org/10.1007/s11356-021-16688-z

Atif B, Hesham A, Fahad S (2021): Biochar coupling with phosphorus fertilization modifies antioxidant activity, osmolyte accumulation and reactive oxygen species synthesis in the leaves and xylem sap of rice cultivars under high-temperature stress. *Physiol Mol Biol Plants* 27(9), 2083–2100 https://doi.org/10.1007/s12298-021-01062-7

Ayman El Sabagh, Akbar Hossain, Celaleddin Barutçular, Muhammad Aamir Iqbal, Sohidul Islam M, Shah Fahad, Oksana Sytar, Fatih Çig, Ram Swaroop Meena, and Murat Erman (2020): Consequences of salinity stress on the quality of crops and its mitigation strategies for sustainable crop production: An outlook of arid and semi-arid regions. In: Fahad S, Hasanuzzaman M, Alam M, Ullah H, Saeed M, Khan AK, Adnan M (Eds.), *Environment, Climate, Plant and Vegetation Growth.* Springer Publ Ltd, Springer Nature Switzerland AG. Part of Springer Nature. pp. 503–534. https://doi.org/10.1007/978-3-030-49732-3

Azad K, Kaminskyj S (2016): A fungal endophyte strategy for mitigating the effect of salt and drought stress on plant growth. *Symbiosis* 68, 73–78.

Aziz K, Daniel KYT, Fazal M, Muhammad ZA, Farooq S, FanW, Fahad S, Ruiyang Z (2017a): Nitrogen nutrition in cotton and control strategies for greenhouse gas emissions: a review. *Environ Sci Pollut Res* 24, 23471–23487 https://doi.org/10.1007/s11356-017-0131-y

Aziz K, Daniel KYT, Muhammad ZA, Honghai L, Shahbaz AT, Mir A, Fahad S (2017b): Nitrogen fertility and abiotic stresses management in cotton crop: a review. *Environ Sci Pollut Res* 24, 14551–14566 https://doi.org/10.1007/s11356-017-8920-x

Bacilio-Jiménez M, Aguilar-Flores S, Ventura-Zapata E, Pérez-Campos E, Bouquelet S, Zenteno E (2003): Chemical characterization of root exudates from rice (Oryza sativa) and their effects on the chemotactic response of endophytic bacteria. *Plant and Soil* 249, 271–277.

Baldan R, Cigana C, Testa F, Bianconi I, De Simone M, Pellin D, Di Serio C, Bragonzi A, Cirillo DM (2014): Adaptation of Pseudomonas aeruginosa in cystic fibrosis airways influences virulence of Staphylococcus aureus in vitro and murine models of co-infection. *PloS One* 9, e89614.

Barnawal D, Bharti N, Tripathi A, Pandey SS, Chanotiya CS, Kalra A (2016): ACC-deaminase-producing endophyte Brachybacterium paraconglomeratum strain SMR20 ameliorates Chlorophytum salinity stress via altering phytohormone generation. *J Plant Growth Regul* 35, 553–564.

Baseer M, Adnan M, Fazal M, Fahad S, Muhammad S, Fazli W, Muhammad A, Jr. Amanullah, Depeng W, Saud S, Muhammad N, Muhammad Z, Fazli S, Beena S, Mian AR, Ishaq AM (2019): Substituting urea by organic wastes for improving maize yield in alkaline soil. *J Plant Nutrition* 2423–2434 doi.org/10.1080/01904167.2019.1659344

Basu S, Kumar A, Benazir I, Kumar G (2021): Reassessing the role of ion homeostasis for improving salinity tolerance in crop plants. *Physiologia Plantarum* 171, 502–519.

Bayram AY, Seher Ö, Nazlican A (2020): Climate change forecasting and odelling for the year of 2050. In: Fahad S, Hasanuzzaman M, Alam M, Ullah H,Saeed M, Khan AK, Adnan M (Eds.), *Environment, Climate, Plant and Vegetation Growth.* Springer Publ Ltd, Springer Nature Switzerland AG. Part of Springer Nature. pp. 109–122. https://doi.org/10.1007/978-3-030-49732-3

Begley TP, Ealick SE (2004): Enzymatic reactions involving novel mechanisms of carbanion stabilization. *Curr Opin Chem Biol* 8, 508–515.

Berg G, Rybakova D, Grube M, Köberl M (2016): The plant microbiome explored: implications for experimental botany. *J Exp Bot* 67, 995–1002.

Bilal S, Shahzad R, Khan AL, Kang S-M, Imran QM, Al-Harrasi A, Yun B-W, Lee I-J (2018): Endophytic microbial consortia of phytohormones-producing fungus *Paecilomyces formosus* LHL10 and bacteria *Sphingomonas* sp. LK11 to *Glycine max* L. regulates physio-hormonal changes to attenuate aluminum and zinc stresses. *Front Plant Sci*, 9, 1273.

Bizos G, Papatheodorou EM, Chatzistathis T, Ntalli N, Aschonitis VG, Monokrousos N (2020): The role of microbial inoculants on plant protection, growth stimulation, and crop productivity of the olive tree (Olea europea L.). *Plants* 9, 743.

Bona E, Lingua G, Manassero P, Cantamessa S, Marsano F, Todeschini V, Copetta A, D'Agostino G, Massa N, Avidano L (2015): AM fungi and PGP pseudomonads increase flowering, fruit production, and vitamin content in strawberry grown at low nitrogen and phosphorus levels. *Mycorrhiza* 25, 181–193.

Bruins MR, Kapil S, Oehme FW (2000): Microbial resistance to metals in the environment. *Ecotoxicol Environ Saf* 45, 198–207.

Bukhari MA, Adnan NS, Fahad S, Javaid I, Fahim N, Abdul M, Mohammad SB (2021): Screening of wheat (Triticum aestivum L.) genotypes for drought tolerance using polyethylene glycol. *Arabian J Geosci* 14, 2808.

Canfora L, Bacci G, Pinzari F, Lo Papa G, Dazzi C, Benedetti A (2014): Salinity and bacterial diversity: to what extent does the concentration of salt affect the bacterial community in a saline soil? *PloS One* 9, e106662.

Chang W, Qiujuan J, Evgenios A, Haitao L, Gezi L, Jingjing Z, Fahad S, Ying J (2021): Hormetic effects of zinc on growth and antioxidant defense system of wheat plants. *Sci Total Environ* 807(2), 150992 https://doi.org/10.1016/j.scitotenv.2021.150992

Chao W, Youjin S, Beibei Q, Fahad S (2022): Effects of asymmetric heat on grain quality during the panicle initiation stage in contrasting rice genotypes. *J Plant Growth Regul* https://doi.org/10.1007/s00344-022-10598-1

Chen P, Greenberg B, Taghavi S, Romano C, van der Lelie D, He C (2005): An exceptionally selective lead (II)-regulatory protein from ralstonia metallidurans: development of a fluorescent lead (II) probe. *Angewandte Chemie* 117, 2775–2779.

Chen Y, Guo Z, Dong L, Fu Z, Zheng Q, Zhang G, Qin L, Sun X, Shi Z, Fahad S, Xie F, Saud S. (2021): Turf performance and physiological responses of native Poa species to summer stress in Northeast China. *Peer J* 9, e12252 http://doi.org/10.7717/peerj.12252

Compant S, Samad A, Faist H, Sessitsch A (2019): A review on the plant microbiome: ecology, functions, and emerging trends in microbial application. *J Adv Res* 19, 29–37.

Dastogeer KM, Li H, Sivasithamparam K, Jones MG, Wylie SJ (2018): Fungal endophytes and a virus confer drought tolerance to Nicotiana benthamiana plants through modulating osmolytes, antioxidant enzymes and expression of host drought responsive genes. *Environ Exp Bot* 149, 95–108.

Deepranjan S, Ardith SO, Siva D, Sonam S, Shikha, Manoj P, Amitava R, Sayyed RZ, Abdul G, Mohammad JA, Subhan D, Fahad S, Rahul D (2021): Optimizing nutrient use efficiency, productivity, energetics, and economics of red cabbage following mineral fertilization and biopriming with compatible rhizosphere microbes. *Sci Rep* 11, 15680 https://doi.org/10.1038/s41598-021-95092-6

Depeng W, Fahad S, Saud S, Muhammad K, Aziz K, Mohammad NK, Hafiz MH, Wajid N (2018): Morphological acclimation to agronomic manipulation in leaf dispersion and orientation to promote "Ideotype" breeding: evidence from 3D visual modeling of "super" rice (Oryza sativa L.). *Plant Physiol Biochem* 135, 499–510 https://doi.org/10.1016/j.plaphy.2018.11.010

Dreischhoff S, Das IS, Jakobi M, Kasper K, Polle A (2020): Local responses and systemic induced resistance mediated by ectomycorrhizal fungi. *Frontiers in Plant Science*, 1908.

Egamberdieva D, Jabborova D, Berg G (2016): Synergistic interactions between Bradyrhizobium japonicum and the endophyte Stenotrophomonas rhizophila and their effects on growth, and nodulation of soybean under salt stress. *Plant and soil* 405, 35–45.

Egamberdieva D, Wirth S, Jabborova D, Räsänen LA, Liao H (2017): Coordination between Bradyrhizobium and Pseudomonas alleviates salt stress in soybean through altering root system architecture. *J. Plant Interact.* 12, 100–107.

Eitinger T, Suhr J, Moore L, Smith JAC (2005): Secondary transporters for nickel and cobalt ions: theme and variations. *Biometals* 18, 399–405.

El Sabagh A, Islam MS, Hossain A, Iqbal MA, Mubeen M, Waleed M, Reginato M, Battaglia M, Ahmed S, Rehman A, Arif M, Athar H-U-R, Ratnasekera D, Danish S, Raza MA, Rajendran K, Mushtaq M, Skalicky M, Brestic M, Soufan W, Fahad S, Pandey S, Kamran M, Datta R. Abdelhamid MT (2022): Phytohormones as growth regulators during abiotic stress tolerance in plants. *Front. Agron.* 4, 765068. doi: 10.3389/fagro.2022.765068

Emre B, Ömer SU, Martín LB, Andre D, Fahad S, Rahul D, Muhammad Z-ul-H, Ghulam SH, Subhan D (2021): Studying soil erosion by evaluating changes in physico-chemical properties of soils under different land-use types. *J Saudi Society Agricultural Sci* 20, 190–197.

Etesami H (2018): Ecotoxicology and environmental safety bacterial mediated alleviation of heavy metal stress and decreased the accumulation of metals in plant tissues: mechanisms and future prospects. *Ecotoxicol Environ Saf* 147, 175–191.

Etesami H, Glick BR (2020): Halotolerant plant growth–promoting bacteria: Prospects for alleviating salinity stress in plants. *Environ Exp Bot* 178, 104124.

Fahad S, Bano A (2012): Effect of salicylic acid on physiological and biochemical characterization of maize grown in saline area. *Pak J Bot* 44, 1433–1438.

Fahad S, Chen Y, Saud S, Wang K, Xiong D, Chen C, Wu C, Shah F, Nie L, Huang J (2013): Ultraviolet radiation effect on photosynthetic pigments, biochemical attributes, antioxidant enzyme activity and hormonal contents of wheat. *J Food, Agri Environ* 11(3&4), 1635–1641.

Fahad S, Hussain S, Bano A, Saud S, Hassan S, Shan D, Khan FA, Khan F, Chen Y, Wu C, Tabassum MA, Chun MX, Afzal M, Jan A, Jan MT, Huang J (2014a): Potential role of phytohormones and plant growth-promoting rhizobacteria in abiotic stresses: consequences for changing environment. *Environ Sci Pollut Res* 22(7):4907–4921. https://doi.org/10.1007/s11356-014-3754-2

Fahad S, Hussain S, Matloob A, Khan FA, Khaliq A, Saud S, Hassan S, Shan D, Khan F, Ullah N, Faiq M, Khan MR, Tareen AK, Khan A, Ullah A, Ullah N, Huang J (2014b): Phytohormones and plant responses to salinity stress: a review. *Plant Growth Regul* 75(2), 391–404. https://doi.org/10.1007/s10725-014-0013-y

Fahad S, Hussain S, Saud S, Tanveer M, Bajwa AA, Hassan S, Shah AN, Ullah A, Wu C, Khan FA, Shah F, Ullah S, Chen Y, Huang J (2015a): A biochar application protects rice pollen from high-temperature stress. *Plant Physiol Biochem* 96, 281–287.

Fahad S, Nie L, Chen Y, Wu C, Xiong D, Saud S, Hongyan L, Cui K, Huang J (2015b): Crop plant hormones and environmental stress. *Sustain Agric Rev* 15, 371–400.

Fahad S, Hussain S, Saud S, Hassan S, Chauhan BS, Khan F et al. (2016a): Responses of rapid viscoanalyzer profile and other rice grain qualities to exogenously applied plant growth regulators under high day and high night temperatures. *PloS One* 11(7), e0159590. https://doi.Org/10.1371/journal.pone.0159590

Fahad S, Hussain S, Saud S, Khan F, Hassan S, Jr A, Nasim W, Arif M, Wang F, Huang J (2016b): Exogenously applied plant growth regulators affect heat-stressed rice pollens. *J Agron Crop Sci* 202, 139–150.

Fahad S, Hussain S, Saud S, Hassan S, Ihsan Z, Shah AN, Wu C, Yousaf M, Nasim W, Alharby H, Alghabari F, Huang J (2016c): Exogenously applied plant growth regulators enhance the morphophysiological growth and yield of rice under high temperature. *Front Plant Sci* 7, 1250. https://doi.org/10.3389/fpls.2016. 01250

Fahad S, Hussain S, Saud S, Hassan S, Tanveer M, Ihsan MZ, Shah AN, Ullah A, Nasrullah KF, Ullah S, AlharbyH NW, Wu C, Huang J (2016d): A combined application of biochar and phosphorus alleviates heat-induced adversities on physiological, agronomical and quality attributes of rice. *Plant Physiol Biochem* 103, 191–198.

Fahad S, Bajwa AA, Nazir U, Anjum SA, Farooq A, Zohaib A, Sadia S, NasimW, Adkins S, Saud S, Ihsan MZ, Alharby H, Wu C, Wang D, Huang J (2017): Crop production under drought and heat stress: plant responses and management options. *Front Plant Sci* 8, 1147. https://doi.org/10.3389/fpls.2017.01147

Fahad S, Muhammad ZI, Abdul K, Ihsanullah D, Saud S, Saleh A, Wajid N, Muhammad A, Imtiaz AK, Chao W, Depeng W, Jianliang H (2018a): Consequences of high temperature under changing climate optima for rice pollen characteristics-concepts and perspectives, *Archives Agron Soil Sci* 64(11), 1473–1488 doi: 10.1080/03650340.2018.1443213

Fahad S, Abdul B, Adnan M. (Eds.) (2018b): Global wheat production. *IntechOpen United Kingdom* 2018. http://dx.doi.org/10.5772/intechopen.72559

Fahad S, Rehman A, Shahzad B, Tanveer M, Saud S, Kamran M, Ihtisham M, Khan SU, Turan V, Rahman MHU (2019a): Rice responses and tolerance to metal/metalloid toxicity. In: Hasanuzzaman, M and Fujita, M and Nahar, K and Biswas, JK (Eds.), *Advances in Rice Research for Abiotic Stress Tolerance.* Woodhead Publ Ltd, pp. 299–312.

Fahad S, Adnan M, Hassan S, Saud S, Hussain S, Wu C, Wang D, Hakeem KR, Alharby HF, Turan V, Khan MA, Huang J (2019b): Rice responses and tolerance to high temperature. In: Hasanuzzaman, M and Fujita, M and Nahar, K and Biswas, JK (Eds.), *Advances in Rice Research for Abiotic Stress Tolerance.* Woodhead Publ Ltd, pp. 201–224.

Fahad S, Hasanuzzaman M, Alam M, Ullah H, Saeed M, Ali Khan I, Adnan M (Eds.) (2020): Environment, climate, plant and vegetation growth. *Springer Nature Switzerland AG 2020.* doi: https://doi.org/10.1007/978-3-030-49732-3

Fahad S, Sönmez O, Saud S, Wang D, Wu C, Adnan M, Turan V (Eds.) (2021a): Plant growth regulators for climate-smart agriculture, First edition, *Footprints of Climate Variability on Plant Diversity.* CRC Press.

Fahad S, Sonmez O, Saud S, Wang D, Wu C, Adnan M, Turan V (Eds.) (2021b): Climate change and plants: biodiversity, growth and interactions, First edition, *Footprints of Climate Variability on Plant Diversity*. CRC Press.

Fahad S, Sonmez O, Saud S, Wang D, Wu C, Adnan M, Turan V (Eds.) (2021c): Developing climate resilient crops: improving global food security and safety, First edition, *Footprints of Climate Variability on Plant Diversity*. CRC Press.

Fahad S, Sönmez O, Turan V, Adnan M, Saud S, Wu C, Wang D (Eds.) (2021d): Sustainable soil and land management and climate change, First edition, *Footprints of Climate Variability on Plant Diversity*. CRC Press.

Fahad S, Sönmez O, Saud S, Wang D, Wu C, Adnan M, Arif M, Amanullah (Eds.) (2021e): Engineering tolerance in crop plants against abiotic stress, First edition, *Footprints of Climate Variability on Plant Diversity*. CRC Press.

Fahad S, Saud S, Yajun C, Chao W, Depeng W (Eds.) (2021f): Abiotic stress in plants. *IntechOpen United Kingdom* 2021. http://dx.doi.org/10.5772/intechopen.91549

Fahad S, Adnan M, Saud S. (Eds.) (2022a): Improvement of plant production in the era of climate change. First edition, *Footprints of Climate Variability on Plant Diversity*. CRC Press.

Fahad S, Adnan M, Saud S, Nie L (Eds.) (2022b): Climate change and ecosystems: challenges to sustainable development. First edition, *Footprints of Climate Variability on Plant Diversity*. CRC Press.

Fakhre A, Ayub K, Fahad S, Sarfraz N, Niaz A, Muhammad AA, Muhammad A, Khadim D, Saud S, Shah H, Muhammad ASR, Khalid N, Muhammad A, Rahul D, Subhan D (2021): Phosphate solubilizing bacteria optimize wheat yield in mineral phosphorus applied alkaline soil. *J Saudi Soc Agric Sci* 21(5), 339–348 https://doi.org/10.1016/j.jssas.2021.10.007

Farah R, Muhammad R, Muhammad SA, Tahira Y, Muhammad AA, Maryam A, Shafaqat A, Rashid M, Muhammad R, Qaiser H, Afia Z, Muhammad AA, Muhammad A, Fahad S (2020): Alternative and non-conventional soil and crop management strategies for increasing water use efficiency. In: Fahad S, Hasanuzzaman M, Alam M, Ullah H, Saeed M, Khan AK, Adnan M (Eds.), *Environment, Climate, Plant and Vegetation Growth*. Springer Publ Ltd, Springer Nature Switzerland AG. Part of Springer Nature. pp. 323–338. https://doi.org/10.1007/978-3-030-49732-3

Farhana G, Ishfaq A, Muhammad A, Dawood J, Fahad S, Xiuling L, Depeng W, Muhammad F, Muhammad F, Syed AS (2020): Use of crop growth model to simulate the impact of climate change on yield of various wheat cultivars under different agro-environmental conditions in Khyber Pakhtunkhwa, Pakistan. *Arabian J Geosci* 13, 112 https://doi.org/10.1007/s12517-020-5118-1

Farhat A, Hafiz MH, Wajid I, Aitazaz AF, Hafiz FB, Zahida Z, Fahad S, Wajid F, Artemi C (2020): A review of soil carbon dynamics resulting from agricultural practices. *J Environ Manage* 268 (2020), 110319.

Farhat UK, Adnan AK, Kai L, Xuexuan X, Muhammad A, Fahad S, Rafiq A, Mushtaq AK, ·Taufiq N, Faisal Z (2022): Influences of long-term crop cultivation and fertilizer management on soil aggregates stability and fertility in the Loess Plateau, Northern China. *J Soil Sci Plant Nutri* 22, 1446–1457 https://doi.org/10.1007/s42729-021-00744-1

Faust K, Raes J (2012): Microbial interactions: from networks to models. *Nat Rev Microbiol* 10, 538–550.

Fazli W, Muhmmad S, Amjad A, Fahad S, Muhammad A, Muhammad N, Ishaq AM, Imtiaz AK, Mukhtar A, Muhammad S, Muhammad I, Rafi U, Haroon I, Muhammad A (2020): Plant–microbes interactions and functions in changing climate. In: Fahad S, Hasanuzzaman M, Alam M, Ullah H, Saeed M, Khan AK, Adnan M (Eds.), *Environment, Climate, Plant and Vegetation Growth*. Springer Publ Ltd, Springer Nature Switzerland AG. Part of Springer Nature. pp. 397–420. https://doi.org/10.1007/978-3-030-49732-3

Feng J, Liu R, Chen P, Yuan S, Zhao D, Zhang J, Zheng Z (2015): Degradation of aqueous 3, 4-dichloroaniline by a novel dielectric barrier discharge plasma reactor. *Environ Sci Pollut Res* 22, 4447–4459.

Garg N, Singla P (2011): Arsenic toxicity in crop plants: physiological effects and tolerance mechanisms. *Environ Chem Lett* 9, 303–321.

Genre A, Lanfranco L, Perotto S, Bonfante P (2020): Unique and common traits in mycorrhizal symbioses. *Nat Rev Microbiol* 18, 649–660.

Ghulam M, Muhammad AA, Donald LS, Sajid M, Muhammad FQ, Niaz A, Ateeq ur R, Shakeel A, Sajjad H, Muhammad A, Summia M, Aqib HAK, Fahad S, Rahul D, Mazhar I, Timothy DS (2021): Formalin fumigation and steaming of various composts differentially influence the nutrient release, growth and yield of muskmelon (Cucumis melo L.). *Sci Rep* 11, 21057.

Glick BR (2014): Bacteria with ACC deaminase can promote plant growth and help to feed the world. *Microbiol Res* 169, 30–39.

Gopakumar L, Bernard NO, Donato V (2020): Soil microarthropods and nutrient cycling. In: Fahad S, Hasanuzzaman M, Alam M, Ullah H, Saeed M, Khan AK, Adnan M (Eds.), *Environment, Climate, Plant and Vegetation Growth.* Springer Publ Ltd, Springer Nature Switzerland AG. Part of Springer Nature. pp. 453–472. https://doi.org/10.1007/978-3-030-49732-3

Guofu L, Zhenjian B, Fahad S, Guowen C, Zhixin X, Hao G, Dandan L, Yulong L, Bing L, Guoxu J, Saud S (2021): Compositional and structural changes in soil microbial communities in response to straw mulching and plant revegetation in an abandoned artificial pasture in Northeast China. *Glob Ecol Conserv* 31 (2021): e01871.

Gururani MA, Upadhyaya CP, Baskar V, Venkatesh J, Nookaraju A, Park SW (2013): Plant growth-promoting rhizobacteria enhance abiotic stress tolerance in Solanum tuberosum through inducing changes in the expression of ROS-scavenging enzymes and improved photosynthetic performance. *J Plant Growth Regul* 32, 245–258.

Habib ur R, Ashfaq A, Aftab W, Manzoor H, Fahd R, Wajid I, Md Aminul I, Vakhtang S, Muhammad A, Asmat U, Abdul W, Syeda RS, Shah S, Shahbaz K, Fahad S, Manzoor H, Saddam H, Wajid N (2017): Application of CSM-CROPGRO-cotton model for cultivars and optimum planting dates: Evaluation in changing semi-arid climate. *Field Crops Res* 238, 139–152 http://dx.doi.org/10.1016/j.fcr.2017.07.007

Hafeez M, Farman U, Muhammad MK, Xiaowei L, Zhijun Z, Sakhawat S, Muhammad I, Mohammed AA, Mandela F-G, Nicolas D, Muzammal R, Fahad S, Yaobin L (2021): Metabolic-based insecticide resistance mechanism and ecofriendly approaches for controlling of beet armyworm Spodoptera exigua:a review. *Environ Sci Pollution Res* 29(2), 1746–1762 https://doi.org/10.1007/s11356-021-16974-w

Hafez Y, Attia K, Alamery S, Ghazy A, Al-Doss A, Ibrahim E, Rashwan E, El-Maghraby L, Awad A, Abdelaal K (2020): Beneficial effects of biochar and chitosan on antioxidative capacity, osmolytes accumulation, and anatomical characters of water-stressed barley plants. *Agronomy* 10, 630.

Hafiz MH, Wajid F, Farhat A, Fahad S, Shafqat S, Wajid N, Hafiz FB (2016): Maize plant nitrogen uptake dynamics at limited irrigation water and nitrogen. *Environ Sci Pollut Res* 24(3), 2549–2557. https://doi.org/10.1007/s11356-016-8031-0

Hafiz MH, Farhat A, Shafqat S, Fahad S, Artemi C, Wajid F, Chaves CB, Wajid N, Muhammad M, Hafiz FB (2018): Offsetting land degradation through nitrogen and water management during maize cultivation under arid conditions. *Land Degrad Dev* 29(5), 1–10. doi: 10.1002/ldr.2933

Hafiz MH, Muhammad A, Farhat A, Hafiz FB, Saeed AQ, Muhammad M, Fahad S, Muhammad A (2019): Environmental factors affecting the frequency of road traffic accidents: a case study of sub-urban area of Pakistan. *Environ Sci Pollut Res* 26, 11674–11685 https://doi.org/10.1007/s11356-019-04752-8

Hafiz MH, Farhat A, Ashfaq A, Hafiz FB, Wajid F, Carol Jo W, Fahad S, Gerrit H (2020a): Predicting kernel growth of maize under controlled water and nitrogen applications. *Int J Plant Prod* 14, 609–320 https://doi.org/10.1007/s42106-020-00110-8

Hafiz MH, Abdul K, Farhat A, Wajid F, Fahad S, Muhammad A, Ghulam MS, Wajid N, Muhammad M, Hafiz FB (2020b): Comparative effects of organic and inorganic fertilizers on soil organic carbon and wheat productivity under arid region. *Commun Soil Sci Plant Anal* 51, 1406–1422 doi: 10.1080/00103624.2020.1763385

Haider SA, Lalarukh I, Amjad SF, Mansoora N, Naz M, Naeem M, Bukhari SA, Shahbaz M, Ali SA, Marfo TD, Subhan D, Rahul D, Fahad S (2021): Drought stress alleviation by potassium-nitrate-containing chitosan/montmorillonite microparticles confers changes in Spinacia oleracea L. *Sustain* 13, 9903. https://doi.org/10.3390/su13179903

Haldar S, Sengupta S (2015): Plant-microbe cross-talk in the rhizosphere: insight and biotechnological potential. *Open Microbiol J* 9, 1.

Hamza SM, Xiukang W, Sajjad A, Sadia Z, Muhammad N, Adnan M, Fahad S et al. (2021): Interactive effects of gibberellic acid and NPK on morpho-physio-biochemical traits and organic acid exudation pattern in coriander (Coriandrum sativum L.) grown in soil artificially spiked with boron. *Plant Physiol Biochem* 167(2021), 884–900.

Hao X, Xie P, Zhu Y-G, Taghavi S, Wei G, Rensing C (2015): Copper tolerance mechanisms of Mesorhizobium amorphae and its role in aiding phytostabilization by Robinia pseudoacacia in copper contaminated soil. *Environm Sci Tech* 49, 2328–2340.

Haoliang Y, Matthew TH, Ke L, Bin W, Puyu F, Fahad S, Holger M, Rui Y, De LL, Sotirios A, Isaiah H, Xiaohai T, Jianguo M, Yunbo Z, Meixue Z (2022): Crop traits enabling yield gains under more frequent extreme climatic events. *Sci Total Environ* 808(2022), 152170.

Hasegawa PM, Bressan RA, Zhu J-K, Bohnert HJ (2000): Plant cellular and molecular responses to high salinity. *Annu Rev Plant Biol* 51, 463–499.

Heidari M, Mousavinik SM, Golpayegani A (2011): Plant growth promoting rhizobacteria (PGPR) effect on physiological parameters and mineral uptake in basil (Ociumum basilicm L.) under water stress. *ARPN J Agric Biol Sci* 6, 6–11.

Hesham FA and Fahad S (2020): Melatonin application enhances biochar efficiency for drought tolerance in maize varieties: Modifications in physio-biochemical machinery. *Agron J* 112(4), 1–22.

Hu Y, Xie G, Jiang X, Shao K, Tang X, Gao G (2020): The relationships between the free-living and particle-attached bacterial communities in response to elevated eutrophication. *Front Microbiol* 11, 423.

Huang Li-Y, Li Xiao-X, Zhang Yun-B, Fahad S, Wang F (2021): dep1 improves rice grain yield and nitrogen use efficiency simultaneously by enhancing nitrogen and dry matter translocation. *J Integrative Agri* 21(11), 3185–3198 doi: 10.1016/S2095-3119(21)63795-4

Hussain MA, Fahad S, Rahat S, Muhammad FJ, Muhammad M, Qasid A, Ali A, Husain A, Nooral A, Babatope SA, Changbao S, Liya G, Ibrar A, Zhanmei J, Juncai H (2020): Multifunctional role of brassinosteroid and its analogues in plants. *Plant Growth Regul* 92, 141–156 https://doi.org/10.1007/s10725-020-00647-8

Ibad U, Dost M, Maria M, Shadman K, Muhammad A, Fahad S, Muhammad I, Ishaq AM, Aizaz A, Muhammad HS, Muhammad S, Farhana G, Muhammad I, Muhammad ASR, Hafiz MH, Wajid N, Shah S, Jabar ZKK, Masood A, Naushad A, Rasheed AkbarM, Shah MK Jan B (2022): Comparative effects of biochar and NPK on wheat crops under different management systems. *Crop Pasture Sci* 74, 31–40 https://doi.org/10.1071/CP21146

Ibrar H, Muqarrab A, Adel MG, Khurram S, Omer F, Shahid I, Fahim N, Shakeel A, Viliam B, Marian B, Sami Al Obaid, Fahad S, Subhan D, Suleyman T, Hanife AKÇA, Rahul D (2021): Improvement in growth and yield attributes of cluster bean through optimization of sowing time and plant spacing under climate change scenario. *Saudi J Bio Sci* 29(2), 781–792 https://doi.org/10.1016/j.sjbs.2021.11.018

Ibrar K, Aneela R, Khola Z, Urooba N, Sana B, Rabia S, Ishtiaq H, Mujaddad Ur Rehman, Salvatore M (2020): Microbes and environment: Global warming reverting the frozen zombies. In: Fahad S, Hasanuzzaman M, Alam M, Ullah H, Saeed M, Khan AK, Adnan M (Eds.), *Environment, Climate, Plant and Vegetation Growth*. Springer Publ Ltd, Springer Nature Switzerland AG. Part of Springer Nature. pp. 607–634. https://doi.org/10.1007/978-3-030-49732-3

Ihsan MZ, Abdul K, Manzer HS, Liaqat A, Ritesh K, Hayssam MA, Amar M, Fahad S (2022): The response of Triticum aestivum treated with plant growth regulators to acute day/night temperature rise. *J Plant Growth Regul* 41(5), 2020–2033 https://doi.org/10.1007/s00344-022-10574-9

Ikram U, Khadim D, Muhammad T, Muhammad S, Fahad S (2021): Gibberellic acid and urease inhibitor optimize nitrogen uptake and yield of maize at varying nitrogen levels under changing climate. *Environ Sci Pollution Res* 29(5), 6568–6577 https://doi.org/10.1007/s11356-021-16049-w

Ilangumaran G, Smith DL (2017): Plant growth promoting rhizobacteria in amelioration of salinity stress: a systems biology perspective. *Front Plant Sci* 8, 1768.

Ilyas M, Nisar M, Khan N, Hazrat A, Khan AH, Hayat K, Fahad S, Khan A, Ullah A (2021): Drought tolerance strategies in plants: a mechanistic approach. *J Plant Growth Regul* 40, 926–944.

Iqra M, Amna B, Shakeel I, Fatima K, Sehrish L, Hamza A, Fahad S (2020): Carbon cycle in response to global warming. In: Fahad S, Hasanuzzaman M, Alam M, Ullah H,Saeed M, Khan AK, Adnan M (Eds.), *Environment, Climate, Plant and Vegetation Growth*. Springer Publ Ltd, Springer Nature Switzerland AG, pp. 1–16. https://doi.org/10.1007/978-3-030-49732-3

Irfan M, Muhammad M, Muhammad JK, Khadim MD, Dost M, Ishaq AM, Waqas A, Fahad S, Saud S et al. (2021): Heavy metals immobilization and improvement in maize (Zea mays L.) growth amended with biochar and compost. *Sci Rep* 11, 18416I.

Izrael-Živković L, Rikalović M, Gojgić-Cvijović G, Kazazić S, Vrvić M, Brčeski I, Beškoski V, Lončarević B, Gopčević K, Karadžić I (2018): Cadmium specific proteomic responses of a highly resistant Pseudomonas aeruginosa san ai. *RSC Adv* 8, 10549–10560.

Jabborova D, Sulaymanov K, Sayyed RZ, Alotaibi SH, Enakiev Y, Azimov A, Jabbarov Z, Ansari MJ, Fahad S, Danish S et al. (2021): Mineral fertilizers improve the quality of turmeric and soil. *Sustain* 13, 9437. https://doi.org/10.3390/su13169437

Jan M, Muhammad Anwar-ul-Haq, Adnan NS, Muhammad Y, Javaid I, Xiuling L, Depeng W, Fahad S (2019): Modulation in growth, gas exchange, and antioxidant activities of salt-stressed rice (Oryza sativa L.) genotypes by zinc fertilization. *Arabian J Geosci* 12, 775 https://doi.org/10.1007/s12517-019-4939-2

Jorge GL, Kisiala A, Morrison E, Aoki M, Nogueira APO, Emery RN (2019): Endosymbiotic Methylobacterium oryzae mitigates the impact of limited water availability in lentil (Lens culinaris Medik.) by increasing plant cytokinin levels. *Environ Exp Bot* 162, 525–540

Kamaran M, Wenwen C, Irshad A, Xiangping M, Xudong Z, Wennan S, Junzhi C, Shakeel A, Fahad S, Qingfang H, Tiening L (2017): Effect of paclobutrazol, a potential growth regulator on stalk mechanical strength, lignin accumulation and its relation with lodging resistance of maize. *Plant Growth Regul* 84, 317–332. https://doi.org/10.1007/ s10725-017-0342-8

Kang S-M, Shahzad R, Bilal S, Khan AL, Park Y-G, Lee K-E, Asaf S, Khan MA, Lee I-J (2019): Indole-3-acetic-acid and ACC deaminase producing Leclercia adecarboxylata MO1 improves Solanum lycopersicum L. growth and salinity stress tolerance by endogenous secondary metabolites regulation. *BMC Microbiol* 19, 1–14.

Karakas S, Dikilitas M, Tıpırdamaz R (2020): Phytoremediation of salt-affected soils using halophytes. *Handbook of Halophytes: From Molecules to Ecosystems towards Biosaline Agriculture*, 1–18.

Kasotia A, Varma A, Tuteja N, Choudhary DK (2016): Microbial-mediated amelioration of plants under abiotic stress: an emphasis on arid and semiarid climate. In: Devendra K. Choudhary, Ajit Varma, Narendra Tuteja (Eds.), *Plant-Microbe Interaction: An Approach to Sustainable Agriculture*. Springer, pp. 155–163.

Kaushal M, Wani SP (2016): Plant-growth-promoting rhizobacteria: drought stress alleviators to ameliorate crop production in drylands. *Annals of Microbiology* 66, 35–42.

Khadim D, Fahad S, Jahangir MMR, Iqbal M, Syed SA, Shah AK, Ishaq AM, Rahul D et al. (2021a): Biochar and urease inhibitor mitigate NH_3 and N_2O emissions and improve wheat yield in a urea fertilized alkaline soil. *Sci Rep* 11, 17413.

Khadim D, Saif-ur-R, Fahad S, Syed SA, Shah AK et al. (2021b): Influence of variable biochar concentration on yield-scaled nitrous oxide emissions, Wheat yield and nitrogen use efficiency. *Sci Rep* 11, 16774.

Khan M, Asaf S, Khan A, Adhikari A, Jan R, Ali S, Imran M, Kim KM, Lee IJ (2020): Plant growth-promoting endophytic bacteria augment growth and salinity tolerance in rice plants. *Plant Biology* 22, 850–862.

Khan MA, Asaf S, Khan AL, Ullah I, Ali S, Kang S-M, Lee I-J (2019): Alleviation of salt stress response in soybean plants with the endophytic bacterial isolate Curtobacterium sp. SAK1. *Annals Microbiol* 69, 797–808.

Khan MHU, Khattak JZK, Jamil M, Malook I, Khan SU, Jan M, Din I, Saud S, Kamran M, Alharby H (2017): Bacillus safensis with plant-derived smoke stimulates rice growth under saline conditions. *Environ Sci Pollut Res* 24, 23850–23863.

Khan MMH, Niaz A, Umber G, Muqarrab A, Muhammad AA, Muhammad I, Shabir H, Shah F, Vibhor A, Shams HA-H, Reham A, Syed MBA, Nadiyah MA, Ali TKZ, Subhan D, Rahul D (2021): Synchronization of Boron application methods and rates is environmentally friendly approach to improve quality attributes of Mangifera indica L. on sustainable basis. *Saudi J Bio Sci* 29(3), 1869–1880 https://doi.org/10.1016/j.sjbs.2021.10.036

Khan N, Ali S, Shahid MA, Mustafa A, Sayyed R, Curá JA (2021): Insights into the interactions among roots, rhizosphere, and rhizobacteria for improving plant growth and tolerance to abiotic stresses: a review. *Cells* 10, 1551.

Khatun M, Sarkar S, Era FM, Islam AKMM, Anwar MP, Fahad S, Datta R, Islam AKMA (2021): Drought stress in grain legumes: effects, tolerance mechanisms and management. *Agron* 11, 2374. https://doi.org/10.3390/agronomy11122374

Kim K, Jang Y-J, Lee S-M, Oh B-T, Chae J-C, Lee K-J (2014): Alleviation of salt stress by Enterobacter sp. EJ01 in tomato and Arabidopsis is accompanied by up-regulation of conserved salinity responsive factors in plants. *Mol Cells* 37, 109.

Kumar M, Mishra S, Dixit V, Agarwal L, Chauhan PS, Nautiyal CS (2016): Synergistic effect of Pseudomonas putida and Bacillus amyloliquefaciens ameliorates drought stress in chickpea (Cicer arietinum L.). *Plant Signaling Behav* 11, e1071004.

Lakshmanan V, Selvaraj G, Bais HP (2014): Functional soil microbiome: belowground solutions to an aboveground problem. *Plant Physiology* 166, 689–700.

Lawson CE, Harcombe WR, Hatzenpichler R, Lindemann SR, Löffler FE, O'Malley MA, García Martín H, Pfleger BF, Raskin L, Venturelli OS (2019): Common principles and best practices for engineering microbiomes. *Nature Rev Microbiol* 17, 725–741.

Liao X, Chen J, Guan R, Liu J, Sun Q (2021): Two arbuscular mycorrhizal fungi alleviate drought stress and improves plant growth in Cinnamomum migao seedlings. *Mycobiol* 49, 396–405.

Lim J-H, Kim S-D (2013): Induction of drought stress resistance by multi-functional PGPR Bacillus licheniformis K11 in pepper. *Plant Pathol J* 29, 201.

Liu F, Jensen CR, Shahanzari A, Andersen MN, Jacobsen S-E (2005): ABA regulated stomatal control and photosynthetic water use efficiency of potato (Solanum tuberosum L.) during progressive soil drying. *Plant Sci* 168, 831–836.

Liu H, Carvalhais LC, Crawford M, Singh E, Dennis PG, Pieterse CM, Schenk PM (2017): Inner plant values: diversity, colonization and benefits from endophytic bacteria. *Front Microbiol* 8, 2552.

Luo W, Xu C, Ma W, Yue X, Liang X, Zuo X, Knapp AK, Smith MD, Sardans J, Dijkstra FA (2018): Effects of extreme drought on plant nutrient uptake and resorption in rhizomatous vs bunchgrass-dominated grasslands. *Oecologia* 188, 633–643.

Luziatelli F, Ficca AG, Cardarelli M, Melini F, Cavalieri A, Ruzzi M (2020): Genome sequencing of Pantoea agglomerans C1 provides insights into molecular and genetic mechanisms of plant growth-promotion and tolerance to heavy metals. *Microorg* 8, 153.

Maggio A, Miyazaki S, Veronese P, Fujita T, Ibeas JI, Damsz B, Narasimhan ML, Hasegawa PM, Joly RJ, Bressan RA (2002): Does proline accumulation play an active role in stress-induced growth reduction? *Plant J* 31, 699–712.

Mahar A, Amjad A, Altaf HL, Fazli W, Ronghua L, Muhammad A, Fahad S, Muhammad A, Rafiullah, Imtiaz AK, Zengqiang Z (2020): Promising technologies for Cd-contaminated soils: Drawbacks and possibilities. In: Fahad S, Hasanuzzaman M, Alam M, Ullah H, Saeed M, Khan AK, Adnan M (Eds.), *Environment, Climate, Plant and Vegetation Growth.* Springer Publ Ltd, Springer Nature Switzerland AG. Part of Springer Nature. pp. 63–92. https://doi.org/10.1007/978-3-030-49732-3

Mahmood Ul H, Tassaduq R, Chandni I, Adnan A, Muhammad A, Muhammad MA, Muhammad H-ur-R, Mehmood AN, Alam S, Fahad S (2021): Linking plants functioning to adaptive responses under heat stress conditions: a mechanistic review. *J Plant Growth Regul* 41, 2596–2613 https://doi.org/10.1007/s00344-021-10493-1

Mallick I, Islam E, Kumar Mukherjee S (2015): Fundamentals and application potential of arsenic-resistant bacteria for bioremediation in rhizosphere: a review. *Soil Sediment Contam: Int J* 24, 704–718.

Manavalan LP, Guttikonda SK, Phan Tran L-S, Nguyen HT (2009): Physiological and molecular approaches to improve drought resistance in soybean. *Plant and Cell Physiology* 50, 1260–1276.

Manivasagan P, Venkatesan J, Sivakumar K, Kim S-K (2014): Pharmaceutically active secondary metabolites of marine actinobacteria. *Microbial Res* 169, 262–278.

Manzer HS, Saud A, Soumya M, Abdullah A, Al-A, Qasi DA, Bander MA. Al-M, Hayssam MA, Hazem MK, Fahad S, Vishnu DR, Om PN (2021): Molybdenum and hydrogen sulfide synergistically mitigate arsenic toxicity by modulating defense system, nitrogen and cysteine assimilation in faba bean (Vicia faba L.) seedlings. *Environ Pollut* 290, 117953.

Mariotti L, Scartazza A, Curadi M, Picciarelli P, Toffanin A (2021): Azospirillum baldaniorum sp245 induces physiological responses to alleviate the adverse effects of drought stress in purple basil. *Plants* 10, 1141.

Marulanda A, Barea J-M, Azcón R (2009): Stimulation of plant growth and drought tolerance by native microorganisms (AM fungi and bacteria) from dry environments: mechanisms related to bacterial effectiveness. *J Plant Growth Regul* 28, 115–124.

Matias SR, Pagano MC, Muzzi FC, Oliveira CA, Carneiro AA, Horta SN, Scotti MR (2009): Effect of rhizobia, mycorrhizal fungi and phosphate-solubilizing microorganisms in the rhizosphere of native plants used to recover an iron ore area in Brazil. *Eur J Soil Biol* 45, 259–266.

Mazhar SH, Herzberg M, Fekih IB, Zhang C, Bello SK, Li YP, Su J, Xu J, Feng R, Zhou S (2020): Comparative insights into the complete genome sequence of highly metal resistant Cupriavidus metallidurans strain BS1 isolated from a gold–copper mine. *Front in Microbiol* 11, 47

Md Enamul H, Shoeb AZM, Mallik AH, Fahad S, Kamruzzaman MM, Akib J, Nayyer S, Mehedi AKM, Swati AS, Md Yeamin A, Most SS (2020): Measuring vulnerability to environmental hazards: qualitative to quantitative. In: Fahad S, Hasanuzzaman M, Alam M, Ullah H, Saeed M, Khan AK, Adnan M (Eds.), *Environment, Climate, Plant and Vegetation Growth.* Springer Publ Ltd, Springer Nature Switzerland AG. Part of Springer Nature. pp. 421–452. https://doi.org/10.1007/978-3-030-49732-3

Md Jakir H, Allah B (2020): Development and applications of transplastomic plants: a way towards eco-friendly agriculture. In: Fahad S, Hasanuzzaman M, Alam M, Ullah H, Saeed M, Khan AK, Adnan M (Eds.), *Environment, Climate, Plant and Vegetation Growth*. Springer Publ Ltd, Springer Nature Switzerland AG. Part of Springer Nature. pp. 285–322. https://doi.org/10.1007/978-3-030-49732-3

Mehmood K, Bao Y, Saifullah, Bibi S, Dahlawi S, Yaseen M, Abrar MM, Srivastava P, Fahad S, Faraj TK (2022): Contributions of open biomass burning and crop straw burning to air quality: current research paradigm and future outlooks. *Front Environ Sci* 10, 852492. doi: 10.3389/fenvs.2022.852492

Mendes R, Raaijmakers JM (2015): Cross-kingdom similarities in microbiome functions. *ISME J* 9, 1905–1907.

Milet A, Chaouche NK, Dehimat L, Kaki AA, Ali MK, Thonart P (2016): Flow cytometry approach for studying the interaction between Bacillus mojavensis and Alternaria alternata. *Afr J Biotechnol* 15, 1417–1428.

Mohammad I Al-Wabel, Munir Ahmad, Adel RA Usman, Mutair Akanji, and Muhammad Imran Rafique (2020a): Advances in pyrolytic technologies with improved carbon capture and storage to combat climate change. In: Fahad S, Hasanuzzaman M, Alam M, Ullah H, Saeed M, Khan AK, Adnan M (Eds.), *Environment, Climate, Plant and Vegetation Growth*. Springer Publ Ltd, Springer Nature Switzerland AG. Part of Springer Nature. pp. 535–576. https://doi.org/10.1007/978-3-030-49732-3

Mohammad I Al-Wabel, Abdelazeem S, Munir A, Khalid E, Adel RAU (2020b): Extent of climate change in Saudi Arabia and its impacts on agriculture: a case study from Qassim region. In: Fahad S, Hasanuzzaman M, Alam M, Ullah H, Saeed M, Khan AK, Adnan M (Eds.), *Environment, Climate, Plant and Vegetation Growth*. Springer Publ Ltd, Springer Nature Switzerland AG. Part of Springer Nature. pp. 635–658. https://doi.org/10.1007/978-3-030-49732-3

Moretti C, Mattos L, Calbo A, Sargent S (2010): Climate changes and potential impacts on postharvest quality of fruit and vegetable crops: a review. *Food Res Int* 43, 1824–1832.

Morsy M, Cleckler B, Armuelles-Millican H (2020): Fungal endophytes promote tomato growth and enhance drought and salt tolerance. *Plants* 9, 877.

Mubeen M, Ashfaq A, Hafiz MH, Muhammad A, Hafiz UF, Mazhar S, Muhammad Sami ul Din, Asad A, Amjed A, Fahad S, Wajid N (2020): Evaluating the climate change impact on water use efficiency of cotton-wheat in semi-arid conditions using DSSAT model. *J Water Climate Change* 11(4), 1661–1675 doi/10.2166/wcc.2019.179/622035/jwc2019179.pdf

Muhammad I, Khadim D, Fahad S, Imran M, Saud A, Manzer HS, Shah S, Jabar ZKK, Shamsher A, Shah H, Taufiq N, Hafiz MH, Jan B, Wajid N (2022): Exploring the potential effect of Achnatherum splendens L.–derived biochar treated with phosphoric acid on bioavailability of cadmium and wheat growth in contaminated soil. *Environ Sci Pollut Res* 29(25), 37676–37684 https://doi.org/10.1007/s11356-021-17950-0.

Muhammad N, Muqarrab A, Khurram S, Fiaz A, Fahim N, Muhammad A, Shazia A, Omaima N, Sulaiman AA, Fahad S, Subhan D, Rahul D (2021): Kaolin and jasmonic acid improved cotton productivity under water stress conditions. *J Saudi Society Agricultural Sci* 28 (2021), 6606–6614 https://doi.org/10.1016/j.sjbs.2021.07.043

Muhammad Tahir ul Qamar, Amna F, Amna B, Barira Z, Xitong Z, Ling-Ling C (2020): Effectiveness of conventional crop improvement strategies vs. omics. In: Fahad S, Hasanuzzaman M, Alam M, Ullah H, Saeed M, Khan AK, Adnan M (Eds.), *Environment, Climate, Plant and Vegetation Growth*. Springer Publ Ltd, Springer Nature Switzerland AG. Part of Springer Nature. pp. 253–284. https://doi.org/10.1007/978-3-030-49732-3

Muhammad Z, Abdul MK, Abdul MS, Kenneth BM, Muhammad S, Shahen S, Ibadullah J, Fahad S (2019): Performance of Aeluropus lagopoides (mangrove grass) ecotypes, a potential turfgrass, under high saline conditions. *Environ Sci Pollut Res* 26(13), 13410–13412 https://doi.org/10.1007/s11356-019-04838-3

Munns R (2002): Comparative physiology of salt and water stress. *Plant, Cell & Environment* 25, 239–250.

Muzammal R, Fahad S, Guanghui D, Xia C, Yang Y, Kailei T, Lijun L, Fei-Hu L, Gang D (2021): Evaluation of hemp (Cannabis sativa L.) as an industrial crop: a review. *Environ Sci Pollution Res* 28(38), 52832–52843 https://doi.org/10.1007/s11356-021-16264-5

Nadal Jimenez P, Koch G, Thompson JA, Xavier KB, Cool RH, Quax WJ (2012): The multiple signaling systems regulating virulence in Pseudomonas aeruginosa. *Microbiol Mol Biol Rev* 76, 46–65.

Nanda M, Kumar V, Sharma D (2019): Multimetal tolerance mechanisms in bacteria: The resistance strategies acquired by bacteria that can be exploited to "clean-up" heavy metal contaminants from water. *Aquat Toxicol* 212, 1–10.

Naranjo-Ortiz MA, Gabaldón T (2020): Fungal evolution: cellular, genomic and metabolic complexity. *Biol Rev* 95, 1198–1232.

Niaz A, Abdullah E, Subhan D, Muhammad A, Fahad S, Khadim D, Suleyman T, Hanife A, Anis AS, Mohammad JA, Emre B, Omer SU, Rahul D, Bernard RG (2022): Mitigation of lead (Pb) toxicity in rice cultivated with either ground water or wastewater by application of acidified carbon. *J Environ Manage* 307, 114521.

Niu X, Song L, Xiao Y, Ge W (2018): Drought-tolerant plant growth-promoting rhizobacteria associated with foxtail millet in a semi-arid agroecosystem and their potential in alleviating drought stress. *Front in Microbiol* 8, 2580.

Noor M, Naveed ur R, Ajmal J, Fahad S, Muhammad A, Fazli W, Saud S, Hassan S (2020): Climate change and coastal plant lives. In: Fahad S, Hasanuzzaman M, Alam M, Ullah H, Saeed M, Khan AK, Adnan M (Eds.), *Environment, Climate, Plant and Vegetation Growth.* Springer Publ Ltd, Springer Nature Switzerland AG. Part of Springer Nature. pp. 93–108. https://doi.org/10.1007/978-3-030-49732-3

Ottesen AR, González Peña A, White JR, Pettengill JB, Li C, Allard S, Rideout S, Allard M, Hill T, Evans P (2013): Baseline survey of the anatomical microbial ecology of an important food plant: Solanum lycopersicum (tomato). *BMC Microbiol* 13, 1–12.

Phelan VV, Liu W-T, Pogliano K, Dorrestein PC (2012): Microbial metabolic exchange – the chemotype-to-phenotype link. *Nature Chem Biol* 8, 26–35.

Phung LT, Trimble WL, Meyer F, Gilbert JA, Silver S (2012): Draft genome sequence of Alcaligenes faecalis subsp. Faecalis NCIB 8687 (CCUG 2071). *Am Soc Microbiol* 194(18), 5153.

Piazza V, Ferioli A, Giacco E, Melchiorre N, Valenti A, Del Prete F, Biandolino F, Dentone L, Frisenda P, Faimali M (2012): A standardization of Amphibalanus (Balanus) mphitrite (Crustacea, Cirripedia) larval bioassay for ecotoxicological studies. *Ecotoxicol Environ Saf* 79, 134–138.

Qamar-uz Z, Zubair A, Muhammad Y, Muhammad ZI, Abdul K, Fahad S, Safder B, Ramzani PMA, Muhammad N (2017): Zinc biofortification in rice: leveraging agriculture to moderate hidden hunger in developing countries. *Arch Agron Soil Sci* 64, 147–161. https://doi.org/10.1080/03650340.2017.1338343

Qin ZH, Nasib ur Rahman, Ahmad A, Yu-pei Wang, Sakhawat S, Ehmet N, Wen-juan Shao, Muhammad I, Kun S, Rui L, Fazal S, Fahad S (2022): Range expansion decreases the reproductive fitness of Gentiana officinalis (Gentianaceae). *Sci Rep* 12, 2461 https://doi.org/10.1038/s41598-022-06406-1

Rajesh KS, Fahad S, Pawan K, Prince C, Talha J, Dinesh J, Prabha S, Debanjana S, Prathibha MD, Bandana B, Akash H, Gupta NK, Rekha S, Devanshu D, Dalpat LS, Ke L, Matthew TH, Saud S, Adnan NS, Taufiq N (2022): Beneficial elements: new players in improving nutrient use efficiency and abiotic stress tolerance. *Plant Growth Regul* 100, 237–265 https://doi.org/10.1007/s10725-022-00843-8

Rashid M, Qaiser H, Khalid SK, Mohammad I. Al-Wabel, Zhang A, Muhammad A, Shahzada SI, Rukhsanda A, Ghulam AS, Shahzada MM, Sarosh A, Muhammad FQ (2020): Prospects of biochar in alkaline soils to mitigate climate change. In: Fahad S, Hasanuzzaman M, Alam M, Ullah H,Saeed M, Khan AK, Adnan M (Eds.), *Environment, Climate, Plant and Vegetation Growth.* Springer Publ Ltd, Springer Nature Switzerland AG. Part of Springer Nature. pp. 133–150. https://doi.org/10.1007/978-3-030-49732-3

Rath KM, Maheshwari A, Rousk J (2019): Linking microbial community structure to trait distributions and functions using salinity as an environmental filter. *Mbio* 10, e01607–19.

Reddy AR, Chaitanya KV, Vivekanandan M (2004): Drought-induced responses of photosynthesis and antioxidant metabolism in higher plants. *J Plant Physiol* 161, 1189–1202.

Rehana S, Asma Z, Shakil A, Anis AS, Rana KI, Shabir H, Subhan D, Umber G, Fahad S, Jiri K, Sami Al Obaid, Mohammad JA, Rahul D (2021): Proteomic changes in various plant tissues associated with chromium stress in sunflower. *Saudi J Bio Sci* 29(4), 2604–2612 https://doi.org/10.1016/j.sjbs.2021.12.042

Rehman M, Fahad S, Saleem MH, Hafeez M, Muhammad Habib ur Rahman, Liu F, Deng G (2020): Red light optimized physiological traits and enhanced the growth of ramie (Boehmeria nivea L.). *Photosynthetica* 58(4), 922–931.

Reinhold-Hurek B, Bünger W, Burbano CS, Sabale M, Hurek T (2015): Roots shaping their microbiome: global hotspots for microbial activity. *Annu Rev Phytopathol* 53, 403–424.

Rodriguez R, White Jr J, Arnold AE, Redman aRa (2009): Fungal endophytes: diversity and functional roles. *New Phytol* 182, 314–330.

Ruiz-Lozano JM, Porcel R, Azcón C, Aroca R (2012): Regulation by arbuscular mycorrhizae of the integrated physiological response to salinity in plants: new challenges in physiological and molecular studies. *J Exp Bot* 63, 4033–4044.

Ruppel S, Franken P, Witzel K (2013): Properties of the halophyte microbiome and their implications for plant salt tolerance. *Funct Plant Biol* 40, 940–951.

Rysiak A, Dresler S, Hanaka A, Hawrylak-Nowak B, Strzemski M, Kováčik J, Sowa I, Latalski M, Wójciak M (2021): High temperature alters secondary metabolites and photosynthetic efficiency in Heracleum sosnowskyi. *Int J Mol Sci* 22, 4756.

Sadam M, Muhammad Tahir ul Qamar, Ghulam M, Muhammad SK, Faiz AJ (2020): Role of biotechnology in climate resilient agriculture. In: Fahad S, Hasanuzzaman M, Alam M, Ullah H, Saeed M, Khan AK, Adnan M (Eds.), *Environment, Climate, Plant and Vegetation Growth.* Springer Publ Ltd, Springer Nature Switzerland AG. Part of Springer Nature. pp. 339–366. https://doi.org/10.1007/978-3-030-49732-3

Safi UK, Ullah F, Mehmood S, Fahad S, Ahmad Rahi A, Althobaiti F et al. (2021): Antimicrobial, antioxidant and cytotoxic properties of Chenopodium glaucum L. *PLoS One* 16(10), e0255502. https://doi.org/10.1371/journal. pone.0255502

Sahrish N, Shakeel A, Ghulam A, Zartash F, Sajjad H, Mukhtar A, Muhammad AK, Ahmad K, Fahad S, Wajid N, Sezai E, Carol Jo W, Gerrit H (2022): Modeling the impact of climate warming on potato phenology. *European J AgroN* 132, 126404.

Sajid H, Jie H, Jing H, Shakeel A, Satyabrata N, Sumera A, Awais S, Chunquan Z, Lianfeng Z, Xiaochuang C, Qianyu J, Junhua Z (2020): Rice production under climate change: Adaptations and mitigating strategies. In: Fahad S, Hasanuzzaman M, Alam M, Ullah H, Saeed M, Khan AK, Adnan M (Eds.), *Environment, Climate, Plant and Vegetation Growth.* Springer Publ Ltd, Springer Nature Switzerland AG. Part of Springer Nature. pp. 659–686. https://doi.org/10.1007/978-3-030-49732-3

Sajjad H, Muhammad M, Ashfaq A, Waseem A, Hafiz MH, Mazhar A, Nasir M, Asad A, Hafiz UF, Syeda RS, Fahad S, Depeng W, Wajid N (2019): Using GIS tools to detect the land use/land cover changes during forty years in Lodhran district of Pakistan. *Environ Sci Pollut Res* 27, 39676–39692 https://doi.org/10.1007/s11356-019-06072-3

Sajjad H, Muhammad M, Ashfaq A, Fahad S, Wajid N, Hafiz MH, Ghulam MS, Behzad M, Muhammad T, Saima P (2021a): Using space–time scan statistic for studying the effects of COVID-19 in Punjab, Pakistan: a guideline for policy measures in regional Agriculture. *Environ Sci Pollut Res* 30, 42495–42508 https://doi.org/10.1007/s11356-021-17433-2

Sajjad H, Muhammad M, Ashfaq A, Nasir M, Hafiz MH, Muhammad A, Muhammad I, Muhammad U, Hafiz UF, Fahad S, Wajid N, Hafiz MRJ, Mazhar A, Saeed AQ, Amjad F, Muhammad SK, Mirza W (2021b): Satellite-based evaluation of temporal change in cultivated land in Southern Punjab (Multan region) through dynamics of vegetation and land surface temperature. *Open Geo Sci* 13, 1561–1577.

Saleem MH, Fahad S, Adnan M, Mohsin A, Muhammad SR, Muhammad K, Qurban A, Inas AH, Parashuram B, Mubassir A, Reem MH (2020a): Foliar application of gibberellic acid endorsed phytoextraction of copper and alleviates oxidative stress in jute (Corchorus capsularis L.) plant grown in highly copper-contaminated soil of China. *Environ Sci Pollution Res* 27, 37121–37133 https://doi.org/10.1007/s11356-020-09764-3

Saleem MH, Rehman M, Fahad S, Tung SA, Iqbal N, Hassan A, Ayub A, Wahid MA, Shaukat S, Liu L, Deng G (2020b): Leaf gas exchange, oxidative stress, and physiological attributes of rapeseed (Brassica napus L.) grown under different light-emitting diodes. *Photosynthetica* 58(3), 836–845.

Saleem MH, Fahad S, Shahid UK, Mairaj D, Abid U, Ayman ELS, Akbar H, Analía L, Lijun L (2020c): Copper-induced oxidative stress, initiation of antioxidants and phytoremediation potential of flax (Linum usitatissimum L.) seedlings grown under the mixing of two different soils of China. *Environ Sci Poll Res* 27, 5211–5221 https://doi.org/10.1007/s11356-019-07264-7

Saman S, Amna B, Bani A, Muhammad Tahir ul Qamar, Rana MA, Muhammad SK (2020): QTL mapping for abiotic stresses in cereals. In: Fahad S, Hasanuzzaman M, Alam M, Ullah H, Saeed M, Khan AK, Adnan M (Eds.), *Environment, Climate, Plant and Vegetation Growth.* Springer Publ Ltd, Springer Nature Switzerland AG. Part of Springer Nature. pp. 229–252. https://doi.org/10.1007/978-3-030-49732-3

Sandhya V, Ali SZ, Grover M, Reddy G, Venkateswarlu B (2010): Effect of plant growth promoting Pseudomonas spp. On compatible solutes, antioxidant status and plant growth of maize under drought stress. *Plant Growth Regul* 62, 21–30.

Sana U, Shahid A, Yasir A, Farman UD, Syed IA, Mirza MFAB, Fahad S, Al-Misned F, Usman A, Xinle G, Ghulam N, Kunyuan W (2022): Bifenthrin induced toxicity in Ctenopharyngodon della at an acute concentration: a multi-biomarkers based study. *J King Saud Uni Sci* 34(2022), 101752.

Sarkar A, Ghosh PK, Pramanik K, Mitra S, Soren T, Pandey S, Mondal MH, Maiti TK (2018): A halotolerant Enterobacter sp. Displaying ACC deaminase activity promotes rice seedling growth under salt stress. *Res Microbiol* 169, 20–32.

Sarma RK, Saikia R (2014): Alleviation of drought stress in mung bean by strain Pseudomonas aeruginosa GGRJ21. *Plant Soil* 377, 111–126.
Saud S, Chen Y, Long B, Fahad S, Sadiq A (2013): The different impact on the growth of cool season turf grass under the various conditions on salinity and drought stress. *Int J Agric Sci Res* 3, 77–84.
Saud S, Li X, Chen Y, Zhang L, Fahad S, Hussain S, Sadiq A, Chen Y (2014): Silicon application increases drought tolerance of Kentucky bluegrass by improving plant water relations and morph physiological functions. *SciWorld J* 2014, 1–10 https://doi.org/10.1155/2014/ 368694
Saud S, Chen Y, Fahad S, Hussain S, Na L, Xin L, Alhussien SA (2016): Silicate application increases the photosynthesis and its associated metabolic activities in Kentucky bluegrass under drought stress and post-drought recovery. *Environ Sci Pollut Res* 23(17), 17647–17655. https://doi.org/10.1007/s11 356-016-6957-x
Saud S, Fahad S, Yajun C, Ihsan MZ, Hammad HM, Nasim W, Amanullah Jr, Arif M and Alharby H (2017): Effects of nitrogen supply on water stress and recovery mechanisms in Kentucky bluegrass plants. *Front. Plant Sci* 8, 983. doi: 10.3389/fpls.2017.00983
Saud S, Fahad S, Cui G, Chen Y, Anwar S (2020): Determining nitrogen isotopes discrimination under drought stress on enzymatic activities, nitrogen isotope abundance and water contents of Kentucky bluegrass. *Sci Rep* 10, 6415 https://doi.org/10.1038/s41598-020-63548-w
Saud S, Fahad S, Hassan S (2022a): Developments in the investigation of nitrogen and oxygen stable isotopes in atmospheric nitrate. *Sustainable Chem Clim Action* 1, 100003.
Saud S, Li X, Jiang Z, Fahad S, Hassan S (2022b): Exploration of the phytohormone regulation of energy storage compound accumulation in microalgae. *Food Energy Secur* 2022, e418.
Schutzendubel A, Polle A (2002): Plant responses to abiotic stresses: heavy metal-induced oxidative stress and protection by mycorrhization. *J Exp Bot* 53, 1351–1365.
Semchenko M, Leff JW, Lozano YM, Saar S, Davison J, Wilkinson A, Jackson BG, Pritchard WJ, De Long JR, Oakley S (2018): Fungal diversity regulates plant-soil feedbacks in temperate grassland. *Sci Adv* 4(11), eaau4578.
Senol C (2020): The effects of climate change on human behaviors. In: Fahad S, Hasanuzzaman M, Alam M, Ullah H, Saeed M, Khan AK, Adnan M (Eds.), *Environment, Climate, Plant and Vegetation Growth.* Springer Publ Ltd, Springer Nature Switzerland AG. Part of Springer Nature. pp. 577–590. https://doi.org/10.1007/978-3-030-49732-3
Shafi MI, Adnan M, Fahad S, Fazli W, Ahsan K, Zhen Y, Subhan D, Zafar-ul-Hye M, Martin B, Rahul D (2020): Application of single superphosphate with humic acid improves the growth, yield and phosphorus uptake of wheat (Triticum aestivum L.) in calcareous soil. *Agron* 10, 1224. doi:10.3390/agronomy10091224
Shah F, Lixiao N, Kehui C, Tariq S, Wei W, Chang C, Liyang Z, Farhan A, Fahad S, Huang J (2013): Rice grain yield and component responses to near 2°C of warming. *Field Crop Res* 157, 98–110.
Shah S, Shah H, Liangbing X, Xiaoyang S, Shahla A, Fahad S (2022): The physiological function and molecular mechanism of hydrogen sulfide resisting abiotic stress in plants. *Brazilian J Botany* 45, 5639–572 https://doi.org/10.1007/s40415-022-00785-5
Shao H-B, Chu L-Y, Jaleel CA, Zhao C-X (2008): Water-deficit stress-induced anatomical changes in higher plants. *Comptes Rendus Biologies* 331, 215–225.
Sidra K, Javed I, Subhan D, Allah B, Syed IUSB, Fatma B, Khaled DA, Fahad S, Omaima N, Ali TKZ, Rahul D (2021): Physio-chemical characterization of indigenous agricultural waste materials for the development of potting media. *J Saudi Society Agricultural Sci* 28(12), 7491–7498 https://doi.org/10.1016/j.sjbs.2021.08.058
Silver S, Phung LT (2005): Genes and enzymes involved in bacterial oxidation and reduction of inorganic arsenic. *Appl Environ Microbiol* 71, 599–608.
Subhan D, Zafar-ul-Hye M, Fahad S, Saud S, Martin B, Tereza H, Rahul D (2020): Drought stress alleviation by ACC deaminase producing Achromobacter xylosoxidans and Enterobacter cloacae, with and without timber waste biochar in maize. *Sustain* 12(15), 6286 doi:10.3390/su12156286
Sun C, Johnson JM, Cai D, Sherameti I, Oelmüller R, Lou B (2010): Piriformospora indica confers drought tolerance in Chinese cabbage leaves by stimulating antioxidant enzymes, the expression of drought-related genes and the plastid-localized CAS protein. *J Plant Physiol* 167, 1009–1017.
Sun L, Lei P, Wang Q, Ma J, Zhan Y, Jiang K, Xu Z, Xu H (2020): The endophyte Pantoea alhagi NX-11 alleviates salt stress damage to rice seedlings by secreting exopolysaccharides. *Front in Microbiol* 10, 3112.

Tariq M, Ahmad S, Fahad S, Abbas G, Hussain S, Fatima Z, Nasim W, Mubeen M, ur Rehman MH, Khan MA, Adnan M (2018). The impact of climate warming and crop management on phenology of sunflower-based cropping systems in Punjab, Pakistan. *Agri and Forest Met.* 15;256, 270–82.

Thomas DJ, Waters SB, Styblo M (2004): Elucidating the pathway for arsenic methylation. *Toxicol Appl Pharmacol* 198, 319–326.

Treseder KK, Lennon JT (2015): Fungal traits that drive ecosystem dynamics on land. *Microbiol Mol Biol Rev* 79, 243–262.

Treseder KK, Berlemont R, Allison SD, Martiny AC (2018): Drought increases the frequencies of fungal functional genes related to carbon and nitrogen acquisition. *PloS One* 13, e0206441.

Tufail MA, Touceda-González M, Pertot I, Ehlers R-U (2021): Gluconacetobacter diazotrophicus Pal5 enhances plant robustness status under the combination of moderate drought and low nitrogen stress in Zea mays L. *Microorg* 9, 870.

Ullah A, Mushtaq H, Fahad S, Shah A, Chaudhary H (2017): Plant growth promoting potential of bacterial endophytes in novel association with Olea ferruginea and Withania coagulans. *Microbiol* 86, 119–127.

Unsar Naeem-U, Muhammad R, Syed HMB, Asad S, Mirza AQ, Naeem I, Muhammad H ur R, Fahad S, Shafqat S (2020): Insect pests of cotton crop and management under climate change scenarios. In: Fahad S, Hasanuzzaman M, Alam M, Ullah H, Saeed M, Khan AK, Adnan M (Eds.), *Environment, Climate, Plant and Vegetation Growth.* Springer Publ Ltd, Springer Nature Switzerland AG. Part of Springer Nature. pp. 367–396. https://doi.org/10.1007/978-3-030-49732-3

van Elsas JD, Chiurazzi M, Mallon CA, Elhottovā D, Krištůfek V, Salles JF (2012): Microbial diversity determines the invasion of soil by a bacterial pathogen. *Proceedings of the National Academy of Sciences* 109, 1159–1164.

Ventosa A, Mellado E, Sanchez-Porro C, Marquez MC (2008): Halophilic and halotolerant micro-organisms from soils. In: Patrice Dion, Chandra Shekhar Nautiyal (Eds.), *Microbiology of Extreme Soils.* Springer, pp. 87–115.

Ventura-Lima J, Bogo MR, Monserrat JM (2011): Arsenic toxicity in mammals and aquatic animals: a comparative biochemical approach. *Ecotoxicol Environ Saf* 74, 211–218.

Verma H, Kumar D, Kumar V, Kumari M, Singh SK, Sharma VK, Droby S, Santoyo G, White JF, Kumar A (2021): The potential application of endophytes in management of stress from drought and salinity in crop plants. *Microorg* 9, 1729.

Vettori L, Russo A, Felici C, Fiaschi G, Morini S, Toffanin A (2010): Improving micropropagation: effect of Azospirillum brasilense Sp245 on acclimatization of rootstocks of fruit tree. *J Plant Interact* 5, 249–259.

Viti C, Marchi E, Decorosi F, Giovannetti L (2014): Molecular mechanisms of Cr (VI) resistance in bacteria and fungi. *FEMS Microbiol Rev* 38, 633–659.

Wahid F, Fahad S, Subhan D, Adnan M, Zhen Y, Saud S, Manzer HS, Martin B, Tereza H, Rahul D (2020): Sustainable management with mycorrhizae and phosphate solubilizing bacteria for enhanced phosphorus uptake in calcareous soils. *Agri* 10(8), 334. doi:10.3390/agriculture10080334

Wajid N, Ashfaq A, Asad A, Muhammad T, Muhammad A, Muhammad S, Khawar J, Ghulam MS, Syeda RS, Hafiz MH, Muhammad IAR, Muhammad ZH, Muhammad Habib ur R, Veysel T, Fahad S, Suad S, Aziz K, Shahzad A (2017): Radiation efficiency and nitrogen fertilizer impacts on sunflower crop in contrasting environments of Punjab. *Pakistan Environ Sci Pollut Res* 25, 1822–1836. https://doi.org/10.1007/s11356-017-0592-z

Wang C, Guo Y, Liu H, Niu D, Wang Y, Guo J (2012): Enhancement of tomato (Lycopersicon esculentum) tolerance to drought stress by plant-growth-promoting rhizobacterium (PGPR) Bacillus cereus AR156. *J Agric Biotechnol* 20, 1097–1105.

Wang F, Zhong H-H, Chen W-K, Liu Q-P, Li C-Y, Zheng Y-F, Peng G-P (2017): Potential hypoglycaemic activity phenolic glycosides from Moringa oleifera seeds. *Natural Product Res* 31, 1869–1874.

Wang H, Wu P, Liu J, Yang S, Ruan B, Rehman S, Liu L, Zhu N (2020): The regulatory mechanism of Chryseobacterium sp. resistance mediated by montmorillonite upon cadmium stress. *Chemosphere* 240, 124851.

Webb HE, Brichta-Harhay DM, Brashears MM, Nightingale KK, Arthur TM, Bosilevac JM, Kalchayanand N, Schmidt JW, Wang R, Granier SA (2017): Salmonella in peripheral lymph nodes of healthy cattle at slaughter. *Front Microbiol* 8, 2214.

Wei G, Fan L, Zhu W, Fu Y, Yu J, Tang M (2009): Isolation and characterization of the heavy metal resistant bacteria CCNWRS33-2 isolated from root nodule of Lespedeza cuneata in gold mine tailings in China. *J Hazard Mater* 162, 50–56.

Wei Y, Zhao Y, Shi M, Cao Z, Lu Q, Yang T, Fan Y, Wei Z (2018): Effect of organic acids production and bacterial community on the possible mechanism of phosphorus solubilization during composting with enriched phosphate-solubilizing bacteria inoculation. *Bioresource Tech* 247, 190–199.

Wiqar A, Arbaz K, Muhammad Z, Ijaz A, Muhammad A, Fahad S (2022): Relative efficiency of biochar particles of different sizes for immobilising heavy metals and improving soil properties. *Crop Pasture Sci* 74(2), 112–120 https://doi.org/10.1071/CP20453.

Wu C, Kehui C, She T, Ganghua L, Shaohua W, Fahad S, Lixiao N, Jianliang H, Shaobing P, Yanfeng D (2020): Intensified pollination and fertilization ameliorate heat injury in rice (Oryza sativa L.) during the flowering stage. *Field Crops Res* 252, 107795.

Wu C, Tang S, Li G, Wang S, Fahad S, Ding Y (2019): Roles of phytohormone changes in the grain yield of rice plants exposed to heat: a review. *Peer J* 7, e7792 doi: 10.7717/peerj.7792

Xue B, Huang L, Li X, Lu J, Gao R, Kamran M, Fahad S (2022): Effect of clay mineralogy and soil organic carbon in aggregates under straw incorporation. *Agron* 12, 534. https://doi.org/10.3390/ agronomy12020534

Yang R, Dai P, Wang B, Jin T, Liu K, Fahad S, Harrison MT, Man J, Shang J, Meinke H, Deli L, Xiaoyan W, Yunbo Z, Meixue Z, Yingbing T, Haoliang Y (2022): Over-optimistic projected future wheat yield potential in the North China plain: the role of future climate extremes. *Agron* 12, 145. https://doi.org/10.3390/ agronomy12010145

Yang Z, Zhang Z, Zhang T, Fahad S, Cui K, Nie L, Peng S, Huang J (2017): The effect of season-long temperature increases on rice cultivars grown in the central and southern regions of China. *Front Plant Sci* 8, 1908. https://doi.org/10.3389/fpls.2017.01908

Yuan Z, Yi H, Wang T, Zhang Y, Zhu X, Yao J (2017): Application of phosphate solubilizing bacteria in immobilization of Pb and Cd in soil. *Environ Sci Pollut Res* 24, 21877–21884.

Zafar-ul-Hye M, Muhammad N, Subhan D, Fahad S, Rahul D, Mazhar A, Ashfaq AR, Martin B, Jiˇrí H, Zahid HT, Muhammad N (2020a): Alleviation of cadmium adverse effects by improving nutrients uptake in bitter gourd through cadmium tolerant rhizobacteria. *Environ* 7(8), 54 doi:10.3390/environments7080054

Zafar-ul-Hye M, Muhammad T ahzeeb-ul-Hassan, Muhammad A, Fahad S, Martin B, Tereza D, Rahul D, Subhan D (2020b): Potential role of compost mixed biochar with rhizobacteria in mitigating lead toxicity in spinach. *Scientific Rep* 10, 12159 https://doi.org/10.1038/s41598-020-69183-9.

Zafar-ul-Hye M, Akbar MN, Iftikhar Y, Abbas M, Zahid A, Fahad S, Datta R, Ali M, Elgorban AM, Ansari MJ et al. (2021): Rhizobacteria inoculation and caffeic acid alleviated drought stress in lentil plants. *Sustain* 13, 9603. https://doi.org/10.3390/su13179603

Zahida Z, Hafiz FB, Zulfiqar AS, Ghulam MS, Fahad S, Muhammad RA, Hafiz MH,Wajid N, Muhammad S (2017): Effect of water management and silicon on germination, growth, phosphorus and arsenic uptake in rice. *Ecotoxicol Environ Saf* 144, 11–18.

Zahir SM, Zheng-HG, Ala Ud D, Amjad A, Ata Ur R, Kashif J, Shah F, Saud S, Adnan M, Fazli W, Saud A, Manzer HS, Shamsher A, Wajid N, Hafiz MH, Fahad S (2021): Synthesis of silver nanoparticles using Plantago lanceolata extract and assessing their antibacterial and antioxidant activities. *Sci Rep* 11, 20754.

Zaman I, Ali M, Shahzad K, Tahir, MS, Matloob A, Ahmad W, Alamri S, Khurshid MR, Qureshi MM, Wasaya A, Khurram SB, Manzer HS, Fahad S, Rahul D (2021): Effect of plant spacings on growth, physiology, yield and fiber quality attributes of cotton genotypes under nitrogen fertilization. *Agron* 11, 2589. https:// doi.org/ 10.3390/agronomy11122589

Zhang H, Kim M-S, Sun Y, Dowd SE, Shi H, Paré PW (2008): Soil bacteria confer plant salt tolerance by tissue-specific regulation of the sodium transporter HKT1. *Mol Plant–Microbe Interactions* 21, 737–744.

Zhang J, Li Q, Zeng Y, Zhang J, Lu G, Dang Z, Guo C (2019): Bioaccumulation and distribution of cadmium by Burkholderia epacian GYP1 under oligotrophic condition and mechanism analysis at proteome level. *Ecotoxicol Environ Saf* 176, 162–169.

Zhang M, Li Z, Haggblom MM, Young L, He Z, Li F, Xu R, Sun X, Sun W (2020): Characterization of nitrate-dependent As (III)-oxidizing communities in arsenic-contaminated soil and investigation of their metabolic potentials by the combination of DNA-stable isotope probing and metagenomics. *Environ Sci Tech* 54, 7366–7377.

Zia-ur-Rehman M (2020): Environment, climate change and biodiversity. In: Fahad S, Hasanuzzaman M, Alam M, Ullah H, Saeed M, Khan AK, Adnan M (Eds.), *Environment, Climate, Plant and Vegetation Growth.* Springer Publ Ltd, Springer Nature Switzerland AG. Part of Springer Nature. pp. 473–502. https://doi.org/10.1007/978-3-030-49732-3

2 Legacy Phosphorus

Tracking Budget, Vulnerability, and Mobility to Ensure Sustainable Management in the Agricultural System

Sammina Mahmood and Umair Ashraf*
Department of Botany, Division of Science and Technology,
University of Education, Lahore, Punjab, Pakistan
* Correspondence: sammina.mahmood@ue.edu.pk

CONTENTS

2.1 INTRODUCTION

The use of phosphorus (P) is critical for plant active growth and development. This essential nutrient has boosted the use of P-based fertilizers in the agriculture sector over the last few decades at a global scale (Sharpley et al. 2018). However, due to a lack of awareness at the farm level, its use has remained inappropriate and unmanaged, with use over the recommended dose in the quest for yield increases from limited land areas over a long period of time. These unplanned activities have pushed the natural P resources towards two uncertain situations: firstly, over- and unmanaged use either leads to runoff of surplus P to water resources or sediment build up leads to "legacy P" at the ground level. Secondly, over-use has exerted huge pressure on limited and scarce P resources (Mekonnen & Hoekstra 2018). The first situation results in environmental degradion of water resources and the second is disturbing the balance between P supply and demand, and due to this, prices of P0based fertilizers will probably increase in the future (Nesme & Withers 2016)

DOI: 10.1201/9781003286233-2

the elemental form of P is found in two physical forms, the white or yellow form (P_4) and red phosphorus, both of which are allotropes having miscellaneous physical and chemical properties. White phosphorus has higher reactivity due to its tetrahedral structure compared to red phosphorous which is polymeric (P_n) in nature and more stable (Pfitzner et al. 2004). However, red P can be transformed into white P by heating at 300°C in the absence of oxygen or by exposing it to direct sunlight. P, due to its reactivity, is not found in elemental form in nature. It is mineralized itself mostly in the form of phosphates (Ca/Fe/Al), predominantly in sedimentary rock and dispersed worldwide (Zhen-Yu et al. 2013). Rocks formed from volcanic eruptions also act as a source of P, however this is rare compared to sedimentary rock.

Sedimentary rocks containing phosphorus minerals are unevenly distributed around the world. Various physical, chemical, and environmental conditions and geological factors are associated with sedimentary rock formation along the coastal upwelling zones, which vary greatly among countries at a global level (Filippelli 2011). The most common group of phosphorus-containing minerals is apatite. The basic constituent of apatite mineral is calcium phosphate attached with F^-, Cl^-, or OH^- and known as fluorapatite, chloapatite, and hydroxyapatite, respectively. Sedimentary rock or phosphate after processing (the process called beneficiation) is known as phosphate rock (PR). According to an estimate, the global total PR amount was 68,776 Mt in 2014. Among this, almost 75% is in Morocco and Western Sahara and 5.4% in China (Jasinski 2011). An investigation into the future of P reserves predicts that "almost 70% of global P production is from P reserves, and if the same situation continues the P reserves will be depleted within 100 years." Morocco, having the biggest resources (about 77%), produces P fertilizers at an increasing rate which will be almost 700% by 2075. This pressure will be due to an increase in the global population with increasing food demand at an alarming rate, accentuated by depleting P reserves at a global level. This situation will create geopolitical crises among countries for P production and distribution (Cooper et al. 2011; Walan et al. 2014)

This suggests that current wasteful practices in the management of P resources and utilization should be modified, keeping in mind the sensitivity and lethality of the issue. As unwise use on one side is putting pressure on resource management, on the other side, it is increasing environmental concerns, particularly about water pollution. Thus, sustainable management of P-fertilizers in agricultural systems is crucial and could be achieved by adopting strategies based on P-cycling and recycling in natural systems, P-use efficiency, minimizing the losses through runoff, and reducing the reliance on synthetic P inputs. This chapter is focused on reviewing the current status of the above-mentioned factors and find the ommissions hindering their utility towards sustainable agriculture in the future.

2.2 PHOSPHORUS CYCLE

The biogeochemical cycle that describes the phosphorus flow/movement through the lithosphere, hydrosphere, and biosphere of the modern agricultural system is called the phosphorus (P) cycle. P does not exist in a stable gaseous form; hence the P cycle is based on sedimentary flow among multiple components of the environment. In the natural P cycle, with no human intervention, phosphate rock (PR) releases P ions in response to weathering of rocks due to environmental influences, into soil and water media. This released P may be categorized in two forms: soluble and insoluble or particulate forms. The soluble form of P in soil is taken up by plants and then animals or humans (directly or indirectly through the food web). P is then returned to soil through animal/human excretion, and decomposition of dead bodies of animal/humans/plants, as P is one of the basic constituents of all life forms on Earth. The insoluble form of P within soil is either leached down into deep ores of the Earth's crust or a fraction of it may be converted to soluble form by the activity of soil microbes to make it absorbable by plants. However, both soluble and insoluble forms of P can be drained into water reserves in response to poor agricultural practices (Liu et al. 2014; Qiu & Turner 2015).

P released in water medium in response to weathering of rocks may also be categorized in two forms: soluble and insoluble or particulate forms. The soluble form of P is quickly be taken up by phytoplankton and aquatic plants. Competition for and depletion of resources among phytoplankton causes their death and decomposition within water bodies. These dead bodies have two fates; they either cause eutrophication or they pollute the water or it sediments along with other dead aquatic life in the deep ocean. These sediments over many years appear as rock and are exposed in response to geological catastrophes or shifts. On the other hand, the particulate or insoluble form of P has only a single fate, i.e. sedimentation. P is either drained from soil into water or directly released into water; in both situations it deteriorates the quality of water reserves (Capodaglio et al. 2005; Capodaglio et al. 2003; Capodaglio et al. 2016; Søndergaard et al. 2003)

Human interference in the natural P cycle has raised several concerns. Humans have used synthetic P fertilizers unwisely through agricultural management practices in order to get better yields from a reduced land area. This practice has disturbed the natural P cycle and reserves in the soil. Surplus P is either drained into water where it intensifies water pollution crises or it adds to deep soil unused reserves (Filippelli 2011). These unused reserves are called "P legacy" and they act as a secondary P source and could be made available by adopting different agricultural and management practices (Kusmer et al. 2019; Qiu & Turner 2015). P "legacy" denotes surplus P used by any means (organic or inorganic) that leaches down into the deep soil profile or sediment into water reservoirs over time. P legacy varies greatly at inter- and intra-country levels based on the mode of use, soil/land characteristics, consumption and decomposition level, prevailing cropping patterns, and expected runoff into water reservoirs (Borbor-Cordova et al. 2006; Kleinman et al. 2011; Kusmer et al. 2019; Qiu & Turner 2015; Adnan et al. 2018a,b; Adnan et al. 2020; Adnan et al. 2019; Atif et al. 2021).

Buffering capacity refers to the ability of a soil or water reservoir to attenuate the P loading to the surface soil or water by retaining P inputs over time, which varies greatly across human-dominated landscapes (Gentry et al. 2007; Ahmad et al. 2017; Kusmer et al. 2019). The history of P use and accumulation, along with these factors in current P-based planning are considered as key P-regulating services for restricting the excessive use of P in farming systems (Doody et al. 2016). A correct estimate between vulnerability and buffering capacity requires a deep knowledge of the following properties, which varies over space and time.

2.3 GLOBAL PHOSPHORUS RESERVES AND GEOLOGICAL VARIATIONS

In soils vs. aquatic environments, P is widely present in different forms and its availability rate also varies greatly. In soils, P is present in soluble, mineral, adsorbed, and organic forms. The soluble form of P (orthophosphate ion; PO_4^{3-}) is the only form which plants can use for their growth directly; however its percentage of occurrence in soil is very low compared to total phosphorus (TP). Insoluble P consists of organic and mineral P, and is the dominant form of P present in soils. This dominant fraction of P is present in the form of iron/aluminium phosphate in acidic soil and calcium phosphate in alkaline soils, which limits its solubility within these soils (Zhen-Yu et al. 2013). Organic phosphates also include the phosphate forms excreted by microorganisms in soil, which upon microbial action during a mineralization process are broken down into inorganic form. However, the activity of microorganisms is highly dependent upon the soil moisture content, soil temperature, and physiochemical properties of soil. Hence, TP is the sum total of mineral, organic, and adsorbed forms with a negligible amount of soluble P. Soil and fertilizer management practices should be based on TP rather than soluble P (Frossard et al. 2011; Khan et al. 2014).

Compared to soil, in aquatic environments the concentration of soluble P is higher due to the lower percentage of Ca/Fe/Al ions. In addition, agricultural sewage, industrial waste, domestic sewage, and detergents are also sources of P which pollute water resources. As P is an essential nutrient for plant growth and metabolism, its presence in water promotes the growth of plants and phytoplankton within water bodies causing eutrophication (Fahad et al 2016; Fahad et al. 2015a,b; Fahad et al. 2020; Fahad et al. 2021a,b,c,d,e,f; Fahad et al. 2022a,b). P is also a vital component of all living organisms, as a basic building block of DNA, RNA (in the form of PO_4^{3-}), ATP, and phospholipids. Also, considerable amounts of P are also present in the teeth and bones of biological bodies (Boskey 2013).

2.4 PHOSPHORUS BUDGET AND VULNERABILITY

The green revolution has boosted agricultural production with the help of commercial fertilizers. P, as an essential nutrient for crop growth and development, is used as a commercial fertilizer at an increasingly alarming rate at the global level. Increasing competition among growers with the objective of obtaining increased yields from reduced land area with an improved nutritional profile has diverted the mind set from organic fertilizer to synthetic/commercial fertilizer use. This in turns has posed several other undesired issues which initially remained unnoticed. (1) The use of organic fertilizers was reduced, which were not only playing a role in biological nutrient cycling by the process of decomposition but also helping to improve the soil physical and chemical properties. (2) The chief source of commercial P fertilizer is rock phosphate/mineral. Increased production demand rapidly depletes the slowly non-renewable finite resources (Zhen-Yu et al. 2013). (3) At the geological scale, rock phosphate/mineral resources are scattered differently. This situation could create a socio-political imbalance between countries in the future (Khan et al. 2014; Alam et al. 2016). (4) P exhibits the property of leaching down into the deep soil profile in both organic and inorganic forms. Soil with extensive application of commercial fertilizer has a soil P legacy. This P legacy has two fates. (i) It can remain in soil unused by chelating with other minerals. This relationship also makes other nutrients unavailable for plants and sometimes surplus P has caused toxicity in soil as well. (ii) Surplus P can drain into water reservoirs and deteriorate the water table quality both physically and chemically (Kusmer et al. 2019; Qiu &Turner 2015). (5) The reduction in organic fertilizer also necessitates the use of commercial fertilizers with other essential nutrient s,e.g. N, S, etc. (6) There is as yet no international organization or regulation that sets the standard for the P dose rate depending upon soil and environmental history, or that manages P resources and provides legal constraints. All these factors further complicate the situation when estimated on the impacts on developing and developed countries.

P available resources at regional or international level, P recycling, and the status of P legacy in soil determine the overall P budget. The P budget helps to manage the balance between availability and demand. the efficacy of the P budget could be enhanced by improving P use efficiency (PUE) which can be defined as the use of P at a decreased rate with increased efficiency. PUE varies greatly at the global as well as regional level depending upon the management practices adopted in its use.

2.5 THE PHOSPHORUS BALANCE AT THE FARM LEVEL AND FERTILIZATION

P application into soil according to the requirements of available crop cover is called the P balance for that site. However, this ideal situation is rarely met. It is now obvious that in the usual routine, applied P into the soil for food production is quite high compared to the actual need and use at the field level (Liu et al. 2008). This disturbs the P balance which is becoming a common issue not only at the regional level but also attaining global attention. Some general facts reported behind this imbalance include the following. (1) At the global scale, variable geological and geographical conditions further enhance this issue. One site may be exposed to erosion and depletion of resources as in the case of deserts compared to other sites where resources become accumulated with the passage of time, such as forests. Due to this variation, the dose rate recommended for one site cannot be applied with equal success rate at another site. (2) In certain regions, manure from farm animals is the main source of fertilizer on crop lands. This practice also has been adopted in some regions as good agricultural practice. This trend is not common at the regional or global level. This factor, along with fertilizer use, strengthens the P reserves already present in the soil. (3) Each crop type has different P requirements based on the root characteristics and the length of the growth period. However, the application rate of phosphorus (either in the form of manure or fertilizer), often exceeds the required levels for crops. This promotes the P status of the soil to higher level. (4) The soil microbial profile for P fixation characteristics is also variable at regional as well as global levels. This also plays a role in variations in P requirements based on site conditions (de Bruijne et al. 2009). (5) Phosphate available in soil may be immobilized through precipitation and adsorption reactions, making phosphorus unavailable for plant uptake (Syers et al. 2008). All these facts divert researchers' attention towards the development of strategies to determine soil P requirements (application rate/dose rate) depending on the soil fertility level (P status) in order to avoid yield depression at the farm level. This could be achieved by considering the on-site soil characteristics, past cropping history with plant characteristics, and current crop type under consideration for plantation (Cordell et al. 2009; Cornish 2009). However, a single survey cannot be applied for all future applications. This needs to be determined on an annual basis in order to avoid yield depression (Roemer 2009; Ilahi et al. 2020). Besides these strategies, public awareness about the scarce resource, application protocol, and enhanced P use efficiency is also crucial at the regional level to ensure the wise use of this scarce resource.

2.6 FACTORS AFFECTING PHOSPHORUS USE EFFICIENCY (PUE)

2.6.1 Role of P Source and Soil Characteristics in Mobility and Bioavailability of P to Soil

The fate of applied P is also called P-cycling in the soil and is determined by P solubility and reactivity characteristics and topographic characteristics of the particular soil which ensure bioavailability to the planned cropping pattern (Liu et al. 2014). Both P source and soil characteristics are interdependent on each other. Plants usually are unable to uptake almost 80% of applied P fertilizers immediately after application due to sorption, precipitation, and microbial immobilization (either through root or mycorrhizal hyphae) regulated by soil acidity, basicity, pH of soil, etc. (Roberts & Johnston 2015), e.g. P associated with iron and its oxides in soil is sensitive to reduced conditions of soil (Beauchemin et al. 2003), while with aluminium (oxides) and calcium phosphate it is more sensitive towards pH fluctuations of soil (Yan et al. 2014). Exudation of phosphatases and organic acids from roots or microorganisms further play significant roles in accelerating the P cycle by releasing P from the P source (either P legacy or applied fertilizer) (Frossard et al. 2011).

Further bioavailability is hugely dependent upon the organic and inorganic nature of the applied P (Khan et al. 2014), e.g. at pH 4–5 $H_2PO_4^-$ becomes dominant while as the pH increases HPO_4^{2-} and PO_4^{3-} become dominant in the soil (Yadav & Verma 2012). Diversity in soil characteristics is further added to by soil texture, soil management history, cropping pattern/vegetation cover, and economic and social factors associated with vegetation in a particular area. All of these factors in combination determine the magnitude of P cycling in the plant–soil system (Frossard et al. 2011).

2.6.2 Soil P Transformation and Mobilization

Transformation and mobilization of legacy P is a complex process governed by multiple factors. Transformation and mobilization depend upon the power of the activation of variable factors, the addition or manipulation of which change the physical and chemical properties of soil (i.e. texture, soil pH, soil particle charge, etc.). These factors are divided into three main groups: (1) organic matter (animal manure, crop residues, biochar, etc.) including organic acids (humic acid, lignin, etc.); (2) bio-inoculants and bio-fertilizers which include P-solubilizing microbes and phosphatase enzymes; and (3) zeolites.

2.6.3 Organic Matter (Animal Manure, Crop Residues, and Biochar)

In developing countries specifically, and developed ones also, organic matter application into cropping fields in the form of organic acids, crop residues, animal manure, etc. is a common practice. Organic matter is used because of its rich nutrient profile, afordability, readily available at rural sites especially, and as a biological source with few harmful effects. Animal manure, crop residues, and biochar (charcoal produced from crop residues) are all biological sources of multiple nutrients. Land application with these to dispose of the agricultural waste and recycle the nutrients is a predominant farming practice in developing and resource-deficit countries (Dai et al. 2016). The availability of nutrients from these practices depends upon two major factors. (1) The addition and decomposition of these materials influences the overall physical and chemical profile of soil. These fluctuations have a huge and variable impact on the solubility, mobility, and availability of various nutrients, again dependent upon nutrient concentration and type of interaction with soil particles. (2) The type of animal and crop residues added into the soil and the receiving soil environment determine the nutrient profile, their magnitude, and direction of availability. However, in all circumstances, these biological sources act as a primary source of nutrients in soil (Damon et al. 2014; Espinosa et al. 2017).

With reference to P being an organic material, all these are rich sources of highly soluble P (particularly orthophosphate) which could be mineralized in a very short period of time to release soluble P. However, including the factors mentioned above, several additional key factors govern the availability of organic matter-derived P in soil. These factors include: (1) quality of organic residue; (2) the activity level of soil microbial entities; and (3) sorption reactions of mineralized P into soil (Damon et al. 2014). Usually, increased organic matter content in soil provides a more favourable environment to microbes in the soil, which in turn accelerates the P cycling and mineralization (Lone et al. 2015). Crop- and animal-based residues showed differential available P contents. General findings predict almost 2–20% P transfer from crop sources, which is relatively low and highly variable. This could be attributed to any of these factors, e.g. the P concentration in animal or crop residue, physiological maturity of crop residue, soil P status, and soil type (Espinosa et al. 2017). In contrast, P availability from animal manure ranged between 12–100%. This variation in range could be attributed to animal age, diet, and the way manure is stored/preserved (Bahl & Toor 2002). The addition of organic matter in all these forms has not only proved to be a primary source of highly soluble and available P but also affects the native P source and changes its mobility positively. This occurs through chelate formation with insoluble Ca, Al, and Fe forms of P and converting it into soluble forms during the decomposition process (Hountin et al. 2000).

The application of biochar mainly affects the ion exchange capacity of soil. In biochar, the P form changes compared to residues due to its bonding with multivalent metal ions. This ability can affect the P availability either positively or negatively (Chathurika et al. 2016). In acidic soils, biochar exhibits anion exchange capacity which competes with metal oxides (Al, Fe) for sorption of P, similarly to humic acid and lignin (DeLuca et al. 2015; Hunt et al. 2007), while this response is reversed in saline and sodic soils which promote precipitation of P into soil (Xu et al. 2016). P solubility in acidic soils is mainly regulated by its interaction with Al^{3+} and Fe^{2+}/Fe^{3+}, while in saline soils it is with Ca^{2+}. As soon as the application of biochar in acidic soils is increased, soil pH increases and P sorption on metal oxides decreases (Chintala et al. 2014). Biochar application also requires accurate calculation of the dose rate/application rate depending on the soil type as being an organic source it is also enriched with other types of mineral nutrients and elements such as Ca and Mg, etc., with pronounced effects on plant growth and development (Parvage et al. 2013; Vandecasteele et al. 2016). The magnitude of the effects of biochar on P cycling is highly dependent on several key factors, i.e. production method of biochar, climate, variability in feedstock, and soil type. Biochar derived from agricultural residues, animal residues, and crop residues overall show positive responses to plant-available P, while biochar derived from wastewater sludge has a poor response. Since the total P produced annually from organic sources is higher than the global production of

P fertilizer, it seems to be a helpful and economical tool to thermochemically convert manure into biochar to minimize the load on overall mineral P fertilizer production (Steinfeld et al. 2006). This recycling of P from biological residues is a continuous source of P for soils and has environmental benefits compared to direct land application of mineral fertilizers (Dai et al. 2016; Manolikaki et al. 2016; Wang et al. 2015b). It has the potential to enhance the plant-available P in biochar-amended soils and could be a sustainable strategy to complement conventional P fertilizers. Unfortunately, sufficient research is not available regarding P contents of biochar, application rate, or its correlation with plant-available P.

2.6.4 Organic Acids (Humic Acid, Lignin, etc.)

Organic acids are further classified into low-molecular-weight organic acids and high-molecular-weight organic acids. Both types are not only act as direct sources of P but also help in the mobilization of reserved or legacy P. These characteristics are further dependent upon their chemical structure and its reactivity, their solubility, affinity towards other soil elements, and dissociation power. All these characteristics change the soil physical properties directly or modify the chemical profile in a cost-effective way. Both have the property to increase the soil fertility level and improve the soil nutritional profile, including P (Çimrin et al. 2010).

The low-molecular-weight organic acids are derived from two main sources: (1) plants excrete under P-deprived conditions into soil or (2) released in response to decomposition (microbial action and activation) of animal and plant residues (Yuan et al. 2016). Examples of low-molecular-weight molecules include oxalic acid, malic acid, fumaric acid, tartaric acid, succinic acid, tartaric acid, etc., which all exhibit the property of having at least one carboxylic group attached. The exact role of organic acid and the mechanism of activation to P source/P legacy or applied resource remains unknown and needs experiments in a targeted direction, however few experiments based on the artificial application of organic acids alone (Tripathi 2005), in combination (Wang et al. 2015a; Wei et al. 2010; Fakhre et al. 2021; Muhammad et al. 2022; Shafi et al. 2020; Wahid et al. 2020; Zahida et al. 2017), or in combination with phosphatase enzyme (Guan et al. 2013) have been conducted. The results overall have revealed that the possible mechanism of action behind this activation might be any of the following. (1) Addition to soil with specificity may change the soil physio-chemical properties in quite a predictable manner, which in returns affects the solubility of P source/P legacy, e.g. variation in soil pH can affect the dissolution of sparingly soluble minerals containing P to a great extent (Andersson et al. 2015). (2) These might form complexes with other elements present in soil and having great affinity for P to form multiple types of complexes and making it unavailable, e.g. Al, Fe, Ca, etc. (Ström et al. 2005) with the enhanced organic anion-driven solubilization of Ca complex observed compared to plant-based hydrolysis and microbial phytases (Patel et al. 2010). (3) An electrostatic competition observed for sorption into soil between organic acid and P. The presence of two types into the same soil profile affects both antagonistically. However, in all situations the exact mechanism still needs to be elucidated.

High-molecular-weight organic acid compounds include humic acid and lignin. Humic acid is derived from the decomposition of plant litter, while lignin is an important precursor in the formation of humic acid as a result of microbial action. Humic acid exist in a range of structures (aliphatic or aromatic) with a variety of functional groups (phenyl, carboxylic, etc.), mainly oxygen containing (Çimrin et al. 2010). Likewise lignin has major functional groups, such as carboxyl, carbonyl, hydroxyl, and methoxyl. Specifically, both ingredients of organic matter have produced success stories with remarkable increases in available P compared to total P of soil in calcareous soils and soils rich in iron oxides and iron-aluminium oxides (Chen et al. 2003; Zhen-Yu et al. 2013). As far as the mechanism of action is concerned, lignin is a precursor of humic acid and they have almost identical structural and functional properties, and both have very similar mechanisms of action (Sun et al. 2011; Tahir et al. 2011). (1) Shift in soil pH: Decomposition of humic acid into soil changes the pH level of soil by producing H^+ ions. This either decreases the Ca-P mineral precipitation or favours the formation of di-calcium phosphate dehydrate which is the most stable form of phosphate with the least solubility properties into soil (Grossl & Inskeep 1991). (2) The formation of organic ligands/colloidal interactions with metal and metalloids: functional groups present in high-molecular-weight organic acids, e.g. phenyl, carboxylic, etc. make complexes with metal ions, specifically Fe and Al, which are mostly associated with phosphate activity and fixation into soil. The affinity of functional groups for these metal ions enhanced their removal from soil, strengthening the competitive phosphate ratio in soil (Antelo et al. 2007; Regelink et al. 2015).

2.6.5 Diversity of Bio-inoculants and Bio-fertilizers

Microbes, particularly bacteria and fungi, have a natural ability to manipulate the organic and inorganic complexes present in soil and recycle the nutrients present in them into soil. These microbes are called bio-inoculants, exhibiting huge potential to replenish soil nutrient levels in an eco-friendly and cost-effective way. Phosphate-solubilizing microbes (PSMs) are those which are capable of hydrolysing organic and inorganic insoluble P compounds into soluble form (plant accessible) (Khan et al. 2009; Owen et al. 2015; Selvi et al. 2017). PSMs may be comprised of phosphate-solubilizing bacteria (PSBs), phosphate-solubilizing fungi (PSFs), phosphate-solubilizing actinomycetes (PSAs), and cyanobacteria. The role of these microbes in the availability and acquisition of P by plants has been described by several researches worldwide. Well-known examples of bacteria which play a significant role in P replenishment in soil includes *Bacillus megaterium* (Satyaprakash et al. 2017), *Bacillus circulans* (Kumar & Patel 2018), *Bacillus subtilis* (Hajjam & Cherkaoui 2017), and *Bacillus cereus* (Yu et al. 2011). General examples of fungi known to be able to solubilize P includes *Aspergillus awamori* (Mittal et al. 2008; Selvi et al. 2017), *Aspergillus versicolor* (Ho et al. 2013), *Aspergillus niger* (Li et al. 2014), *Aspergillus clavatus* (Alam et al. 2002), *Penicillium citrinum* (Mittal et al. 2008), and *Arthrobotrys oligospora* (Duponnois et al. 2006). Besides bacteria and fungi, actinomycetes and cyanobacteria also have exhibited influential roles in P recycling; examples include *Acinetobacter rhizosphaerae* (Gulati et al. 2010), *Streptomyces albus*, *Streptoverticillium album* (Kumar &Patel 2018), and *Calothrix braunii* (Sharma et al. 2013), respectively.

Soil is the basal medium which provides ideal conditions for microbial growth and replenishment under natural conditions. According to one estimate, one gram of fertile soil contains 10^1–10^{10} microbial load which includes 1–50% bacteria and 0.1–0.5% fungi, respectively (Khan et al. 2009). However, significant variation is observed in the ability of microbes overall, bacteria vs. fungi and among other microbes also, to solubilize insoluble P in soil (Jones & Oburger 2011). The variation in this potential may be attributed to species diversity of particular microbes and plant type, prevailing soil and environmental conditions/factors (biotic and abiotic) of a particular site, efficacy of microbes to produce organic acids and phytohormones, and the ability of microbes to survive under fluctuating environmental conditions. Generally, PSFs produce more acids than bacteria and consequently, irrespective of the lower percentage of their occurrence, they exhibit greater P-solubilizing potential (Sharma et al. 2013).

Besides the ability of actinomycetes to produce phytohormones and survive under harsh environmental conditions (biotic and abiotic), their role in P availability and acquisition by plants also has attracted attention worldwide. The properties and biological role of actinomycetes in soil is paving the way to prove them as more influential microbial entities than bacteria and fungi, particularly under harsh environmental conditions (Hamdali et al. 2008). The overall mechanism of action of microbes may likely include the production of organic acids after inoculation, changing soil physical and chemical properties, ion chelation, promoting root growth, penetration and permeability in establishing a mycorrhizal association among roots and microbes, and strengthening the microbial population with the passage of time. The whole mechanism establishes a microbe–soil–plant triangle and the role after microbial inoculation in P recycling depends directly or indirectly on the properties of the components of this triangle (Suri et al. 2011).

Organic acid production by microbes lowers the soil pH (Kumar & Patel 2018; Selvi et al. 2017). Acidic and basic soils react differently. In basic/alkaline soils, phosphate can precipitate with calcium to form calcium phosphate which is insoluble into soil. an increase in pH is associated with the production of divalent and trivalent forms of inorganic P (HPO_4^{-2} and HPO_4^{-3}), making P unavailable for plants. However, their solubility can be increased by lowering the soil pH. Organic acids produced by microbes play a role in lowering the pH and consequently improving the P solubility in the soil (Satyaprakash et al. 2017).

A positive correlation exists between organic acid concentration and P solubility in soil regulated by the pH profile. pH is lowered with the evolution of CO_2 and soil becomes acidic, which in turn acts as a positive regulator of P solubility (Alam et al. 2002; Selvi et al. 2017; Walpola & Yoon 2012; Yousefi et al. 2011). The efficacy of organic acid to lower soil pH is directly dependent on its type and concentration, which determines the strength of the acid. For example, di- and tri-carboxylic acids are more effective than monocarboxylic and aromatic acids. Aliphatic acids have been found to be more effective than citric, fumaric, and phenolic acids (Mahidi et al. 2011; Walpola & Yoon 2012). The main organic acids include citric, oxalic, lactic, glyconic, gluconic, malic, succinic, acetic, tartaric acid, malonic, glutaric, butyric, and adipic acids (Ahmed & Shahab 2011; Kumar & Patel 2018; Satyaprakash et al. 2017; Selvi et al. 2017; Walpola & Yoon 2012; Yousefi et al. 2011), with each exhibiting variable potential in P solubilization. These organic and inorganic acids compete with phosphate for the sorption site by chelation with cations (Al and Ca oxides) present in soil, and due to this characteristic, they are also known as "chelators." The hydroxyl and carboxyl group of acid chelates with cation initially bound to phosphate, thereby converting it into a soluble P form. These acids stabilize Al and Ca oxides in reaction with them and compete for their reaction sites (Khan et al. 2007; Walpola & Yoon 2012).

2.7 BIO-FERTILIZERS

Various fertilizers are constituted by adding microbes to them, and are called "bio-fertilizers" as the role of microbes is very clearly established in P recycling in an environmentally friendly way without any harmful effects. Hence, the development of bio-fertilizers also serves the same role as microbes in soil. The trend to use bio-fertilizers as an alternative to chemical fertilizers is accelerating currently due to their cost-effectiveness and improved efficiency in P recycling. Microbes used in bio-fertilizer preparation include bacteria (*Bacillus* ***spp.,*** *Pseudomonas* spp., and *Rhizobium* spp.), fungi (*Aspergillus* spp., *Penicillium* spp., and *Rhozopus* spp.) and actinomycetes (*Staphylomyces* spp. and *Streptomyces*) (Mahajan & Gupta 2009). Bio-fertilizers, compared to chemical or synthetic phosphate fertilizers, are based on mycorrhizal infection at root level which not only promotes the conversion and mobilization of P from soil to plant but also serves to improve plant growth and development by secreting a range of exudates (e.g. metabolites, H^+, organic acids and several enzymes including phosphatase, etc. depending upon the species and strain used in the synthesis of bio-fertilizer). The use of dissolved P mixed microbe fertilizers has significantly reduced the dependency on chemical P fertilizers in modern agricultural practices (Behera et al. 2014; Mukherjee & Sen 2015).

The mechanism of action is similar to bio-inoculants by means of the release of organic acids, phosphatase enzymes, and H^+ and finally converting the insoluble forms of P [(tricalcium phosphate($Ca_3(PO_4)_2$, iron phosphate ($FePO_4$), and aluminium phosphate($AlPO_4$)] (Selvi et al. 2017) into soluble forms (Khan et al. 2014). Particularly, the enzymes secreted by PSMs which include a broad range (e.g. C-P lyase, phytase, phosphatse, phosphonatase, and phosphohydrolase) catalyse the hydrolysis of both esters and anhydrides of phosphates (Jones & Oburger 2011). This enzyme range is particularly important in soils where plant- or animal-based manure is applied as the base organic fertilizer. As a consequence, huge amounts of sugars, sugar phosphates, and nucleotides are available in those soils, which serve as huge raw sources of P. However, the reactivity cycles of each enzyme and consequently metabolic products released from the hydrolysis of biological compounds are variable (Khan et al. 2010; Sharma et al. 2013). The effectiveness of bio-fertilizers is dependent upon the type of plant, microbial species and strain used, growth stage of plants, and environment including the soil conditions (biotic and abiotic factors). Hence, plants also act as biocontrol agents for microbes in soil (Behera et al. 2014).

In addition, mechanical agricultural practices also have a huge impact on P solubilization and availability through PSMs. It has been observed that deep tillage affects positively microbial growth and phosphatase activity in clay and loam soils. Deep tillage promotes the growth conditions for microbes and helps to increase their growth rate in soil as well as promoting phosphatase activity. This may be because deep tillage loosens the soil particles, aerates the soil, and mixes organic matter to deep layers of soil, thereby increasing the microbial profile deep in soil. Clay soils are reported to exhibit more PSM activity compared to loam, which may be because clay soils offer a greater surface area and small pores for incorporation, mixing, and manipulation of organic matter compared to loam soils. Both situations reveal that deep tillage practice has been found to have a positive correlation with PSM activity (Ji et al. 2014). However, by increasing the soil depth several factors also fluctuate, including the temperature of the soil. In some reports high temperature has been found to increase phosphatase activity. However, there is a complicated relationship between temperature and phosphatase activity, which cannot be sufficiently explained with only a few examples, as several other factors like soil type, cropping history, environmental conditions, microbe type, and soil chemical profile can complicate the situation and impede prediction of the exact relationship between temperature and phosphatase activity (Das et al. 2014; Štursová & Baldrian 2011).

Abiotic stresses, e.g. salinity and heavy metals, are also found to interact negatively with phosphatase activity (Fahad et al 2021f). At increased salinity and sodicity levels of soil, alkaline phosphatase activity is reduced due to denaturation of enzyme protein and a reduction in microbial growth (Rietz & Haynes 2003). Likewise, heavy metals interact with protein functional groups and bring conformational changes to the structure (Karaca et al. 2010). To mitigate these stresses, the addition of P activators may help in enzyme activity, but insufficient literature is currently available on this topic to describe the mechanisms and effects in detail. Intensive research work is required in this direction in future so that a successful strategy can be developed to promote P resource utilization through biological means before their depletion. All these are organic amendments to enhance P utilization.

2.8 ZEOLITES

Zeolites are microporous, hydrated, and crystaline calcium, sodium, potassium, and barium aluminosilicates used as an inorganic amendment in soil to improve P fixation and utilization. The uses of inorganic zeolites however vary. Zeolites are formed in nature by the chemical reaction of volcanic lava with saline water. It is characterized by a large sorption site, ion exchange capacity and selectivity, catalytic activities, and most importantly thermal stability for a temperature range of up to 700–750°C (Yangyuoru et al. 2006). These properties impart strength to zeolite, making it difficult to break down. It can be retained in soil for a long period of time as a source of minerals required by plants for their growth, and hence plays a significant role in soil management. The hydrated and amorphous silica skeleton helps to retain water and nutrients at the root zone level, which ultimately reduces the water and fertilizer costs (Polat et al. 2004).

Zeolites are found to have an effect in both acidic and saline soils with a range of soil textures, i.e. sand to clay, however, under differential soil chemical and textural conditions, zeolite exhibit variable catalytic activity to ameliorate the stress situation and enable a balance between nutrient availability and uptake (Aainaa et al. 2014; Ahmed et al. 2017; Ahmed et al. 2010; Al-Busaidi et al. 2008; Omar & Ahmed 2011). A change in soil nutrient availability chiefly depends on three factors: (1) P addition from the applied source; (2) shift in pH-dependent soil chemical equilibrium; and (3) the effect on microbial growth and activity (Demeyer et al. 2001). In terms of their abundance, low cost, and ecofriendly nature, zeolites are considered as a good source that reduces the load on commercial fertilizers needed for plant in a cost-effective way.

Another aspect of zeolites which has attracted attention is that natural zeolites can be modified to enhance their ion exchange capacity and ion exchange selectivity. These modifications have been found to have a significant effect on P availability (Yang et al. 2015). Acid-modified zeolites in soil can reduce pH, and alkali-modified zeolites can enhance cation exchange through alkali metal ions, with both situations promoting the release of available P into soil (Fadaeerayeni et al. 2015). The combination of rock phosphate (RP) with zeolite acts as an exchange-fertilizer (exchanging Ca^{2+} onto the zeolite). Also, plant uptake of NH_4^+ or K^+ further frees the exchange sites which can be occupied by Ca^{2+}, which ultimately lowers the soil Ca^{2+} concentration and enhances further dissolution of RP. Both conditions strengthen the available P concentration from source P into soil and promote its uptake by plants (Pickering et al. 2002):

$$RP + NH_4^{+-}\text{Zeolite} \rightleftharpoons Ca^{2+} - \text{Zeolite} + NH_4^+ + PO_4^{3}$$

2.9 CONCLUSION

The importance of phosphorus in agricultural production systems cannot be denied. The P issue is clearly a double-edged one. It concerns both the depletion of phosphorus reserves and environmental pollution, particularly water pollution. Water scarcity is already a global issue, which is getting worse due to factors which are deteriorating the quality of this already scarce resource. Human interference is breaking the natural P cycle and releasing more P-containing wastes into water. This acceleration in the natural cycle is posing disastrous effects on water quality and aquatic life. It is changing not only the physiochemical properties of water but also biological processes in the marine environment, causing eutrophication. As almost 95% of P pollutants enter water from poorly managed agricultural practices and the rest comes from other sources including industries, detergents, etc., it is clear that current agricultural practices waste a large amount of these resources, as only 16% of the total is used effectively. By adopting best management practices (BMPs) in the agriculture sector the potential pressure on P reserves and threats to water sources can be reduced. On the other hand, strikingly, no international organization currently exists to devise rules and regulations about the use of P keeping in view the scarcity of these resources. Legislation at the international level chiefly concentrates more on the environmental aspects of P use than on its finiteness and the recovery of resources. It therefore seems that environmental issues are higher on the global agenda than the development of strategies to enhance P use efficiency or reuse, keeping in view the future scarcity. P use is determined by keeping in focus the rapid growth in population, land use, yield quality parameters, etc. rather than taking into account the divergent plant phosphorus availability explicitly on an eco-regional basis. The increase in the world's population is also not being matched by expanding agricultural land. These facts accentuate the need to consider the P status of soil as an important factor for estimation of its impact on food quality and quantity under divergent eco-regional conditions, i.e. comparable ecological conditions at a regional level are important to determine P status rather than at a global level. The focus henceforth should be on technologies and approaches allowing the least use of synthetic P fertilizers, increasing PUE, and reduced P leaching and runoff, to ensure the sustainability of this element.

REFERENCES

Aainaa HN, Ahmed OH, Kasim S, Majid NMA (2014): Reducing Egypt rock phosphate use in Zea mays cultivation on an acid soil using clinoptilolite zeolite. *Sustainable Agric Res* 4, 56–66.

Adnan M, Fahad S, Khan IA, Saeed M, Ihsan MZ, Saud S, Riaz M, Wang D, Wu C (2019): Integration of poultry manure and phosphate solubilizing bacteria improved availability of Ca bound P in calcareous soils. *3 Biotech.* 9(10), 368.

Adnan M, Fahad S, Muhammad Z, Shahen S, Ishaq AM, Subhan D, Zafar-ul-Hye M, Martin LB, Raja MMN, Beena S, Saud S, Imran A, Zhen Y, Martin B, Jiri H, Rahul D (2020): Coupling phosphate-solubilizing bacteria with phosphorus supplements improve maize phosphorus acquisition and growth under lime induced salinity stress. *Plants* 9(7), 900. doi: 10.3390/plants9070900

Adnan M, Zahir S, Fahad S, Arif M, Mukhtar A, Imtiaz AK, Ishaq AM, Abdul B, Hidayat U, Muhammad A, Inayat-Ur R, Saud S, Muhammad ZI, Yousaf J, Amanullah, Hafiz MH, Wajid N (2018a): Phosphate-solubilizing bacteria nullify the antagonistic effect of soil calcification on bioavailability of phosphorus in alkaline soils. *Sci Rep* 8, 4339. https://doi.org/10.1038/s41598-018-22653-7

Adnan M, Shah Z, Sharif M, Rahman H (2018b): Liming induces carbon dioxide (CO_2) emission in PSB inoculated alkaline soil supplemented with different phosphorus sources. *Environ Sci Poll Res.* 25(10), 9501–9509.

Ahmad I, Jadoon SA, Said A, Adnan M, Mohammad F, Munsif F (2017): Response of sunflower varieties to NPK fertilization. *Pure Appl Biol* 6(1), 272–277.

Ahmed N, Shahab S (2011): Phosphate solubilization: their mechanism genetics and application. *Int JMicrobiol* 9, 4408–4412.

Ahmed O, Aainaa H, Majid N (2017): *Zeolites to Unlock Phosphorus Fixation in Agriculture*. Universiti Putra Malaysia: Serdang, Malaysia, p. 188.

Ahmed O, Sumalatha G, Muhamad AN (2010): Use of zeolite in maize (Zea mays) cultivation on nitrogen, potassium and phosphorus uptake and use efficiency. *Int J Phys Sci* 5, 2393–2401.

Al-Busaidi A, Yamamoto T, Inoue M, Eneji AE, Mori Y, Irshad M (2008): Effects of zeolite on soil nutrients and growth of barley following irrigation with saline water. *J Plant Nutr* 31, 1159–1173..

Alam A, Baig J, Din SU, Arshad M, Adnan M, Khan A (2016): Relative abundance of benthic macro-invertebrates in relation to abiotic environment in Hussainbadnallah, Hunza, Gilgit Baltistan, Pakistan. *Int J Biosci* 9(3), 185–193.

Alam S, Khalil S, Ayub N, Rashid M (2002): In vitro solubilization of inorganic phosphate by phosphate solubilizing microorganisms (PSM) from maize rhizosphere. *Int J Agric Biol* 4, 454–458

Andersson KO, Tighe MK, Guppy CN, Milham PJ, McLaren TI (2015): Incremental acidification reveals phosphorus release dynamics in alkaline vertic soils. *Geoderma* 259, 35–44.

Antelo J, Arce F, Avena M, Fiol S, López R, Macías F (2007): Adsorption of a soil humic acid at the surface of goethite and its competitive interaction with phosphate. *Geoderma* 138, 12–19.

Atif B, Hesham A, Fahad S (2021): Biochar coupling with phosphorus fertilization modifies antioxidant activity, osmolyte accumulation and reactive oxygen species synthesis in the leaves and xylem sap of rice cultivars under high-temperature stress. *Physiol Mol Biol Plants* 27(9), 2083–2100. https://doi.org/10.1007/s12298-021-01062-7

Bahl G, Toor G (2002): Influence of poultry manure on phosphorus availability and the standard phosphate requirement of crop estimated from quantity–intensity relationships in different soils. *Bioresour Technol* 85, 317–322.

Beauchemin S, Hesterberg D, Chou J, Beauchemin M, Simard RR, Sayers DE (2003): Speciation of phosphorus in phosphorus-enriched agricultural soils using X-ray absorption near-edge structure spectroscopy and chemical fractionation. *J Environ Qual* 32, 1809–1819.

Behera B, Singdevsachan SK, Mishra R, Dutta S, Thatoi H (2014): Diversity, mechanism and biotechnology of phosphate solubilising microorganism in mangrove—a review. *Biocatal Agric Biotechnol* 3, 97–110.

Borbor-Cordova MJ, Boyer EW, McDowell WH, Hall CA (2006): Nitrogen and phosphorus budgets for a tropical watershed impacted by agricultural land use: Guayas, Ecuador. *Biogeochem* 79, 135–161.

Boskey AL (2013): Bone composition: relationship to bone fragility and antiosteoporotic drug effects. *BoneKEy Reports* 2.

Capodaglio A, Boguniewicz J, Llorens E, Salerno F, Copetti D, Legnani E, Buraschi E, Tartari G (2005): Integrated lake/catchment approach as a basis for the implementation of the WFD in the Lake Pusiano watershed. In: John Lawson (Ed.), *River Basin Management*. CRC Press, pp. 77–86.

Capodaglio A, Muraca A, Becchi G (2003): Accounting for water quality effects of future urbanization: Diffuse pollution loads estimates and control in Mantua's Lakes (Italy). *Water Sci Technol* 47, 291–298.

Capodaglio AG, Ghilardi P, Boguniewicz-Zablocka J (2016): New paradigms in urban water management for conservation and sustainability. *Water Pract Technol* 11, 176–186.

Chathurika JS, Kumaragamage D, Zvomuya F, Akinremi OO, Flaten DN, Indraratne SP, Dandeniya WS (2016): Woodchip biochar with or without synthetic fertilizers affects soil properties and available phosphorus in two alkaline, chernozemic soils. *Can J Soil Sci* 96, 472–484.

Chen Q, Mu H, Huang Y, Yang W (2003): Influence of lignin on transformation of phosphorus fractions and its validity. *J Agro-Environ Sci* 22, 745–748.

Chintala R, Schumacher TE, McDonald LM, Clay DE, Malo DD, Papiernik SK, Clay SA, Julson JL (2014): Phosphorus sorption and availability from biochars and soil/biochar mixtures. *CLEAN–Soil, Air, Water* 42, 626–634.

Çimrin KM, Türkmen Ö, Turan M, Tuncer B (2010): Phosphorus and humic acid application alleviate salinity stress of pepper seedling. *African J Biotechnol* 9, 5845–5851.

Cooper J, Lombardi R, Boardman D, Carliell-Marquet C (2011): The future distribution and production of global phosphate rock reserves. *Resour, Conserv Recycl* 57, 78–86.

Cordell D, Drangert J-O, White S (2009): The story of phosphorus: global food security and food for thought. *Global Environ Change* 19, 292–305

Cornish PS (2009): Research directions: improving plant uptake of soil phosphorus, and reducing dependency on input of phosphorus fertiliser. *Crop Pasture Sci* 60, 190–196.

Dai L, Li H, Tan F, Zhu N, He M, Hu G (2016): Biochar: a potential route for recycling of phosphorus in agricultural residues. *Gcb Bioenergy* 8, 852–858.

Damon PM, Bowden B, Rose T, Rengel Z (2014): Crop residue contributions to phosphorus pools in agricultural soils: a review. *Soil Biol Biochem* 74, 127–137.

Das S, Jana TK, De TK (2014): Vertical profile of phosphatase activity in the Sundarban mangrove forest, North East Coast of Bay of Bengal, India. *Geomicrobiol J* 31, 716–725.

de Bruijne G, Caldwell I, Rosemarin A (2009): Peak phosphorus–The next inconvenient truth. The Broker.

DeLuca TH, Gundale MJ, MacKenzie MD, Jones DL (2015): Biochar effects on soil nutrient transformations. In: Johannes Lehmann, Stephen Joseph (Eds.), *Biochar for Environmental Management*. Routledge, pp. 453–486.

Demeyer A, Nkana JV, Verloo M (2001): Characteristics of wood ash and influence on soil properties and nutrient uptake: an overview. *Bioresour Technol* 77, 287–295.

Doody DG, Withers PJ, Dils RM, McDowell RW, Smith V, McElarney YR, Dunbar M, Daly D (2016): Optimizing land use for the delivery of catchment ecosystem services. *Front Ecol Environ* 14, 325–332.

Duponnois R, Kisa M, Plenchette C (2006): Phosphate-solubilizing potential of the nematophagous fungus Arthrobotrys oligospora. *J Plant Nutr Soil Sci* 169, 280–282.

Espinosa D, Sale P, Tang C (2017): Effect of soil phosphorus availability and residue quality on phosphorus transfer from crop residues to the following wheat. *Plant Soil* 416, 361–375.

Fadaeerayeni S, Sohrabi M, Royaee S (2015): Kinetic modeling of MTO process applying ZSM-5 zeolite modified with phosphorus as the reaction catalyst. *Pet Sci Technol* 33, 1093–1100.

Fahad S, Adnan M, Saud S (Eds.) (2022a): Improvement of plant production in the era of climate change. First edition, *Footprints of Climate Variability on Plant Diversity*. CRC Press.

Fahad S, Adnan M, Saud S, Nie L. (Eds.) (2022b): Climate change and ecosystems: challenges to sustainable development. First edition, *Footprints of Climate Variability on Plant Diversity*. CRC Press.

Fahad S, Hasanuzzaman M, Alam M, Ullah H, Saeed M, Ali Khan I, Adnan M (Eds.) (2020): *Environment, Climate, Plant and Vegetation Growth*. Springer Nature Switzerland AG 2020. doi: https://doi.org/10.1007/978-3-030-49732-3

Fahad S, Hussain S, Saud S, Hassan S, Tanveer M, Ihsan MZ, Shah AN, Ullah A, Nasrullah KF, Ullah S, AlharbyH NW, Wu C, Huang J (2016): A combined application of biochar and phosphorus alleviates heat-induced adversities on physiological, agronomical and quality attributes of rice. *Plant Physiol Biochem* 103, 191–198.

Fahad S, Hussain S, Saud S, Tanveer M, Bajwa AA, Hassan S, Shah AN, Ullah A,Wu C, Khan FA, Shah F, Ullah S, Chen Y, Huang J (2015a): A biochar application protects rice pollen from high-temperature stress. *Plant Physiol Biochem* 96, 281–287.

Fahad S, Nie L, Chen Y, Wu C, Xiong D, Saud S, Hongyan L, Cui K, Huang J (2015b): Crop plant hormones and environmental stress. *Sustain Agric Rev* 15, 371–400.

Fahad S, Sönmez O, Saud S, Wang D, Wu C, Adnan M, Turan V (Eds.) (2021a): Plant growth regulators for climate-smart agriculture, First edition, *Footprints of Climate Variability on Plant Diversity*. CRC Press.

Fahad S, Sonmez O, Saud S, Wang D, Wu C, Adnan M, Turan V (Eds.) (2021b): Climate change and plants: biodiversity, growth and interactions, First edition, *Footprints of Climate Variability on Plant Diversity*. CRC Press.

Fahad S, Sonmez O, Saud S, Wang D, Wu C, Adnan M, Turan V (Eds.) (2021c): Developing climate resilient crops: improving global food security and safety, First edition, *Footprints of Climate Variability on Plant Diversity*. CRC Press.

Fahad S, Sönmez O, Turan V, Adnan M, Saud S, Wu C, Wang D (Eds.) (2021d): Sustainable soil and land management and climate change, First edition, *Footprints of Climate Variability on Plant Diversity*. CRC Press.

Fahad S, Sönmez O, Saud S, Wang D, Wu C, Adnan M, Arif M, Amanullah (Eds.) (2021e): Engineering tolerance in crop plants against abiotic stress, First edition, *Footprints of Climate Variability on Plant Diversity*. CRC Press.

Fahad S, Saud S, Yajun C, Chao W, Depeng W (Eds.) (2021f): Abiotic stress in plants. *IntechOpen United Kingdom* 2021. http://dx.doi.org/10.5772/intechopen.91549

Fakhre A, Ayub K, Fahad S, Sarfraz N, Niaz A, Muhammad AA, Muhammad A, Khadim D, Saud S, Shah H, Muhammad ASR, Khalid N, Muhammad A, Rahul D, Subhan D (2021): Phosphate solubilizing bacteria optimize wheat yield in mineral phosphorus applied alkaline soil. *J Saudi Soc Agric Sci* 21(5), 339–348 https://doi.org/10.1016/j.jssas.2021.10.007

Filippelli GM (2011): Phosphate rock formation and marine phosphorus geochemistry: the deep time perspective. *Chemosphere* 84, 759–766.

Frossard E, Achat DL, Bernasconi SM, Bünemann EK, Fardeau J-C, Jansa J, Morel C, Rabeharisoa L, Randriamanantsoa L, Sinaj S (2011): The use of tracers to investigate phosphate cycling in soil–plant systems. *Phosphorus in Action*. Springer, pp. 59–91.

Gentry L, David M, Royer T, Mitchell C, Starks K (2007): Phosphorus transport pathways to streams in tile-drained agricultural watersheds. *J Environ Qual* 36, 408–415.

Grossl PR, Inskeep WP (1991): Precipitation of dicalcium phosphate dihydrate in the presence of organic acids. *Soil Sci Soc Am J* 55, 670–675.

Guan R, Su J, Meng X, Li S, Li Y, Xu J, et al. (2015): Multi-layered regulation of ethylene induction plays a positive role in *Arabidopsis* resistance against *Pseudomonas syringae*. *Plant Physiol* 169, 299–312. doi: 10.1104/pp.15.00659

Gulati A, Sharma N, Vyas P, Sood S, Rahi P, Pathania V, Prasad R (2010): Organic acid production and plant growth promotion as a function of phosphate solubilization by Acinetobacter rhizosphaerae strain BIHB 723 isolated from the cold deserts of the trans-Himalayas. *Arch Microbiol* 192, 975–983.

Hajjam Y, Cherkaoui S (2017): The influence of phosphate solubilizing microorganisms on symbiotic nitrogen fixation: perspectives for sustainable agriculture. *J Mater* 8, 801–808.

Hamdali H, Hafidi M, Virolle MJ, Ouhdouch Y (2008): Rock phosphate-solubilizing Actinomycetes: screening for plant growth-promoting activities. *World J Microbiol Biotechnol* 24, 2565–2575.

Ho B, Huang E-h, Ho TK, Ho TW (2013): Microorganisms, microbial phosphate fertilizers and methods for preparing such microbial phosphate fertilizers. *Google Patents.*

Hountin J, Karam A, Couillard D, Cescas M (2000): Use of a fractionation procedure to assess the potential for P movement in a soil profile after 14 years of liquid pig manure fertilization. *Agric, Ecosyst Environ* 78, 77–84.

Hunt JF, Ohno T, He Z, Honeycutt CW, Dail DB (2007): Inhibition of phosphorus sorption to goethite, gibbsite, and kaolin by fresh and decomposed organic matter. *Biol Fertil Soils* 44, 277–288.

Ilahi H, Wahid F, Ullah R, Adnan M, Ahmad J, Azeem M, ... Amin H (2020): Evaluation of bacterial and fertility status of Rawalpindi cultivated soil with special emphasis on fluoride content. *Int J Agric Environ Res* 6(3), 177–184.

Jasinski SM (2011): Phosphate rock. *Commodity Summaries* 118–119.

Ji BY, Hu H, Zhao YL, Mu XY, Liu K, Li CH (2014): Effects of deep tillage and straw returning on soil microorganism and enzyme activities. *Sci World J* 2014, 451493. doi:10.1155/2014/451493

Jones DL, Oburger E (2011): Solubilization of phosphorus by soil microorganisms. In: Else Bünemann, Astrid Oberson, Emmanuel Frossard (Eds.), *Phosphorus in Action*. Springer, pp. 169–198.

Karaca A, Cetin SC, Turgay OC, Kizilkaya R (2010): Effects of heavy metals on soil enzyme activities. In: Brian J. Alloway (Ed.), *Soil Heavy Metals*. Springer, pp. 237–262.

Khan AA, Jilani G, Akhtar MS, Naqvi SMS, Rasheed M (2009): Phosphorus solubilizing bacteria: occurrence, mechanisms and their role in crop production. *J Agric Biol Sci* 1, 48–58.

Khan MS, Zaidi A, Wani PA (2007): Role of phosphate-solubilizing microorganisms in sustainable agriculture—a review. *Agron Sustainable Dev* 27, 29–43.

Khan MS, Zaidi A, Ahemad M, Oves M, Wani PA (2010): Plant growth promotion by phosphate solubilizing fungi–current perspective. *Arch Agron Soil Sci* 56, 73–98.

Khan MS, Zaidi A, Ahmad E (2014): Mechanism of phosphate solubilization and physiological functions of phosphate-solubilizing microorganisms. In: Mohammad Saghir Khan, Almas Zaidi, Javed Musarrat (Eds.), *Phosphate Solubilizing Microorganisms*. Springer, pp. 31–62.

Kleinman P, Sharpley A, Buda A, McDowell R, Allen A (2011): Soil controls of phosphorus in runoff: Management barriers and opportunities. *Can J Soil Sci* 91, 329–338.

Kumar A, Patel H (2018): Role of microbes in phosphorus availability and acquisition by plants. *Int J Curr Microbiol Appl Sci* 7, 1344–1347.

Kusmer A, Goyette J-O, MacDonald G, Bennett E, Maranger R, Withers P (2019): Watershed buffering of legacy phosphorus pressure at a regional scale: a comparison across space and time. *Ecosyst* 22, 91–109.

Li N, Qiao Z, Hong J, Xie Y, Zhang P (2014): Effect of soluble phosphorus microbial mixed fertilizers on phosphorus nutrient and phosphorus adsorption-desorption characteristics in calcareous cinnamon soil. *Chin J Appl Environ Biol* 20, 662–668.

Liu J, Yang J, Liang X, Zhao Y, Cade-Menun BJ, Hu Y (2014): Molecular speciation of phosphorus present in readily dispersible colloids from agricultural soils. *Soil Science Soc Am J* 78, 47–53.

Liu Y, Villalba G, Ayres RU, Schroder H (2008): Global phosphorus flows and environmental impacts from a consumption perspective. *J Ind Ecol* 12, 229–247.

Lone A, Najar G, Ganie M, Sofi J, Ali T (2015): The impact of biochar on sustainable soil health: a review of exploration and hotspots. *Pedosphere* 25, 639–653.

Mahajan A, Gupta R (2009): Bio-fertilizers: Their kinds and requirement in India. In: Anil Mahajan, R. D. Gupta (Eds.), *Integrated Nutrient Management (INM) in a Sustainable Rice-Wheat Cropping System*. Springer, pp. 75–100.

Mahidi S, Hassan G, Hussain A, Rasool F (2011): Phosphorus availability issue-Its fixation and role of phosphate solubilizing bacteria in phosphate solubilization-Case study. *Agric Sci Res J* 2, 174–179.

Manolikaki II, Mangolis A, Diamadopoulos E (2016): The impact of biochars prepared from agricultural residues on phosphorus release and availability in two fertile soils. *J Environ Manage* 181, 536–543.

Mekonnen MM, Hoekstra AY (2018): Global anthropogenic phosphorus loads to freshwater and associated grey water footprints and water pollution levels: a high-resolution global study. *Water Resour Res* 54, 345–358.

Mittal V, Singh O, Nayyar H, Kaur J, Tewari R (2008): Stimulatory effect of phosphate-solubilizing fungal strains (Aspergillus awamori and Penicillium citrinum) on the yield of chickpea (Cicer arietinum L. cv. GPF2). *Soil Bio Biochem* 40, 718–727.

Muhammad I, Khadim D, Fahad S, Imran M, Saud A, Manzer HS, Shah S, Jabar ZKK, Shamsher A, Shah H, Taufiq N, Hafiz MH, Jan B, Wajid N (2022): Exploring the potential effect of Achnatherum splendens L.—derived biochar treated with phosphoric acid on bioavailability of cadmium and wheat growth in contaminated soil. *Environ Sci Pollut Res* 29(25), 37676–37684. https://doi.org/10.1007/s11356-021-17950-0.

Mukherjee S, Sen SK (2015): Exploration of novel rhizospheric yeast isolate as fertilizing soil inoculant for improvement of maize cultivation. *J Sci Food Agric* 95, 1491–1499.

Nesme T, Withers PJ (2016): *Sustainable Strategies towards a Phosphorus Circular Economy*. Springer.

Omar L, Ahmed OH (2011): Effect of mixing urea with zeolite and sago waste water on nutrient use efficiency of maize (Zea mays L.). *African J Microbiol Res* 5, 3462–3467.

Owen D, Williams AP, Griffith GW, Withers PJ (2015): Use of commercial bio-inoculants to increase agricultural production through improved phosphrous acquisition. *Appl Soil Ecol* 86, 41–54.

Parvage MM, Ulén B, Eriksson J, Strock J, Kirchmann H (2013): Phosphorus availability in soils amended with wheat residue char. *Biol Fertil Soils* 49, 245–250.

Patel KJ, Singh AK, Nareshkumar G, Archana G (2010): Organic-acid-producing, phytate-mineralizing rhizobacteria and their effect on growth of pigeon pea (Cajanus cajan). *Appl Soil Ecol* 44, 252–261.

Pfitzner A, Bräu MF, Zweck J, Brunklaus G, Eckert H (2004): Phosphorus nanorods—Two allotropic modifications of a long-known element. *Angewandte Chemie International Edition* 43, 4228–4231.

Pickering HW, Menzies NW, Hunter MN (2002): Zeolite/rock phosphate—a novel slow release phosphorus fertiliser for potted plant production. *Sci Hortic* 94, 333–343.

Polat E, Karaca M, Demir H, Onus AN (2004): Use of natural zeolite (clinoptilolite) in agriculture. *J Fruit Ornamental Plant Res* 12, 183–189.

Qiu J, Turner MG (2015): Importance of landscape heterogeneity in sustaining hydrologic ecosystem services in an agricultural watershed. *Ecosphere* 6, 1–19.

Regelink I, Weng L, Lair G, Comans R (2015): Adsorption of phosphate and organic matter on metal (hydr) oxides in arable and forest soil: a mechanistic modelling study. *Eur J Soil Sci* 66, 867–875.

Rietz D, Haynes R (2003): Effects of irrigation-induced salinity and sodicity on soil microbial activity. *Soil Biol Biochem* 35, 845–854.

Roberts TL, Johnston AE (2015): Phosphorus use efficiency and management in agriculture. *Resour, Conserv Recycl* 105, 275–281.

Roemer W (2009): Concepts for a more efficient use of phosphorus based on experimental observations. *Berichte über Landwirtschaft* 87, 5–30.

Satyaprakash M, Nikitha T, Reddi E, Sadhana B, Vani SS (2017): Phosphorous and phosphate solubilising bacteria and their role in plant nutrition. *Int J Curr Microbiol Appl Sci* 6, 2133–2144.

Selvi K, Paul J, Vijaya V, Saraswathi K (2017): Analyzing the efficacy of phosphate solubilizing microorganisms by enrichment culture techniques. *Biochem Mol Biol J* 3, 1–7.

Shafi MI, Adnan M, Fahad S, Fazli W, Ahsan K, Zhen Y, Subhan D, Zafar-ul-Hye M, Martin B, Rahul D (2020): Application of single superphosphate with humic acid improves the growth, yield and phosphorus uptake of wheat (Triticum aestivum L.) in calcareous soil. *Agron* 10, 1224. doi:10.3390/agronomy10091224

Sharma SB, Sayyed RZ, Trivedi MH, Gobi TA (2013): Phosphate solubilizing microbes: sustainable approach for managing phosphorus deficiency in agricultural soils. *SpringerPlus* 2, 1–14.

Sharpley A, Jarvie H, Flaten D, Kleinman P (2018): Celebrating the 350th anniversary of phosphorus discovery: a conundrum of deficiency and excess. *J Environ Qual* 47, 774–777.

Søndergaard M, Jensen JP, Jeppesen E (2003): Role of sediment and internal loading of phosphorus in shallow lakes. *Hydrobiologia* 506, 135–145.

Steinfeld H, Gerber P, Wassenaar T, Castel V, Rosales M, de Haan C (2006): Livestock's long shadow: environmental issues and options. Food & Agriculture Org.

Ström L, Owen AG, Godbold DL, Jones DL (2005): Organic acid behaviour in a calcareous soil: implications for rhizosphere nutrient cycling. *Soil Biol Biochem* 37, 2046–2054.

Štursová M, Baldrian P (2011): Effects of soil properties and management on the activity of soil organic matter transforming enzymes and the quantification of soil-bound and free activity. *Plant Soil* 338, 99–110.

Sun G, Jin J, Shi Y (2011): Advances in the effect of humic acid and modified lignin on availability to crops. *Chin J Soil Sci* 42, 1003–1009.

Suri V, Choudhary AK, Chander G, Verma T, Gupta M, Dutt N (2011): Improving phosphorus use through co-inoculation of vesicular arbuscular mycorrhizal fungi and phosphate-solubilizing bacteria in maize in an acidic Alfisol. *Commun Soil Sci Plant Anal* 42, 2265–2273.

Syers J, Johnston A, Curtin D (2008): Efficiency of soil and fertilizer phosphorus use. *FAO Fertilizer and Plant Nutrition Bulletin* 18, 108.

Tahir M, Khurshid M, Khan M, Abbasi M, Kazmi M (2011): Lignite-derived humic acid effect on growth of wheat plants in different soils. *Pedosphere* 21, 124–131.

Tripathi K (2005): Effect of organic acid on phosphorus-use efficiency by clusterbean (Cyamopsis tetragonoloba) in arid soil of Rajasthan. *Indian J Agric Sci* 75, 651–653.

Vandecasteele B, Sinicco T, D'Hose T, Nest TV, Mondini C (2016): Biochar amendment before or after composting affects compost quality and N losses, but not P plant uptake. *J Environ Manage* 168, 200–209.

Wahid F, Fahad S, Subhan D, Adnan M, Zhen Y, Saud S, Manzer HS, Martin B, Tereza H, Rahul D (2020): Sustainable management with mycorrhizae and phosphate solubilizing bacteria for enhanced phosphorus uptake in calcareous soils. *Agri* 10(8), 334. doi:10.3390/agriculture10080334

Walan P, Davidsson S, Johansson S, Höök M (2014): Phosphate rock production and depletion: regional disaggregated odelling and global implications. *Resour, Conserv Recycl* 93, 178–187.

Walpola BC, Yoon M-H (2012): Prospectus of phosphate solubilizing microorganisms and phosphorus availability in agricultural soils: A review. *African J Microbiol Res* 6, 6600–6605.

Wang Y, Chen X, Whalen JK, Cao Y, Quan Z, Lu C, Shi Y (2015a): Kinetics of inorganic and organic phosphorus release influenced by low molecular weight organic acids in calcareous, neutral and acidic soils. *J Plant Nutr Soil Sci* 178, 555–566.

Wang Y, Lin Y, Chiu PC, Imhoff PT, Guo M (2015b): Phosphorus release behaviors of poultry litter biochar as a soil amendment. *Sci Total Environ* 512, 454–463.

Wei L, Chen C, Xu Z (2010): Citric acid enhances the mobilization of organic phosphorus in subtropical and tropical forest soils. *Biol Fertil Soils* 46, 765–769.

Xu G, Zhang Y, Sun J, Shao H (2016): Negative interactive effects between biochar and phosphorus fertilization on phosphorus availability and plant yield in saline sodic soil. *Sci Total Environ* 568, 910–915.

Yadav B, Verma A (2012): Phosphate solubilization and mobilization in soil through microorganisms under arid ecosystems. In: Mahamane Ali (Ed.), *The Functioning of Ecosystems*. IntechOpen, pp. 93–108.

Yan Y, Liu Jr F, Li W, Liu F, Feng X, Sparks D (2014): Sorption and desorption characteristics of organic phosphates of different structures on aluminium (oxyhydr) oxides. *Eur J Soil Sci* 65, 308–317.

Yang M, Lin J, Zhan Y, Zhu Z, Zhang H (2015): Immobilization of phosphorus from water and sediment using zirconium-modified zeolites. *Environ Sci Pollut Res* 22, 3606–3619.

Yangyuoru M, Boateng E, Adiku S, Acquah D, Adjadeh T, Mawunya F (2006): Effects of natural and synthetic soil conditioners on soil moisture retention and maize yield. *West African J Appl Ecol* 9, 1–8.

Yousefi AA, Khavazi K, Moezi AA, Rejali F, Nadian HA (2011): Phosphate solubilizing bacteria and arbuscular mycorrhizal fungi impacts on inorganic phosphorus fractions and wheat growth. *World Appl Sci J* 15, 1310–1318.

Yu X, Liu X, Zhu TH, Liu GH, Mao C (2011): Isolation and characterization of phosphate-solubilizing bacteria from walnut and their effect on growth and phosphorus mobilization. *Biol Fert Soils* 47, 437–446.

Yuan H, Blackwell M, Mcgrath S, George T, Granger S, Hawkins J, Dunham S, Shen J (2016): Morphological responses of wheat (Triticum aestivum L.) roots to phosphorus supply in two contrasting soils. *J Agric Sci* 154, 98–108.

Zahida Z, Hafiz FB, Zulfiqar AS, Ghulam MS, Fahad S, Muhammad RA, Hafiz MH, Wajid N, Muhammad S (2017): Effect of water management and silicon on germination, growth, phosphorus and arsenic uptake in rice. *Ecotoxicol Environ Saf* 144, 11–18.

Zhen-Yu D, Qing-Hua W, Fang-Chun L, Hai-Lin M, Bing-Yao M, Malhi S (2013): Movement of phosphorus in a calcareous soil as affected by humic acid. *Pedosphere* 23, 229–235.

3 Bio-fertilizers for Improving Micronutrient Availability

Nazish Huma Khan,[*1] *Muhammad Sufyan Khan,*[1]
Tooba Saeed,[2] *Fazli Zuljalal,*[1] *and Muhammad Adnan*[3]
[1] Department of Environmental Sciences, University of Swabi, Pakistan
[2] National Centre of Excellence in Physical Chemistry, University of Peshawar, Pakistan
[3] Department of Agriculture, University of Swabi, Pakistan
* Correspondence: humakhan@uoswabi.edu.pk

CONTENTS

3.1 INTRODUCTION

The global population continues to grow, and by 2050 it is expected to reach 9.7 billion people (Ehrlich and Harte 2015). Intensive industry, urbanization, and crop yields are closely linked to this rapid expansion (Saman et al. 2020; Muhammad Tahir et al. 2020; Jakirand Allah 2020; Mahmood et al. 2021; Farah et al. 2020; Sadam et al. 2020; Unsar et al. 2020; Fazli et al. 2020; Enamul et al. 2020; Gopakumar et al. 2020; Zia-ur-Rehman 2020; Ayman et al. 2020; Mohammad I. Al-Wabel

DOI: 10.1201/9781003286233-3

et al. 2020a,b; Senol 2020; Amjad et al. 2020; Ibrar et al. 2020; Sajid et al. 2020; Muhammad et al. 2021; Sidra et al. 2021; Zahir et al. 2021; Sahrish et al. 2022). Agricultural production is essential to meet the nutrient needs of humanity, which were expected to reach, by 2020, 321 million tons of edible cereals (Al-Zahrani et al. 2022; Rajesh et al. 2022; Anam et al. 2021; Deepranjan et al. 2021; Haider et al. 2021; Amjad et al. 2021; Sajjad et al. 2021a,b; Fakhre et al. 2021; Khatun et al. 2021). Traditional farming methods, on the other hand, rely heavily on the extensive use of synthetic pesticides and fertilizers to feed crops and prevent infections (Vesile et al. 2015; Afzal et al. 2017; Ibrar et al. 2021; Bukhari et al. 2021; Haoliang et al. 2022; Sana et al. 2022; Abid et al. 2021; Zaman et al. 2021; Sajjad et al. 2021a,b; Rehana et al. 2021; Yang et al. 2022; Ahmad et al. 2022; Shah et al. 2022; Muhammad et al. 2022; Wiqar et al. 2022). The proper application of synthetic chemicals has undeniable benefits not only for plant growth, yield, and quality, but also the farmer's income is also important. However, the increasing use of synthetic supplies may end up polluting the air, water, and soil, posing a significant threat to the natural environment (Rehman and Zhang 2018; Farhat et al. 2020; Farhat et al. 2022; Niaz et al. 2022; Ihsan et al. 2022; Chao et al. 2022; Qin et al. 2022; Xue et al. 2022; Ali et al. 2022; Mehmood et al. 2022; El Sabagh et al. 2022; Ibad et al. 2022). The thoughtless use of agrochemicals and their inability to biodegrade leads to their accumulation in the soil, leading to negative changes in soil properties such as structure, fertility, and water-holding capacity. Excessive use of artificial fertilizers is also linked to algal blooms in water supplies, greenhouse gases, and the buildup of toxic substances like As and Cd (Atafar et al. 2010; Deepranjan et al. 2021; Haider et al. 2021; Huang Li et al. 2021; Ikram et al. 2021; Jabborova et al. 2021; Khadim et al. 2021a,b; Manzer et al. 2021; Muzammal et al. 2021; Abdul et al. 2021a,b; Ashfaq et al. 2021; Amjad et al. 2021; Atif et al. 2021).

Organic farming is a viable alternative to traditional farming, as it helps us to provide quality food without adversely affecting the soil's health or the environment. It incorporates environmentally friendly agronomic approaches and enables contamination-free food production while ensuring soil quality and biodiversity (Niggli 2015; Athar et al. 2021; Adnan et al. 2018a,b; Adnan et al. 2019; Akram et al. 2018a,b; Aziz et al. 2017a,b; Chang et al. 2021; Chen et al. 2021; Emre et al. 2021; Habib et al. 2017; Hafiz et al. 2016; Hafiz et al. 2019; Ghulam et al. 2021; Guofu et al. 2021). The-above mentioned dangers of overloading the soil with chemically synthesized agrochemicals, along with increased consumer awareness of the importance of protecting the natural environment and human health, have prompted researchers to seek alternatives that would be just as effective while posing no risk to the environment (Geiger et al. 2010; Hafeez et al. 2021; Khan et al. 2021; Kamaran et al. 2017; Muhmmad et al. 2019; Safi et al. 2021; Sajjad et al. 2019; Saud et al. 2013; Saud et al. 2014; Saud et al. 2017; Saud et al. 2016; Shah et al. 2013; Saud et al. 2020; Saud et al. 2022a,b; Fahad and Bano 2012; Fahad et al. 2017; Fahad et al. 2013; Fahad et al. 2014a,b;Fahad et al. 2016a,b,c,d; Fahad et al. 2015a,b; Fahad et al. 2018a,b; Fahad et al. 2019a,b; Fahad et al. 2020; Fahad et al. 2021a,b,c,d,e,f; Fahad et al. 2022a,b; Hesham and Fahad 2020). These microbial strains have great potential in agriculture as they promote crop growth by increasing the supply of natural nutrients (Tyota and Watenabe 2013). Compared to artificial chemicals, bio-fertilizers are an economical, ecologically benign, and sustainable supplier of nutrients for crops, and therefore, they are gaining attractiveness and importance in agricultural production (Swapna et al. 2016; Qamar et al. 2017; Hamza et al. 2021; Irfan et al. 2021;Wajid et al. 2017; Yang et al. 2017; Zahida et al. 2017; Depeng et al. 2018; Hussain et al. 2020; Hafiz et al. 2020 a,b; Shafi et al. 2020; Wahid et al. 2020; Subhan et al. 2020; Zafar-ul-Hye et al. 2020a,b; Zafar et al. 2021).

3.2 BIO-FERTILIZERS

The term "bio-fertilizer" has several meanings. Seaweed extracts, compound municipal waste, microbiological mixtures with undetermined ingredients, organic fertilizer products, as well as mineral fertilizer products that have been augmented with organic compounds are all examples of

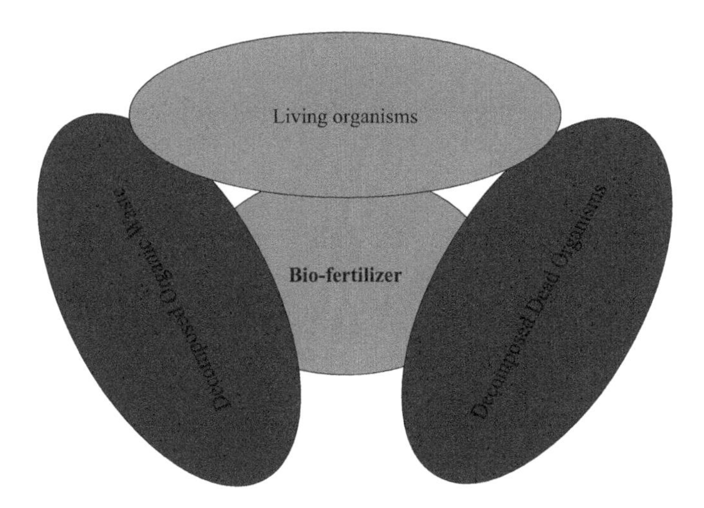

FIGURE 3.1 Sources of bio-fertilizers.

bio-fertilizers. In 2003, Vesey defined bio-fertilizer as "a product containing microbes that colonizes the rhizosphere (interior) of the plant and stimulates development when administered to the surface of roots, seeds or soil, by improving the supply or availability of essential minerals to the host plant" (Vesey 2003; Ali et al. 2018). Another study (Fuentes-Ramirez and Caballero-Mellado 2005) described the fact that bio-fertilizer contains live microbes that exert a favourable impact on crop yield and agriculture through several methods (Figure 3.1). Products that include beneficial microorganisms used to reduce plant pathogens can be classified as bio-fertilizers, although they are more commonly referred to as bio-pesticides (Fuantes-Ramireez and Cabellero-Mellado 2005).

In science, bio-fertilizer is a single microorganism with plant growth-promoting abilities, but in agronomy the term refers to a product composed of many beneficial strains, which are useful for nutrient mobilization, are enclosed in a container, have properties allowing conservation at the time specified by the producer, and are ready for planting or application to the soil. Bio-fertilizer can also allow for the addition of chemicals to help microorganisms work better. The term "bio-fertilizer" should not even be confused with terms like plant or animal compost, intercropping, or fertilizers, which refer to a mixture of mineral and organic chemicals, as well as to bio-stimulants from microorganisms (Reddy 2014). The main objective of using bio-fertilizers is to promote plant development while minimizing negative environmental consequences and improving crop yields. They also improve plant–water relationships, protect plants from dehydration, reduce the incidence of insect pests, and make plants more resistant to several soilborne diseases, such as those caused by fungi that also create mycotoxins (Dey et al. 2014). In addition, microbial fertilizers need to be in a greater quantity to supply sufficient nutrient content to plants, their efficacy is dependent on the application zone's soil conditions (Macik et al. 2020).

3.3 NEED FOR BIO-FERTILIZERS

Synthetic fertilizers used haphazardly contaminate soil, pollute water basins, eliminate beneficial microorganisms and insects, increase disease susceptibility and decrease soil fertility (Mishra et al. 2013; Adnan et al. 2020; Ilyas et al. 2020; Saleem et al. 2020a,b,c; Rehman 2020; Frahat et al. 2020; Wu et al. 2020; Mubeen et al. 2020; Farhana 2020; Jan et al. 2019; Wu et al. 2019; Ahmad et al. 2019; Baseer et al. 2019; Hafiz et al. 2018; Tariq et al. 2018). The demand is far greater than the supply. By 2020, 28.8 million tons of minerals were needed to accomplish the 321 million tons of grain that needed to be produced, while only 21.6 million tons were available, resulting in a deficit of 7.2 million tons (Arun 2007). Depletion of raw materials/fossil fuels (energy crisis) and increased fertilizer costs result. Besides the above, their long-term use means it costs more to use than bio-fertilizers, which are environmentally friendly, efficient, and work well, while synthetic fertilizers are more expensive and inaccessible to the wider agricultural community (SubbaRoa 2001).

3.4 TYPES

Nitrogen fixers and phosphate absorbers are two types of bio-fertilizers.

3.4.1 Nitrogen Fixers

3.4.1.1 *Rhizobium*

Rhizobium is a symbiotic bacterium that exclusively fixes nitrogen with legumes (50–100 kg/ha). *Rhizobium*'s ability to nodulate leguminous crops is primarily dependent on the availability of suitable strains for each legume. It takes up residence in the roots of some legumes, producing root nodules, which are tumour-like growths that operate as NH_3 manufacturers. In a symbiotic association with legumes, *Rhizobium* may improve soil fertility and growth of non-legumes such as *Parasomnia*. Some *Rhizobium* strains have been shown to operate as bio-control agents and stimulate crop development (Kumar et al. 2011).

3.4.1.2 *Azospirillum*

Azospirillum is a fungus that is both heterotrophic and associative, and it belongs to the Spirilaceae family. It is a growth-regulator, in addition to having an ability to fix nitrogen at a rate of 20–40 kg/ha, and it can also produce compounds. There are several varieties in this genus, such as *A. amazonense*, *A. lipoferum*, and *A. brasilense* which have been shown to have a worldwide range and provide inoculation benefits. The *Azotobacter* that colonizes roots does not just stay on the surface; a significant fraction enters the cellular tissues of the roots and coexists with the crop (Chavan 2019).

3.4.1.3 *Azotobacter*

Azotobacter is a type of bacteria that lives in aerobic, free-living environments; they are heterotrophic bacteria from the Azotobacteriaceae family. *Azotobacter* can be found in both neutral and alkaline soils, with *A. chrococcum* being the most common species in arable soils. Other species include *A. vinalandii*, *A. beijerinckii*, *A. insignis*, and *A. macro-cytogenes*. Antifungal antibiotics are made by this bacterium which minimizes seedling mortality by stifling the growth of a large number of organisms susceptible to the presence of fungus in the root zone (Subba Rao 2001).

3.4.1.4 Blue Green Algae

This is also known as "paddy organism" being plentiful in paddies. For lowland rice production, considerable amounts of nitrogen are required, which are provided by this type of algae. Sustainability is important for food production; BNF must gradually replace industrial nitrogen fixation as a source of fixed nitrogen. The 50–60% N condition is fulfilled by a hybrid of unrestricted and rice-growing plant-related microbes mineralizing soil organic N and BNF (Roger and Ladha 1992).

3.4.2 Mycorrhiza

Mycorrhiza means "fungus roots." Host plants have a symbiotic connection and a type of fungus found in the roots, in which the fungus partner reaps the rewards by obtaining its carbon requirements from the photosynthetic activity of the host. As a result, the host benefits by getting much-needed nutrients such as P, Ca, Cu, Zn, and other elements with the help of the fungus' fine absorbing hyphae, as it may penetrate areas that are ordinarily inaccessible. *Glomus*, *Gigaspora*, *Acaulospora*,

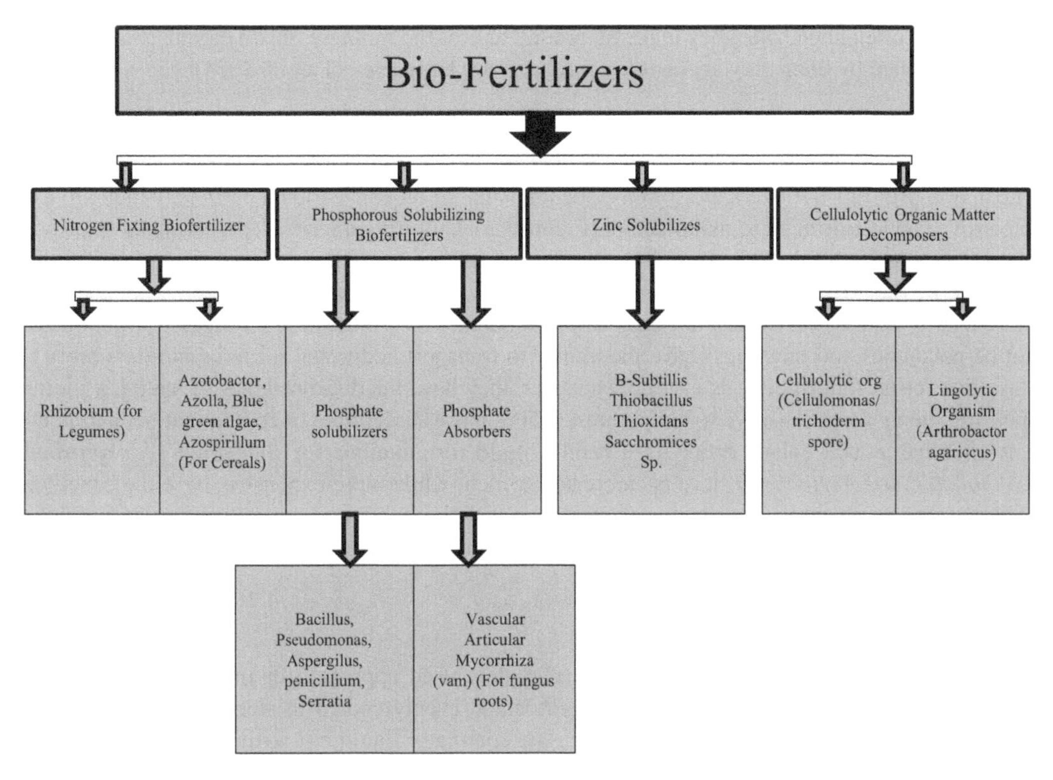

FIGURE 3.2 Agricultural impacts of bio-fertilizers.

Sclerocysts, and *Endogone* are the most prevalent genera employed in the manufacturing of bio-fertilizers (Bagyaraj 1992).

3.5 IMPACTS OF BIO-FERTILIZERS ON AGRICULTURE

Bio-fertilizers are crucial in agricultural techniques to boost crop yields. Fertilizers are required to suit the needs of plants. Nitrogen-fixing fertilizers (*Rhizobium*) are responsible for fixing 50–100 kg of nitrogen per acre in vegetables. Beat vegetables, red gram, and pea can all benefit from these fertilizers. Chickpeas, lentils, green gram, black gram, and other legumes are also beneficial to oil-seed legumes like soybean and groundnut (Mahdi et al. 2010; Khan et al. 2016). Similarly, agriculture is interested in *Azotobacter* because of its potential to fix nitrogen. It can create vitamins and plant hormones like thiamine and riboflavin. In addition to helping with nitrogen fixation, plant development hormones are produced by phosphate-solubilizing bio-fertilizers which promote plants growth (Figure 3.2; Madhav and Hina 2020).

A soil solution is used to raise the phosphate level in soils. Plants are able to absorb sufficient amounts of dissolvable phosphorus, resulting in a 10–20% increase in yield.

The combination of *G. fasciculatum* + *A. chroococcum* + half of the planned P rate produced the best root length, plant stature, bulb size, new weight, root colonization, and P uptake in a nursery pot experiment with onion (Carter 1967). K-solubilizing bacteria, such as *Frateuria aurantia*, are responsible for converting a potassium (K) mixture into a useful structure for plants, and is applied to crops in conjunction with various bio-fertilizers with no adverse effects. Despite the fact that a wide range of agricultural primers have been shown to be effective, there is a notable disparity in responses among yields, geographies, and situations (Ghosh 2004). In farming, *Arbuscular mycorrhiza* (AM) parasites are often abundant. They make up around half of the biomass of soil organisms, and some products formed by them may account for another 3000 kg (Lovelock et al. 2004).

3.6 APPLICATIONS OF BIO-FERTILIZERS

The majority of bio-fertilizers are sold as inoculants based on conventional carriers, which are less expensive and less difficult to manufacture (Figure 3.3). Culturing microbes, preparing the transport ingredient, combining it with the culture broth, and packing are all steps in the mass production of bio-fertilizers. The ideal carrier materials for bio-fertilizer preparation must be less expensive, more readily accessible, as well as being easier to handle; they must have an organic structure and not be poisonous and they must have the ability to transport additional bacteria and the ability to stay alive for prolonged periods of time. However, they have the disadvantages of having a shorter shelf life, being more sensitive to temperature, being more susceptible to contamination, and being less effective as cell counts drop. As a result, liquid formulations for *Rhizobium*, *Azospirillum*, *Azotobacter*, and *Acetobacter* have been created, which, while more expensive, have the benefit of being more convenient to manufacture as well as providing increased soil efficiency (Ngampimol and Kunathigan 2008). The practices listed below are crucial.

3.6.1 Seed Treatment

This is the most successful, cost-efficient, and widely used approach for inoculants of all types (Sethi et al. 2014). The seeds are mixed together and evenly coated in slurry before even being shade-dried for 24 hours before they are used. The coating of liquid bio-fertilizers is possible to do in a plastic bag or bucket, depending on the amount of seeds. Two or more microorganisms can be used to treat seeds without causing an antagonistic reaction, and the greatest amount of bacteria on each seed is necessary for the best outcomes (Chen 2006).

3.6.2 Seedling Root Dipping

Water is used to soak newly planted crop roots in a mixture of bio-fertilizer, such as grains, plants, berries, trees, sisal, silk, grapes, bananas, and tobacco. Various crops require different treatment times; for example, vegetable crops are sprayed for 20–30 minutes, and paddy is sprayed for 8–12 hours before transplantation (Barea and Brown 1974).

3.6.3 Soil Application

In this procedure, bio-fertilizer is applied directly to the roots, either individually or jointly. Cow dung is a phosphate-solubilizing microbial bio-fertilizer and rock phosphate mixture is maintained in the shade overnight and applied to the soil with a moisture level of 50%. *Rhizobium* and *Azotobacter* are two examples of bio-fertilizers that are applied to soil (Hayat et al. 2010).

3.7 ENVIRONMENTAL STRESS AND BIO-FERTILIZERS

Microbial physiology is influenced significantly by environmental stressors (Iqra et al. 2020; Akbar et al. 2020; Mahar et al. 2020; Noor et al. 2020; Bayram et al. 2020; Amanullah, Fahad 2018a,b; Amanullah, Fahad 2017; Amanullah et al. 2020; Amanullah et al. 2021; Rashid et al. 2020; Arif et al. 2020; Amir et al. 2020). Radiation exposure has sufficient energy to allow electrons to escape from a cell's molecules, resulting in the formation of free radicals that can oxidize DNA or RNA, and non-ionizing radiation (ultraviolet radiation), by activating electrons in molecules, has a carcinogenic impact, resulting in pyrimidine formation dimers, which frequently change the structure of DNA in cells, which can lead to complications during pregnancy. Ionizing radiation penetrates cells and endospores easily because of its high frequency. UV light has a negative impact on cyanobacteria's expansion, continued existence, photosynthesis, CO_2 uptake, and nitrogen metabolism, among other things (Sinha 2005). Many researchers believe that UV-B radiation destroys cellular constituents that absorb light between 280 and 315 nm, influencing the porosity of cellular membranes and protein breakdown, finally leading to the death of the cephalopod (Vincent 1993). When exposed to UV-B, reactive oxygen species (ROS) decrease gene expression of critical photosynthetic proteins. Reduced photosynthetic activity (Fv/Fm) and elevated reactive oxygen species were found in *Microcoleus vaginatusa* formation after being exposed with UV-Chen B radiation (Chen et al. 2009). The assimilatory enzyme nitrate reduction is found in almost all bio-fertilizers. It has been proposed that nitrate reductase is linked to membrane fractions that contain chlorophyll, and thus inhibiting chlorophyll directly affects nitrate reductase function. Because their aromatic amino acids absorb UV-B, ultraviolet radiation disrupts the complex organization of phycobilisomes. In cyanobacteria, chlorophyll-binding proteins and phycobilisomes capture and over 99% of UV-B (Sinha 2005). Exposure of *Synechocystis* cells (phycoerythrin-deficient) to moderate UV-B (1.8 Wm^{-2}) causes phycocyanin loss, which could be attributed to the two bilins present in α-phycocyanin versus single bilins in other biliproteins. Both α- and β-phycocyanin were photodegraded when *Synechococcus* sp. PCC 7942 phycobilisomes were subjected to UV-B light (Sah et al. 1998). DNA is one of the most significant targets of solar UV radiation in all living things (Jans 2005).

3.8 BIO-FERTILIZERS IN THE ECOSYSTEM

Despite the fact that bio-fertilizers have been routinely employed in agriculture for a number of years, information on their colonization and ecology is lacking. Furthermore, the process underlying their interactions with plants and the local microbial population continues to pique people's interest. Microflora that is native to the area in the rhizosphere is one of the primary elements that influences a bio-fertilizer's efficiency in a living ecosystem. This highly competitive rhizosphere community with a varied range of organisms may have an impact on the bio-fertilizer's survival and qualities that encourage crop growth (Hibbing et al. 2010). Furthermore, seeds and seedlings are bacterized, and also soil additives may cause local microflora to change structure, which must be taken into account when determining the risk of microorganisms being released into the environment (Dey et al. 2012). Finally, microbial bio-fertilizers' effect on non-target species, the biogeochemical processes affected, changes in soil texture, soil qualities such as permeability, richness, and water-holding ability, as well as erosion avoidance should all be carefully studied (Pereg and McMillan 2015). As a result, prior to the release of bio-fertilizers into the surroundings, it is critical to evaluate their unintended consequences for local microbiota communities and, as a result, ecosystems, as well as a thorough examination of bio-fertilizers and their effects before changing agricultural methods. Although the impacts of bio-fertilizers on non-target members of the soil rhizosphere and food web have been examined to some extent, the effects of bio-fertilizers on non-target members of the soil rhizosphere and food web have yet to be fully investigated. Despite the fact that most research found quantifiable changes as a result of introducing bio-inoculants into the rhizosphere, the extent of the alterations and their implications for ecological functions have yet to be published (Martinez-Viveros et al. 2010). According to reports, the amount of influence in residential areas, of bio-fertilizers being implemented is dependent on a variety of elements such as soil characteristics, bio-fertilizer application technique, environmental circumstances, and so on (Dey et al. 2012). However, relatively few trials have been conducted over a longer length of time, which is critical for drawing conclusions about the efficacy and risk aspects of bio-inoculants. As a result, more thorough research on the longer term effects of non-target creatures and bio-fertilizers are needed prior to the use of bio-inoculants that can be classified as "safe" for commercial use. Additionally, bio-fertilizer's impact on non-target groups have been investigated using methods that are both culture-dependent and culture-independent, as well as genetic and physiological assays. Both plating and cytochemical methods for assessing microbial community structures are appropriate approaches to investigate the effects of bio-fertilizers on the microbiota of the local environment. The application of new technology must be done while evaluating the influence on resident microbiota as well as soil function. Although DNA is a trustworthy as a useful criterion for evaluating the diversity and potential of a community, recent successes in extracting messenger RNA (mRNA) from soil are significant enough to include the experiment in bio-fertilizer risk and efficiency assessments. Furthermore, an mRNA-based method would take into account the real functional variety of bio-fertilizers at any given time in the system under study. As a result, high-throughput, high-resolution methodologies must be used in conjunction with established multi-dimensional inspection approaches to bio-fertilizer performance, diversity, and risk assessment studies before they are released into the environment (Sharma et al. 2012).

3.9 ADVANTAGES OF BIO-FERTILIZERS

Bio-fertilizers aid in the production of high crop yields by adding nutrients to the soil and beneficial microbes required for crop yield. Chemical fertilizers are no longer used as they are harmful to plants. Chemical fertilizers stunt plant development and release hazardous chemicals into the environment, polluting the environment. Bio-fertilizers are used because they include natural ingredients that do not damage but rather benefit plants, and crop yield can be accelerated. They are environmentally beneficial because they safeguard the environment from contaminants. The soil will preserve its fertility if it is free of chemicals, which will benefit both the plants and the ecosystem, as the crops will be disease-free and there will be no pollution in the ecosystem. Bio-fertilizers eliminate the toxic components in soil that cause plant illnesses. Plants can be safeguarded in the same way from famine and other situations by employing bio-fertilizers. Bio-fertilizers are affordable, and they can be used by even low-income farmers (Chavan 2019).

3.10 DISADVANTAGES OF BIO-FERTILIZERS

Because bio-fertilizers are alive, they require careful handling when storing them for lengthy periods of time. They have to be utilized before they expire. Bio-fertilizers can be used in conjunction with traditional fertilizers, but they cannot completely replace them. Bio-fertilizers must be used at the correct temperature and pH range to be effective; otherwise, they are ineffective and inhibit the growth of helpful microorganisms. They are less efficient if the carrier medium is polluted with other microorganisms or if the grower uses the incorrect strain of bacterium. For organisms to thrive and work, the soil in which the bio-fertilizer is placed must possess sufficient nutrients in a suitable form and amount. Due to the slow release rate of nutrients, nutrient deficiency cannot be remedied quickly with the use of a bio-fertilizer. Because bio-fertilizers supply fewer nutrients than traditional fertilizers, a greater quantity is required to meet the required quantity of nutrients for better plant production (Kumar et al. 2017).

3.11 BIO-FERTILIZERS: A WAY TOWARDS SUSTAINABILITY

Bio-fertilizers are environmentally benign since they are created from readily available natural sources such as rice husk, leftover vegetables, and organic debris. Bio-fertilizers are less expensive to manufacture than chemical fertilizers, especially when nitrogen and phosphorus are used.

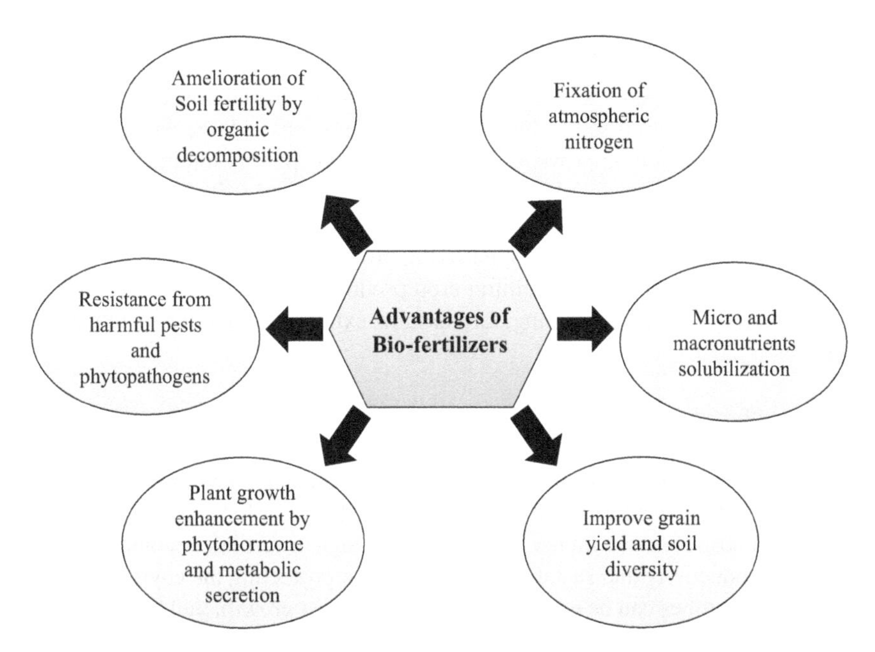

FIGURE 3.3 Advantages of bio-fertilizer.

Bio-fertilizers are inexpensive and simple to create; in fact, farmers can prepare them themselves. Bio-fertilizers improve soil fertility by mobilizing nutrients from organic materials and encourage long-term agricultural production. Bio-fertilizers also boost soil biological activity, which boosts nutrient mobilization from natural and synthetic sources as well as harmful material breakdown. Bio-fertilizers increase soil organic matter content, and boost water retention, exchangeable cations, and improve soil and buffering ability against soil acidity, salinity, alkalinity, pesticides, and toxic heavy chemicals. Bio-fertilizers provide nourishment for beneficial microorganisms and earthworms, while also encouraging their growth. Bio-fertilizers boost crop yields by 10–20%. The release of growth-promoting hormones by bio-fertilizers promotes plant root multiplication. Fungi, bacteria, and other parasites in the soil are suppressed by the presence of bio-fertilizers. Plants are more resistant to stress and survive longer. Antibiotic compounds are released by some bio-fertilizers, such as *Azotobacter*, in response to biotic stressors. In the soil, cyanobacteria release proteins, amino acids, vitamins, and other compounds that enhance organic carbon and aid nitrogen fixation.

3.12 CONSTRAINTS IN BIO-FERTILIZER TECHNOLOGY

Regardless of the fact that bio-fertilizer innovation is an ecologically friendly option, it is not without its drawbacks, and due to a variety of issues, its application or installation is restricted. These include issues with infrastructures, such as a lack of major equipment, energy, and so on. Other restrictions include financial constraints and difficulties acquiring bank loans. Seasonal requirement for bio-fertilizers, simultaneous cropping activities, and a short period of sowing/planting in a specific location are all examples of environmental restrictions. Due to challenges in farmers implementing this device, there is an unawareness of the technology's benefits due to diverse techniques of inoculation, when compared to inorganic fertilizers, and there is no discernible difference in crop growth. For manufacturers, there aren't enough retail shops or a market network, as well as the lacking of a correct time and place when the right inoculant is administered, which are all marketing obstacles. Different restrictions influence manufacturing, marketing, and usage techniques in one way or another (Kumar et al. 2017).

3.13 BIO-FERTILIZER STRATEGIES

To get updated knowledge about soil and climatic conditions in relation to a specific crop, it is necessary to strengthen research and technology. The applications of biotechnological technologies and procedures are important to improve nitrogen fixation, phosphorus solubilization, and other bio-fertilizers. Bio-fertilizer dose standardization in a specific crop and soil and finding a better carrier medium for strains to extend their shelf life are needed. Monitoring of bio-fertilizer manufacturing units on a regular basis to ensure proper viable count, production process, storage, and other factors is also required. Bio-fertilizers for agricultural crop production have received a lot of attention. In this regard, the efforts of scientific training, farmer fairs, exhibitions, and the media are to be highly praised (Singh et al. 2016).

3.14 CONCLUSION

Bio-fertilizers play an important role in integrated nutrition management and a renewable supply of plant nutrients that can be used to augment chemical fertilizers in long-term agricultural systems. As ecofriendly and cost-effective inputs for farmers, biological fertilizers could play a major role in increasing soil productivity and sustainability, while also protecting the environment. In an apple orchard, beneficial microbes can be employed as a tool to boost growth, yield, and fruit quality. The goal of biological pest control is to limit the use of insecticides while also maintaining a clean environment, food safety, and ultimately human health.

REFERENCES

Abdul S, Muhammad AA, Shabir H, Hesham A El E, Sajjad H, Niaz A, Abdul G, RZ Sayyed, Fahad S, Subhan D, Rahul D (2021a) Zinc nutrition and arbuscular mycorrhizal symbiosis effects on maize (Zea mays L.) growth and productivity. *J Saudi Soc Agri Sci* 28(11), 6339–6351. https://doi.org/10.1016/j.sjbs.2021.06.096

Abdul S, Muhammad AA, Subhan D, Niaz A, Fahad S, Rahul D, Mohammad JA, Omaima N, Muhammad Habib ur R, Bernard RG (2021b) Effect of arbuscular mycorrhizal fungi on the physiological functioning of maize under zinc-deficient soils. *Sci Rep* 11, 18468.

Abid M, Khalid N, Qasim A, Saud A, Manzer HS, Chao W, Depeng W, Shah S, Jan B, Subhan D, Rahul D, Hafiz MH, Wajid N, Muhammad M, Farooq S, Fahad S (2021) Exploring the potential of moringa leaf extract as bio stimulant for improving yield and quality of black cumin oil. *Sci Rep* 11, 24217. https://doi.org/10.1038/s41598-021-03617-w

Adnan M, Fahad S, Khan IA, Saeed M, Ihsan MZ, Saud S, Riaz M, Wang D, Wu C (2019) Integration of poultry manure and phosphate solubilizing bacteria improved availability of Ca bound P in calcareous soils. *3 Biotech* 9(10), 368.

Adnan M, Fahad S, Muhammad Z, Shahen S, Ishaq AM, Subhan D, Zafar-ul-Hye M, Martin LB, Raja MMN, Beena S, Saud S, Imran A, Zhen Y, Martin B, Jiri H, Rahul D (2020) Coupling phosphate-solubilizing bacteria with phosphorus supplements improve maize phosphorus acquisition and growth under lime induced salinity stress. *Plants* 9(7), 900. doi: 10.3390/plants9070900

Adnan M, Shah Z, Sharif M, Rahman H (2018b) Liming induces carbon dioxide (CO_2) emission in PSB inoculated alkaline soil supplemented with different phosphorus sources. *Environ Sci Poll Res* 25(10), 9501–9509.

Adnan M, Zahir S, Fahad S, Arif M, Mukhtar A, Imtiaz AK, Ishaq AM, Abdul B, Hidayat U, Muhammad A, Inayat-Ur R, Saud S, Muhammad ZI, Yousaf J, Amanullah, Hafiz MH, Wajid N (2018a) Phosphate-solubilizing bacteria nullify the antagonistic effect of soil calcification on bioavailability of phosphorus in alkaline soils. *Sci Rep* 8, 4339. https://doi.org/10.1038/s41598-018-22653-7

Afzal F, Adnan M, Rehman IU, Noor M, Khan A, Shah JA, Khan MA, Roman M, Wahid F, Nawaz S, Perveez R (2017) Growth response of olive cultivars to air layering. *Pure Appl Biol* 6(4), 1403–1409.

Ahmad N, Hussain S, Ali MA, Minhas A, Waheed W, Danish S, Fahad S, Ghafoor U, Baig, KS, Sultan H, Muhammad IH, Mohammad JA, Theodore DM (2022) Correlation of soil characteristics and citrus leaf nutrients contents in current scenario of Layyah District. *Hortic* 8, 61. https://doi.org/10.3390/horticulturae8010061

Ahmad S, Kamran M, Ding R, Meng X, Wang H, Ahmad I, Fahad S, Han Q (2019) Exogenous melatonin confers drought stress by promoting plant growth, photosynthetic capacity and antioxidant defense system of maize seedlings. *PeerJ* 7, e7793. http://doi.org/10.7717/peerj.7793

Akbar H, Timothy JK, Jagadish T, Golam M, Apurbo KC, Muhammad F, Rajan B, Fahad S, Hasanuzzaman M (2020) Agricultural land degradation: Processes and problems undermining future food security. In: Fahad S, Hasanuzzaman M, Alam M, Ullah H, Saeed M, Khan AK, Adnan M (Eds.), *Environment, Climate, Plant and Vegetation Growth.* Springer Publ Ltd, Springer Nature Switzerland AG. Part of Springer Nature. pp. 17–62. https://doi.org/10.1007/978-3-030-49732-3

Akram R, Turan V, Hammad HM, Ahmad S, Hussain S, Hasnain A, Maqbool MM, Rehmani MIA, Rasool A, Masood N, Mahmood F, Mubeen M, Sultana SR, Fahad S, Amanet K, Saleem M, Abbas Y, Akhtar HM, Waseem F, Murtaza R, Amin A, Zahoor SA, ul Din MS, Nasim W (2018a) Fate of organic and inorganic pollutants in paddy soils. In: Hashmi, MZ and Varma, A (Eds.), *Environmental Pollution of Paddy Soils, Soil Biology.* Springer International Publishing, pp. 197–214.

Akram R, Turan V, Wahid A, Ijaz M, Shahid MA, Kaleem S, Hafeez A, Maqbool MM, Chaudhary HJ, Munis, MFH, Mubeen M, Sadiq N, Murtaza R, Kazmi DH, Ali S, Khan N, Sultana SR, Fahad S, Amin A, Nasim W (2018b) Paddy land pollutants and their role in climate change. In: Hashmi, MZ and Varma, A (Eds.), *Environmental Pollution of Paddy Soils, Soil Biology.* Springer International Publishing, pp. 113–124.

Ali S, Hameed G, Muhammad A, Depeng W, Fahad S (2022) Comparative genetic evaluation of maize inbred lines at seedling and maturity stages under drought stress. *J Plant Growth Regul* 42, 989–1005. https://doi.org/10.1007/s00344-022-10608-2

Ali S, Xu Y, Jia Q, Ma X, Ahmad I, Adnan M, Gerard R, Ren X, Zhang P, Cai T, Zhang J (2018) Interactive effects of plasticfilm mulching with supplemental irrigation on winter wheat photosynthesis, chlorophyllfluorescence and yield under simulated precipitation conditions. *Agric Water Manag* 207, 1–14.

Al-Zahrani HS, Alharby HF and Fahad S (2022) Antioxidative defense system, hormones, and metabolite accumulation in different plant parts of two contrasting rice cultivars as influenced by plant growth regulators under heat stress. *Front Plant Sci* 13, 911846. doi: 10.3389/fpls.2022.911846

Amanullah, Fahad S (Eds.) (2017) Rice – technology and production. *IntechOpen Croatia* 2017 http://dx.doi.org/10.5772/64480

Amanullah, Fahad S (Eds.) (2018a) Corn – Production and human health in changing climate. *IntechOpen United Kingdom* 2018. http://dx.doi.org/10.5772/intechopen.74074

Amanullah, Fahad S (Eds.) (2018b) Nitrogen in agriculture – updates. *IntechOpen Croatia* 2018. http://dx.doi.org/10.5772/65846

Amanullah, Muhammad I, Haider N, Shah K, Manzoor A, Asim M, Saif U, Izhar A, Fahad S, Adnan M et al. (2021) Integrated foliar nutrients application improve wheat (Triticum aestivum L.) productivity under calcareous soils in drylands. *Commun Soil Sci Plant Anal* 52(21), 2748–2766. https://Doi.Org/10.1080/00103624.2021.1956521

Amanullah, Shah K, Imran, Hamdan AK, Muhammad A, Abdel RA, Muhammad A, Fahad S, Azizullah S, Brajendra P (2020) Effects of climate change on irrigation water quality. In: Fahad S, Hasanuzzaman M, Alam M, Ullah H,Saeed M, Khan AK, Adnan M (Eds.), *Environment, Climate, Plant and Vegetation Growth.* Springer Publ Ltd, Springer Nature Switzerland AG, pp. 123–132. https://doi.org/10.1007/978-3-030-49732-3

Amir M, Muhammad A, Allah B, Sevgi Ç, Haroon ZK, Muhammad A, Emre A (2020) Bio fortification under climate change: The fight between quality and quantity. In: Fahad S, Hasanuzzaman M, Alam M, Ullah H, Saeed M, Khan AK, Adnan M (Eds.), *Environment, Climate, Plant and Vegetation Growth.* Springer Publ Ltd, Springer Nature Switzerland AG, pp. 173–228. https://doi.org/10.1007/978-3-030-49732-3

Amjad I, Muhammad H, Farooq S, Anwar H (2020) Role of plant bioactives in sustainable agriculture. In: Fahad S, Hasanuzzaman M, Alam M, Ullah H, Saeed M, Khan AK, Adnan M (Eds.), *Environment, Climate, Plant and Vegetation Growth.* Springer Publ Ltd, Springer Nature Switzerland AG, pp. 591–606. https://doi.org/10.1007/978-3-030-49732-3

Amjad SF, Mansoora N, Din IU, Khalid IR, Jatoi GH, Murtaza G, Yaseen S, Naz M, Danish S, Fahad S et al. (2021) Application of zinc fertilizer and mycorrhizal inoculation on physio-biochemical parameters of wheat grown under water-stressed wnvironment. *Sustainability* 13, 11007. https://doi.org/10.3390/su131911007

Anam I, Huma G, Ali H, Muhammad K, Muhammad R, Aasma P, Muhammad SC, Noman W, Sana F, Sobia A, Fahad S (2021) Ameliorative mechanisms of turmeric-extracted curcumin on arsenic (As)-induced biochemical alterations, oxidative damage, and impaired organ functions in rats. *Environ Sci Pollut Res* 28(46), 66313–66326 https://doi.org/10.1007/s11356-021-15695-4

Arif M, Talha J, Muhammad R, Fahad S, Muhammad A, Amanullah, Kawsar A, Ishaq AM, Bushra K, Fahd R (2020) Biochar: a remedy for climate change. In: Fahad S, Hasanuzzaman M, Alam M, Ullah H, Saeed M, Khan AK, Adnan M (Eds.), *Environment, Climate, Plant and Vegetation Growth.* Springer Publ Ltd, Springer Nature Switzerland AG, pp. 151–172. https://doi.org/10.1007/978-3-030-49732-3

Arun KS (2007) *Bio-Fertilizers for Sustainable Agriculture.* Mechanism of P-solubilization. Agribios Publishers, pp. 196, 197.

Ashfaq AR, Uzma Y, Niaz A, Muhammad AA, Fahad S, Haider S, Tayebeh Z, Subhan D, Süleyman T, Hesham AElE, Pramila T, Jamal MA, Sulaiman AA, Rahul D (2021) Toxicity of cadmium and nickel in the context of applied activated carbon biochar for improvement in soil fertility. *Saudi Society Agric Sci* 29(2), 743–750. https://doi.org/10.1016/j.sjbs.2021.09.035

Atafar Z, Mesdaghinia A, Nouri J, Homaee M, Yunesian M, Ahmadimoghaddam M, Mahvi AH (2010). Effect of fertilizer application on soil heavy metal concentration. *Environ Monit Assess* 160(1), 83–89.

Athar M, Masood IA, Sana S, Ahmed M, Xiukang W, Sajid F, Sher AK, Habib A, Faran M, Zafar H, Farhana G, Fahad S (2021) Bio-diesel production of sunflower through sulphur management in a semi-arid subtropical environment. *Environ Sci Pollution Res* 29(9), 13268–13278. https://doi.org/10.1007/s11356-021-16688-z

Atif B, Hesham A, Fahad S (2021) Biochar coupling with phosphorus fertilization modifies antioxidant activity, osmolyte accumulation and reactive oxygen species synthesis in the leaves and xylem sap of rice cultivars under high-temperature stress. *Physiol Mol Biol Plants* 27(9), 2083–2100. https://doi.org/10.1007/s12298-021-01062-7

Ayman EL Sabagh, Akbar Hossain, Celaleddin Barutçular, Muhammad Aamir Iqbal, Sohidul Islam M, Shah Fahad, Oksana Sytar, Fatih Çig, Ram Swaroop Meena, Murat Erman (2020) Consequences of salinity stress on the quality of crops and its mitigation strategies for sustainable crop production: An outlook of arid and semi-arid regions. In: Fahad S, Hasanuzzaman M, Alam M, Ullah H, Saeed M, Khan AK, Adnan M (Eds.), *Environment, Climate, Plant and Vegetation Growth.* Springer Publ Ltd, Springer Nature Switzerland AG, pp. 503–534. https://doi.org/10.1007/978-3-030-49732-3

Aziz K, Daniel KYT, Fazal M, Muhammad ZA, Farooq S, FanW, Fahad S, Ruiyang Z (2017a) Nitrogen nutrition in cotton and control strategies for greenhouse gas emissions: A review. *Environ Sci Pollut Res* 24, 23471–23487. https://doi.org/10.1007/s11356-017-0131-y

Aziz K, Daniel KYT, Muhammad ZA, Honghai L, Shahbaz AT, Mir A, Fahad S (2017b) Nitrogen fertility and abiotic stresses management in cotton crop: A review. *Environ Sci Pollut Res* 24, 14551–14566. https://doi.org/10.1007/s11356-017-8920-x

Bagyaraj DJ (1992) 19 Vesicular-arbuscular mycorrhiza: Application in agriculture. *Methods in Microbiology* 24, 359–373.

Barea JM, Brown ME (1974) Effects on plant growth produced by Azotobacterpaspali related to synthesis of plant growth regulating substances. *J Appl Bacteriol* 37(4), 583–593.

Baseer M, Adnan M, Fazal M, Fahad S, Muhammad S, Fazli W, Muhammad A, Jr Amanullah, Depeng W, Saud S, Muhammad N, Muhammad Z, Fazli S, Beena S, Mian AR, Ishaq AM (2019) Substituting urea by organic wastes for improving maize yield in alkaline soil. *J Plant Nutrition* 42(19), 2423–2434. doi.org/10.1080/01904167.2019.1659344

Bayram AY, Seher Ö, Nazlican A (2020) Climate change forecasting and modeling for the year of 2050. In: Fahad S, Hasanuzzaman M, Alam M, Ullah H,Saeed M, Khan AK, Adnan M (Eds.), *Environment, Climate, Plant and Vegetation Growth.* Springer Publ Ltd, Springer Nature Switzerland AG, pp. 109–122. https://doi.org/10.1007/978-3-030-49732-3

Bukhari MA, Adnan NS, Fahad S, Javaid I, Fahim N, Abdul M, Mohammad SB (2021) Screening of wheat (*Triticum aestivum* L.) genotypes for drought tolerance using polyethylene glycol. *Arabian J Geosci* 14, 2808.

Carter OG (1967). The effect of chemical fertilizers on seedling establishment. *Aus J Exp Agric* 7(25), 174–180.

Chang W, Qiujuan J, Evgenios A, Haitao L, Gezi L, Jingjing Z, Fahad S, Ying J (2021) Hormetic effects of zinc on growth and antioxidant defense system of wheat plants. *Sci Total Environ* 807, 150992. https://doi.org/10.1016/j.scitotenv.2021.150992

Chao W, Youjin S, Beibei Q, Fahad S (2022) Effects of asymmetric heat on grain quality during the panicle initiation stage in contrasting rice genotypes. *J Plant Growth Regul* 42, 630–636 https://doi.org/10.1007/s00344-022-10598-1

Chen JH (2006, October). The combined use of chemical and organic fertilizers and/or bio-fertilizer for crop growth and soil fertility. In: *International Workshop on Sustained Management of the Soil–Rhizosphere System for Efficient Crop Production and Fertilizer Use*. Land Development Department Bangkok Thailand, Vol. 16, No. 20, pp. 1–11.

Chen LZ, Wang GH, Hong S, Liu A, Li C, Liu YD (2009) UV-B-induced oxidative damage and protective role of exopolysaccharides in desert cyanobacterium Microcoleusvaginatus. *J Integr Plant Biol* 51(2), 194–200.

Chen Y, Guo Z, Dong L, Fu Z, Zheng Q, Zhang G, Qin L, Sun X, Shi Z, Fahad S, Xie F, Saud S (2021) Turf performance and physiological responses of native Poa species to summer stress in Northeast China. *PeerJ* 9, e12252. http://doi.org/10.7717/peerj.12252

Deepranjan S, Ardith SO, Siva D, Sonam S, Shikha, Manoj P, Amitava R, Sayyed RZ, Abdul G, Mohammad JA, Subhan D, Fahad S, Rahul D (2021) Optimizing nutrient use efficiency, productivity, energetics, and economics of red cabbage following mineral fertilization and biopriming with compatible rhizosphere microbes. *Sci Rep* 11, 15680. https://doi.org/10.1038/s41598-021-95092-6

Depeng W, Fahad S, Saud S, Muhammad K, Aziz K, Mohammad NK, Hafiz MH, Wajid N (2018) Morphological acclimation to agronomic manipulation in leaf dispersion and orientation to promote "Ideotype" breeding: Evidence from 3D visual modeling of "super" rice (*Oryza sativa* L.). *Plant Physiol Biochem* 135, 499–510. https://doi.org/10.1016/j.plaphy.2018.11.010

Dey R, Pal KK, Tilak KVBR (2012) Influence of soil and plant types on diversity of rhizobacteria. *Proc Natl Acad Sci India Sect B Biol Sci* 82(3), 341–352.

Dey R, Pal KK, Tilak KVBR (2014) Plant growth promoting rhizobacteria in crop protection and challenges. In *Future Challenges in Crop Protection Against Fungal Pathogens*. Springer, New York, pp. 31–58.

Ehrlich PR, Harte J (2015) Opinion: To feed the world in 2050 will require a global revolution. *Proc Natl Acad Sci*, 112(48), 14743–14744.

EL Sabagh A, Islam MS, Hossain A, Iqbal MA, Mubeen M, Waleed M, Reginato M, Battaglia M, Ahmed S, Rehman A, Arif M, Athar H-U-R, Ratnasekera D, Danish S, Raza MA, Rajendran K, Mushtaq M, Skalicky M, Brestic M, Soufan W, Fahad S, Pandey S, Kamran M, Datta R, Abdelhamid MT (2022) Phytohormones as growth regulators during abiotic stress tolerance in plants. *Front Agron* 4, 765068. doi: 10.3389/fagro.2022.765068

Emre B, Ömer SU, Martín LB, Andre D, Fahad S, Rahul D, Muhammad Z-ul-H, Ghulam SH, Subhan D (2021) Studying soil erosion by evaluating changes in physico-chemical properties of soils under different land-use types. *J Saudi Society Agric Sci* 20, 190–197.

Fahad S, Bano A (2012) Effect of salicylic acid on physiological and biochemical characterization of maize grown in saline area. *Pak J Bot* 44, 1433–1438.

Fahad S, Chen Y, Saud S, Wang K, Xiong D, Chen C, Wu C, Shah F, Nie L, Huang J (2013) Ultraviolet radiation effect on photosynthetic pigments, biochemical attributes, antioxidant enzyme activity and hormonal contents of wheat. *J Food, Agri Environ* 11(3&4), 1635–1641.

Fahad S, Hussain S, Bano A, Saud S, Hassan S, Shan D, Khan FA, Khan F, Chen Y, Wu C, Tabassum MA, Chun MX, Afzal M, Jan A, Jan MT, Huang J (2014a) Potential role of phytohormones and plant growth-promoting rhizobacteria in abiotic stresses: Consequences for changing environment. *Environ Sci Pollut Res* 22(7), 4907–4921. https://doi.org/10.1007/s11356-014-3754-2

Fahad S, Hussain S, Matloob A, Khan FA, Khaliq A, Saud S, Hassan S, Shan D, Khan F, Ullah N, Faiq M, Khan MR, Tareen AK, Khan A, Ullah A, Ullah N, Huang J (2014b) Phytohormones and plant responses to salinity stress: A review. *Plant Growth Regul* 75(2), 391–404. https://doi.org/10.1007/s10725-014-0013-y

Fahad S, Hussain S, Saud S, Tanveer M, Bajwa AA, Hassan S, Shah AN, Ullah A, Wu C, Khan FA, Shah F, Ullah S, Chen Y, Huang J (2015a) A biochar application protects rice pollen from high-temperature stress. *Plant Physiol Biochem* 96, 281–287.

Fahad S, Nie L, Chen Y, Wu C, Xiong D, Saud S, Hongyan L, Cui K, Huang J (2015b) Crop plant hormones and environmental stress. *Sustain Agric Rev* 15, 371–400.

Fahad S, Hussain S, Saud S, Hassan S, Chauhan BS, Khan F et al. (2016a) Responses of rapid viscoanalyzer profile and other rice grain qualities to exogenously applied plant growth regulators under high day and high night temperatures. *PLoS One* 11(7), e0159590. https://doi.org/10.1371/journal.pone.0159590

Fahad S, Hussain S, Saud S, Khan F, Hassan S, Jr A, Nasim W, Arif M, Wang F, Huang J (2016b) Exogenously applied plant growth regulators affect heat-stressed rice pollens. *J Agron Crop Sci* 202, 139–150.

Fahad S, Hussain S, Saud S, Hassan S, Ihsan Z, Shah AN, Wu C, Yousaf M, Nasim W, Alharby H, Alghabari F, Huang J (2016c) Exogenously applied plant growth regulators enhance the morphophysiological growth and yield of rice under high temperature. *Front Plant Sci* 7, 1250. https://doi.org/10.3389/fpls.2016. 01250

Fahad S, Hussain S, Saud S, Hassan S, Tanveer M, Ihsan MZ, Shah AN, Ullah A, Nasrullah KF, Ullah S, AlharbyH NW, Wu C, Huang J (2016d) A combined application of biochar and phosphorus alleviates heat-induced adversities on physiological, agronomical and quality attributes of rice. *Plant Physiol Biochem* 103, 191–198.

Fahad S, Bajwa AA, Nazir U, Anjum SA, Farooq A, Zohaib A, Sadia S, NasimW, Adkins S, Saud S, Ihsan MZ, Alharby H, Wu C, Wang D, Huang J (2017) Crop production under drought and heat stress: Plant responses and management options. *Front Plant Sci* 8, 1147. https://doi.org/10.3389/fpls.2017.01147

Fahad S, Muhammad ZI, Abdul K, Ihsanullah D, Saud S, Saleh A, Wajid N, Muhammad A, Imtiaz AK, Chao W, Depeng W, Jianliang H (2018a) Consequences of high temperature under changing climate optima for rice pollen characteristics-concepts and perspectives. *Archives Agron Soil Sci* 64(11), 1473–1488. doi: 10.1080/03650340.2018.1443213

Fahad S, Abdul B, Adnan M (Eds.) (2018b) Global wheat production. *IntechOpen United Kingdom* 2018. http://dx.doi.org/10.5772/intechopen.72559

Fahad S, Rehman A, Shahzad B, Tanveer M, Saud S, Kamran M, Ihtisham M, Khan SU, Turan V, Rahman MHU (2019a) Rice responses and tolerance to metal/metalloid toxicity. In: Hasanuzzaman, M, Fujita, M, Nahar, K, and Biswas, JK (Eds.), *Advances in Rice Research for Abiotic Stress Tolerance*. Woodhead Publishing Ltd, UK, pp. 299–312.

Fahad S, Adnan M, Hassan S, Saud S, Hussain S, Wu C, Wang D, Hakeem KR, Alharby HF, Turan V, Khan MA, Huang J (2019b) Rice responses and tolerance to high temperature. In: Hasanuzzaman, M, Fujita, M, Nahar, K and Biswas, JK (Eds.), *Advances in Rice Research for Abiotic Stress Tolerance*. Woodhead Publishing Ltd, UK, pp. 201–224.

Fahad S, Hasanuzzaman M, Alam M, Ullah H, Saeed M, Ali Khan I, Adnan M (Eds.) (2020) *Environment, Climate, Plant and Vegetation Growth*. Springer Nature Switzerland AG 2020. doi: https://doi.org/ 10.1007/978-3-030-49732-3

Fahad S, Sönmez O, Saud S, Wang D, Wu C, Adnan M, Turan V (Eds.) (2021a) Plant growth regulators for climate-smart agriculture, First edition, *Footprints of Climate Variability on Plant Diversity*. CRC Press.

Fahad S, Sonmez O, Saud S, Wang D, Wu C, Adnan M, Turan V (Eds.) (2021b) Climate change and plants: biodiversity, growth and interactions, First edition, *Footprints of Climate Variability on Plant Diversity*. CRC Press.

Fahad S, Sonmez O, Saud S, Wang D, Wu C, Adnan M, Turan V (Eds.) (2021c) Developing climate resilient crops: improving global food security and safety, First edition, *Footprints of Climate Variability on Plant Diversity*. CRC Press.

Fahad S, Sönmez O, Turan V, Adnan M, Saud S, Wu C, Wang D (Eds.) (2021d) Sustainable soil and land management and climate change, First edition, *Footprints of Climate Variability on Plant Diversity*. CRC Press.

Fahad S, Sönmez O, Saud S, Wang D, Wu C, Adnan M, Arif M, Amanullah (Eds.) (2021e) Engineering tolerance in crop plants against abiotic stress, First edition, *Footprints of Climate Variability on Plant Diversity*. CRC Press.

Fahad S, Saud S, Yajun C, Chao W, Depeng W (Eds.) (2021f) Abiotic stress in plants. *IntechOpen United Kingdom* 2021. http://dx.doi.org/10.5772/intechopen.91549

Fahad S, Adnan M, Saud S (Eds.) (2022a) Improvement of plant production in the era of climate change. First edition, *Footprints of Climate Variability on Plant Diversity*. CRC Press.

Fahad S, Adnan M, Saud S, Nie L (Eds.) (2022b) Climate change and ecosystems: Challenges to sustainable development. First edition, *Footprints of Climate Variability on Plant Diversity*. CRC Press.

Fakhre A, Ayub K, Fahad S, Sarfraz N, Niaz A, Muhammad AA, Muhammad A, Khadim D, Saud S, Shah H, Muhammad ASR, Khalid N, Muhammad A, Rahul D, Subhan D (2021) Phosphate solubilizing bacteria optimize wheat yield in mineral phosphorus applied alkaline soil. *J Saudi Soc Agric Sci* 21(5), 339–348. https://doi.org/10.1016/j.jssas.2021.10.007

Farah R, Muhammad R, Muhammad SA, Tahira Y, Muhammad AA, Maryam A, Shafaqat A, Rashid M, Muhammad R, Qaiser H, Afia Z, Muhammad AA, Muhammad A, Fahad S (2020) Alternative and non-conventional soil and crop management strategies for increasing water use efficiency. In: Fahad S, Hasanuzzaman M, Alam M, Ullah H, Saeed M, Khan AK, Adnan M (Eds.), *Environment, Climate, Plant and Vegetation Growth*. Springer Publ Ltd, Springer Nature Switzerland AG, pp. 323–338. https://doi.org/10.1007/978-3-030-49732-3

Farhana G, Ishfaq A, Muhammad A, Dawood J, Fahad S, Xiuling L, Depeng W, Muhammad F, Muhammad F, Syed AS (2020) Use of crop growth model to simulate the impact of climate change on yield of various wheat cultivars under different agro-environmental conditions in Khyber Pakhtunkhwa, Pakistan. *Arabian J Geosci* 13, 112. https://doi.org/10.1007/s12517-020-5118-1

Farhat A, Hafiz MH, Wajid I, Aitazaz AF, Hafiz FB, Zahida Z, Fahad S, Wajid F, Artemi C (2020) A review of soil carbon dynamics resulting from agricultural practices. *J Environ Manage* 268(2020), 110319.

Farhat UK, Adnan AK, Kai L, Xuexuan X, Muhammad A, Fahad S, Rafiq A, Mushtaq AK, Taufiq N, Faisal Z (2022) Influences of long-term crop cultivation and fertilizer management on soil aggregates stability and fertility in the Loess Plateau, Northern China. *J Soil Sci Plant Nutri* 22, 1446–1457. https://doi.org/10.1007/s42729-021-00744-1

Fazli W, Muhmmad S, Amjad A, Fahad S, Muhammad A, Muhammad N, Ishaq AM, Imtiaz AK, Mukhtar A, Muhammad S, Muhammad I, Rafi U, Haroon I, Muhammad A (2020) Plant–microbes interactions and functions in changing climate. In: Fahad S, Hasanuzzaman M, Alam M, Ullah H, Saeed M, Khan AK, Adnan M (Eds.), *Environment, Climate, Plant and Vegetation Growth*. Springer Publ Ltd, Springer Nature Switzerland AG, pp. 397–420. https://doi.org/10.1007/978-3-030-49732-3

Fuentes-Ramirez LE, Caballero-Mellado J (2005) Bacterial bio-fertilizers. In: Siddiqui ZA (Ed.), *PGPR: Biocontrol and Biofertilization*. Springer-Verlag, Heidelberg, Berlin. pp. 143–172. doi:10.1007/1-4020-4152-7_5

Geiger F, Bengtsson, J, Berendse F, Weisser WW, Emmerson M, Morales MB ... Inchausti P (2010) Persistent negative effects of pesticides on biodiversity and biological control potential on European farmland. *Basic Appl Ecol*, 11(2), 97–105.

Ghosh, N (2004) Promoting biofertilisers in Indian agriculture. *Economic and Political Weekly*, 5617–5625.

Ghulam M, Muhammad AA, Donald LS, Sajid M, Muhammad FQ, Niaz A, Ateeq ur R, Shakeel A, Sajjad H, Muhammad A, Summia M, Aqib HAK, Fahad S, Rahul D, Mazhar I, Timothy DS (2021) Formalin fumigation and steaming of various composts differentially influence the nutrient release, growth and yield of muskmelon (*Cucumis melo* L.). *Sci Rep* 11, 21057.

Gopakumar L, Bernard NO, Donato V (2020) Soil microarthropods and nutrient cycling. In: Fahad S, Hasanuzzaman M, Alam M, Ullah H, Saeed M, Khan AK, Adnan M (Eds.), *Environment, Climate, Plant and Vegetation Growth*. Springer Publ Ltd, Springer Nature Switzerland AG, pp. 453–472. https://doi.org/10.1007/978-3-030-49732-3

Guofu L, Zhenjian B, Fahad S, Guowen C, Zhixin X, Hao G, Dandan L, Yulong L, Bing L, Guoxu J, Saud S (2021) Compositional and structural changes in soil microbial communities in response to straw mulching and plant revegetation in an abandoned artificial pasture in Northeast China. *Glob Ecol Conserv* 31(2021), e01871.

Habib ur R, Ashfaq A, Aftab W, Manzoor H, Fahd R, Wajid I, Md Aminul I, Vakhtang S, Muhammad A, Asmat U, Abdul W, Syeda RS, Shah S, Shahbaz K, Fahad S, Manzoor H, Saddam H, Wajid N (2017) Application of CSM-CROPGRO-Cotton model for cultivars and optimum planting dates: Evaluation in changing semi-arid climate. *Field Crops Res* 238, 139–152. http://dx.doi.org/10.1016/j.fcr.2017.07.007

Hafeez M, Farman U, Muhammad MK, Xiaowei L, Zhijun Z, Sakhawat S, Muhammad I, Mohammed AA, Mandela F-G, Nicolas D, Muzammal R, Fahad S, Yaobin L (2021) Metabolic-based insecticide resistance mechanism and ecofriendly approaches for controlling of beet armyworm Spodoptera exigua: a review. *Environ Sci Pollution Res* 29, 1746–1762. https://doi.org/10.1007/s11356-021-16974-w

Hafiz MH, Wajid F, Farhat A, Fahad S, Shafqat S, Wajid N, Hafiz FB (2016) Maize plant nitrogen uptake dynamics at limited irrigation water and nitrogen. *Environ Sci Pollut Res* 24(3), 2549–2557. https://doi.org/10.1007/s11356-016-8031-0

Hafiz MH, Farhat A, Shafqat S, Fahad S, Artemi C, Wajid F, Chaves CB, Wajid N, Muhammad M, Hafiz FB (2018) Offsetting land degradation through nitrogen and water management during maize cultivation under arid conditions. *Land Degrad Dev* 29(5), 1–10. doi: 10.1002/ldr.2933

Hafiz MH, Muhammad A, Farhat A, Hafiz FB, Saeed AQ, Muhammad M, Fahad S, Muhammad A (2019) Environmental factors affecting the frequency of road traffic accidents: A case study of sub-urban area of Pakistan. *Environ Sci Pollut Res* 26, 11674–11685. https://doi.org/10.1007/s11356-019-04752-8

Hafiz MH, Farhat A, Ashfaq A, Hafiz FB, Wajid F, Carol Jo W, Fahad S, Gerrit H (2020a) Predicting kernel growth of maize under controlled water and nitrogen applications. *Int J Plant Prod* 14, 609–620. https://doi.org/10.1007/s42106-020-00110-8

Hafiz MH, Abdul K, Farhat A, Wajid F, Fahad S, Muhammad A, Ghulam MS, Wajid N, Muhammad M, Hafiz FB (2020b) Comparative effects of organic and inorganic fertilizers on soil organic carbon and wheat productivity under arid region. *Commun Soil Sci Plant Anal* 29(2), 1746–1762. doi: 10.1080/00103624.2020.1763385

Haider SA, Lalarukh I, Amjad SF, Mansoora N, Naz M, Naeem M, Bukhari SA, Shahbaz M, Ali SA, Marfo TD, Subhan D, Rahul D, Fahad S (2021) Drought stress alleviation by potassium-nitrate-containing chitosan/montmorillonite microparticles confers changes in Spinacia oleracea L. *Sustain* 13, 9903. https://doi.org/10.3390/su13179903

Hamza SM, Xiukang W, Sajjad A, Sadia Z, Muhammad N, Adnan M, Fahad S et al. (2021) Interactive effects of gibberellic acid and NPK on morpho-physio-biochemical traits and organic acid exudation pattern in coriander (Coriandrum sativum L.) grown in soil artificially spiked with boron. *Plant Physiol Biochem* 167, 884–900.

Haoliang Y, Matthew TH, Ke L, Bin W, Puyu F, Fahad S, Holger M, Rui Y, De LL, Sotirios A, Isaiah H, Xiaohai T, Jianguo M, Yunbo Z, Meixue Z (2022) Crop traits enabling yield gains under more frequent extreme climatic events. *Sci Total Environ* 808, 152170.

Hayat R, Ali S, Amara U, Khalid R, Ahmed I (2010) Soil beneficial bacteria and their role in plant growth promotion: A review. *Annals Microbiol* 60(4), 579–598.

Hesham FA and Fahad S (2020) Melatonin application enhances biochar efficiency for drought tolerance in maize varieties: Modifications in physio-biochemical machinery. *Agron J* 112(4), 1–22.

Hibbing ME, Fuqua C, Parsek MR, Peterson SB (2010) Bacterial competition: Surviving and thriving in the microbial jungle. *Nature Rev Microbiol* 8(1), 15–25.

Huang Li-Y, Li Xiao-X, Zhang Yun-B, Fahad S, Wang F (2021) dep1 improves rice grain yield and nitrogen use efficiency simultaneously by enhancing nitrogen and dry matter translocation. *J Integrative Agri* 21(11), 3185–3198. doi: 10.1016/S2095-3119(21)63795-4

Hussain MA, Fahad S, Rahat S, Muhammad FJ, Muhammad M, Qasid A, Ali A, Husain A, Nooral A, Babatope SA, Changbao S, Liya G, Ibrar A, Zhanmei J, Juncai H (2020) Multifunctional role of brassinosteroid and its analogues in plants. *Plant Growth Regul* 92, 141–156. https://doi.org/10.1007/s10725-020-00647-8

Ibad U, Dost M, Maria M, Shadman K, Muhammad A, Fahad S, Muhammad I, Ishaq AM, Aizaz A, Muhammad HS, Muhammad S, Farhana G, Muhammad I, Muhammad ASR, Hafiz MH, Wajid N, Shah S, Jabar ZKK, Masood A, Naushad A, Rasheed Akbar M, Shah MK Jan B (2022) Comparative effects of biochar and NPK on wheat crops under different management systems. *Crop Pasture Sci* 74, 31–40. https://doi.org/10.1071/CP21146

Ibrar H, Muqarrab A, Adel MG, Khurram S, Omer F, Shahid I, Fahim N, Shakeel A, Viliam B, Marian B, Sami Al Obaid, Fahad S, Subhan D, Suleyman T, Hanife AKÇA, Rahul D (2021) Improvement in growth and yield attributes of cluster bean through optimization of sowing time and plant spacing under climate change Scenario. *Saudi J Bio Sci* 29(2), 781–792. https://doi.org/10.1016/j.sjbs.2021.11.018

Ibrar K, Aneela R, Khola Z, Urooba N, Sana B, Rabia S, Ishtiaq H, Mujaddad Ur Rehman, Salvatore M (2020) Microbes and environment: Global warming reverting the frozen zombies. In: Fahad S, Hasanuzzaman M, Alam M, Ullah H, Saeed M, Khan AK, Adnan M (Eds.), *Environment, Climate, Plant and Vegetation Growth*. Springer Publ Ltd, Springer Nature Switzerland, pp. 607–634. https://doi.org/10.1007/978-3-030-49732-3

Ihsan MZ, Abdul K, Manzer HS, Liaqat A, Ritesh K, Hayssam MA, Amar M, Fahad S (2022) The response of Triticum aestivum treated with plant growth regulators to acute day/night temperature rise. *J Plant Growth Regul* 41, 2020–2033. https://doi.org/10.1007/s00344-022-10574-9

Ikram U, Khadim D, Muhammad T, Muhammad S, Fahad S (2021) Gibberellic acid and urease inhibitor optimize nitrogen uptake and yield of maize at varying nitrogen levels under changing climate. *Environ Sci Pollution Res* 29(5), 6568–6577. https://doi.org/10.1007/s11356-021-16049-w

Ilyas M, Mohammad N, Nadeem K, Ali H, Aamir HK, Kashif H, Fahad S, Aziz K, Abid U (2020) Drought tolerance strategies in plants: A mechanistic approach. *J Plant Growth Regulation* 40, 926–944. https://doi.org/10.1007/s00344-020-10174-5

Iqra M, Amna B, Shakeel I, Fatima K,Sehrish L, Hamza A, Fahad S (2020) Carbon cycle in response to global warming. In: Fahad S, Hasanuzzaman M, Alam M, Ullah H,Saeed M, Khan AK, Adnan M (Eds.), *Environment, Climate, Plant and Vegetation Growth*. Springer Publ Ltd, Springer Nature Switzerland, pp. 1–16. https://doi.org/10.1007/978-3-030-49732-3

Irfan M, Muhammad M, Muhammad JK, Khadim MD, Dost M, Ishaq AM, Waqas A, Fahad S, Saud S et al. (2021) Heavy metals immobilization and improvement in maize (Zea mays L.) growth amended with biochar and compost. *Sci Rep* 11, 18416.

Jabborova D, Sulaymanov K, Sayyed RZ, Alotaibi SH, Enakiev Y, Azimov A, Jabbarov Z, Ansari MJ, Fahad S, Danish S et al. (2021) Mineral fertilizers improve the quality of turmeric and soil. *Sustain* 13, 9437. https://doi.org/10.3390/su13169437

Jan M, Muhammad Anwar-ul-Haq, Adnan NS, Muhammad Y, Javaid I, Xiuling L, Depeng W, Fahad S (2019) Modulation in growth, gas exchange, and antioxidant activities of salt-stressed rice (Oryza sativa L.) genotypes by zinc fertilization. *Arabian J Geosci* 12, 775. https://doi.org/10.1007/s12517-019-4939-2

Jans J, Schul W, Sert YG, Rijksen Y, Rebel H, Eker AP, ... Van Der Horst GT (2005) Powerful skin cancer protection by a CPD-photolyase transgene. *Curr Biol* 15(2), 105–115.

Kamaran M, Wenwen C, Irshad A, Xiangping M, Xudong Z, Wennan S, Junzhi C, Shakeel A, Fahad S, Qingfang H, Tiening L (2017) Effect of paclobutrazol, a potential growth regulator on stalk mechanical strength, lignin accumulation and its relation with lodging resistance of maize. *Plant Growth Regul* 84, 317–332. https://doi.org/10.1007/ s10725-017-0342-8

Khadim D, Fahad S, Jahangir MMR, Iqbal M, Syed SA, Shah AK, Ishaq AM, Rahul D et al. (2021a) Biochar and urease inhibitor mitigate NH_3 and N_2O emissions and improve wheat yield in a urea fertilized alkaline soil. *Sci Rep* 11, 17413.

Khadim D, Saif-ur-R, Fahad S, Syed SA, Shah AK et al. (2021b) Influence of variable biochar concentration on yield-scaled nitrous oxide emissions, Wheat yield and nitrogen use efficiency. *Sci Rep* 11, 16774.

Khan I, Hussain H, Shah B, Ullah W, Naeem A, Ali W, Khan N, Adnan M, Junaid K, Shah SRA, Ahmed N, Iqbal M (2016) Evaluation of phytobiocides and different culture media for growth, isolation and control of Rhizoctonia solani in vitro. *J Ento & Zool Studies* 4(2), 417–420.

Khan MMH, Niaz A, Umber G, Muqarrab A, Muhammad AA, Muhammad I, Shabir H, Shah F, Vibhor A, Shams HA-H, Reham A, Syed MBA, Nadiyah MA, Ali TKZ, Subhan D, Rahul D (2021) Synchronization of boron application methods and rates is environmentally friendly approach to improve quality attributes of *Mangifera indica* L. on sustainable basis. *Saudi J Bio Sci* 29(3), 1869–1880. https://doi.org/10.1016/j.sjbs.2021.10.036

Khatun M, Sarkar S, Era FM, Islam AKMM, Anwar MP, Fahad S, Datta R, Islam AKMA (2021) Drought stress in grain legumes: Effects, tolerance mechanisms and management. *Agron* 11, 2374. https://doi.org/10.3390/agronomy11122374

Kumar H, Dubey RC, Maheshwari DK (2011) Effect of plant growth promoting rhizobia on seed germination, growth promotion and suppression of Fusarium wilt of fenugreek (Trigonellafoenum-graecum L.). *Crop Prot* 30(11), 1396–1403.

Kumar R, Kumawat N, Sahu YK (2017) Role of bio-fertilizers in agriculture. *Pop Kheti* 5(4), 63–66.

Lovelock CE, Wright SF, Clark DA, Ruess RW (2004) Soil stocks of glomalin produced by arbuscularmycorrhizal fungi across a tropical rain forest landscape. *J Ecol* 92(2), 278–287.

Macik M, Gryta A, Frac M (2020) Bio-fertilizers in agriculture: An overview on concepts, strategies and effects on soil microorganisms. *Adv Agron* 162, 31–87.

Mahar A, Amjad A, Altaf HL, Fazli W, Ronghua L, Muhammad A, Fahad S, Muhammad A, Rafiullah, Imtiaz AK, Zengqiang Z (2020) Promising technologies for Cd-contaminated soils: Drawbacks and possibilities. In: Fahad S, Hasanuzzaman M, Alam M, Ullah H,Saeed M, Khan AK, Adnan M (Eds.), *Environment, Climate, Plant and Vegetation Growth*. Springer Publ Ltd, Springer Nature Switzerland, pp. 63–92. https://doi.org/10.1007/978-3-030-49732-3

Mahdi SS, Hassan GI, Samoon SA, Rather HA, Dar SA, Zehra B (2010) Bio-fertilizers in organic agriculture. *J Phytol* 2(10), 42–54.

Mahmood Ul H, Tassaduq R, Chandni I, Adnan A, Muhammad A, Muhammad MA, Muhammad H-ur-R, Mehmood AN, Alam S, Fahad S (2021) Linking plants functioning to adaptive responses under heat stress conditions: A mechanistic review. *J Plant Growth Regul* 41, 2596–2613. https://doi.org/10.1007/s00344-021-10493-1

Manzer HS, Saud A, Soumya M, Abdullah A. Al-A, Qasi DA, Bander M. Al, Hayssam MA, Hazem MK, Fahad S, Vishnu DR, Om PN (2021) Molybdenum and hydrogen sulfide synergistically mitigate arsenic toxicity by modulating defense system, nitrogen and cysteine assimilation in faba bean (*Vicia faba* L.) seedlings. *Environ Pollut* 290, 117953.

Martínez-Viveros O, Jorquera MA, Crowley DE, Gajardo GMLM, Mora ML (2010). Mechanisms and practical considerations involved in plant growth promotion by rhizobacteria. *J Soil Sci Plant Nutr* 10(3), 293–319.

Md Enamul H, Shoeb AZM, Mallik AH, Fahad S, Kamruzzaman MM, Akib J, Nayyer S, Mehedi KM, Swati AS, Md Yeamin A, Most SS (2020) Measuring vulnerability to environmental hazards: Qualitative to quantitative. In: Fahad S, Hasanuzzaman M, Alam M, Ullah H, Saeed M, Khan AK, Adnan M (Eds.), *Environment, Climate, Plant and Vegetation Growth.* Springer Publ Ltd, Springer Nature Switzerland, pp. 421–452. https://doi.org/10.1007/978-3-030-49732-3

Md Jakir H, Allah B (2020) Development and applications of transplastomic plants: A way towards eco-friendly agriculture. In: Fahad S, Hasanuzzaman M, Alam M, Ullah H, Saeed M, Khan AK, Adnan M (Eds.), *Environment, Climate, Plant and Vegetation Growth.* Springer Publ Ltd, Springer Nature Switzerland, pp. 285–322. https://doi.org/10.1007/978-3-030-49732-3

Mehmood K, Bao Y, Saifullah, Bibi S, Dahlawi S, Yaseen M, Abrar MM, Srivastava P, Fahad S, Faraj TK (2022) Contributions of open biomass burning and crop straw burning to air quality: Current research paradigm and future outlooks. *Front Environ Sci* 10, 852492. doi: 10.3389/fenvs.2022.852492

Mishra D, Rajvir S, Mishra U, Kumar SS (2013) Role of bio-fertilizer in organic agriculture: A review. *Res J Recent Sci* 2, 39–41. ISSN, 2277, 2502.

Mohammad I Al-Wabel, Ahmad M, Adel RA Usman, Mutair Akanji, and Muhammad Imran Rafique (2020a) Advances in pyrolytic technologies with improved carbon capture and storage to combat climate change. In: Fahad S, Hasanuzzaman M, Alam M, Ullah H, Saeed M, Khan AK, Adnan M (Eds.), *Environment, Climate, Plant and Vegetation Growth.* Springer Publ Ltd, Springer Nature Switzerland, pp. 535–576. https://doi.org/10.1007/978-3-030-49732-3

Mubeen M, Ashfaq A, Hafiz MH, Muhammad A, Hafiz UF, Mazhar S, Muhammad Sami ul Din, Asad A, Amjed A, Fahad S, Wajid N (2020) Evaluating the climate change impact on water use efficiency of cotton-wheat in semi-arid conditions using DSSAT model. *J Water Clim Change* 11(4), 1661–1675. doi/10.2166/wcc.2019.179/622035/jwc2019179.pdf

Muhammad I, Khadim D, Fahad S, Imran M, Saud A, Manzer HS, Shah S, Jabar ZKK, Shamsher A, Shah H, Taufiq N, Hafiz MH, Jan B, Wajid N (2022) Exploring the potential effect of *Achnatherum splendens* L.-derived biochar treated with phosphoric acid on bioavailability of cadmium and wheat growth in contaminated soil. *Environ Sci Pollut Res* 29(25), 37676–37684. https://doi.org/10.1007/s11356-021-17950-0.

Muhammad N, Muqarrab A, Khurram S, Fiaz A, Fahim N, Muhammad A, Shazia A, Omaima N, Sulaiman AA, Fahad S, Subhan D, Rahul D (2021) Kaolin and jasmonic acid improved cotton productivity under water stress conditions. *J Saudi Society Agricultural Sci* 28(2021) 6606–6614. https://doi.org/10.1016/j.sjbs.2021.07.043

Muhammad Tahir ul Qamar, Amna F, Amna B, Barira Z, Xitong Z, Ling-Ling C (2020) Effectiveness of conventional crop improvement strategies vs. omics. In: Fahad S, Hasanuzzaman M, Alam M, Ullah H, Saeed M, Khan AK, Adnan M (Eds.), *Environment, Climate, Plant and Vegetation Growth.* Springer Publ Ltd, Springer Nature Switzerland, pp. 253–284. https://doi.org/10.1007/978-3-030-49732-3

Muhammad Z, Abdul MK, Abdul MS, Kenneth BM, Muhammad S, Shahen S, Ibadullah J, Fahad S (2019) Performance of Aeluropus lagopoides (mangrove grass) ecotypes, a potential turfgrass, under high saline conditions. *Environ Sci Pollut Res* 26(13), 13410–13412. https://doi.org/10.1007/s11356-019-04838-3

Muzammal R, Fahad S, Guanghui D, Xia C, Yang Y, Kailei T, Lijun L, Fei-Hu L, Gang D (2021) Evaluation of hemp (Cannabis sativa L.) as an industrial crop: A review. *Environ Sci Pollut Res* 28(38), 52832–52843 https://doi.org/10.1007/s11356-021-16264-5

Ngampimol H, Kunathigan V (2008) The study of shelf life for liquid bio-fertilizer from vegetable waste. *AU JT* 11(4), 204–208.

Niaz A, Abdullah E, Subhan D, Muhammad A, Fahad S, Khadim D, Suleyman T, Hanife A, Anis AS, Mohammad JA, Emre B, Ömer SU, Rahul D, Bernard RG (2022) Mitigation of lead (Pb) toxicity in rice cultivated with either ground water or wastewater by application of acidified carbon. *J Environ Manage* 307, 114521.

Niggli U (2015) Sustainability of organic food production: challenges and innovations. *Proc Nutr Soc*, 74(1), 83–88.

Noor M, Naveed ur R, Ajmal J, Fahad S, Muhammad A, Fazli W, Saud S, Hassan S (2020) Climate change and coastal plant lives. In: Fahad S, Hasanuzzaman M, Alam M, Ullah H,Saeed M, Khan AK, Adnan M (Eds.), *Environment, Climate, Plant and Vegetation Growth*. Springer Publ Ltd, Springer Nature Switzerland, pp. 93–108. https://doi.org/10.1007/978-3-030-49732-3

Pereg L, McMillan M (2015) Scoping the potential uses of beneficial microorganisms for increasing productivity in cotton cropping systems. *Soil Biol Biochem* 80, 349–358.

Qamar-uz Z, Zubair A, Muhammad Y, Muhammad ZI, Abdul K, Fahad S, Safder B, Ramzani PMA, Muhammad N (2017) Zinc biofortification in rice: Leveraging agriculture to moderate hidden hunger in developing countries. *Arch Agron Soil Sci* 64, 147–161. https://doi.org/10.1080/03650340.2017.1338343

Qin ZH, Nasib ur Rahman, Ahmad A, Yu-pei Wang, Sakhawat S, Ehmet N, Wen-juan Shao, Muhammad I, Kun S, Rui L, Fazal S, Fahad S (2022) Range expansion decreases the reproductive fitness of *Gentiana officinalis* (*Gentianaceae*). *Sci Rep* 12, 2461. https://doi.org/10.1038/s41598-022-06406-1

Rahman KM, Zhang D (2018) Effects of fertilizer broadcasting on the excessive use of inorganic fertilizers and environmental sustainability. *Sustainability* 10(3), 759.

Rajesh KS, Fahad S, Pawan K, Prince C, Talha J, Dinesh J, Prabha S, Debanjana S, Prathibha MD, Bandana B, Akash H, Gupta NK, Rekha S, Devanshu D, Dalpat LS, Ke L, Matthew TH, Saud S, Adnan NS, Taufiq N (2022) Beneficial elements: New players in improving nutrient use efficiency and abiotic stress tolerance. *Plant Growth Regulation* 100, 237–265. https://doi.org/10.1007/s10725-022-00843-8

Rashid M, Qaiser H, Khalid SK, Mohammad I Al-Wabel, Zhang A, Muhammad A, Shahzada SI, Rukhsanda A, Ghulam AS, Shahzada MM, Sarosh A, Muhammad FQ (2020) Prospects of biochar in alkaline soils to mitigate climate change. In: Fahad S, Hasanuzzaman M, Alam M, Ullah H,Saeed M, Khan AK, Adnan M (Eds.), *Environment, Climate, Plant and Vegetation Growth*. Springer Publ Ltd, Springer Nature Switzerland AG, pp. 133–150. https://doi.org/10.1007/978-3-030-49732-3

Reddy PP (2014) *Plant Growth Promoting Rhizobacteria for Horticultural Crop Protection*. Springer India, Vol. 10, pp. 978–981.

Rehana S, Asma Z, Shakil A, Anis AS, Rana KI, Shabir H, Subhan D, Umber G, Fahad S, Jiri K, Sami Al Obaid, Mohammad JA, Rahul D (2021) Proteomic changes in various plant tissues associated with chromium stress in sunflower. *Saudi J Bio Sci* 29(4), 2604–2612. https://doi.org/10.1016/j.sjbs.2021.12.042

Rehman M, Fahad S, Saleem MH, Hafeez M, Muhammad Habib ur Rahman, Liu F, Deng G (2020) Red light optimized physiological traits and enhanced the growth of ramie (Boehmeria nivea L.). *Photosynthetica* 58(4), 922–931.

Roger PA, Ladha JK (1992) Biological N_2 fixation in wetland rice fields: Estimation and contribution to nitrogen balance. In: Ladha JK, George T, Bohlool BB (Eds.), *Biological Nitrogen Fixation for Sustainable Agriculture*. Springer, pp. 41–55.

Sadam M, Muhammad Tahir ul Qamar, Ghulam M, Muhammad SK, Faiz AJ (2020) Role of biotechnology in climate resilient agriculture. In: Fahad S, Hasanuzzaman M, Alam M, Ullah H, Saeed M, Khan AK, Adnan M (Eds.), *Environment, Climate, Plant and Vegetation Growth*. Springer Publ Ltd, Springer Nature Switzerland, pp. 339–366. https://doi.org/10.1007/978-3-030-49732-3

Safi UK, Ullah F, Mehmood S, Fahad S, Ahmad Rahi A, Althobaiti F et al. (2021) Antimicrobial, antioxidant and cytotoxic properties of Chenopodium glaucum L. *PLoS One* 16(10), e0255502. https://doi.org/10.1371/journal. pone.0255502

Sah JF, Krishna KB, Srivastava M, Mohanty P (1998) Effects of ultraviolet-B radiation on phycobilisomes of Synechococcus PCC 7942: Alterations in conformation and energy transfer characteristics. *IUBMB Life* 44(2), 245–257.

Sahrish N, Shakeel A, Ghulam A, Zartash F, Sajjad H, Mukhtar A, Muhammad AK, Ahmad K, Fahad S, Wajid N, Sezai E, Carol Jo W, Gerrit H (2022) Modeling the impact of climate warming on potato phenology. *European J Agro N* 132, 126404.

Sajid H, Jie H, Jing H, Shakeel A, Satyabrata N, Sumera A, Awais S, Chunquan Z, Lianfeng Z, Xiaochuang C, Qianyu J, Junhua Z (2020) Rice production under climate change: Adaptations and mitigating strategies. In: Fahad S, Hasanuzzaman M, Alam M, Ullah H, Saeed M, Khan AK, Adnan M (Eds.), *Environment,*

Climate, Plant and Vegetation Growth. Springer Publ Ltd, Springer Nature Switzerland, pp. 659–686. https://doi.org/10.1007/978-3-030-49732-3

Sajjad H, Muhammad M, Ashfaq A, Waseem A, Hafiz MH, Mazhar A, Nasir M, Asad A, Hafiz UF, Syeda RS, Fahad S, Depeng W, Wajid N (2019) Using GIS tools to detect the land use/land cover changes during forty years in Lodhran district of Pakistan. *Environ Sci Pollut Res* 27, 39676–39692. https://doi.org/10.1007/s11356-019-06072-3

Sajjad H, Muhammad M, Ashfaq A, Fahad S, Wajid N, Hafiz MH, Ghulam MS, Behzad M, Muhammad T, Saima P (2021a) Using space–time scan statistic for studying the effects of COVID-19 in Punjab, Pakistan: A guideline for policy measures in regional Agriculture. *Environ Sci Pollut Res* 30, 42495–42508. https://doi.org/10.1007/s11356-021-17433-2

Sajjad H, Muhammad M, Ashfaq A, Nasir M, Hafiz MH, Muhammad A, Muhammad I, Muhammad U, Hafiz UF, Fahad S, Wajid N, Hafiz MRJ, Mazhar A, Saeed AQ, Amjad F, Muhammad SK, Mirza W (2021b) Satellite-based evaluation of temporal change in cultivated land in Southern Punjab (Multan region) through dynamics of vegetation and land surface temperature. *Open Geo Sci* 13, 1561–1577.

Saleem MH, Fahad S, Adnan M, Mohsin A, Muhammad SR, Muhammad K, Qurban A, Inas AH, Parashuram B, Mubassir A, Reem MH (2020a) Foliar application of gibberellic acid endorsed phytoextraction of copper and alleviates oxidative stress in jute (Corchorus capsularis L.) plant grown in highly copper-contaminated soil of China. *Environ Sci Pollution Res* 27, 37121–37133. https://doi.org/10.1007/s11356-020-09764-3

Saleem MH, Rehman M, Fahad S, Tung SA, Iqbal N, Hassan A, Ayub A, Wahid MA, Shaukat S, Liu L, Deng G (2020b) Leaf gas exchange, oxidative stress, and physiological attributes of rapeseed (Brassica napus L.) grown under different light-emitting diodes. *Photosynthetica* 58 (3), 836–845.

Saleem MH, Fahad S, Shahid UK, Mairaj D, Abid U, Ayman ELS, Akbar H, Analía L, Lijun L (2020c) Copper-induced oxidative stress, initiation of antioxidants and phytoremediation potential of flax (Linum usitatissimum L.) seedlings grown under the mixing of two different soils of China. *Environ Sci Poll Res* 27, 5211–5221. https://doi.org/10.1007/s11356-019-07264-7

Saman S, Amna B, Bani A, Muhammad Tahir ul Qamar, Rana MA, Muhammad SK (2020) QTL mapping for abiotic stresses in cereals. In: Fahad S, Hasanuzzaman M, Alam M, Ullah H, Saeed M, Khan AK, Adnan M (Eds.), *Environment, Climate, Plant and Vegetation Growth*. Springer Publ Ltd, Springer Nature Switzerland, pp. 229–252. https://doi.org/10.1007/978-3-030-49732-3

Sana U, Shahid A, Yasir A, Farman UD, Syed IA, Mirza MFAB, Fahad S, Al-Misned F, Usman A, Xinle G, Ghulam N, Kunyuan W (2022) Bifenthrin induced toxicity in Ctenopharyngodon della at an acute concentration: A multi-biomarkers based study. *J King Saud Uni – Sci* 34 (2022) 101752.

Saud S, Chen Y, Long B, Fahad S, Sadiq A (2013) The different impact on the growth of cool season turf grass under the various conditions on salinity and drought stress. *Int J Agric Sci Res* 3,77–84.

Saud S, Li X, Chen Y, Zhang L, Fahad S, Hussain S, Sadiq A, Chen Y (2014) Silicon application increases drought tolerance of Kentucky bluegrass by improving plant water relations and morph physiological functions. *Sci World J* 2014, 1–10. https://doi.org/10.1155/2014/ 368694

Saud S, Chen Y, Fahad S, Hussain S, Na L, Xin L, Alhussien SA (2016) Silicate application increases the photosynthesis and its associated metabolic activities in Kentucky bluegrass under drought stress and post-drought recovery. *Environ Sci Pollut Res* 23(17), 17647–17655. https://doi.org/10.1007/s11356-016-6957-x

Saud S, Fahad S, Yajun C, Ihsan MZ, Hammad HM, Nasim W, Amanullah Jr, Arif M and Alharby H (2017) Effects of nitrogen supply on water stress and recovery mechanisms in Kentucky bluegrass plants. *Front Plant Sci* 8, 983. DOI: 10.3389/fpls.2017.00983

Saud S, Fahad S, Cui G, Chen Y, Anwar S (2020) Determining nitrogen isotopes discrimination under drought stress on enzymatic activities, nitrogen isotope abundance and water contents of Kentucky bluegrass. *Sci Rep* 10, 6415. https://doi.org/10.1038/s41598-020-63548-w

Saud S, Fahad S, Hassan S (2022a) Developments in the investigation of nitrogen and oxygen stable isotopes in atmospheric nitrate. *Sustainable Chem Clim Action* 1, 100003.

Saud S, Li X, Jiang Z, Fahad S, Hassan S (2022b) Exploration of the phytohormone regulation of energy storage compound accumulation in microalgae. *Food Energy Secur* 2022, e418.

Senol C (2020) The effects of climate change on human behaviors. In: Fahad S, Hasanuzzaman M, Alam M, Ullah H, Saeed M, Khan AK, Adnan M (Eds.), *Environment, Climate, Plant and Vegetation Growth*. Springer Publ Ltd, Springer Nature Switzerland, pp. 577–590. https://doi.org/10.1007/978-3-030-49732-3

Sethi SK, Adhikary SP (2014) Growth response of region specific Rhizobium strains isolated from Arachis hypogea and Vigna radiata to different environmental variables. *African J Biotechnol* 13(34), 3496–3504.

Shafi MI, Adnan M, Fahad S, Fazli W, Ahsan K, Zhen Y, Subhan D, Zafar-ul-Hye M, Martin B, Rahul D (2020) Application of single superphosphate with humic acid improves the growth, yield and phosphorus uptake of wheat (Triticum aestivum L.) in calcareous soil. *Agron* (10), 1224. DOI:10.3390/agronomy10091224

Shah F, Lixiao N, Kehui C, Tariq S, Wei W, Chang C, Liyang Z, Farhan A, Fahad S, Huang J (2013) Rice grain yield and component responses to near 2°C of warming. *Field Crop Res* 157, 98–110.

Shah S, Shah H, Liangbing X, Xiaoyang S, Shahla A, Fahad S (2022) The physiological function and molecular mechanism of hydrogen sulfide resisting abiotic stress in plants. *Brazilian J Botany* 45, 563–572. https://doi.org/10.1007/s40415-022-00785-5

Sharma S, Gupta R, Dugar G, Srivastava AK (2012) Impact of application of bio-fertilizers on soil structure and resident microbial community structure and function. In: *Bacteria in Agrobiology: Plant Probiotics*. Springer, Berlin, Heidelberg, pp. 65–77.

Sidra K, Javed I, Subhan D, Allah B, Syed IUSB, Fatma B, Khaled DA, Fahad S, Omaima N, Ali TKZ, Rahul D (2021) Physio-chemical characterization of indigenous agricultural waste materials for the development of potting media. *J Saudi Society Agricultural Sci* 28(12), 7491–7498. https://doi.org/10.1016/j.sjbs.2021.08.058

Singh M, Dotaniya ML, Mishra A, Dotaniya CK, Regar KL, Lata M (2016) Role of bio-fertilizers in conservation agriculture. In: Muhammad Farooq, Kadambot HM Siddique (Eds.) *Conservation Agriculture*. Springer. pp. 113–134.

Sinha RP, Kumar A, Tyagi MB, Hader D (2005) Ultraviolet-B-induced destruction of phycobiliproteins in cyanobacteria. *Physiol Mol Biol Plants* 11(2), 313.

SubbaRoa NS (2001) An appraisal of bio-fertilizers in India. The biotechnology of bio-fertilizers (Doctoral dissertation, ed.) S. Kannaiyan, Narosa Pub. House, New.

Subhan D, Zafar-ul-Hye M, Fahad S, Saud S, Martin B, Tereza H, Rahul D (2020) Drought stress alleviation by ACC deaminase producing *Achromobacter xylosoxidans* and *Enterobacter cloacae*, with and without timber waste biochar in maize. *Sustain* 12(15), 6286 DOI:10.3390/su12156286

Swapna G, Divya M, Brahmaprakash GP (2016) Survival of microbial consortium in granular formulations, degradation and release of microorganisms in soil. *Ann Plant Sci* 5(5), 1348–1352.

Tariq M, Ahmad S, Fahad S, Abbas G, Hussain S, Fatima Z, Nasim W, Mubeen M, ur Rehman MH, Khan MA, Adnan M (2018) The impact of climate warming and crop management on phenology of sunflower-based cropping systems in Punjab, Pakistan. *Agri and Forest Met* 256, 270–282.

Toyota K, Watanabe T (2013) Recent trends in microbial inoculants in agriculture. *Microbes Environ* 28(4), 403–404.

Unsar Naeem-U, Muhammad R, Syed HMB, Asad S, Mirza AQ, Naeem I, Muhammad H ur R, Fahad S, Shafqat S (2020) Insect pests of cotton crop and management under climate change scenarios. In: Fahad S, Hasanuzzaman M, Alam M, Ullah H, Saeed M, Khan AK, Adnan M (Eds.), *Environment, Climate, Plant and Vegetation Growth*. Springer Publ Ltd, Springer Nature Switzerland, pp. 367–396. https://doi.org/10.1007/978-3-030-49732-3

Vasile AJ, Popescu C, Ion RA, Dobre I (2015) From conventional to organic in Romanian agriculture – Impact assessment of a land use changing paradigm. *Land Use Policy*, 46, 258–266.

Vessey JK (2003) Plant growth promoting rhizobacteria as bio-fertilizers. *Plant Soil* 255(2), 571–586.

Vincent WF, Roy S (1993) Solar ultraviolet-B radiation and aquatic primary production: damage, protection, and recovery. *Environm Rev* 1(1), 1–12.

Wahid F, Fahad S, Subhan D, Adnan M, Zhen Y, Saud S, Manzer HS, Martin B, Tereza H, Rahul D (2020) Sustainable management with mycorrhizae and Phosphate Solubilizing Bacteria for Enhanced Phosphorus Uptake in Calcareous Soils. *Agri* 10(8), 334. DOI:10.3390/agriculture10080334

Wajid N, Ashfaq A, Asad A, Muhammad T, Muhammad A, Muhammad S, Khawar J, Ghulam MS, Syeda RS, Hafiz MH, Muhammad IAR, Muhammad ZH, Muhammad Habib ur R, Veysel T, Fahad S, Suad S, Aziz K, Shahzad A (2017) Radiation efficiency and nitrogen fertilizer impacts on sunflower crop in contrasting environments of Punjab. *Pakistan Environ Sci Pollut Res* 25, 1822–1836. https://doi.org/10.1007/s11356-017-0592-z

Wiqar A, Arbaz K, Muhammad Z, Ijaz A, Muhammad A, Fahad S (2022) Relative efficiency of biochar particles of different sizes for immobilising heavy metals and improving soil properties. *Crop Pasture Sci* 42(2), 112–120 https://doi.org/10.1071/CP20453.

Wu C, Kehui C, She T, Ganghua L, Shaohua W, Fahad S, Lixiao N, Jianliang H, Shaobing P, Yanfeng D (2020) Intensified pollination and fertilization ameliorate heat injury in rice (Oryza sativa L.) during the flowering stage. *Field Crops Res* 252, 107795.

Wu C, Tang S, Li G, Wang S, Fahad S, Ding Y (2019) Roles of phytohormone changes in the grain yield of rice plants exposed to heat: A review. *PeerJ* 7, e7792. DOI: 10.7717/peerj.7792

Xue B, Huang L, Li X, Lu J, Gao R, Kamran M, Fahad S (2022) Effect of clay mineralogy and soil organic carbon in aggregates under straw incorporation. *Agron* 12, 534. https://doi.org/10.3390/ agronomy12020534

Yang R, Dai P, Wang B, Jin T, Liu K, Fahad S, Harrison MT, Man J, Shang J, Meinke H, Deli L, Xiaoyan W, Yunbo Z, Meixue Z, Yingbing T, Haoliang Y (2022) Over-optimistic projected future wheat yield potential in the North China plain: The role of future climate extremes. *Agron* 12, 145. https://doi.org/10.3390/ agronomy12010145

Yang Z, Zhang Z, Zhang T, Fahad S, Cui K, Nie L, Peng S, Huang J (2017) The effect of season-long temperature increases on rice cultivars grown in the central and southern regions of China. *Front Plant Sci* 8, 1908. https://doi.org/10.3389/fpls.2017.01908

Zafar-ul-Hye M, Muhammad N, Subhan D, Fahad S, Rahul D, Mazhar A, Ashfaq AR, Martin B, Ji˘rí H, Zahid HT, Muhammad N (2020a) Alleviation of cadmium adverse effects by improving nutrients uptake in bitter gourd through cadmium tolerant Rhizobacteria. *Environ* 7(8), 54. DOI:10.3390/environments7080054

Zafar-ul-Hye M, Muhammad Tahzeeb-ul-Hassan, Muhammad A, Fahad S, Martin B, Tereza D, Rahul D, Subhan D (2020b) Potential role of compost mixed biochar with rhizobacteria in mitigating lead toxicity in spinach. *Scientific Rep* 10, 12159. https://doi.org/10.1038/s41598-020-69183-9.

Zafar-ul-Hye M, Akbar MN, Iftikhar Y, Abbas M, Zahid A, Fahad S, Datta R, Ali M, Elgorban AM, Ansari MJ et al. (2021) Rhizobacteria inoculation and caffeic acid alleviated drought stress in lentil plants. *Sustain* 13, 9603. https://doi.org/10.3390/su13179603

Zahida Z, Hafiz FB, Zulfiqar AS, Ghulam MS, Fahad S, Muhammad RA, Hafiz MH, Wajid N, Muhammad S (2017) Effect of water management and silicon on germination, growth, phosphorus and arsenic uptake in rice. *Ecotoxicol Environ Saf* 144,11–18.

Zahir SM, Zheng-HG, Ala Ud D, Amjad A, Ata Ur R, Kashif J, Shah F, Saud S, Adnan M, Fazli W, Saud A, Manzer HS, Shamsher A, Wajid N, Hafiz MH, Fahad S (2021) Synthesis of silver nanoparticles using Plantago lanceolata extract and assessing their antibacterial and antioxidant activities. *Sci Rep* 11, 20754.

Zaman I, Ali M, Shahzad K, Tahir, MS, Matloob A, Ahmad W, Alamri S, Khurshid MR, Qureshi MM, Wasaya A, Khurram SB, Manzer HS, Fahad S, Rahul D (2021) Effect of plant spacings on growth, physiology, yield and fiber quality attributes of cotton genotypes under nitrogen fertilization. *Agron* 11, 2589. https://doi.org/10.3390/agronomy11122589

Zia-ur-Rehman M (2020) Environment, climate change and biodiversity. In: Fahad S, Hasanuzzaman M, Alam M, Ullah H, Saeed M, Khan AK, Adnan M (Eds.), *Environment, Climate, Plant and Vegetation Growth*. Springer Publ Ltd, Springer Nature Switzerland, pp. 473–502. https://doi.org/10.1007/ 978-3-030-49732-3

4 Role of Plant Bio-stimulants and Their Classification

Waseem Ahmed,[1] *Adnan Noor Shah,*[2*] *Asad Abbas,*[3] *Muhammad Nawaz,*[2] *Abdul Qayyum,*[4] *Muhammad Umair Hassan,*[5] *Muther Mansoor Qaisrani,*[6] *and Jallat Khan*[7]
[1] Department of Horticulture, The University of Haripur, Pakistan
[2] Department of Agricultural Engineering, Khwaja Fareed University of Engineering and Information Technology, Punjab, Pakistan
[3] School of Horticulture, Anhui Agricultural University, Hefei, China
[4] Department of Agronomy, The University of Haripur, Pakistan
[5] Research Center on Ecological Sciences, Jiangxi Agricultural University, Nanchang, China
[6] Department of Bioscience & Technology Khwaja Fareed University of Engineering and Information Technology, Punjab, Pakistan
[7] Department of Chemistry, Khwaja Fareed University of Engineering & Information Technology, Punjab, Pakistan
*Correspondence: ans.786@yahoo.com

CONTENTS

DOI: 10.1201/9781003286233-4

4.1 INTRODUCTION

Agriculture faces numerous obstacles and issues, including scarcity of natural resources and destruction caused by over usage of heavy metals (Le Mire et al. 2016). Furthermore, it has been forecast that, in the year 2050, the world's population will have grown to 9.7 billion people (Fahad et al. 2016). As the population depends on agriculture to meet their daily food demands, the best way to fulfill these demands is to make efficient use of available resources to grow healthy products and maintain a healthy diet. Over the past few years, different types of technological revolutions have been proposed to boost immunity to the production system. This is due to the significant reduction of agriculture chemicals. According to the report of environment forums, one in a million deaths occur per year as a result of fatal diseases caused by pesticide poisoning. Fertilizer application is associated with pollution, aberrant plants, groundwater contamination, microbe development, acidity, and soil mineral depletion (Rahman and Zhang 2018; Adnan et al. 2018a,b; Adnan et al. 2020; Adnan et al. 2019; Atif et al. 2021). In this regard, there is a need for new farming methods that are eco-friendly, flexible, and efficient in the production of food (Rouphael et al. 2018; Xu and Geelen 2018; Fahad et al. 2015a,b; Fahad et al. 2020; Fahad et al. 2021a,b,c,d,e,f; Fahad et al. 2022a,b). For this purpose, the use of chemicals or microbe development is a promising method to accelerate plant growth, increase to the prevention of damage to the environment or soil conditions, and increase the efficiency of resources to be used in a more effective way for betterment of plant growth and development (Fakhre et al. 2021; Muhammad et al. 2022; Shafi et al. 2020; Wahid et al. 2020; Zahida et al. 2017). The term "bio-stimulants" was proposed for these chemicals and microorganisms (Zhang and Schmidt 1997). In agriculture, bio-stimulants are natural substances that improve crop growth and quality, and helps the crop to pass through conditions of biotic and abiotic stress (Parađiković et al. 2019). Bio-stimulants will encourage growth of a plant from seed germination to its maturity by increasing its nutrient absorption, intensive growth, improvement in the quality of the product and its features (such as sugar content, colour, frui,t and seed), a complex of soil–microbial interactions, improvements in physical and chemical characteristics of the soil, and increased tolerance to extensive abiotic stress and stress-related disorders.

Humic substances, plant extracts, smoke-derived plants, hydrolysed protein and amino acids, chitosan, and polysaccharides (agar, alginates, and carrageenans), as well as inorganic compounds (Al, Co, Na, Se, and Si) or microorganisms (bacteria or fungi) can all be used as plant bio-stimulants for seed, plants, and soil (Pilon-Smits et al. 2009; Rouphael and Colla 2018). Bio-stimulants, which are created in the form of natural or synthetic ingredients and comprise hormones or plant hormone precursors, are the most common. They have a direct impact on physiological processes in the delivery of possible advantages to growth, development, and responses to stress, water, salt, and toxic substances, such as toxic aluminium, when employed in different cultures (du Jardin 2012).

In the scientific literature, Kauffman et al. (2007) was the first to define the term "bio-stimulants" as bio-stimulants being materials that are different from the fertilizers that help crops to develop when used in tiny amounts. Russo and Berlyn (1991) described bio-stimulants as "the substances or products that are not fertilizers but have a favorable impact on plant growth". Most bio-stimulants are made from organic commodities which contain no additional chemicals or laboratory-manufactured growth regulators. According to Basak (2008), bio-stimulants can be divided into various categories depends on the mechanism of action and the origin of the actively used ingredients. Bulgari et al. (2015) stated that bio-stimulants can also be ranked on the basis of their action or impact on the physiological parameters of plants other than their state of composition. Yakhin et al. (2017) defined bio-stimulants as genetic agents that boost agricultural production by new or derivative capabilities of complicated substances, rather than solely through the presence of known important mineral elements, bio-fertilizers, or plant chemicals.

The Association of American Plant Food Control Officials (2019) defined bio-stimulants as compounds other than primary, secondary, and micronutrients for plants which may be proposed by scientific researchers for the benefit of many species of plant when applied to the plant or soil. The European Biostimulants Industry Council (2012) proposed that they are plant-based products containing chemicals or microbes whose goal is to promote biological responses in plants and soil to improve nutrient absorption, nutritional efficiency, abiotic stress tolerance, and crop productivity.

Plant bio-stimulants are commonly used to improve the productivity and quality of crops having great economic value such as greenhouse crops, fruit trees, plants, vegetables, flowers, and ornamental plants. Also, their beneficial effects have been realized on cereals, grains, oilseeds, pulses, and other crops. However, there is a lack of information on the method of application and their expiry. Bio-stimulants do not have a direct effect on pests and are therefore not included in the regulation of pesticides (European Biostimulants Industry Council and Ertani 2012). The mechanism triggered by bio-stimulants is difficult to measure and still being analysed (Paul et al. 2019).

In past years, the expression "bio-stimulator" has gradually begun to be utilized in scientific publications along with the expansion of the range of organic materials and microorganisms. The bio-stimulants market is increasing daily. Figure 4.1 shows the market size of bio-stimulants.

The current issue addresses the latest development of bio-stimulants in horticulture from various experiments showing that the utilization of materials and useful microbes can be both a beneficial and viable tool for horticultural cultivation. A unique release of bio-stimulants in horticulture includes 10 revisions on the plant bio-stimulants' concept, strategy, basic classes, and management (Du Jardin 2015), humorous and fulvic acids (Canellas et al. 2015), protein hydrolysates (Canellas and Olivares 2014), marine plants release (Battacharyya et al. 2015), chitosan (Pichyangkura and Chadchawan 2015), silicon (Adnan et al. 2015; Savvas and Ntatsi 2015), phosphites (Gómez-Merino and Trejo-Téllez 2015), arbuscular mycorrhizal mould (Rouphael et al. 2015), *Trichoderma* (López-Bucio et al. 2015), and rhizobacteria that promote plant development (Ruzzi and Aroca 2015).

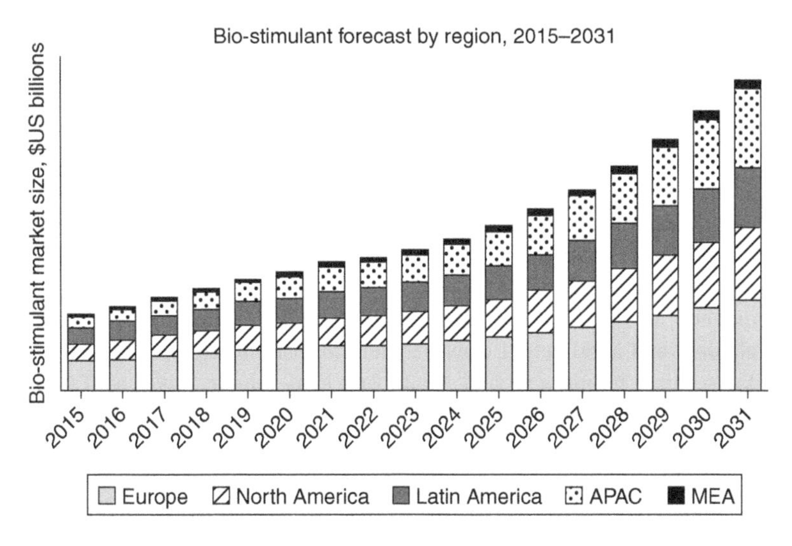

FIGURE 4.1 Market size of bio-stimulants globally.

4.2 MAJOR CLASSIFICATION OF BIO-STIMULANTS

Apart from recent initiatives to define the regulating nature of bio-stimulants, no official or legislative concept of plant bio-stimulants exists elsewhere in the world, including in the European Union and the United States. These circumstances prevent comprehensive description or ranking of certain materials as well as microorganisms. Despite this, bio-stimulants are divided into various classes according to their role or action on plant growth and development (Calvo et al. 2014; Rouphael et al. 2017; Ahmad et al. 2021) giving certain groups of bio-stimulants that include:

- Organic compounds (humic and fulvic acid, animal, vegetal protein hydrolysates, seaweed extracts, and smoke-derived compounds)
- Chiton and chitanose derivatives
- Beneficial microorganisms (N_2-fixing bacteria, mycorrhizal fungi, and plant growth-promoting rhizobacteria)
- Inorganic compounds (copper, silicon, sodium, selenium, and aluminium).
- They are discussed below.

Humic acid and fulvic acid:

Humic substances (HS) are naturally occurring nutrients in the soil that are produced by the decomposition of plant, animal, and microbial wastes, as well as the metabolic process of soil-dwelling insects that consume certain subterranean parts. HS is a group belonging to contrasting substances, divided by the number of molecules and their stability into humins, humic acid, and also fulvic acid. Humic substances are found to affect microbial metabolism as well as chemical and biological modifications of living organisms (including mainly plant and animal waste) that exist also in the atmosphere and water (Canellas et al. 2015; Graber and Rudich 2006; Salma et al. 2010). The most comprehensive content of HS comes from prehistoric deposits when flora and fauna were submerged in wetlands. Within the Earth's atmosphere, humic organisms contain large (about 60%) compounds of organic matter and seem to be largely responsible for soil quality by affecting physio-chemical soil boundaries such as form and permeability (Bronick and Lal 2005; Du Jardin 2015; Trevisan et al. 2010; Ali et al. 2016). Macromolecular formations of diverse materials (such as sugars, essential fats, peptides, hydrocarbon chains, and aromatic rings) are held together by hydrocarbons (such as van der Waals and ion–dipole forces) and ionic bonds (Piccolo 2001; Trevisan et al. 2010). Humic material and its structures in soil are therefore affected by interactions between microorganisms and plant roots. Any experiment to use numerous materials to promote plant growth and crop harvesting requires the expansion of these interactions to obtain the desired results. The major constituent of humic material is humus (containing carbon compounds), which is created by a series of chemical as well as biological fertilizers for soil and organic matter under controlled temperature, time period, water, aeration, and moisture (Canellas and Olivares 2014). Humus is the main part of organic matter found in soil (65–75%) and can remain in the soil for hundreds of years. A recent randomized conceptual analysis of humic substances sprayed on plants showed that dry growth of 24% of shoots and 26% of roots was achieved (Rose et al. 2014).

Humic compounds are classified as alkaline (high pH) or acid (low pH) based on soluble differences in alkaline and alkaline pH (low pH) humin-insoluble (carbon-based macromolecular substances); alkali-soluble-humic acid and alkali- and acid-soluble-fulvic acid (humic and fulvic acid are the final elements in the decomposition of animals along with plants fossils) (Trevisan et al. 2010). Humic acid contains molecules of dark brown cells, has a high molecular weight, and dissolves in alkaline solutions. It is produced by microorganisms and can break down dead organisms. Fulvic acid is a category of chemicals that are formed by the decomposition of plants and animals. It is found in humus. Fulvic acids help to absorb essential nutritional elements because they can quickly enter via the cellular membrane. While humic acids can enhance nutrient absorption from magnificent properties of removing heavy metals and water loss. Humic acids have also been noticed to raise soil structure and microbial activity.

4.2.1 Effect of HS on Soil

Soil organisms, including humic substances, are accepted as a major component in soil fertility, including having physical and chemical functions (Bronick and Lal 2005; Khan et al. 2017). Therefore, HS causes the formation of stable soil compounds, soil aeration, and hydration for the absorption and discovery of nutrients. Canellas et al. (2015) stated that humic substances work on the carbon exchange capacity in chemical reactions and neutralize the pH of the soil. It has been shown that plants that grow in soil containing enough humin, humic acid, and fulvic acid are healthy with high production along with providing good-quality nutritious food. HS can also work to obtain phosphate in plants by storing phosphorus solution in the plant but this useful effect depends on the pH of the soil and the concentration of calcium (Delgado et al. 2002). The activation of cytoplasmic membrane H±ATPs, which convert available energy supplied by hydrolysis of ATPs to electrical and chemical energy utilized to acquire nitrates and other nutrients, is another contribution of HS to root feeding.

4.2.2 Effect of HS on Plant Physiology

Humic substances directly affect plant physiology which means that the impacts are not related to soil characteristics and the availability of nutrients but include regulation of cellular activities such as mutations, genetic modification, and hormonal actions (Canellas et al. 2008). The processes by which some humic chemicals behave as external plant hormones that alter root expansion are unknown. HS are reported to have a significant impact on the formation of the root system (Canellas et al. 2002; Khristeva 1957; Ramos et al. 2015; Zandonadi et al. 2007). Zandonadi et al. (2007) revealed that humic substances are not involved in root elongation at lateral root emergence sites (Kulkarni et al. 2011). Humic substances are also reported to form genital roots with identified plasma lemma and the function of H+ tonoplast pumps (Zandonadi et al. 2007). Jannin et al. (2012) explained that humic acid utilization in *Brassica napus* brought about a resurgence of root development and increase in N and S intake due to the development of genomes found in roots including a sulphate carrier and a nitrate carrier. As claimed by various researchers, humic substances can also have a significant impact on important metabolic processes and in reducing stress levels (De Pascale et al. 2017; Khan et al. 2016). It has been reported that humic substances contain hormone-like substances such as auxins, gibberellins, cytokinins, and polyamines (Nardi et al. 2009). They also have been noticed to have a beneficial impact on some other plant body procedures such as the chlorophyll level. It is also reported that humic substances trigger several cellular actions in plant cells and increase plant resistance to many types of abiotic stresses, e.g. in tomatoes and okra, and HS has been already proven to improve plant salt stress (Paksoy et al. 2010; Türkmen et al. 2004; Van Oosten et al. 2017).

4.3 PROTEIN HYDROLYSATE

Protein hydrolysates (PHs) are a fusion of polypeptides, oligopeptides, and amino acids generated by the incomplete breakdown of plant proteins and common bio-stimulants (Schaafsma 2009). Amino acid and peptide composition are acquired by protein breakdown through biochemical means via agricultural and industrial materials derived from plants (plant wastes) and animal faeces (e.g. collagen, epithelial tissue) (Calvo et al. 2014; du Jardin 2012; Halpern et al. 2015; Zeeshan et al. 2016). PHs have obtained importance as bio-stimulants due to their potential to boost the development, yield, and durability of a variety of horticulture crops. The use of PHs can also reduce the negative outcomse of pressure on abiotic plants due to salinity, drought, and heavy metals.

4.3.1 Classification of Protein Hydrolysates (PHs)

The chemical properties of PHs are determined by the protein supply and the technique of protein breakdown in the presence of water (chemical, thermal or enzymatic hydrolysis) (Colla et al. 2015a; Du Jardin 2015; Halpern et al. 2015). Chemical degradation using H_2SO_4 or HCl at extreme temperatures (> 121°C) and compression (> 220.6 kPa) is used to oxidize animal-derived proteins in acidic circumstances (protein processing continued by specified heating has incorporation of alkaline agents such as Ca, Na, or KOH) (Colla et al. 2015b; Pasupuleti and Braun 2008). Some researchers have described a decreased growth of fruit plants as a result of PHs found in animals that may be due to unequal amino acid formation or high density of amino acids and high salts (Lisiecka et al. 2011). Companies should carefully inspect the stability of mammal PHs before releasing them onto the market (company safety testing). European Regulation (No. 354/2014) restricts the usage of PHs found in animals in edible parts of living plants due to growing concerns about food safety, however they can be used in the inedible part of plants.

- Enzymatic hydrolysis is appropriate for the generation of protease, proteinase, or peptidase via animals (pancreatin, pepsin), plants (papain, ficin, bromelain), or microbes (alcalase, flavourzyme). Protein hydrolysates are present in powder form and with water added can be used in leaves such as foliar applications or extracts that are taken in proximity to plant roots (Colla et al. 2017).

The emergence of degradation of certain molecules and enzymes can conserve vitality and maintain the formation of amino molecules (Niculescu et al. 2009).

4.3.2 Effect of Protein Hydrolysates (PHs)

4.3.2.1 Direct Effect of Protein Hydrolysates (PHs)

The latest studies show that protein hydrolysates affect plants directly by stimulating the mechanism of nitrogen and carbon by regulating the enzymes involved in the formation of N, its genes, and by acting in a signature form of N in roots. The elimination of oxidants by other organic matter substances, involving glycine betaine or proline, can provide antioxidant action that helps to reduce environmental stress. Some organic molecules like proline have solubilizing properties that can protect flora against toxic chemicals while providing the movement and availability of micronutrients.

4.3.2.2 Indirect Effect of Protein Hydrolysates (PHs)

In the farming industry, secondary impacts on crop nutritive value are also significant where protein hydrolysate is used in plants and soils. PHs can also play a role in increasing nutrient availability in low-growing plant areas, increasing nutrient intake and nutrient utilization of plants. They boost microbiological biomass production cellular processes and total crop production. Certain amino molecules and peptides are believed to help with nutrient uptake along with root access due to their chelating and complex activities.

Hydrolysate proteins are involved in the formation of macro-chelate (such as K, Ca, and Mg) as well as micronutrients (Cu, Fe, Mn, and Zn) utilized in industry for increasing potent nutrient intake (De Pascale et al. 2017). They contributeg to plant enlargement by amendments in the phytohormone balance (Colla and Rouphael 2020; Ertani et al. 2009). They are also reported to promote phytonutrients (such as polyphenols, carotenoids, flavonoids) with the antioxidant action of portable herbs (Ertani et al. 2009; Tsouvaltzis et al. 2014). Regulation Commission (EU) No. 354/2014 lays down health rules as regards animal by-products and derived products in consumable plant components not intended for human consumption. Therefore, bio-stimulants based on protein hydrolysate have positive effects and enable productive methods for organic farming. Therefore, the formation of PHs must be inhibited by the utilization of basic resources, substrate catalysts, and degradation situations.

4.4 SEAWEED EXTRACTS

Seaweeds are a large group of species divided into distinct phyla compromising of brown, red, and green macroalgae and are utilized as manure in agriculture. *Ascophyllum nodosum*, *Ecklonia maxima*, *Durvillea potatorum*, and *Macrocystis pyrifera* are the most common sources of seaweed extract. They are mostly obtained from carbohydrates (such as laminar, fucoid, and alginates) (Battacharyya et al. 2015), phenolic compounds (Audibert et al. 2010), phlorotannins (Rengasamy et al. 2015), ACC (1-aminocyclopropane-1-carboxylic acid) (Nelson and Van Staden 1985), osmolytes (Reed et al. 1985), and phytohormones (such as abscisic acid, auxin, cytokinin, and gibberellins) for the enhancement of plants (Stirk and Van Staden 2014). They can be given to the plant as a foliar treatment. Many of these compounds are distinct from their algal source, so they are defined by the interests of the scientific community and industries into these groups of species.

4.4.1 Application of Seaweeds

- The use of seaweed bio-stimulant is known to enhance seed sprouting (Masondo et al. 2018), photosynthetic components (Di Stasio et al. 2017; Patel et al. 2018), nutrient absorption (Di Stasio et al. 2017), blooming of flowers and fruits (Pohl et al. 2019), tolerance to abiotic and biotic suppression (Bajpai et al. 2019; D'Addabbo et al. 2019; Jayaraj et al. 2008), and increased production and life span of fruit juices (Kamel 2014; Renaut et al. 2019).
- The positive effect on soil microflora is also explained by pathogens being encouraged to increase plant development and pathogen resistance in repressive soils (du Jardin 2012).
- The use of liquid seaweed bio-stimulant has been shown to encourage flower initiation (flower onset) and growth in many plant species. The phytohormone content in bio-stimulants may affect the setting of flowers and is also associated with early fruits.
- Useful effects and defensive substances found in seaweed include antioxidants as well as a genetic regression response that may be involved (Calvo et al. 2014).
- The exceptional effects of gelation, fluid retention, and moisture content in soil are due to seaweed polysaccharides present in soil (du Jardin 2012).
- The polyionic compounds of seaweeds take part in the preparation and exchange of cations which are also of interest in heavy metal complexes and soil preparation (du Jardin 2012).

Commercial seaweed extraction from agrochemical companies has gained widespread acceptance in agriculture as natural or plant bio-stimulants over the past few years. Commercial liquid extraction such as Stimplex (a marine commodity product) has shown a significant increase in the range of ornamental *Capsicum annuum* (Ozbay and Demirkiran 2019). Stimplex enabled responses such as cytokinin in *A. thaliana* (Khan et al. 2011). The extraction of seaweed (Acadian, Acadian Seaplants Ltd Canada) with *Ascophyllum nodosum* increased root growth and resistance to water in pumpkin, tomatoes, peppers, apricot, petunia, pansy, lettuce, and cosmos (Neily et al. 2010). Seaweeds are reported to have many useful effects on a variety of plants but there is a need to better understand how they work and to better utilize them in agriculture.

4.4.2 Microalgae

Recently microalgae, which includes single-celled eukaryotic and prokaryotic (cyanobacteria) blue algae are has become well-known as a bio-stimulant due to its ability to boost flourishing, plant expansion, production, mineral utilization, and tolerance to various abiotic stressors. The effective use of bio-stimulants of plant-based microalgae is limited even though the polyunsaturated fatty acids, carotenoids, betaines, polyamines, phycobiliproteins, sterols, vitamins, and polysaccharides are some of the bioactive organisms.

4.4.2.1 Activity of Microalgae as Bio-stimulants

In some review papers the authors have described the dominant types of microalgae that have stimulatory action as: *Calothrix elenkinii*, *Acutodesmus dimorphus*, *Chlorella ellipsoida*, *Scenedesmus platensis*, *Scenedesmus quadricauda*, *Chlorella vulgaris*, *Dunaliella salina*, *Chlorella infusionum*, and *Scenedesmus platensis* (Colla and Rouphael 2020). The number of key substances (protein, lipids, carbohydrates), vitamins, amino molecules (tryptophan, arginine), polysaccharides (glucan), and osmolytes (proline and glycine betaine) is linked to the bio-stimulant scheme of microalgal extracts (Chiaiese et al. 2018). A polysaccharide found in microalgae has been revealed to contain a variety of biological agents which are used in farming to boost plant productivity, nutrient availability, and biological and eco-physiological stress resistance. The microalgal exopolysaccharides are well-known to reduce salt pressure and can be utilized under severe environmental circumstances. Green alga in *Solanum lycopersicum* have lowered salt pressure as well as reduced the size and loss mass of plant seeds and also roots (El Arroussi et al. 2018). Screening and selecting microalgal isolates containing organic plant hormones, especially cytokinins and auxin, regarded as major sources for developing the possibility of algal economization in increasing plant maturation, productivity, and immunological responses especially in abiotic stress (Ronga et al. 2019).

It is vital to remember that algae species, growth circumstances, and extraction processes all influence the active chemicals identified in microalgal fragments. Physical (mashing, autoclaving, liquid nitrogen), chemical (HCL, NaOH, and osmotic shock), and enzymatic (cellulase and protease) approaches are used to remove different kinds of toxins from microalgae (Chiaiese et al. 2018). In addition, bio-stimulants based on microalgae are commonly distributed to plants and the rhizosphere by various means such as soil wetness, composting, fertilizer spray, and seed treatment (Colla and Rouphael 2020). The microalgae's bio-stimulant action can be attributed to their ability to alter the number of microbes in the ground, rhizosphere, and biosphere (Renuka et al. 2018; Xiao and Zheng 2016). The effect of the incorporation of cyanobacteria *Calothrix* into the genus on the growth of rice plants and with the pathogen culture is supported by research (Priya et al. 2015). They found that *Calothrix* in the region upgraded the growth and formation of rice and revived the growing number of bacteria. Chiaiese et al. (2018) and Ronga et al. (2019) described the methods of bio-stimulant use based on microalgae that were provided with various explicit and implicit placement methods, such as:

- Biochemical (stimulating carotenoids and chlorophyll synthesis) and physical modification (prevented senescence)
- Increase in the number of key genes involved in plant metabolism
- Increased activity around bacterial communities (rhizobacteria and mycorrhizae) by releasing low- and high-molecular-weight organic solute is associated with the rhizosphere
- Variations in root structure and as a result an increase in macro- and micronutrient efficiency and effectiveness.

4.5 PLANT-DERIVED SMOKE

Plant smoke is a complex collection of compounds of chemical substances based on heating a natural blend of flowering plants to encourage germinating seeds and vegetation in a variety of plant species (Simorangkir 2007). Smoke contains life-forming compounds that dissolve easily into freshwater so its extract or smoke water (SW) can be produced to enhance plant health (De

Lange and Boucher 1990; Gupta et al. 2020). This can reduce the issue of air pollution caused by smoke. Plant smoke has been proven to be a stimulant for several plant-related conditions including seed breakdown, accelerated germination, the percentage of seedlings, and health of seedlings (Aslam et al. 2017). It has been found that SW has a dual control effect, where a low SW concentration is very effective and there is a negative effect with high concentrations (Daws et al. 2007; Gupta et al. 2019; Light et al. 2002). This is because smoke water contains both stimulating and suppressive compounds (Flematti et al. 2004; Van Staden et al. 2004). A high volume of inhibitors (trimethylbutane gold) in unmodified SW has been noticed because SW exhibits an inhibitory effect on high aggregation (Gupta et al. 2020; Light et al. 2002). It acts to reduce the inhibition of heavy metals, dryness, salinity, and extremes in temperature on growing plants (Akhtar et al. 2017).

Like the extraction of seaweed, the major effects of bio-stimulants from plant-based fumes such as smoke water (SW) and karrikin gold (KAR1) (Gupta et al. 2020; Kulkarni et al. 2011; Light et al. 2002; Masondo et al. 2018) have been recorded in many agricultural plants. Smoke water is used in unusually small areas until it attains growth-promoting tasks and post-emergence growth (Gupta et al. 2020). An aspect of plant-derived smoke it that while creating various stress resistances it qualifies for inclusion in the category of bio-stimulants. Among the many functions of plant smoke in plant life are developing stress resistance which helps to eliminate salinity stress in *Zea mays* and *Oryza sativa*, reducing heat stress in *Solanum lycopersicum*, reducing heavy rain stress in *Glycine max*, and reversing abscisic acid stress in *Lactuca sativa*, which have strongly supported its use as a bio-stimulant. Smoke contains antibacterial agents that can help to protect young plants from dangerous microorganisms (Holley and Patel 2005). Many plant species have shown hormone-like responses to chemicals found in SW and interrelated plant hormones in photoblastic as well as thermos-dormant seeds have been reported (Gupta et al. 2019; Van Staden et al. 2004).

These findings suggest that this smoking technology could be useful in commercially cultivated plants (Van Staden et al. 2004). As a result, plants inoculated with plant-derived smoke are safe to consume. It has shown that SW and smoke compounds (KAR1) do not show any mutagenic genotoxic effects. Smoke influences the expansion and advancement of various plant species and the use of smoked substances (KAR1) and water from the smoke can provide a new way to improve crop production and harvest.

4.6 CHITIN AND CHITOSAN DERIVATIVES

Chitosan is a commercially available form of the natural and industrial biopolymer chitin without an acetyl group. The physical responses of chitosan oligomers in plants are the outcome of this polycationic substance's ability to hold together many cellular elements, including genetic material on the cell surface along with cellular fragments, as well as to attach specific receptors associated with immune system activation, in a similar way providing plant protection (Hadwiger 2013; Katiyar et al. 2015). Crustaceans (king crab, shrimp, lobster), fungi (filamentous fungus, black mould), insects (ladybird, insect wax, butterfly), mole oyster, and squid pen are all naturally occurring sources of chitin and chitosan (Yadav et al. 2019). Microbial proteases have been the most extensively used method for chitin and chitosan synthesis (Philibert et al. 2017). Crustacean shells are a significant source of chitin, and chitin improvement involves three stages including demineralization, deproteination, and removal of pigments and lipids (Holley and Patel 2005; Philibert et al. 2017).

In agriculture, chitosan is utilized as a fertilizer in seed, leaf, fruit, and vegetable crops to increase crop production and provide protection against various microorganisms.

Because of the polymer's capacity to break down chitin slowly it is commonly using it as a biofertilizer, extracting essential nutritional elements from plants without the risk of over-fertilization.

4.6.1 Chitin (CH)-based Polymers

The chemical structure of chitin is given below:

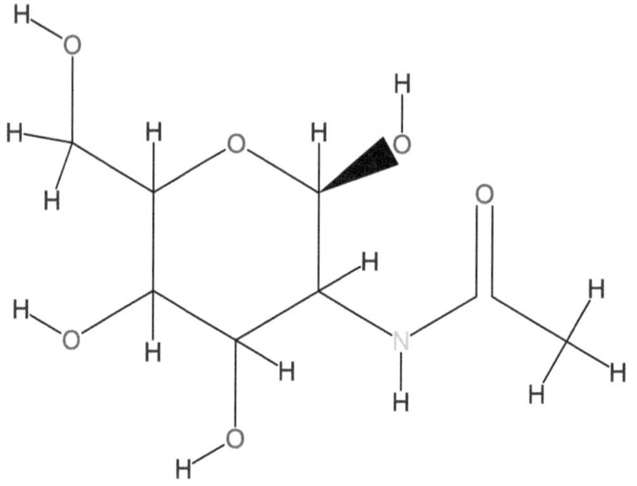

Chitin is the second most frequent carbohydrate found in the plant body, and is a key component of the formation of fungal cells and bones of invertebrates (Pusztahelyi 2018). According to the latest market research, the global chitin trade is predicted to reach USD 2900 million by 2027, with significant segments including health maintenance, waste, water treatment, and biochemical (Market and Chitin 2017). It is a highly concentrated compound of b-1,4-N-acetylglucosamine that is stretchable, therefore, its structural alteration, including chemical variants, is critical to the optimal use and validity of this biopolymer.

4.6.1.1 Chitin as a Bio-stimulant

Chitin can be employed instantly as a bio-fertilizer to promote plant expansion because it contains a high nitrogen concentration and a weak C/N ratio. Furthermore, adding chitin to soil enhances the population and composition of microbes. It is identified as a pathogen-associated molecular receptor pattern (PAMP) in plants by specific receptors found in the plasma cell membrane (Chandanie et al. 2009). It follows that chitin could be used as a PAMP-induced immunity which can create immune responses to fungi, bacteria, and viruses (Pusztahelyi 2018). The commonly produced chitin is insoluble in most typical liquid chemicals due to its high density.
The main methods of applying chitin to plants include

- Foliar spraying
- Direct application to soil.

Foliar application as a useful impact of CH on plants is linked with its direct action as an obstacle for several microbial diseases and also by an incidental action which triggers antibodies simultaneously with molecular signaling mechanisms (Eckardt 2008; Ramírez et al. 2010; Wan et al. 2008).

The effects of applying chitin directly to soil are more complicated than those of foliar application and involve the following:

a. Enhanced nitrogen availability because of a higher chitin concentration
b. Enhanced action of the chitinolytic compound which will adversely affect microbes in the soil

c. The positive impact of soil microbes such as mycorrhiza that work in conjunction with planting and improving plant performance (Li et al. 2007).

In addition, chitin use in consumable coats in agricultural produce can impart a definitive hurdle to gases interchange and can delay maturation while also reducing water loss and respiratory levels (Dhall 2013).

The elements of chitin have activity on the exterior surfaces of living cells such as polycationic and lipid molecules. This leads to the application of water purification but the emergence of chitosan films on the waxy surface of the plant leaves give rise to its use as an anti-transpirant take out CH from yeast and fungi have various benefits (Sun et al. 2018).

- Cannot make humans or animals more susceptible to illnesses, such as CH from vertebrates
- Have physiochemical qualities like eco-friendly, bio-stability together with non-toxicity, which facilitate its agricultural use.

Some procedures recommend using CH with a low concentration to facilitate melting. An example of chitin oligosaccharide produced in *Arabidopsis thaliana* is a genetic expression of vegetable growth, nitrogen, and carbon survival. Plants medicated with CH oligosaccharide demonstrated improved weight gain (10%), maximum root length, and overall carbon and nitrogen content when compared to organic plants (Winkler et al. 2017).

4.6.1.2 Bio-stimulatory Effect of Chitin in Vegetables (Table 4.1)

TABLE 4.1
The Consequences of Applying Chitin to Vegetables

Vegetables	Scientific Name	Effect of Chitin	References
Cabbage	*Brassica oleracea*	Complex infections such as root knot nematode and cottony soft rot have been minimized due to the composition of chitin and *Trichoderma*	Loganathan et al. 2010
Chili pepper	*Capsicum annum*	Chitin and salicylic acid treatment, as well as opponents (fluorescent pseudomonads SE21 and RD41), are well regulated for damping off (*Thanatephorus cucumeris*) seeds	Rajkumar et al. 2008
Eggplant	*Solanum melongena*	Chitin-containing soil supplements found in crabs reduce verticillium pressure in plants	Inderbitzin et al. 2018
Lettuce	*Lactuca sativa*	Soil use of chitin is associated with the use of betaine paper for improved plant performance under conditions of water stress	Lin et al. 2020
Tobacco	*Nicotiana tabacum*	Nanochitin enhanced seed sprouting and development, as well as *Fusarium* species tolerance	Zhou et al. 2017
Wheat	*Triticun aestivum*	Nanochitin whiskers develop resistance to *Fusarium* crown rot of wheat and also *Gibberella zeae*	Liang et al. 2018
Wheat	*Triticun aestivum*	Nanochitin works on the enhancement of crop production, and protein, ferrous, and zinc contents	Xu and Geelen 2018

4.6.2 Chitosan (CHT)-based Polymer

The chemical structure of chitosan is:

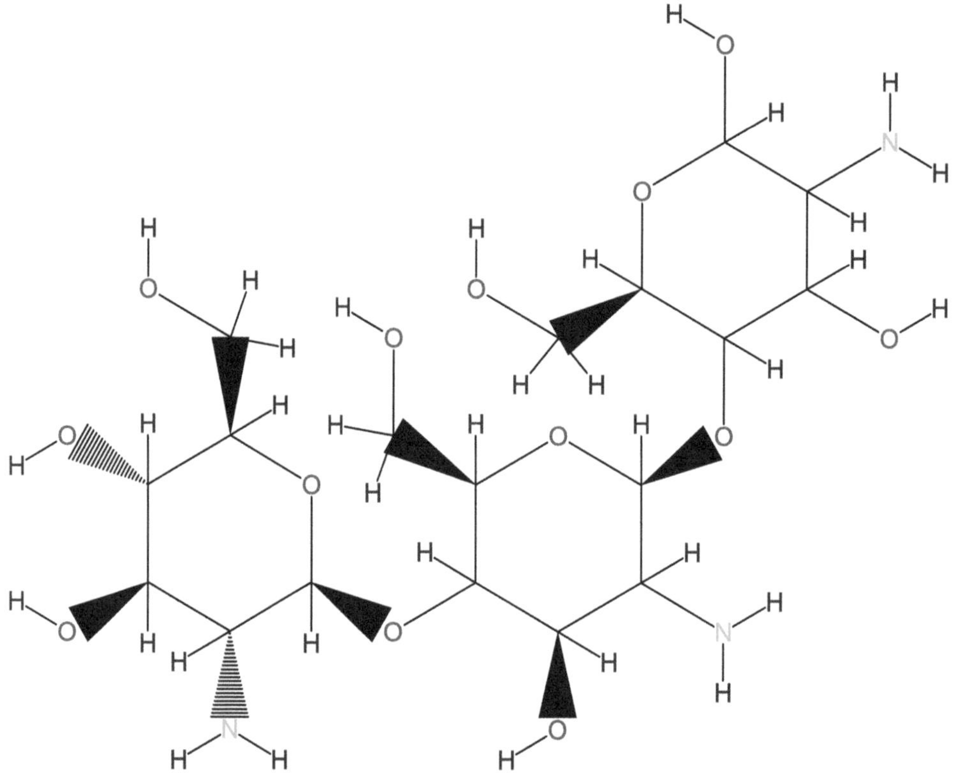

Excessive usage of chemical compounds to boost production and germ tolerance have the ability to trigger permanent environmental damage because of its assembling in the atmosphere around the plant and affecting biodiversity. To look for new ways to resolve this problem, the use of nanotechnology is very favourable. Chitosan (CHT) of biomaterials is very helpful in nanotechnology because of their biological degradation, compatibility, and non-harmful nature. Moreover, CHT can be easily altered without compromising its reproductive capacity compared to other biopolymers (Chakraborty et al. 2020). The advantages of the use of chitosan are very dominant because it is described GRAS (generally recognized as safe) and is easily soaked, cheap, accessible, and easy to use (Bellich et al. 2016).

4.6.2.1 Chitosan as a Bio-stimulant

- The key functions of CHT have been strongly linked with improved photosynthetic action and resistance to abiotic pressures such as drying and salt as well as low or high temperatures, and increased activity of the antioxidant enzyme and its expression protective genes (Pichyangkura and Chadchawan 2015).
- Furthermore, the use of chitosan can lead to increased plant growth, especially with an increase in nitrogen and nutrients, and also can be used as an additional source of carbon in biosynthetic plant processes (Mondal et al. 2013; Pirbalouti et al. 2017).
- Chitosan's major agricultural goals rely on its promising impacts on macromolecule production in insects and diseases (Chakraborty et al. 2020; Patel et al. 2020).

TABLE 4.2
The Consequences of Applying Chitosan to Vegetables

Vegetables	Scientific Name	Effect of Chitosan	References
Ginger	*Zingiber officinale*	Chitosan and oligochitosan improve immune enzyme action on ginger	Liu et al. 2016
Lettuce	*Lactuca sativa*	Applying chitosan at 2% to Ni-contaminated soil can importantly control Ni bioavailability	Turan 2019
Eggplant	*Solanum melongena*	Combined nanocomposites enhance nematocidal activity with a standardized immune response	Attia et al. 2021
Okra	*Hibiscus esculentus*	Nanochitosan has a positive effect on plant morphogenesis, growth, and physiology	Mondal et al. 2012
Potato	*Solanum tuberosum*	Chitosan spray mixed with humic acid led to extensive tuber production and yield	Harfoush et al. 2017
Tomato	*Solanum lycopersicum*	Foliar use of salicylic acid and chitosan at 75 mg L^{-1} can be used at the beginning of high growth, acquiring maximum fruit yield in summer tomatoes.	Mondal et al. 2016
		Application of chitosan + compost + arbuscular mycorrhizal fungus has improved tomato growth	El Amerany et al. 2020

- In plants, several antimicrobial and control functions have been discovered through chitosan molecules usage. For example, the reproductive elicitor of *Oryza sativa* L. is chitosan nanoparticles (CHTNP), which contain CHT structures and nanoparticle properties (such as interface), surface impact, tiny size effect, and quantum size (Divya et al. 2019), which influence the germination of seeds and maturation of wheat seed positively (Li et al. 2019).
- It can also expand the roots and lower the respiration rate, leading to improved water absorption and better water use in plants (Bittelli et al. 2001; Mondal et al. 2016).
- It can be used in a variety of ways including seed coverage, foliar application, along with being cover-up factor for vegetable or fruit protection after harvesting (Gómez-Merino and Trejo-Téllez 2015; Pandey et al. 2018; Toscano et al. 2018).

4.6.2.2 Bio-stimulatory Effect of Chitosan in Vegetables (Table 4.2)

Chitin and chitosan are natural biopolymers that have many functions in plants. The use of these compounds has shown good results in protecting the biodiversity of horticultural crops as well as in crop production and growth, especially under natural barriers that emphasize their positive role in crop cultivation under arid and non-arid conditions.

4.7 BENEFICIAL MICROBES AS BIO-STIMULANTS

Microbial bio-stimulants include microbes or an association of microbes that can be employed to boost yield, absorption of nutrients, abiotic stress tolerance, crop productivity, and yield in seeds, young plants, and land. This indludes plant-growth ptomoting Rhizobacteria (PGPR, soil-dwelling), plant growth-promoting fungi (PGPF, specifically mycorrhizal fungi), and endophytes (both fungi and bacteria) (Fiorentino et al. 2018; Rouphael et al. 2017).

4.7.1 Beneficial Fungi

Species of plants, as well as fungus, are developed independently from soil, plants, and the mutualistic theory. This continuation contributes to the understanding of the diverse range of relationships that emerge during evolutionary periods (Bonfante and Genre 2010; Johnson and Graham 2013). Fungi interconnect with the root systems of plants in a variety of different mutualistic symbioses (meaning that when two species live near each other they communicate and develop beneficial partnerships) to parasitic in the living cytoplasm (Behie and Bidochka 2014). Mycorrhizal fungi are the most numerous genus in symbioses, with more than 90% of all plant species. Plants that promote fungal growth, such as arbuscular mycorrhiza fungi (AMF; *Rhizophagus intraradices* and endomycorrhizal fungus), and *Trichoderma* species (Teleomorph Hypocrea) can encourage plant growth by increasing the plant mineral content (Fe, Cu, Zn) as well as minerals in the soil (Fe^{3+}, Mg^{2+}, K^{+}, etc.) through improved formation of stimulant hormones (such as indole acetic acid) (López-Bucio et al. 2015; Rouphael et al. 2015; Smith and Read 2008). AMF is formed between the cortical cells of the root system and a category of fungus that sets up distinctive structures called arbuscules and vesicles. AMF requires host plants to form an external hyphae network that can expand over 40 times in space (Krüger et al. 2012). There is growing interest in utilizing mycorrhiza to upgrade the agribusiness, given the common advantages of efficiently high nutrient markers, water balancing, and protection from biotic and abiotic plant pressures. Metagenomics is an important tool for tracking and studying the bacterial interrelation in the space around plants. A fusion of plant particles and soil capacity proceeds these processes (Colla et al. 2015a; Savvas and Ntatsi 2015). *Trichoderma* species, *Sebacinales* and *Piriformospora indica*, as model organisms are different from mycorrhiza fungi because they can spend a portion of their life span elsewhere than with their host plant, attaching roots, transferring nutrients to the moderator, and using various methods (Behie and Bidochka 2014). *Trichoderma*-based bio-stimulants have reportedly enhanced plant growth, yield, healthy food quality, and biotic and abiotic stress tolerance (López-Bucio et al. 2015; Lorito and Woo 2015). *Trichoderma* has efficacy against 30 *Tricoderma asperellum* samples as well as four different *Fusarium oxysporum* species (El Komy et al. 2015). Some classes of *Trichoderma* have a prominent bio-stimulant effect, which distinguishes them from their widespread use in agriculture. They are non-toxic to himans, animals, and crops, and also to native habitats, where they invade plant roots without causing harm. The advantageous outcomes of *Rhizophagus intraradices* and *Trichoderma atroviride* application are suitable for the production of auxin-containing substances with the release of flexible organic chemicals that improve the branching strength, and the solubility of essential nutrients (Zn, Fe, P, Mn, etc). (Rouphael et al. 2017). The combination of *Hypocrea lixii*, *Hypocrea rufa*, and *Hypocrea virens* increases the growth parameters, biomass production, as well as nitrogen and phosphorus roots assimilation in *Cicer arietinum* (chickpea) in glasshouse and field experiments (Rudresh et al. 2005).

The accomplishment of AMF and *Trichoderma* sp. improved the plant performance under water stress conditions (Augé et al. 2015; Chitarra et al. 2016). It has been noted that the inclusion of both AMF and *Trichoderma* species can have a beneficial impact on plant relationships and growth suppression (Chandanie et al. 2009; Colla et al. 2015a). These fungal endophytes can be considered bio-stimulants, although crops have been generally assisted by the use of bio-pesticides.

4.7.2 Beneficial Bacteria

Studies show that the use of advantageous bacteria can provide farmers with a solution to reduce fertilizer application without compromising crop quality and making existing crop management systems economically and environmentally viable. Bacteria such as *Pseudomonas* in plant roots form beneficial relationships with plants that encourage nutrient absorption as well as plant development. Successful colonization should reduce the need to re-install microbial inoculants. Nitrification, siderophore manufacturing, dissolving phosphate, formation of plant hormones (auxins, cytokinin, and gibberellin), the activity of ACC, and presence of organic molecules and organic polymer agents are all reported to promote maturation and production in various plants (Bharti and Barnawal 2019; Vessey 2003). Bacteria affiliate to the flora in many ways (Ahmad et al. 2008):

I. Mould, where there is a continuity between parasitism and mutualism.
II. Bacterial strips travel from soil to inside tissues, with rhizosphere and rhizoplane regions.
III. Some bacteria are transported upwards by seeds, and this relationship may last for a short or long time.
IV. Activities affecting plant life span in the biogeochemical cycle, provide nutrients, increase nutrient efficiency, increase resistance to disease, improve abiotic stress tolerance, and cause changes in plant morphology through plant growth hormones.

PGPR can also stimulate plant growth in a variety of forms such as increasing nutrient uptake, and maximizing the root zone and yield of plants (Paungfoo-Lonhienne et al. 2019). Indirect enhancements include a decrease in the unfavourable effects of phytopathogenic organisms like bacteria and fungi through a process called induced systematic resistance (ISR). ISR regulates phytopathogenic fungi by forming antifungal substances like hydrogen cyanide, phenazines, pyrrolnitrin, tensin, etc. (Bhattacharyya and Jha 2012). Cakmakci et al. (2007) described PGPR as enhancing the effects of glutathione reductase (reduces oxidized glutathione for reuse as an antioxidant compound) and activity, and improving plant growth in wheat. *Bacillus lentimorbus* has also been reported to increase antioxidant activity in edible components of lettuce, spinach, and carrot plants (Niculescu et al. 2009). Through the use of bio-stimulants in cultivation within botanical variants, two key classes of functional and natural categories should be considered:

(i) Intersection endosymbionts of the *Rhizobium* type
(ii) Interactions of *Rhizobium* plant growth-promoting rhizobacteria (PGPR).

Rhizobium and related genera are sold as organic fertilizers, which are bacterial inocula that help plants absorb nutrients.

Phosphorus is the second most important macronutrient for the growth of plants and a lack of it reduces crop production. It is naturally available in the soil but it is unavailable to plants. Through a process known as mineral phosphate solubilization, phosphate solubilizing rhizobacteria (PSRB) release chemical compounds and phosphatases to transform soluble phosphates into soluble monobasic (H_2PO_{4-}) and dibasic (HPO_{42-}) ions (Gyaneshwar et al. 2002).

Fe is an important micronutrient in plants, however, it is present mainly in the form of Fe^{3+} ions and forms hydroxides and oxyhydroxides, but is needed in the form of Fe^{2+} by plants. By releasing moderate chelating agents known as siderophores, PGPR sequesters iron as Fe^{2+} ions (Fe^{3+} is reduced to Fe^{2+} in the siderophore membrane and transmitted to the cell wall). Crops directly incorporate Fe^{2+} from siderophores by exploiting Fe-siderophore characteristics or by alternating reactions with a suitable ligand (Novo et al. 2018). In addition, the global economy for bacterial bio-stimulants is expanding, and PGPR inocula are increasingly regarded as a type of "probiotic" plants, that is, they contribute actively to nourishment and immune defence (Berendsen et al. 2012).

4.8 INORGANIC COMPOUNDS

Inorganic compounds are minerals like silica, selenium, cobalt, and others that enhance the development of plants, purity of plant foods, and abiotic stress resistance. Out of five beneficial properties, Si is most widely used as a bio-stimulant; it reduces salt, drought, and nutrient stress, as well as climate-related stress, and also reducing metal toxicity (Yan et al. 2020). Useful chemicals are synthetic components that aid soil fertility which can be significant for certain taxons but may not be essential for all species (Pilon-Smits et al. 2009). Aluminium, copper, sodium, selenium, and silicon are the five most important nutrients, and they can be found in a variety of forms in soil as well as in crops, including undissolved structures such as amorphous silica ($SiO_2.nH_2O$) (Du Jardin 2015).

Such favourable roles may synthesize, such as nourishing the cellular structure by silica crystals, or be exposed to specific ecological circumstances, such as selenium microbe attacks and osmotic stress for sodium. These involve cell wall stiffening, heat reduction by crystallization, radioactivity-regulated heat, composition-influenced functions of enzymes, connections between flora and foods, as well as other aspects throughout absorption, movability, enzyme preservation, and mutually beneficial contact between two organisms, microbe and herbivore actions, protection against metal toxicity, hormone production, and signaling (Pilon-Smits et al. 2009).

Chlorides, phosphates, phosphites, silicates, and carbonates are synthetic compounds and important chemicals employed as antifungal agents (Deliopoulos et al. 2010). Many synthetic chemicals affect pH, osmosis, and oxidation–reduction stability as well as the biological stress signatures of hormones and enzymes (e.g. peroxidase). Their roles as bio-stimulants for plant development show they are effective in nutritional efficiency and resistance to abiotic pressure, which differ from their fungicidal activity and their role as fertilizers, which requires much attention (Du Jardin 2015).

4.9 CONCLUSION

Bio-stimulants are not confined to a single substance or microbe. The component can be a synthetic form of chemical or perhaps a class of chemical with approved biological action, e.g. extruded plants but not complete arrangements. In this case they correspond to the current definition of a substance. This substance involves the European REACH Regulation (EC No 1907/2006) relating to the certification, testing, authorization, and limitations of chemicals, that acknowledge the group of flexible materials: "UVCB materials can be recorded as a single material within this Act, regardless of their distinct structure". There is a European Commission guideline document on effective compounds that protects plant produce. Agricultural plant materials include any compounds identified in plants and obtained by compressing, digesting, granulating, and/or eliminating plants or sections of plants of the same type. To be seen as a bio-stimulant, a substance has to appear to alter the plant structure, making it more systematic at using limited water and nutrients resources, or protecting it from harmful factors, such as reactive oxygen produced by environmental stress, pests, or microbes.

REFERENCES

Adnan M, Basir A, Arif M, Shah SRA, Khan M, Jamal Y (2015) Impact of grazing on wheat yield and associated weeds. *Pak J Weed Sci Res* 21(3):351–358.

Adnan M, Fahad S, Khan IA, Saeed M, Ihsan MZ, Saud S, Riaz M, Wang D, Wu C (2019) Integration of poultry manure and phosphate solubilizing bacteria improved availability of Ca bound P in calcareous soils. *3 Biotech* 9(10):368.

Adnan M, Fahad S, Muhammad Z, Shahen S, Ishaq AM, Subhan D, Zafar-ul-Hye M, Martin LB, Raja MMN, Beena S, Saud S, Imran A, Zhen Y, Martin B, Jiri H, Rahul D (2020) Coupling phosphate-solubilizing bacteria with phosphorus supplements improve maize phosphorus acquisition and growth under lime induced salinity stress. *Plants* 9(7):900. doi: 10.3390/plants9070900

Adnan M, Zahir S, Fahad S, Arif M, Mukhtar A, Imtiaz AK, Ishaq AM, Abdul B, Hidayat U, Muhammad A, Inayat-Ur R, Saud S, Muhammad ZI, Yousaf J, Amanullah, Hafiz MH, Wajid N (2018a) Phosphate-solubilizing bacteria nullify the antagonistic effect of soil calcification on bioavailability of phosphorus in alkaline soils. *Sci Rep* 8:4339. https://doi.org/10.1038/s41598-018-22653-7

Adnan M, Shah Z, Sharif M, Rahman H (2018b) Liming induces carbon dioxide (CO_2) emission in PSB inoculated alkaline soil supplemented with different phosphorus sources. *Environ Sci Poll Res.* 25(10):9501–9509.

Ahmad I, Pichtel J, Hayat S (2008) *Plant-Bacteria Interactions: Strategies and Techniques to Promote Plant Growth.* WILEY-VCH Verlag GmbH and Co., KGaA, Weinheim.

Ahmad K, Aslam M, Saleem MH, Ijaz M, Ul-Allah SAMI, El-Sheikh AHM, ... Ali S (2021) Genetic diversity and characterization of salt stress tolerance traits in maize (Zea mays L.) under normal and saline conditions. *Pak J Bot*, 54:3.

Akhtar N, Khan S, Malook I, Rehman SU, Jamil M (2017) Pb-induced changes in roots of two cultivated rice cultivars grown in lead-contaminated soil mediated by smoke. *Environ Sci Poll Res* 24(26):21298–21310.

Ali M, Ullah Z, Mian IA, Khan M, Khan N, Adnan M, Saeed M (2016) Response of maize to nitrogen levels and seed priming. *Pure Appl Biol* 5(3):578–587.

Aslam MM, Jamil M, Khatoon A, Hendawy S, Al-Suhaibani NA, Malook I, Rehman S (2017) Physiological and biochemical responses of maize (*Zea mays* L.) to plant derived smoke solution. *Pak J Bot* 49(2):435–443.

Atif B, Hesham A, Fahad S (2021) Biochar coupling with phosphorus fertilization modifies antioxidant activity, osmolyte accumulation and reactive oxygen species synthesis in the leaves and xylem sap of rice cultivars under high-temperature stress. *Physiol Mol Biol Plants* 27(9):2083–2100. https://doi.org/10.1007/s12298-021-01062-7

Attia MS, El-Sayyad GS, Abd Elkodous M, Khalil WF, Nofel MM, Abdelaziz AM, Farghali AA, El-Batal AI, El Rouby WM (2021) Chitosan and EDTA conjugated graphene oxide antinematodes in eggplant: toward improving plant immune response. *Int J Biolo Macromole* 179:333–344.

Audibert L, Fauchon M, Blanc N, Hauchard D, Ar Gall E (2010) Phenolic compounds in the brown seaweed Ascophyllum nodosum: distribution and radical-scavenging activities. *Phytochem Analy* 21(5):399–405.

Augé RM, Toler HD, Saxton AM (2015) Arbuscular mycorrhizal symbiosis alters stomatal conductance of host plants more under drought than under amply watered conditions: A meta-analysis. *Mycorrhiza* 25(1):13–24.

Bajpai S, Shukla PS, Asiedu S, Pruski K, Prithiviraj B (2019) A biostimulant preparation of brown seaweed Ascophyllum nodosum suppresses powdery mildew of strawberry. *The Plant Pathol J* 35(5):406.

Basak A (2008) Effect of preharvest treatment with seaweed products, Kelpak® and Goëmar BM 86®, on fruit quality in apple. *Int J Fruit Sci* 8(1–2):1–14.

Battacharyya D, Babgohari MZ, Rathor P, Prithiviraj B (2015) Seaweed extracts as biostimulants in horticulture. *Sci Hort* 196:39–48.

Behie SW, Bidochka MJ (2014) Nutrient transfer in plant–fungal symbioses. *Trends Plant Sci* 19(11):734–740.

Bellich B, D'Agostino I, Semeraro S, Gamini A, Cesàro A (2016) "The good, the bad and the ugly" of chitosans. *Mar Drugs* 14(5):99.

Berendsen RL, Pieterse CM, Bakker PA (2012) The rhizosphere microbiome and plant health. *Trends Plant Sci* 17(8):478–486.

Bharti N, Barnawal D (2019) Amelioration of salinity stress by PGPR: ACC deaminase and ROS scavenging enzymes activity PGPR amelioration in sustainable agriculture. *Food Security and Environmental Management,* Woodhead Publishing, United Kingdom. Elsevier. pp. 85–106.

Bhattacharyya PN, Jha DK (2012) Plant growth-promoting rhizobacteria (PGPR): emergence in agriculture. *World J Microb Biotech* 28(4):1327–1350.

Bittelli M, Flury M, Campbell GS, Nichols EJ (2001) Reduction of transpiration through foliar application of chitosan. *Agri Forest Meteorol* 107(3):167–175.

Bonfante P, Genre A (2010) Mechanisms underlying beneficial plant–fungus interactions in mycorrhizal symbiosis. *Nat Comm* 1(1):1–11.

Bronick CJ, Lal R (2005) Soil structure and management: a review. *Geoderma* 124(1–2):3–22.

Bulgari R, Cocetta G, Trivellini A, Vernieri P, Ferrante A (2015) Biostimulants and crop responses: a review. *Biol Agric Hort* 31(1):1–17.

Cakmakci R, Dönmez MF, Erdoğan Ü (2007) The effect of plant growth promoting rhizobacteria on barley seedling growth, nutrient uptake, some soil properties, and bacterial counts. *Turk J Agric Fores* 31(3): 189–199.

Calvo P, Nelson L, Kloepper JW (2014) Agricultural uses of plant biostimulants. *Plant Soil* 383(1):3–41.

Canellas L, Teixeira Junior L, Dobbss L, Silva C, Medici L, Zandonadi D, Façanha A (2008) Humic acids crossinteractions with root and organic acids. *Ann App Biol* 153(2):157–166.

Canellas LP, Olivares FL, Okorokova-Façanha AL, Façanha AR (2002) Humic acids isolated from earthworm compost enhance root elongation, lateral root emergence, and plasma membrane H+-ATPase activity in maize roots. *Plant Physiol* 130(4):1951–1957.

Canellas LP, Olivares FL (2014) Physiological responses to humic substances as plant growth promoter. *Chem Biol Technol Agric* 1(1):1–11.

Canellas LP, Olivares FL, Aguiar NO, Jones DL, Nebbioso A, Mazzei P, Piccolo A (2015) Humic and fulvic acids as biostimulants in horticulture. *Sci Hort* 196:15–27.

Chakraborty M, Hasanuzzaman M, Rahman M, Khan M, Rahman A, Bhowmik P, Mahmud NU, Tanveer M, Islam T (2020) Mechanism of plant growth promotion and disease suppression by chitosan biopolymer. *Agric* 10(12):624.

Chandanie W, Kubota M, Hyakumachi M (2009) Interactions between the arbuscular mycorrhizal fungus Glomus mosseae and plant growth-promoting fungi and their significance for enhancing plant growth and suppressing damping-off of cucumber (*Cucumis sativus* L.). *App Soil Ecol* 41(3):336–341.

Chiaiese P, Corrado G, Colla G, Kyriacou MC, Rouphael Y (2018) Renewable sources of plant biostimulation: microalgae as a sustainable means to improve crop performance. *Front Plant Sci* 9:1782.

Chitarra W, Pagliarani C, Maserti B, Lumini E, Siciliano I, Cascone P, Schubert A, Gambino G, Balestrini R, Guerrieri E (2016) Insights on the impact of arbuscular mycorrhizal symbiosis on tomato tolerance to water stress. *Plant Physiol* 171(2):1009–1023.

Colla G, Nardi S, Cardarelli M, Ertani A, Lucini L, Canaguier R, Rouphael Y (2015a) Protein hydrolysates as biostimulants in horticulture. *Sci Hort* 196:28–38.

Colla G, Rouphael Y, Di Mattia E, El-Nakhel C, Cardarelli M (2015b) Co-inoculation of Glomus intraradices and Trichoderma atroviride acts as a biostimulant to promote growth, yield and nutrient uptake of vegetable crops. *J Sci Foo Agric* 95(8):1706–1715.

Colla G, Hoagland L, Ruzzi M, Cardarelli M, Bonini P, Canaguier R, Rouphael Y (2017) Biostimulant action of protein hydrolysates: unraveling their effects on plant physiology and microbiome. *Front Plant Sci* 8:2202.

Colla G, Rouphael Y (2020) *Microalgae: New Source of Plant Biostimulants*. vol 10. Multidisciplinary Digital Publishing Institute, p. 1240.

D'Addabbo T, Laquale S, Perniola M, Candido V (2019) Biostimulants for plant growth promotion and sustainable management of phytoparasitic nematodes in vegetable crops. *Agronomy* 9(10):616.

Daws MI, Davies J, Pritchard HW, Brown NA, Van Staden J (2007) Butenolide from plant-derived smoke enhances germination and seedling growth of arable weed species. *Plant Growth Reg* 51(1):73–82.

De Lange J, Boucher C (1990) Autecological studies on Audouinia capitata (Bruniaceae). I. Plant-derived smoke as a seed germination cue. *South Afri J Bot* 56(6):700–703.

De Pascale S, Rouphael Y, Colla G (2017) Plant biostimulants: Innovative tool for enhancing plant nutrition in organic farming. *Eur J Hortic Sci* 82(6):277–285.

Delgado A, Madrid A, Kassem S, Andreu L, del Carmen del Campillo M (2002) Phosphorus fertilizer recovery from calcareous soils amended with humic and fulvic acids. *Plant Soil* 245(2):277–286.

Deliopoulos T, Kettlewell PS, Hare MC (2010) Fungal disease suppression by inorganic salts: a review. *Crop Protect* 29(10):1059–1075.

Dhall R (2013) Advances in edible coatings for fresh fruits and vegetables: a review. *Critical Rev Food Sci Nutrit* 53(5):435–450.

Di Stasio E, Rouphael Y, Colla G, Raimondi G, Giordano M, Pannico A, El-Nakhel C, De Pascale S (2017) The influence of Ecklonia maxima seaweed extract on growth, photosynthetic activity and mineral composition of Brassica rapa L. subsp sylvestris under nutrient stress conditions. *Euro J Hort Sci* 82(6):286–293.

Divya K, Vijayan S, Nair SJ, Jisha M (2019) Optimization of chitosan nanoparticle synthesis and its potential application as germination elicitor of *Oryza sativa* L. *Int J. Biol Macrom* 124:1053–1059.

du Jardin P (2012) The Science of Plant Biostimulants—A bibliographic analysis, Ad hoc study report. European Commission DG ENTR. 546, 20.23.

Du Jardin P (2015) Plant biostimulants: definition, concept, main categories and regulation. *Scientia Hort* 196:3–14.

Eckardt NA (2008) Chitin signaling in plants: insights into the perception of fungal pathogens and rhizobacterial symbionts. *Plant Cell* 20: 241–243.

El Amerany F, Rhazi M, Wahbi S, Taourirte M, Meddich A (2020) The effect of chitosan, arbuscular mycorrhizal fungi, and compost applied individually or in combination on growth, nutrient uptake, and stem anatomy of tomato. *Scientia Hort* 261:109015.

El Arroussi H, Benhima R, Elbaouchi A, Sijilmassi B, El Mernissi N, Aafsar A, Meftah-Kadmiri I, Bendaou N, Smouni A (2018) Dunaliella salina exopolysaccharides: a promising biostimulant for salt stress tolerance in tomato (*Solanum lycopersicum*). *J App Phycol* 30(5):2929–2941.

El Komy MH, Saleh AA, Eranthodi A, Molan YY (2015) Characterization of novel Trichoderma asperellum isolates to select effective biocontrol agents against tomato Fusarium wilt. *The Plant Pathol J* 31(1):50.

Ertani A, Cavani L, Pizzeghello D, Brandellero E, Altissimo A, Ciavatta C, Nardi S (2009) Biostimulant activity of two protein hydrolyzates in the growth and nitrogen metabolism of maize seedlings. *J Plant Nutri Soil Sci* 172(2):237–244.

European Biostimulants Industry Council Eabibhwbe, Ertani A, Pizzeghello D, Francioso O, Sambo P, Sanchez-Cortes S, Nardi SL (2012). EBIC and biostimulants in brief. www.biostimulants.eu

Fahad S, Hussain S, Saud S, Tanveer M, Bajwa AA, Hassan S, Shah AN, Ullah A,Wu C, Khan FA, Shah F, Ullah S, Chen Y, Huang J (2015a) A biochar application protects rice pollen from high-temperature stress. *Plant Physiol Biochem* 96:281–287.

Fahad S, Nie L, Chen Y, Wu C, Xiong D, Saud S, Hongyan L, Cui K, Huang J (2015b) Crop plant hormones and environmental stress. *Sustain Agric Rev* 15:371–400.

Fahad S, Hussain S, Saud S, Hassan S, Tanveer M, Ihsan MZ, Shah AN, Ullah A, Nasrullah KF, Ullah S, AlharbyH NW, Wu C, Huang J (2016) A combined application of biochar and phosphorus alleviates heat-induced adversities on physiological, agronomical and quality attributes of rice. *Plant Physiol Biochem* 103:191–198.

Fahad S, Hasanuzzaman M, Alam M, Ullah H, Saeed M, Ali Khan I, Adnan M (Eds.) (2020) *Environment, Climate, Plant and Vegetation Growth.* Springer Nature Switzerland AG 2020. doi: https://doi.org/10.1007/978-3-030-49732-3

Fahad S, Sönmez O, Saud S, Wang D, Wu C, Adnan M, Turan V (Eds.) (2021a) Plant growth regulators for climate-smart agriculture. First edition. *Footprints of Climate Variability on Plant Diversity.* CRC Press, Boca Raton.

Fahad S, Sonmez O, Saud S, Wang D, Wu C, Adnan M, Turan V (Eds.) (2021b) Climate change and plants: biodiversity, growth and interactions. First edition. *Footprints of Climate Variability on Plant Diversity.* CRC Press, Boca Raton.

Fahad S, Sonmez O, Saud S, Wang D, Wu C, Adnan M, Turan V (Eds.) (2021c) Developing climate resilient crops: improving global food security and safety. First edition. *Footprints of Climate Variability on Plant Diversity.* CRC Press, Boca Raton.

Fahad S, Sönmez O, Turan V, Adnan M, Saud S, Wu C, Wang D (Eds.) (2021d) Sustainable soil and land management and climate change. First edition. *Footprints of Climate Variability on Plant Diversity.* CRC Press, Boca Raton.

Fahad S, Sönmez O, Saud S, Wang D, Wu C, Adnan M, Arif M, Amanullah (Eds.) (2021e) Engineering tolerance in crop plants against abiotic stress. First edition. *Footprints of Climate Variability on Plant Diversity.* CRC Press, Boca Raton.

Fahad S, Saud S, Yajun C, Chao W, Depeng W (Eds.) (2021f) Abiotic stress in plants. *IntechOpen United Kingdom* 2021. http://dx.doi.org/10.5772/intechopen.91549

Fahad S, Adnan M, Saud S (Eds.) (2022a) Improvement of plant production in the era of climate change. First edition. *Footprints of Climate Variability on Plant Diversity.* CRC Press, Boca Raton.

Fahad S, Adnan, M, Saud S, Nie L (Eds.) (2022b) Climate change and ecosystems: challenges to sustainable development. First edition. *Footprints of Climate Variability on Plant Diversity.* CRC Press, Boca Raton.

Fakhre A, Ayub K, Fahad S, Sarfraz N, Niaz A, Muhammad AA, Muhammad A, Khadim D, Saud S, Shah H, Muhammad ASR, Khalid N, Muhammad A, Rahul D, Subhan D (2021) Phosphate solubilizing bacteria optimize wheat yield in mineral phosphorus applied alkaline soil. *J Saudi Soc Agric Sci* 21(5):339–348. https://doi.org/10.1016/j.jssas.2021.10.007

Fiorentino N, Ventorino V, Woo SL, Pepe O, De Rosa A, Gioia L, Romano I, Lombardi N, Napolitano M, Colla G (2018) Trichoderma-based biostimulants modulate rhizosphere microbial populations and improve N uptake efficiency, yield, and nutritional quality of leafy vegetables. *Fronti Plant Sci* 9:743.

Flematti GR, Ghisalberti EL, Dixon KW, Trengove RD (2004) A compound from smoke that promotes seed germination. *Science* 305(5686):977.

Gómez-Merino FC, Trejo-Téllez LI (2015) Biostimulant activity of phosphite in horticulture. *Sci Hort* 196:82–90.

Graber E, Rudich Y (2006) Atmospheric HULIS: How humic-like are they? A comprehensive and critical review. *Atm Chemi Phys* 6(3):729–753.

Gupta S, Hrdlička J, Ngoroyemoto N, Nemahunguni NK, Gucký T, Novák O, Kulkarni MG, Doležal K, Van Staden J (2020) Preparation and standardisation of smoke-water for seed germination and plant growth stimulation. *J Plant Growth Reg* 39(1):338–345.

Gupta S, Plačková L, Kulkarni MG, Doležal K, Van Staden J (2019) Role of smoke stimulatory and inhibitory biomolecules in phytochrome-regulated seed germination of *Lactuca sativa*. *Plant Physiol* 181(2):458–470.

Gyaneshwar P, Naresh Kumar G, Parekh L, Poole P (2002) Role of soil microorganisms in improving P nutrition of plants. *Plant Soil* 245(1):83–93.

Hadwiger LA (2013) Multiple effects of chitosan on plant systems: solid science or hype. *Plant Sci* 208:42–49.

Halpern M, Bar-Tal A, Ofek M, Minz D, Muller T, Yermiyahu U (2015) The use of biostimulants for enhancing nutrient uptake. *Adv Agro* 130:141–174.

Harfoush E, Abdel-Razzek A, El-Adgham F, El-Sharkawy A (2017) Effects of humic acid and chitosan under different levels of nitrogen and potassium fertilizers on growth and yield potential of potato plants (*Solanum tuberosum*, L.). *Alex J Agri Sci* 62(1):135–148.

Holley RA, Patel D (2005) Improvement in shelf-life and safety of perishable foods by plant essential oils and smoke antimicrobials. *Food Microb* 22(4):273–292.

Inderbitzin P, Ward J, Barbella A, Solares N, Izyumin D, Burman P, Chellemi DO, Subbarao KV (2018) Soil microbiomes associated with Verticillium wilt-suppressive broccoli and chitin amendments are enriched with potential biocontrol agents. *Phytopathology* 108(1):31–43.

Jannin L, Arkoun M, Ourry A, Laîné P, Goux D, Garnica M, Fuentes M, Francisco SS, Baigorri R, Cruz F (2012) Microarray analysis of humic acid effects on Brassica napus growth: involvement of N, C and S metabolisms. *Plant and Soil* 359(1):297–319.

Jayaraj J, Wan A, Rahman M, Punja Z (2008) Seaweed extract reduces foliar fungal diseases on carrot. *Crop Protect* 27(10):1360–1366.

Johnson NC, Graham JH (2013) The continuum concept remains a useful framework for studying mycorrhizal functioning. *Plant Soil* 363(1):411–419.

Kamel H (2014) Impact of garlic oil, seaweed extract and imazalil on keeping quality of Valencia orange fruits during cold storage. *J Hortic Sci Ornam Plants* 6:116–125.

Katiyar D, Hemantaranjan A, Singh B (2015) Chitosan as a promising natural compound to enhance potential physiological responses in plant: a review. *Ind J Plant Physiol* 20(1):1–9.

Kauffman GL, Kneivel DP, Watschke TL (2007) Effects of a biostimulant on the heat tolerance associated with photosynthetic capacity, membrane thermostability, and polyphenol production of perennial ryegrass. *Crop Sci* 47(1):261–267.

Khan MA, Basir A, Adnan M, Shah AS, Noor M, Khan A, Shah JA, Ali Z, Rehman A (2017) Wheat phenology and density and fresh and dry weight of weeds as affected by potassium sources levels and tillage practices. *Pak J Weed Sci Res* 23(4):451–462.

Khan N, Shah Z, Adnan M, Ali M, Khan B, Mian IA, Ali A, Zahoor M, Roman M, Ullah L, Khaliq A, Khan WA, Alam A (2016) Evaluation of soil for important properties and chromium concentration in the basin of Chromite Hills in Lower Malakand. *Adv Environ Biol* 10(7):141–147.

Khan W, Hiltz D, Critchley AT, Prithiviraj B (2011) Bioassay to detect *Ascophyllum nodosum* extract-induced cytokinin-like activity in *Arabidopsis thaliana*. *J App Phycol* 23(3):409–414.

Khristeva L (1957) Stimulating activity of humic acid towards higher plants and the nature of this phenomenon. In: *Humic Fertilizers. Theory and Practice of Their Application*. Kharkiv, Gorky State University. pp. 75–93.

Krüger M, Krüger C, Walker C, Stockinger H, Schüßler A (2012) Phylogenetic reference data for systematics and phylotaxonomy of arbuscular mycorrhizal fungi from phylum to species level. *New Phytol* 193(4):970–984.

Kulkarni M, Light M, Van Staden J (2011) Plant-derived smoke: old technology with possibilities for economic applications in agriculture and horticulture. *South Afr J Bot* 77(4):972–979.

Le Mire G, Nguyen M, Fassotte B, du Jardin P, Verheggen F, Delaplace P, Jijakli H (2016) Implementing biostimulants and biocontrol strategies in the agroecological management of cultivated ecosystems. *Biotechnol Agro Société Environ* 1381 (20): 299–313.

Li R, He J, Xie H, Wang W, Bose SK, Sun Y, Hu J, Yin H (2019) Effects of chitosan nanoparticles on seed germination and seedling growth of wheat (*Triticum aestivum* L.). *Int J. Biol Macromo* 126:91–100.

Li Y, Chen XG, Liu N, Liu CS, Liu CG, Meng XH, Kenendy JF (2007) Physicochemical characterization and antibacterial property of chitosan acetates. *Carboh Poly* 67(2):227–232.

Liang R, Li X, Yuan W, Jin S, Hou S, Wang M, Wang H (2018) Antifungal activity of nanochitin whisker against crown rot diseases of wheat. *J Agric Food Chem* 66(38):9907–9913.

Light M, Gardner M, Jäger A, Van Staden J (2002) Dual regulation of seed germination by smoke solutions. *Plant Growth Reg* 37(2):135–141.

Lin F-W, Lin K-H, Wu C-W, Chang Y-S (2020) Effects of betaine and chitin on water use efficiency in lettuce (lactuca sativa var. capitata). *Hort Sci* 55(1):89–95.

Lisiecka J, Knaflewski M, Spizewski T, Fraszczak B, Kaluzewicz A, Krzesinski W (2011) The effect of animal protein hydrolysate on quantity and quality of strawberry daughter plants cv.'Elsanta'. *Acta Sci Pol Hortorum Cultus* 10(1):31–40.

Liu Y, Wisniewski M, Kennedy JF, Jiang Y, Tang J, Liu J (2016) Chitosan and oligochitosan enhance ginger (Zingiber officinale Roscoe) resistance to rhizome rot caused by Fusarium oxysporum in storage. *Carbohydr Polym* 151:474–479. doi: 10.1016/j.carbpol.2016.05.103.

Loganathan M, Sible G, Maruthasalam S, Saravanakumar D, Raguchander T, Sivakumar M, Samiyappan R (2010) Trichoderma and chitin mixture based bioformulation for the management of head rot (Sclerotinia sclerotiorum (Lib.) deBary)–root-knot (Meloidogyne incognita Kofoid and White; Chitwood) complex diseases of cabbage. *Arch Phytop Plant Pro* 43(10):1011–1024.

López-Bucio J, Pelagio-Flores R, Herrera-Estrella A (2015) Trichoderma as biostimulant: exploiting the multilevel properties of a plant beneficial fungus. *Sci Hort* 196:109–123.

Lorito M, Woo SL (2015) Discussion agronomic. In: Lugtenberg B (Ed.). *Principles of Plant-Microbe 1400 Interactions*, Berlin, Springer International Publishing. pp. 345–353.

Market, Chitin (2017) Chitin Market: Agrochemical End Use Industry Segment Inclined Towards High Growth—Moderate Value during the Forecast Period: Global Industry Analysis (2012–2016) and Opportunity Assessment (2017–2027).

Masondo NA, Kulkarni MG, Finnie JF, Van Staden J (2018) Influence of biostimulants-seed-priming on Ceratotheca triloba germination and seedling growth under low temperatures, low osmotic potential and salinity stress. *Ecotoxicol Environ Safety* 147:43–48.

Mondal MM, Malek M, Puteh A, Ismail M, Ashrafuzzaman M, Naher L (2012) Effect of foliar application of chitosan on growth and yield in okra. *Aus J Crop Sci* 6(5):918–921.

Mondal M, Puteh A, Dafader N, Rafii M, Malek M (2013) Foliar application of chitosan improves growth and yield in maize. *J Food Agric Environ* 11(2):520–523.

Mondal M, Puteh AB, Dafader NC (2016) Foliar application of chitosan improved morphophysiological attributes and yield in summer tomato (*Solanum lycopersicum*). *Pak J Agric Sci* 53:339–344. 53(2).

Muhammad I, Khadim D, Fahad S, Imran M, Saud A, Manzer HS, Shah S, Jabar ZKK, Shamsher A, Shah H, Taufiq N, Hafiz MH, Jan B, Wajid N (2022) Exploring the potential effect of *Achnatherum splendens* L.–derived biochar treated with phosphoric acid on bioavailability of cadmium and wheat growth in contaminated soil. *Environ Sci Pollut Res* 29(25), 37676–37684. https://doi.org/10.1007/s11356-021-17950-0.

Nardi S, Carletti P, Pizzeghello D, Muscolo A (2009) Biophysico-chemical processes involving natural nonliving organic matter in environmental systems. Vol 2, part 1: fundamentals and impact of mineral-organic biota interactions on the formation, transformation, turnover, and storage of natural nonliving organic matter (NOM). In: Senesi N, Xing B, Huang PM (Eds.). *Biological Activities of Humic Substances*, Hoboken, Wiley, pp. 305–339.

Neily W, Shishkov L, Nickerson S, Titus D, Norrie J (2010) Commercial extract from the brown seaweed Ascophyllum nodosum (Acadian (R)) improves early establishment and helps resist water stress in vegetable and flower seedlings. *HortScience* 45:S105–S106.

Nelson WR, Van Staden J (1985) 1-Aminocyclopropane-l-carboxylic acid in seaweed concentrate. *Bot Mar* 28:415–417.

Niculescu M, Bajenaru S, Gaidau C, Simion D, Filipescu L (2009) Extraction of the protein components as amino-acids hydrolysates from chrome leather wastes through hydrolytic processes. *Rev Chim* 60:1070–1078.

Novo LA, Castro PM, Alvarenga P, da Silva EF (2018) Plant growth-promoting rhizobacteria-assisted phytoremediation of mine soils. In: Prasad MNV, de Campos Favas PJ, Maiti SK (Eds.) *Bio-Geotechnologies for Mine Site Rehabilitation*, Elsevier Inc., Amsterdam, pp. 281–295.

Ozbay N, Demirkiran A (2019) Enhancement of growth in ornamental pepper (*Capsicum annuum* L.) Plants with application of a commercial seaweed product, stimplex®. *Appl Ecol Environ Res* 17:4361–4375.

Paksoy M, Türkmen Ö, Dursun A (2010) Effects of potassium and humic acid on emergence, growth and nutrient contents of okra (*Abelmoschus esculentus* L.) seedling under saline soil conditions. *Afr J Biotech* 9(33):5343–5346.

Pandey P, Verma MK, De N (2018) Chitosan in agricultural context—A review. *Bull Environ Pharmacol Life Sci* 7:87–96.

Parchiković N, Teklić T, Zeljković S, Lisjak M, Špoljarević M (2019) Biostimulants research in some horticultural plant species—A review. *Food and Ener Se* 8(2):e00162.

Pasupuleti VK, Braun S (2008) State of the art manufacturing of protein hydrolysates. In: Pasupuleti V, Demain A (Eds.) *Protein Hydrolysates in Biotechnology*. Springer, Dordrecht, pp. 11–32. doi: 10.1007/978-1-4020-6674-0_2

Patel JS, Selvaraj V, Gunupuru LR, Rathor PK, Prithiviraj B (2020) Combined application of Ascophyllum nodosum extract and chitosan synergistically activates host-defense of peas against powdery mildew. *BMC Plant Biol* 20(1):1–10.

Patel K, Agarwal P, Agarwal PK (2018) Kappaphycus alvarezii sap mitigates abiotic-induced stress in Triticum durum by modulating metabolic coordination and improves growth and yield. *J App Phycol* 30(4):2659–2673.

Paul K, Sorrentino M, Lucini L, Rouphael Y, Cardarelli M, Bonini P, Reynaud H, Canaguier R, Trtílek M, Panzarová K (2019) Understanding the biostimulant action of vegetal-derived protein hydrolysates by high-throughput plant phenotyping and metabolomics: A case study on tomato. *Front Plant Sci* 10:47.

Paungfoo-Lonhienne C, Redding M, Pratt C, Wang W (2019) Plant growth promoting rhizobacteria increase the efficiency of fertilisers while reducing nitrogen loss. *J Environ Manag* 233:337–341.

Philibert T, Lee BH, Fabien N (2017) Current status and new perspectives on chitin and chitosan as functional biopolymers. *App Biochem Biotech* 181(4):1314–1337.

Piccolo A (2001) The supramolecular structure of humic substances. *Soil Science* 166(11):810–832.

Pichyangkura R, Chadchawan S (2015) Biostimulant activity of chitosan in horticulture. *Sci Hort* 196:49–65.

Pilon-Smits EA, Quinn CF, Tapken W, Malagoli M, Schiavon M (2009) Physiological functions of beneficial elements. *Curr Opinion Plant Biol* 12(3):267–274.

Pirbalouti AG, Malekpoor F, Salimi A, Golparvar A (2017) Exogenous application of chitosan on biochemical and physiological characteristics, phenolic content and antioxidant activity of two species of basil (*Ocimum ciliatum* and *Ocimum basilicum*) under reduced irrigation. *Sci hort* 217:114–122.

Priya H, Prasanna R, Ramakrishnan B, Bidyarani N, Babu S, Thapa S, Renuka N (2015) Influence of cyanobacterial inoculation on the culturable microbiome and growth of rice. *Microbiol Res* 171:7–889.

Pusztahelyi T (2018) Chitin and chitin-related compounds in plant–fungal interactions. *Mycology* 9(3):189–201.

Rahman K, Zhang D (2018) Effects of fertilizer broadcasting on the excessive use of inorganic fertilizers and environmental sustainability. *Sustainability* 10(3):759.

Rajkumar M, Lee K, Freitas H (2008) Effects of chitin and salicylic acid on biological control activity of Pseudomonas spp. against damping off of pepper. *South Afr J of Bot* 74(2):268–273.

Ramírez MA, Rodríguez AT, Alfonso L, Peniche C (2010) Chitin and its derivatives as biopolymers with potential agricultural applications. *Biotecn Apl* 27(4):270–276.

Ramos AC, Dobbss LB, Santos LA, Fernandes MS, Olivares FL, Aguiar NO, Canellas LP (2015) Humic matter elicits proton and calcium fluxes and signaling dependent on Ca^{2+}-dependent protein kinase (CDPK) at early stages of lateral plant root development. *Chem Biol Tech Agric* 2(1):1–12.

Reed R, Davison I, Chudek J, Foster R (1985) The osmotic role of mannitol in the Phaeophyta: an appraisal. *Phycologia* 24(1):35–47.

Renaut S, Masse J, Norrie JP, Blal B, Hijri M (2019) A commercial seaweed extract structured microbial communities associated with tomato and pepper roots and significantly increased crop yield. *Microbial Biotech* 12(6):1346–1358.

Rengasamy KR, Kulkarni MG, Stirk WA, Van Staden J (2015) Eckol-a new plant growth stimulant from the brown seaweed Ecklonia maxima. *J App phycol* 27(1):581–587.

Renuka N, Guldhe A, Prasanna R, Singh P, Bux F (2018) Microalgae as multi-functional options in modern agriculture: current trends, prospects and challenges. *Biotec Adv* 36(4):1255–1273.

Ronga D, Biazzi E, Parati K, Carminati D, Carminati E, Tava A (2019) Microalgal biostimulants and biofertilisers in crop productions. *Agronomy* 9(4):192.

Rose MT, Patti AF, Little KR, Brown AL, Jackson WR, Cavagnaro TR (2014) A meta-analysis and review of plant-growth response to humic substances: practical implications for agriculture. *Adv Agro* 124:37–89.
Rouphael Y, Franken P, Schneider C, Schwarz D, Giovannetti M, Agnolucci M, De Pascale S, Bonini P, Colla G (2015) Arbuscular mycorrhizal fungi act as biostimulants in horticultural crops. *Sci Hort* 196:91–108.
Rouphael Y, Cardarelli M, Bonini P, Colla G (2017) Synergistic action of a microbial-based biostimulant and a plant derived-protein hydrolysate enhances lettuce tolerance to alkalinity and salinity. *Front Plant Sci* 8:131.
Rouphael Y, Colla G (2018) Synergistic biostimulatory action: designing the next generation of plant biostimulants for sustainable agriculture. *Front Plant Sci* 9:1655.
Rouphael Y, Kyriacou MC, Petropoulos SA, De Pascale S, Colla G (2018) Improving vegetable quality in controlled environments. *Sci Hort* 234:275–289.
Rudresh D, Shivaprakash M, Prasad R (2005) Effect of combined application of Rhizobium, phosphate solubilizing bacterium and Trichoderma spp. on growth, nutrient uptake and yield of chickpea (*Cicer aritenium* L.). *App Soil Eco* 28(2):139–146.
Russo RO, Berlyn GP (1991) The use of organic biostimulants to help low input sustainable agriculture. *J Sustainable Agric* 1(2):19–42.
Ruzzi M, Aroca R (2015) Plant growth-promoting rhizobacteria act as biostimulants in horticulture. *Sci Hort* 196:124–134.
Salma I, Mészáros T, Maenhaut W, Vass E, Majer Z (2010) Chirality and the origin of atmospheric humic-like substances. *Atm Chem Ph* 10(3):1315–1327.
Savvas D, Ntatsi G (2015) Biostimulant activity of silicon in horticulture. *Sci Hort* 196:66–81.
Schaafsma G (2009) Safety of protein hydrolysates, fractions thereof and bioactive peptides in human nutrition. *Eur J Cli Nutrit* 63(10):1161–1168.
Shafi MI, Adnan M, Fahad S, Fazli W, Ahsan K, Zhen Y, Subhan D, Zafar-ul-Hye M, Martin B, Rahul D (2020) Application of single superphosphate with humic acid improves the growth, yield and phosphorus uptake of wheat (Triticum aestivum L.) in calcareous soil. *Agron* (10):1224. doi:10.3390/agronomy10091224
Simorangkir D (2007) Fire use: is it really the cheaper land preparation method for large-scale plantations? *Mitig Ada Strat Global Change* 12(1):147–164.
Smith S, Read D (2008) *Mycorrhizal Symbiosis*. Third Edition. London. Academic Press.
Stirk WA, Van Staden J (2014) Plant growth regulators in seaweeds: occurrence, regulation and functions. *Adv Bot Res* 71:125–159
Sun C, Fu D, Jin L, Chen M, Zheng X, Yu T (2018) Chitin isolated from yeast cell wall induces the resistance of tomato fruit to Botrytis cinerea. *Carbohydr Polym* 199:341–352.
Toscano S, Romano D, Massa D, Bulgari R, Franzoni G, Ferrante A (2018) Biostimulant applications in low input horticultural cultivation systems. *Italus Hortus* 25(2):27–36. doi: 10.26353/j.itahort/2018.1.2736
Trevisan S, Francioso O, Quaggiotti S, Nardi S (2010) Humic substances biological activity at the plant–soil interface: from environmental aspects to molecular factors. *Plant Sign Beh* 5(6):635–643.
Tsouvaltzis P, Koukounaras A, Siomos AS (2014) Application of amino acids improves lettuce crop uniformity and inhibits nitrate accumulation induced by the supplemental inorganic nitrogen fertilization. *Int J Agricul Biol* 16(5):951–955.
Türkmen Ö, Dursun A, Turan M, Erdinç Ç (2004) Calcium and humic acid affect seed germination, growth, and nutrient content of tomato (*Lycopersicon esculentum* L.) seedlings under saline soil conditions. *Acta Agriculturae Scandinavica, Section B-Soil Plant Sci* 54(3):168–174.
Van Oosten MJ, Pepe O, De Pascale S, Silletti S, Maggio A (2017) The role of biostimulants and bioeffectors as alleviators of abiotic stress in crop plants. *Chem Biol Tech Agricult* 4(1):1–12.
Van Staden J, Jager A, Light M, Burger B (2004) Isolation of the major germination cue from plant-derived smoke. *South Afr J Bot* 70:654–659.
Vessey JK (2003) Plant growth promoting rhizobacteria as biofertilizers. *Plant and Soil* 255(2):571–586.
Wahid F, Fahad S, Subhan D, Adnan M, Zhen Y, Saud S, Manzer HS, Martin B, Tereza H, Rahul D (2020) Sustainable management with mycorrhizae and phosphate solubilizing bacteria for enhanced phosphorus uptake in calcareous soils. *Agri* 10(8), 334. doi:10.3390/agriculture10080334
Wan J, Zhang X, Stacey G (2008) Chitin signaling and plant disease resistance. *Plant Signal Behav* 3: 831–833.
Winkler AJ, Dominguez-Nuñez JA, Aranaz I, Poza-Carrión C, Ramonell K, Somerville S, Berrocal-Lobo M (2017) *Short-chain chitin oligomers. Prom Plant Growth Mar Drugs* 15(2):40.

Xiao R, Zheng Y (2016) Overview of microalgal extracellular polymeric substances (EPS) and their applications. *Biotech Ad* 34(7):1225–1244.

Xu L, Geelen D (2018) Developing biostimulants from agro-food and industrial by-products. *Front Plant Sci* 9:1567.

Yadav M, Goswami P, Paritosh K, Kumar M, Pareek N, Vivekanand V (2019) Seafood waste: a source for preparation of commercially employable chitin/chitosan materials. *Bior Biop* 6(1):1–20.

Yakhin OI, Lubyanov AA, Yakhin IA, Brown PH (2017) Biostimulants in plant science: a global perspective. *Front Plant Sci* 7:2049.

Yan G, Fan X, Peng M, Yin C, Xiao Z, Liang Y (2020) Silicon improves rice salinity resistance by alleviating ionic toxicity and osmotic constraint in an organ-specific pattern. *Front Plant Sci* 11:260.

Zahida Z, Hafiz FB, Zulfiqar AS, Ghulam MS, Fahad S, Muhammad RA, Hafiz MH,Wajid N, Muhammad S (2017) Effect of water management and silicon on germination, growth, phosphorus and arsenic uptake in rice. *Ecotoxicol Environ Saf* 144:11–18.

Zandonadi DB, Canellas LP, Façanha AR (2007) Indolacetic and humic acids induce lateral root development through a concerted plasmalemma and tonoplast H+ pumps activation. *Planta* 225(6):1583–1595.

Zeeshan M, Ahmad, Khan I, Shah B, Naeem A, Khan N, Ullah W, Adnan M, Shah SRA, Junaid K, Iqbal M (2016) Study on the management of *Ralstoniasolanacearum*(Smith) with spent mushroom compost. *J Ento & Zool Studies* 4(3):114–121.

Zhang X, Schmidt RE (1997) The impact of growth regulators on alpha-tocopherol status of water-stressed Poa pratensis L. *Int Turfgrass Soc Res J* 8:1364–1371.

Zhou Y, Jiang S, Jiao Y, Wang H (2017) Synergistic effects of nanochitin on inhibition of tobacco root rot disease. *Int J Biol Macro* 99:205–212.

5 Approaches for Using Bio-fertilizers as a Substitute for Chemical Fertilizers to Improve Soil Health and Crop Yields in Pakistan

Anas Iqbal,[1] Muhammad Izhar Shafi,[2] Mazhar Rafique,[3] Maid Zaman,[4] Izhar Ali,[1] Waqar-un-Nisa,[3] Ayesha Jabeen,[3] Sofia Asif,[5] Bushra Gul,[6] Muhammad Adnan,[7] Rafi Ullah,[7] Nazeer Ahmed,[7] Muhammad Haroon,[7] Muhammad Romman,[8] and Ligeng Jiang[1]*

[1] College of Agriculture, Guangxi University, Nanning, China
[2] Department of Soil and Environmental Sciences, Faculty of Crop Production Sciences, The University of Agriculture, Peshawar, Pakistan
[3] Department of Soil & Climate Sciences, The University of Haripur, Khyber Pakhtunkhwa, Pakistan
[4] Department of Entomology, The University of Haripur, Khyber Pakhtunkhwa, Pakistan
[5] Department of Food Science and Technology, The University of Haripur, Khyber Pakhtunkhwa, Pakistan
[6] Department of Biosciences, University of Wah, Punjab, Pakistan
[7] Department of Agriculture, University of Swabi, Khyber Pakhtunkhwa, Pakistan
[8] Department of Botnay, University of Chitral, Khyber Pakhtunkhwa, Pakistan
* Correspondence: Ligeng Jiang: jiang@gxu.edu.cn

CONTENTS

DOI: 10.1201/9781003286233-5

5.1 INTRODUCTION

Soils are one of the world's most significant natural resources, and protecting, maintaining, and improving them is crucial for the survival of life on Earth. Soil's fertility allows for the supply of critical chemical elements in the quantities and ratios required for the growth of plants (Itelima et al. 2018; Al-Zahrani et al. 2022; Rajesh et al. 2022; Anam et al. 2021; Deepranjan et al. 2021; Haider et al. 2021; Amjad et al. 2021; Sajjad et al. 2021a,b). It is critical for crop production, yet poor soils and runoff remain a management concern in many regions of the world. The main reason for this is that researchers and farmers commonly assess soil fertility using different theories and ambiguous literature findings (Yageta et al. 2019; Fakhre et al. 2021; Khatun et al. 2021; Ibrar et al. 2021; Bukhari et al. 2021; Haoliang et al. 2021; Sana et al. 2022; Abid et al. 2021; Zaman et al. 2021). As a result, understanding soil fertility is critical for enhanced soil production and appropriate land management strategies. Soil researchers have created numerous chemical, physical, and biological methods for measuring soil fertility, but the assessment is not confined to scientific measurements and is based on farmers' qualitative judgements (Ali et al. 2020; Iqbal et al. 2021a; Iqbal et al. 2020; Karlen et al. 2003; Sajjad et al. 2021a,b; Rehana et al. 2022; Yang et al. 2022; Ahmad et al. 2022; Shah et al. 2022; Muhammad et al. 2022; Wiqar et al. 2022; Farhat et al. 2022; Niaz et al. 2022).

Disparagement of the ineffectiveness of major technology implementations and scientific allocation of material by extension facilities has increased interest in the importance and incorporation of farmers' understanding (Berazneva et al. 2018; Guzman et al. 2018; Ihsan et al. 2022; Chao et al. 2022, Qin et al. 2022; Xue et al. 2022; Ali et al. 2022; Mehmood et al. 2022; El Sabagh et al. 2022; Ibad et al. 2022). Farmers apply their local soil skills to make day-to-day land managerial decisions by observing and evaluating them (Bado & Bationo 2018; Deepranjan et al. 2021; Haider et al. 2021; Huang Li et al. 2021; Ikram et al. 2021; Jabborova et al. 2021; Khadim et al. 2021a,b; Manzer et al. 2021; Muzammal et al. 2021; Abdul et al. 2021a,b; Ashfaq et al. 2021; Amjad et al.

2021; Atif et al. 2021). Incorporating indigenous data assists extension staff in matching their energies to native requirements and may result in increased uptake of co-produced technologies (Khan et al. 2016; Ingram et al. 2018). Farmers' assessments of soil health are widely reported as "regional" or "farmer's soil awareness" in much ethno-pedological research (Barrera-Bassols & Zinck 2003; Athar et al. 2021; Adnan et al. 2018a,b; Adnan et al. 2019; Akram et al. 2018a,b; Aziz et al. 2017a,b; Chang et al. 2021; Chen et al. 2021; Emre et al. 2021; Habib et al. 2017; Hafiz et al. 2016; Hafiz et al. 2019; Ghulam et al. 2021; Guofu et al. 2021), demonstrating that farmers may be aware of the mechanism and scientific attributes of soil type but use different connotations or conceptions to interact and plan their soil productivity. As a result of how local information systems differ from scientific knowledge systems, shared understanding among farmers and researchers is difficult to achieve (Agrawal 1995). According to Barrios et al. (2006), while both systems share indispensable concepts, such as the importance of water in plant growth, each information system comprises gaps that are filled by others. They also claimed that attempting to strike a balanced scientific precision and local relevance broadens common information, resulting in a new, hybrid knowledge base. Farmers and agronomists both begin their appraisal of soil fertility with the same aim: crop growth efficiency (Murage et al. 2000; Hafeez et al. 2021; Khan et al. 2021; Kamaran et al. 2017; Muhmmad et al. 2019; Safi et al. 2021; Sajjad et al. 2019; Sajjad et al. 2021a,b; Saud et al. 2013; Saud et al. 2014; Saud et al. 2017; Saud et al. 2016; Shah et al. 2013; Saud et al. 2020; Saud et al. 2022a,b; Qamar et al. 2017; Hamza et al. 2021; Irfan et al. 2021;Wajid et al. 2017; Yang et al. 2017; Zahida et al. 2017; Depeng et al. 2018). In addition, growers also define the qualities of healthy or unfertile topsoil, primarily through physical and morphological traits like colour and texture, which are regarded as universal soil fertility criteria (Mairura et al. 2007; Iqbal et al. 2016; Hussain et al. 2020; Hafiz et al. 2020 a,b; Shafi et al. 2020; Wahid et al. 2020; Subhan et al. 2020; Zafar-ul-Hye et al. 2020a,b; Zafar et al. 2021; Adnan et al. 2020; Ilyas et al. 2020; Saleem et al. 2020a,b,c; Rehman 2020; Farhat et al. 2020; Wu et al. 2020; Mubeen et al. 2020; Farhana 2020; Jan et al. 2020; Wu et al. 2019). Soil scientists use quantitative analysis to assess soil as a natural resource, whereas growers assess soils as part of their day-to-day work in the field. Producers have more knowledge or "technical experience" of soil, whereas scientists have more scientific expertise or understanding of soil (Ingram et al. 2010; Ahmad et al. 2019; Baseer et al. 2019; Hafiz et al. 2018;Tariq et al. 2018; Fahad and Bano 2012). Such distinctions can be classified into three parts: awareness of additional environmental knowledge, spatial scale, and timing. Examining the various approaches taken by growers and researchers reveals the potential worth of increased consciousness regarding indigenous descriptions of soil quality, which indicate full forms of information and livelihood knowledge and have implications for developing an integrated soil approach to management (Yageta et al. 2019).

Sustainable development in the agricultural system could be accomplished without affecting future generations' environmental resources or capacity to meet their own needs (Umesha et al. 2018; Fahad et al. 2017; Fahad et al. 2013; Fahad et al. 2014a,b; Fahad et al. 2016a,b,c,d; Fahad et al. 2015a,b; Fahad et al. 2018a,b; Fahad et al. 2019a,b; Fahad et al. 2020; Fahad et al. 2021a,b,c,d,e,f; Fahad et al. 2022a,b; Hesham and Fahad 2020). Excessive use of synthetic fertilizers depletes favourable living circumstances since residues that act as secondary contaminants might infiltrate food chains and eventually people (Kumar et al. 2019; Iqra et al. 2020; Akbar et al. 2020; Mahar et al. 2020; Noor et al. 2020; Bayram et al. 2020; Amanullah, Fahad 2018a,b; Amanullah, Fahad 2017; Amanullah et al. 2020; Amanullah et al. 2021). Secondary pollutants can linger in the ecosystem for an extended time, posing a health risk (Uosif et al. 2014). The use of bio-fertilizers rather than agrochemicals may usher in a new era of industry. Bio-fertilizers could help plants obtain the nutrients they need while not harming the environment (Mishra & Dash 2014). This section could assist as a helpful guide for developing bio-fertilizers and using them to accomplish agricultural sustainability.

5.1.1 Current Fertility Status of Pakistani Soils

Optimum crop yield is dependent on good soil fertility. Soil analysis over time is essential and provides details on the basic and current soil quality. For several reasons, soils in arid and semi-arid parts of the globe are often infertile (Vanlauwe et al. 2011). Nutrient loss reduces soil fertility, while restoration with organic or inorganic inputs impacts crop development and production (Chukwuka 2009; Luo et al. 2020; Ullah et al. 2020). Deteriorations in soil quality are thought to be a major contributor to the low productivity of crops such as rice, wheat, sugarcane, maize, and tobacco (Rahman et al. 2016; Belachew & Abera 2010, Yuan et al. 2022; Rashid et al. 2020; Arif et al. 2020; Amir et al. 2020; Saman et al. 2020; Muhammad Tahir et al. 2020; Md Jakir and Allah 2020; Mahmood et al. 2021; Farah et al. 2020; Sadam et al. 2020; Unsar et al. 2020; Fazli et al. 2020; Md. Enamul et al. 2020; Gopakumar et al. 2020; Zia-ur-Rehman 2020; Ayman et al. 2020; Mohammad I. Al-Wabel et al. 2020a,b).

Pakistan is primarily a dryland region, with 80% of its land area classified as desert or semi-arid, 12% classified as sub-humid, and 8% classified as humid (Khan et al. 2013). As a result, soils in arid and semi-arid locations are subjected to various degradation processes. The most significant reasons for soil deterioration and desertification, as well as reduced agronomic productivity, are salinization, drought stress, soil erosion, and reduction of soil fertility and soil organic matter (SOM) contents (Smith et al. 2020; Senol 2020; Amjad et al. 2020; Ibrar et al. 2020; Sajid et al. 2020; Muhammad et al. 2021; Sidra et al. 2021; Zahir et al. 2021; Sahrish et al. 2022). Therefore, understanding the climate–soil–productivity nexus is critical for satisfying the expanding population's food and nutrition needs. Pakistan's population grew from approximately 30 million to 201 million from 1947 to 2018 and is expected to reach 244 million in 2030 and 352 million in 2100 (Lal 2018). However, the current annual growth rate of approximately 2.0% is falling and is anticipated to reach 0.3% by 2100. As evidenced by the rapid growth of the populations of particular cities, the rise in global population is indicative of Pakistan's strong urbanization tendency (Alam et al. 2007). From 1960 to 2018, Pakistan's population grew by 4.5 million, while overall cereal production (wheat, sorghum, maize, rice, millet, etc.) increased by 6.5 million metric tons (from 6.6 to 43.0 million metric tons). Therefore, per capita cereal crop yields increased significantly between 1961 and 1980, but remained stable between 1980 and 2016 at 220 kg per person. Despite these tremendous improvements, there is no reason to be complacent because much greater difficulties are already apparent for the near future. It is not only the populations that doubled between now and 2100, but nutritional tastes may move towards animal-based goods because of increasing wealth and overall economic success. The issue of promoting food security and nutrition is exacerbated further by the ever-increasing hazards of soil pollution, expanding suburbanization, global warming, and decreasing aquifers (Lal 2018).

The Indus plains in Pakistan have the lowest soil organic carbon (SOC) contents, ranging from 0.5 to 0.1% in the root zone. A low SOC content has an impact on agronomic production and input performance (Lal 2018), particularly in Pakistan's rice–wheat as well as other crop cultivation. However, by implementing effective management techniques, SOC concentration can be regained. The goal is to improve the soil/ecosystem by expanding the use of bio-fertilizers. As a result, site-specific best management practices, such as cover crops, irrigation tillage, conservation tillage, mulches, integrated nutrient management (INM) incorporating manure/compost input, usage of biochar, bio-fertilizers, contour farming, and crop interaction with livestock and plants are recommended (Sarfaraz et al. 2020).

5.2 BIO-FERTILIZERS

Bio-fertilizers are organic in nature and include metabolites derived from microbes or bacteria (Mishra & Dash 2014). Microorganisms extracted from soil (rhizosphere), air, and water are used to make bio-fertilizers, then purified for use in the field. Microorganisms begin creating agriculturally important metabolites in response to particular environmental conditions, and plants may use these

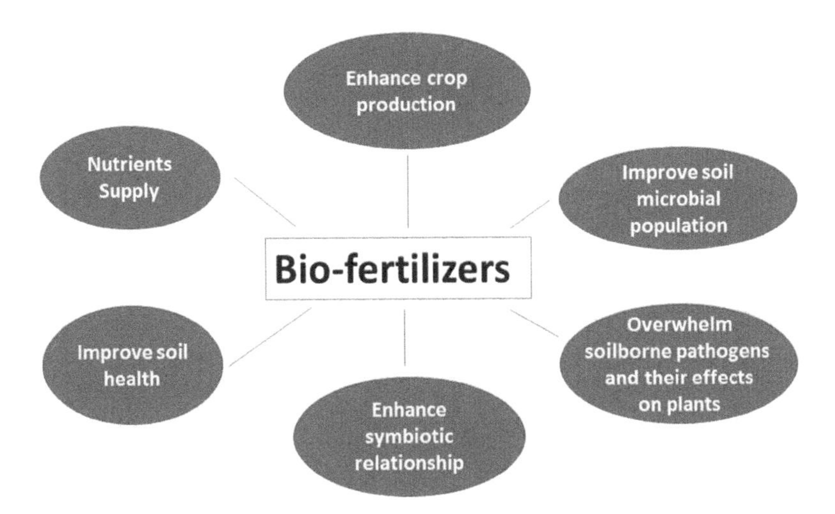

FIGURE 5.1 A schematic diagram of bio-fertilizers with various functions enlightened.

metabolites to support numerous biochemical processes (Salar et al. 2017). Microbes and microbial metabolites facilitate the breakdown of complicated soil minerals/particles into simpler forms and the resulting forms work as a growth stimulator for specific crops. Bio-fertilizers can be applied for various purposes (Kaur & Purewal 2019; Xie et al. 2021) (Figure 5.1).

5.2.1 Types of Bio-fertilizers

Bio-fertilizers are amongst the most effective current agricultural fertility contributors. Organic fertilizers are used in agriculture as an alternative to traditional fertilizers, including compost, domestic garbage, and green manure (Khan et al. 2017; Ali et al. 2021; Iqbal et al. 2021c; Mishra et al. 2013). However, synthetic fertilizers are more successful in this regard. As a result, farmers frequently use chemical fertilizers for crop production, but on the other hand, their excessive use is harmful to the ecosystem by polluting water, air, and soil (Iqbal et al. 2019; Iqbal et al. 2021b). Furthermore, they have the potential to deplete soil health in the long term (Itelima et al. 2018; Wu et al. 2021). Bio-fertilizers comprise microorganisms that encourage appropriate nutrient supply to the host plants and maintain optimal growth and physiological regulation. Organic fertilizers are made using several living microorganisms (Xie et al. 2021). Only microorganisms with specialized functions to improve plant growth and reproduction are employed (Gupta et al. 2015). Bio-fertilizers, as fundamental constituents of organic agriculture, develop the quality and stability of soil classified into several types based on their kind, action, and availability (Kaur & Purewal 2019).

5.2.2 Phosphate-solubilizing Microbe Bio-fertilizers (PSBs)

Phosphorus is an important macronutrient because it influences root growth, protein synthesis, signal transduction, respiration, and N fixation in plants (Ahmad et al. 2019; Izhar Shafi et al. 2020). However, plants cannot utilize it directly as it is present in unavailable forms in the soil. Therefore, it must be converted to plant-available forms from unavailable forms (Shafi & Sharif 2019). There are many strains of useful bacteria that can reduce phosphorus into its most basic form, allowing it to be easily absorbed by root systems. Phosphate-solubilizing microbes (PSMs), although naturally common, vary in number depending on the soil type and location (Awais et al. 2017). In developing nations, PSBs, in combination with rock phosphate (RP) of poor quality, could be a substitute for expensive phosphate fertilizer (Mahanta et al. 2018; Rafique et al. 2017). In this regard, research activities are being carried out worldwide to identify microbes that may be important in maintaining agricultural sustainability. According to various researchers, bacterial strains such as *Micrococcus*, *Achromobactin*, *Erwinia*, *Pseudomonas*, and *Aerobacter*, etc. play a prominent role in solubilization of unavailable insoluble complexed forms of phosphate (Chen et al. 2006). Aerobic and anaerobic microbes coexist in the rhizospheric soil. Bacterial strains or spores have different degrees of P solubilization depending on the location from which they are collected, and among all, the spores isolated from the rhizosphere have the highest P solubilization capacity. Phosphorus can bind with Fe, Al, and K to generate complex compounds (Wahid et al. 2019). The entire conversion process is made up of a series of biochemical processes involving the action of several enzymes caused by bacterial strains. The conversion of strongly bound P into organic and inorganic acids takes place in the first stage, which reduces the soil pH and maximizes the accessibility of phosphorus to growing plants.

5.2.3 *Rhizobium* Bio-fertilizers

In developing countries, critical nutrient deficiencies in food crops are more difficult to overcome (Kumari et al. 2018). To solve these issues, there is a strong focus on the employment of microbial consortia, particularly PGPR, for continuous plant growth and meeting food requirements in the future (Khatoon et al. 2020). *Rhizobium* is a nitrogen-fixing continually evolving member of the Rhizobiaceae family. *Rhizobium* infects plant roots, causing the production of particular rhizosphere soil (Gouda et al. 2018). According to Kumari et al. (2018), the more common *Rhizobium* isolates BHU-M and BHU-B13-398 were extracted from mung bean roots. These strains enhance the shoot and root growth, and the height and yield of the plants as they are associated with plant roots and capture nutrients for plant growth. Moreover, *Rhizobium* inoculation was reported to be involved in the regulation of phytochelatin-related gene expressions in *Medicago sativa* and protects plants against excessive Cu-stress (Chen et al. 2018). Their findings revealed that *Rhizobium* strain inoculation enhanced plant growth through higher N uptake by the plants. When untreated and *Rhizobium*-inoculated treated plants were compared, a significant increase in Cu uptake was noted. Several scientific studies have found that inoculating chickpeas with efficient microbial strains at the time of planting increases the total grain yield (Funga et al. 2016).

Microorganisms in root nodules degrade molecular nitrogen to ammonia, which is then used by the plant system to synthesize proteins, vitamins, and other N-containing substances (Belhadi et al. 2018). The use of *Rhizobium* in particular legumes and other host plants aids in the maintenance of major agricultural benefits (Sahu et al. 2019). These bacteria are harmless and have shown no negative environmental impact (Singh et al. 2011). Despite their occurrence in leguminous plant nodules, several artificially created *Rhizobium* formulations are also available in the market.

5.2.4 Arbuscular Mycorrhizal Bio-fertilizers

Natural resources are constantly subjected to abiotic stressors at various growth and development phases which soil microbes are able cope with (Wahid et al. 2019). Plants begin manufacturing a particular type of minor metabolites when stressed to combat the excessive production of reactive oxygen species (Kaur et al. 2018a; Kaur et al. 2018b). To some extent, the creation of certain ingredients aids the plant's survival under severe conditions. One of the essential factors contributing to crop plant health is the symbiosis interaction. Arbuscular mycorrhizal fungi (AMF) are essential symbionts with roots that aid in nutrient uptake and numerous enzymatic activities in most plants (Ortas et al. 2021; Yang et al. 2018). The AMF connections with plant rhizospheres give a variety of growth-promoting effects such as improved nutrition, increased resistance, drought tolerance, and modified soil composition (Berruti et al. 2016; Rafique & Ortas 2018). Water-soluble chemical fertilizers are avoided in organic farming, and involve a variety of crop rotations. According to scientific investigations, this increases AMF infection in soils with maximum nutrient uptake (Ortaş et al. 2017). As a result, AMF may be a viable alternative to chemical fertilizers.

5.2.5 *Azotobacter* Bio-fertilizers

Azotobacter are anaerobic bacteria from the family Azotobacteraceae (Sethi & Adhikary 2012). They are non-symbiont, Gram-positive diazotrophs, that give numerous benefits to plants and their interaction with growing crops enables them to maintain a stable growth with enhanced production. The use of *Azotobacter* as a bio-fertilizer to maximize the production and cropping yield has been recommended by several researchers. It also helps in improving plant dry matter, yield, and secondary metabolite synthesis (Damir et al. 2011). *Azotobacter* strains with imperative practical qualities (enhancing the health of soil and N-fixation, promoting growth and production of crops, protecting plants against drought and pathogens) could be a boon for sustainable farming techniques (Shirinbayan et al. 2019). In certain conditions, *Azotobacter* and related bacteria begin to develop cysts – a normal defensive mechanism against various environmental factors (Socolofsky & Wyss 1962). The strains commence the production of pigments from a deep brown to yellowish-green and purple colour throughout the N-fixation process. The fundamental reason strains produce pigment during the nitrogen fixation process is to shield nitrogenase from the destructive impact of oxygen (Shivprasad & Page 1989). *Azotobacter* is now produced using a fermenter and mixer on a commercial scale. The use of a fermenter is a scientific and automated method for the proliferation of microbes. Specific nutritional media essential to maintain microorganism development are created and pasteurized, and the pH of the medium may be controlled to commence appropriate microbial populations. A mother culture (1–2% of the total) may be employed to enhance growth. Other significant needs include a constant supply of oxygen and the ability to maintain a constant temperature. Depending on the required demand, growth can be accelerated by utilizing a shaker, which increases the rate of nutrient absorption in a brief period.

5.2.6 *Azospirillum* Bio-fertilizers

Azospirillum is another type of bio-fertilizer that aids crops in maintaining different biochemical reactions essential for agricultural production (Llorente et al. 2016). It is an essential member of the group Rhodospirillales and is closely associated with grasses and occasionally with monocots, particularly rice and corn (Ruíz-Sánchez et al. 2011). Their interaction is directly related to nitrification, the release of particular fungicides, and plant hormones (Gonzalez et al. 2015). *Azospirillum* is capable of producing phytohormones such as salicylic acid (Sahoo et al. 2014), auxins (Spaepen & Vanderleyden 2015), and indole-3-acetic acid (Fukami et al. 2018). *Azospirillum* improves moisture and nutrient retention by plants and protects plants against environmental stress, resulting in higher total production (Fukami et al. 2018). *Azospirillum* inoculation in plants results in dramatic morpho-physiological alterations, including shoots and grains with increased nitrogen content. When *Azospirillum* is used on the field, it requires less synthetic fertilizer than in fields without its application (Cassán & Diaz-Zorita 2016).

5.2.7 *Azolla* and Blue-green Algae Bio-fertilizers

Azolla is a member of the Salviniaceae family, which includes seven different species of duckweed phototrophic ferns (Roger & Ladha 1992). Depending on numerous circumstances, including soil properties, *Azolla* can develop to generate massive biomass in as little as 10 days. *Azolla* is a tiny free-floating plant having rough leaves and flowing roots. It is well known for its N-fixing symbiotic relationship with *Anabaena azollae* in developing and underdeveloped nations (Emrooz et al. 2018). Rice crops are widely recognized for their high-water use, and growers use *Azolla* to prevent extreme weed development. It can deliver up to 10 tons of proteins and other critical nutrients to rice crops in cultivation (Yao et al. 2018). Blue-green algae (BGA) are N-fixing microorganisms filamentous by nature and have a type of cell called a heterocyst (micronodules). Heterocysts demonstrate nitrogen fixation process functioning. These microorganisms form symbiotic partnerships with fungal strains, ferns, and flowering plants for nitrogen fixation (Soma et al. 2018). Blue-green algae are particularly important in agriculture because of their fast activity and effective nitrogen fixation. As well as N-fixing, they also fix P, Zn, K, S, and other nutrients (Adeniyi et al. 2018).

5.2.8 Silicon-solubilizing Microbe Bio-fertilizers

Disintegration of silica and silica-based rocks and minerals can lead to a change in the soil layer (Vasanthi et al. 2018). Microbial consortia of several types play a significant part in the decay, transformation, and activation of silicon and its derivatives. A microbial consortia's action is determined by the availability of condensation, pH conditions, and growth regulators in soil. These are involved in synthesizing various enzymes and metabolic products that may be useful in mineralization (Gadd 2010). Biological methods of converting tough silicon derivative products into the simplest eatable forms have gained significance over chemical and physical methods. Biological methods include microbial activities, which are self-manageable and inexpensive and can result in conversion in a small period. *Thiobacillus thiooxidans* and *Bacillus globisporus* showed the greatest ability to leach silicon (Friedrich et al. 1991; Sheng et al. 2008).

5.3 BIO-FERTILIZER EFFECT ON QUALITATIVE AND QUANTITATIVE ATTRIBUTES: A CASE STUDY OF CUCUMBER

Several studies have been conducted on studying the effects of bio-fertilizers on soils for different crops and vegetables. When applied to various crops in combination with synthetic or other fertilizers, it showed promising results in meeting the plant's nutritional demand in an eco-friendly manner. A detailed study of the efficiency of bio-fertilizers in cucurbits is available (Kumar et al. 2018). The useful insight and use of bio-fertilizers and their effects on cucumber as a case study are outlined in Table 5.1.

5.4 MARKET CHARACTERISTICS NEEDED FOR THE RELEASE OF BIO-FERTILIZER

Farmers' use of bio-fertilizers for increased crop production is one of the foremost constraints in the farming sector. Although various bio-fertilizers are now commercially available, their quality and quantity may fluctuate based on the manufacturing division. Bio-fertilizers should have the following characteristics before being released to the market.

5.4.1 Availability

Bio-fertilizers should be widely available in the marketplace. Farmers benefit from reduced transportation costs and saved time.

5.4.2 Solubility and Mode of Action

The formulation should be water-soluble to decrease costs and allow for spray application in larger field areas.

TABLE 5.1
Use of Bio-fertilizer and Integrated Nutrient Management (INM) Practices in Cucumber Crop

S. no.	Treatments	Characters Enhanced in Cucumber Crop	References
1	Application of minerals (25%) and organic N (75%)	Increase in plant growth, yield, and quality	Mahmoud et al. 2009
2	Use of bio-fertilizers	Increase in the fruit count, fruit length, average fruit weight, and fruit yield	Jilani et al. 2009
3	Use of FYM/vermicompost	An increase in the yield was observed	Narayanamma et al. 2010
4	Use of bio-fertilizers	Enhanced yield and yield attributing characters	Isfahani & Besharati 2012, Saeed et al. 2015
5	Use of vermicompost	An increase in yield and fruit weight was noted	Ghasem et al. 2014
6	Use of poultry manure with NPK	A significant increase in the weight, number of leaves, fruit count, and size with quality and yield were found	Okoli & Nweke 2015, Solaiman et al. 2020
7	Use of bio-fertilizers	Significant increase in the fruit length and diameter, fruit count, average fruit weight, and yield	Kanaujia & Daniel 2016
8	Use of poultry manure at 20 ton/ha	An increase in yield was noted	Khan et al. 2017

5.4.3 Stability of Storage

Bio-fertilizer formulations must be reliable in a broad range of climate circumstances. The strength of the preparation must not deteriorate over time.

5.4.4 Effectiveness

Bio-fertilizers should be used in small quantities in the field and must be successful in providing a balanced mix of nutrients to plants. The formulation should provide crops with an immediate supply of nutrients while causing no adverse effects. It should be simple to use and have no negative effects on the health of growers. It must be affordable to growers, as this impacts crop prices. It should be season-independent and accessible to farmers throughout the year.

5.5 PAKISTAN AND BIO-FERTILIZERS

Presently, Pakistan spends a significant amount of money importing and producing 8.41 million nutrient tons of synthetic fertilizers. On the other hand, a significant opportunity exists for the enhancement bio-fertilizers use in sustainable agriculture. In Pakistan, saving of 10.0 billion rupees annually is possible by the addition of a 10% contribution of bio-fertilizers to the total fertilizer consumption (Ali et al. 2012). Various groups/organizations are engaged in bio-fertilizers research and innovation in Pakistan. They have stated substantial rises in yield and yield components of important crops due to microorganism inoculation (Alam et al. 2007, Zahir et al. 2005) . The extent to which this bio-fertilizers benefit depends on their quantity and efficiency, which is ruled by a diversity of environmental and soil elements. In comparison to chemical fertilizer plants, the system used for bio-fertilizers production is much simpler and the costs for its installation are very negotiable. Furthermore, the use of bio-fertilizers for a long time is efficient, more cost-effective, eco-friendly, effective, and readily available to growers. A list of major problems, limitations, and recommendations regarding the production of bio-fertilizers on large-scale and future technologies in the county also have been discussed in detail.

5.5.1 History of Bio-fertilizers in Pakistan

Rhizobium is the world's oldest bio-fertilizer for leguminous plants and soil quality enricher "*Theophrastus*, 372 287 BC" as observed by (Danso 1992). J.B. Boussingault, a French chemist and agronomist, proposed the classic concept of biological nitrogen fixation (BNF) in 1834 and later Hellriegel and Wilfarth (1888) confirmed it. Beijerinck isolated N-fixing *Rhizobium* organisms in 1888, *Azotobacter* in 1901, and *Azospirillum* in 1925. *Rhizobium* is an N-fixer, which was first commercialized in the United States in 1895 under the trade name "Nitragin" and was developed by Noble and Hiltner in 1896 (Chandra & Sharma 2017). Stalstrom (1903) was the first to report microbial P-solubilization and Pikovskaya isolated microbes in 1948 and Sperber in 1957.

Before the establishment of Pakistan, in early 1920, India's first Agricultural College, named the Punjab Agricultural College and Research Institute Lyallpur, began research on biological N fixation. After six years, in 1926, these research activities were boosted when an independent post of "Agricultural Bacteriologist" was established at the Institute. The microbiological centre was developed in 1927 in Lyallpur (Naveed et al. 2015). The laboratories were developed, and field trials were conducted at a larger scale in the Lyallpur and Gurdaspur areas to evaluate the effectiveness of synthetic inoculum on chickpea, Egyptian and Persian clover, alfalfa, sweet clover, mash beans, mung bean, and cluster bean. It was concluded from the early research that seeds treated with inoculum generated greater yields of higher quality than untreated seeds (Naveed et al. 2015). After this, the commercial production of *Rhizobium* inocula began in 1956 in the region.

5.5.2 Bio-fertilizer Research and Development in Pakistan

The uniqueness and capabilities of microbes, particularly in specific cultural and environmental conditions, have shown that they have the potential to resolve the food security issues in agriculture and other areas of life. Several organizations, research groups, and institutes in Pakistan are working on the research and development of bio-fertilizers to overcome food scarcity and increase the country's agricultural production. As summarized in the following sections, research and development efforts are underway to expand the role of bio-fertilizers in Pakistan.

5.5.3 AARI (Ayyub Agricultural Research Institute), Faisalabad

The Ayub Agricultural Research Institute in Faisalabad, formerly known as the Punjab Agricultural Research Institute Lyallpur, is a parallel research institute of the Punjab Agricultural College. Lyallpur was the country's first and earliest BNF and bio-fertilizers research institute. Work on research and innovation began in the early 1920s and was focused on at various times. Since 1956, AARI scientists have provided organic fertilizers with the trade name "Associative Diazotrophs". The fruitful and steady approval of legume *Rhizobium* cultures in the field (Naveed et al. 2015) prompted the AARI's "Soil Bacteriology" section to collaborate for useful microbial associations prevalent in different crops in the early 1990s, and *Azospirillum* and *Azotobacter* inoculants were introduced. Their consortia were released as a commercial product in the mid-1990s under the trade name "Fasloon ka jarasimi teeka". This contains familiarized phosphate-solubilizing microbes (PSMs), which grabbed the attention of many growers who were struggling with P-fertilizer scarcity market prices. Data from the field experiments resulted in a 20% increase in the yield of leguminous and non-leguminous crops by applying rhizobial, diazotrophic, and PSM inoculants. On a limited scale, the AARI's Soil Bacteriology Section was manufacturing and providing 38,800 carrier-based

250.0 g inoculum culture bags in the region. This was adequate for the inoculation of 14,000 ha of arable land during 2000–2011 (Naveed et al. 2015).

5.5.4 NIAB (Nuclear Institute of Agriculture and Biology) and NIBGE (National Institute for Biotechnology and Genetic Engineering), Faisalabad

In 1972, NIAB established a very energetic BNF research centrr in the Department of Soil Biology, having published work at the national and international levels. They have conducted research using *Azolla anabaena* as N-fixing BGA used as symbiont on rice bio-fertilizers "Azolla" a water-fern. The severe hot environmental conditions of Punjab did not respond according to its potential on a larger scale. In contrast, this technology provided the best in rice production in northern areas where the environmental conditions were mild and humid (Malik et al. 2002).

With the foundation of NIBGE in 1992, Dr Kausar Abdullah Malik led the "Bio-fertilizers Division" by securing funding from several donor agencies such as the International Centre for Nuclear Research, International Atomic Energy Agency, International Centre for Genetic Engineering and Biotechnology, and the Islamic Development Bank for the development of a Bio-fertilizers Resource Centre in the South Asian region. In 1996, they successfully introduced the commercial organic fertilizer "BioPower". *Rhizobium* species were used for chickpea, mash bean, soybean, mung bean, cowpea, and alfalfa as legume bio-fertilizers, whereas implicit nitrogen-fixing and PGPR were used in crops such as wheat and maize. The research conducted revealed that bio-fertilizer use could meet 40–70% of crop plant nitrogen requirements, which also improves crop yield by 60–80% (Hafeez et al. 1998). After pot and field trials, "BioPower" was used commercially on an area of 11,000 ha with different testing crops in Punjab, and a 50–70% reduction in nitrogen fertilizer costs with a 20% increase in crop production was claimed by the research team (Naveed et al. 2015). It was revealed that using half the recommended dose of NPK with "BioPower" produced the same results as with the full recommended dose of NPK alone. NIBGE joined public and private sector entrepreneurs to popularize the bio-fertilizers, transfer manufacturing capabilities, and provide proper training to farmers. Farmers were able to save a significant amount of money (up to \$292 USD ha^{-1}) by using "BioPower" in several crops, as per the benefit–cost ratio of the technology (Naveed et al. 2015). The NIBGE has a fully established bio-fertilizers pilot production unit to scale up bio-fertilizer production to meet rising demand. "BioPower" has been supplied to between 9000–12,000 hectares (Naveed et al. 2015).

5.5.5 NARC (National Agricultural Research Centre), Islamabad

During the early 1980s, the Soil Biology and Biochemistry Department of NARC's Land Resources Research Program began investigating N-fixation in legumes. They investigated the effects of imported rhizobial strains (from NifTAL, Hawaii) on legume production in Pakistan. Later, a local *Rhizobium* spp. was isolated and used to inoculate important crop legume crops (lentil, chickpea, mash bean, mung bean, groundnut, soybean, Egyptian clover, pea, alfalfa, and sesbania). The NARC "Rhizobium Gene Bank" contains over 200 isolates of various rhizobia. In 1990, the Centre introduced "Biozote", a bio-fertilizer product. The efficacy of "Biozote" was tested commercially during a combined project of PARC (Pakistan Agricultural Research Council), Islamabad and Engro-Chemical Pakistan Ltd. This project ran for three years to test different leguminous crops, and approximately 60,000 packets of "Biozote" were provided to growers. Data from 300 growers' fields revealed a 20–50% improvement in crop production using "Biozote". Additionally, it was added that the benefit–cost ratio of this technology was 30:1, and if applied to a 50% leguminous area, it has the potential to improve the national economy by enhancing crop yield (Naveed et al. 2015). The Centre could produce 150,000 culture bags per year and currently, it is supplying about 2000 culture bags to growers each year.

5.5.6 ISES (Institute of Soil and Environmental Sciences), University of Agriculture, Faisalabad

In 2003, the University of Agriculture, Faisalabad's Department of Soil Science was upgraded to the status of "Institute of Soil and Environmental Sciences". The Institute is vigorously involved in basic and applied research on soil microbiology and biotechnology etc., by isolating soil microbes with various beneficial strains and using them as bio-fertilizers. The researchers are not using only living cells of inoculants but also proposed using microbial metabolites or plant growth regulators, which could be the best approach to improve crop growth (Khalid et al. 2009). In 2002, they created a liquid preparatory work of microbial metabolite-based bio-fertilizers called "Rice-Biofert". Data collected from multi-location experimental fields for three years indicated an increase of 20% in rice production (Zahir & Arshad 2004). The Soil Microbiology and Biochemistry Group has also isolated various cultures of *Azotobacter* from various soils, and their performance in increasing crop production has been extensively studied. Many PGPR, such as *Burkholderia*, *Pseudomonas*, *Serratia*, *Bacillus*, and others, have been isolated and demonstrated their value as plant-growth promoters. ACC-deaminase is an enzyme that hydrolyses ACC (ethylene precursor) into ammonia and α-ketobutyrate in various PGPR strains. The PGPR having ACC-deaminase act as ACC reservoirs when colonized with plant roots and lower the plant ethylene concentration. This mechanism has the potential to inhibit the impact of high ethylene concentrations in plants and promote stronger root structure and function. These plants also develop anti-environmental stressed qualities such as protecting against drought, salinity, heavy metals, etc. (Ahmad et al. 2011; Nadeem et al. 2010). A series of field trials were conducted by ISES on growers' land to demonstrate the potential impact of PGPR-based bio-fertilizer "Uni Grow" for the purpose of encouraging the use of bio-fertilizers in the farmer community and received highly encouraging results (Shahzad et al. 2008). According to the literature, the combined application of chemical, bio-, and organic fertilizers has the potential to increase the crop yield and meet the food demands of the country. Under field conditions, inoculation of rhizobia in leguminous as well as in non-leguminous crops produced excellent results (Hussain et al. 2009; Mehboob et al. 2011). The ISES recently developed a combined culture of ACC-deaminase containing PGPR and *Rhizobium* called "Rhizogold" (RG), which enhanced 40–45% yield of legumes. Another multi-strain bio-fertilizer called "RhizogoldPlus" (RG+) was obtained from effective strains of PGPR having ACC-deaminase with the purpose of mitigating salinity stress on cereal crops (Khan et al. 2013; Naveed et al. 2015).

5.5.7 NFRDF (Nature Farming Research and Development Foundation)

The effective microorganism (EM) technique was introduced by a former scientist of the Soil Science Department of the University of Agriculture, Faisalabad, who brought it from a Japanese scientist, Dr. Teruo Higa, and used it as biological input for sustainable yield. After his tremendous work, the foundation of the Nature Farming Research Centre was laid at UAF to work on this technology. Further, the soil fertility and productivity were enhanced by using beneficial microorganisms (BM) s in combination with manure, crop residues, waste from industries, green manures, and composts from various sources. This technology reduced the costly application of chemical fertilizers. A new type of BM fermenter/superfermenter has been developed to achieve the minimum number of organisms available and to use salt water to irrigate with BM technology (Hussain et al. 2009). This centre has carried out many experimental projects in growers' fields to assess the efficacy of this technology for preserving soil fertility and productivity, encouraging the sustainable use of soil, proliferating soil biological activities, reducing pollution, and recycling waste of plants and animals. It was concluded that using the technology improved the soil's biological activities, increased crop yield and profit per hectare, and improved the quality of soil and water resources.

Many of the products of EM technology are in use by farmers in Punjab, e.g., for crop production and fish farming, EM-BIOAAB is used, whereas for animal and poultry production, EM-BIOVET is preferably used. EM-BIOCONTROL, which is not a pesticide or insecticide, is used to control insect/pest diseases in crops, vegetables, and fruits (Hussain et al. 2009).

5.5.8 Bio-fertilizer Study in Higher Education Institutes

At various higher education institutions across the nation, researchers are studying soil–microbial prospects and plant–microbe relations to understand how they affect the health of soil and plants. Various higher institutes such as Quaid-i-Azam University Islamabad, Comsat University, Islamabad; Karachi University, Karachi; PMAS-Arid Agriculture University, Rawalpindi; Punjab University, Lahore; The University of Agriculture, Peshawar; AJK University, University of Poonch, Rawalakot; and many others have made outstanding contributions. An extensive list of research work published in national and international journals has been documented. Relationships among *Rhizobium* and leguminous and non-leguminous crops, and the isolation and identification of various microorganisms for disease management (Hussain et al. 2009; Mehboob et al. 2011) have been studied in depth. Microflora supplying resistance to various stresses (Arshad et al. 2008; Saleem et al. 2007), microorganism production of phytohormones (Qureshi et al. 2013; Shakir et al. 2012), exploitation of bacterial and fungal populations for improved health of soil and plants, assessment of variations and development of markers for maintaining and evaluating microbial efficacies (Malusà et al. 2016), and phytoremediation of soil and environment (Naveed et al. 2015) are amongst the research areas. Research is being conducted at various universities and higher education research centres, providing applied research strategies. Interactions among research and educational institutes can lead to the translation of scientific concepts into practice. Academia and industrial linkages for the cheap, sustainable, and easy supply of the end product to the consumers (farmers) are the need of the time.

5.6 PROBLEMS IN MASS PRODUCTION AND COMMERCIALIZATION OF BIO-FERTILIZERS IN PAKISTAN

Although microbe technology has shown its value when used in various agricultural and environmental issues with remarkable success over the past 50 years, it has not been widely accepted. It is often difficult to replicate its positive effects in different areas. These conditions are most common in the upper and lower extremities. The following are the major barriers to mass production and technological advancement in pakistan.

1. Lack of bio-fertilizer regulation and standards. Regulations for the production and selling of bio-fertilizers have yet to be established at the national level in Pakistan. As a result, sub-standard inoculants are among the significant limitations.
2. An insufficient community of growers that undestand microbial inoculants.
3. As most bio-fertilizers are environment and ecology specific, they do not produce the required results always, and so, eventually growers may lose faith in this technology.
4. The communication difference between marketing, extension work, and end-users.
5. Lack of qualified labour and the excessive cost of making high-quality organic fertilizers.
6. The country lacks transportation and storage facilities to prevent contamination.
7. Extreme climatic conditions frequently cause bio-fertilizer results to be inconsistent.
8. A low amount of soil organic matter prevents beneficial microorganisms from surviving and interacting positively with plants.
9. An insufficient supply of appropriate excipients for bio-fertilizer production.
10. Poor labelling and packaging of bio-fertilizers damage their reliability.

5.7 FUTURE SCENARIOS AND RECOMMENDATIONS

There are several concerns that need to be addressed by the government in future studies for more comprehensive production and application of bio-fertilizers.

1. Necessary legislation to monitor bio-fertilizers, their quality, and any harmful effects on humans and plant species. This serious concern must be evaluated and necessitates collaboration among the government and private sectors.
2. The government should sponsor the production of bio-fertilizers, or there should be the availability of loans from the government to produce bio-fertilizers on a small scale, e.g., seed money, agri-preneur start ups, etc.
3. The country is in desperate need of microbial strain banks. All characterized microbes/potential bio-fertilizer candidates from various institutes and independent scientists should be collected, conserved, molecular tagging internationally, and validated chemotaxonomically if necessary.
4. Farmer communities and stakeholders should be trained by adopting intensive training and extension workshops to use bio-fertilizer technology to its full potential.
5. Development of bio-fertilizers by using microbial consortia having active, competitive, and stress-tolerant microbial strains.
6. The ability of bio-fertilizers to provide micronutrients and bio-fortify food plants should be investigated.
7. Phosphate-solubilizing microorganisms (PSMs) and P mobilizers such as vesicular-arbuscular mycorrhizae (VAM), which are less commonly used bio-fertilizers, show promising results in providing P and other micronutrients. Therefore, the laboratory-produced strains of these symbionts will allow testing of their performance in the field. The genetic basis for a competitive advantage must still be determined.
8. Selection of a low-cost synthetic carrier capable of maintaining a high viable count and developments in inoculation procedures to guarantee soil establishment and perseverance.
9. Creation of polymicrobial bio-fertilizers such as PGPR, *Rhizobium*, PSM, and VAM.
10. Locally available organic wastes should be converted into value-added bio-fertilizers.
11. Endophyte molecular breeding is also required to improve endophyte–host plant interactions. Endophytic bacteria genetic engineering should be a much simpler process than crop genetic engineering. Endophytes that have been genetically modified by using helpful genes will introduce new characteristics to host plants that have been inoculated with these strains.
12. Synthetic fertilizers coated with promising microbial strains may mark the start of a new understanding of synthetic/natural sources of nutrition, potentially providing knowledge of "microbial-enhanced" fertilizer use efficiency.

5.8 CONCLUSIONS

Understanding the production and application of bio-fertilizers is needed for a country's economic growth. Knowing the basic sustainability principles in agriculture requires understanding the design, method of production, utilization, and storage conditions. Sustainability in agriculture is extremely beneficial in resolving the actual problems in the agriculture sector with crop production. Furthermore, marginal farmers in developing countries must be trained in the biotechnological features of bio-fertilizers in agricultural system planning. This chapter provides an in-depth examination of the efficacy of bio-fertilizers in achieving sustainable agriculture. Bio-fertilizers can meet agro-industry challenges and create novel prospects for the benefit of growers in the agriculture sector and business, and for the research, academia, and other government sectors.

REFERENCES

Abdul S, Muhammad AA, Shabir H, Hesham A El E, Sajjad H, Niaz A, Abdul G, RZ Sayyed, Fahad S, Subhan D, Rahul D (2021a) Zinc nutrition and arbuscular mycorrhizal symbiosis effects on maize (Zea mays L.) growth and productivity. *J Saudi Soc Agri Sci* 28(11), 6339–6351. https://doi.org/10.1016/j.sjbs.2021.06.096

Abdul S, Muhammad AA, Subhan D, Niaz A, Fahad S, Rahul D, Mohammad JA, Omaima N, Muhammad Habib ur R, Bernard RG (2021b) Effect of arbuscular mycorrhizal fungi on the physiological functioning of maize under zinc-deficient soils. *Sci Rep* 11, 18468.

Abid M, Khalid N, Qasim A, Saud A, Manzer HS, Chao W, Depeng W, Shah S, Jan B, Subhan D, Rahul D, Hafiz MH, Wajid N, Muhammad M, Farooq S, Fahad S (2021) Exploring the potential of moringa leaf extract as bio stimulant for improving yield and quality of black cumin oil. *Sci Rep* 11, 24217. https://doi.org/10.1038/s41598-021-03617-w

Adeniyi OM, Azimov U, Burluka A (2018) Algae biofuel: current status and future applications. *Renewable Sustainable Energy Reviews* 90, 316–335.

Adnan M, Zahir S, Fahad S, Arif M, Mukhtar A, Imtiaz AK, Ishaq AM, Abdul B, Hidayat U, Muhammad A, Inayat-Ur R, Saud S, Muhammad ZI, Yousaf J, Amanullah, Hafiz MH, Wajid N (2018a) Phosphate-solubilizing bacteria nullify the antagonistic effect of soil calcification on bioavailability of phosphorus in alkaline soils. *Sci Rep* 8, 4339. https://doi.org/10.1038/s41598-018-22653-7

Adnan M, Shah Z, Sharif M, Rahman H (2018b) Liming induces carbon dioxide (CO_2) emission in PSB inoculated alkaline soil supplemented with different phosphorus sources. *Environ Sci Poll Res* 25(10), 9501–9509.

Adnan M, Fahad S, Khan IA, Saeed M, Ihsan MZ, Saud S, Riaz M, Wang D, Wu C (2019) Integration of poultry manure and phosphate solubilizing bacteria improved availability of Ca bound P in calcareous soils. *3 Biotech* 9(10), 368.

Adnan M, Fahad S, Muhammad Z, Shahen S, Ishaq AM, Subhan D, Zafar-ul-Hye M, Martin LB, Raja MMN, Beena S, Saud S, Imran A, Zhen Y, Martin B, Jiri H, Rahul D (2020) Coupling phosphate-solubilizing bacteria with phosphorus supplements improve maize phosphorus acquisition and growth under lime induced salinity stress. *Plants* 9(7), 900. doi: 10.3390/plants9070900

Agrawal A (1995) Dismantling the divide between indigenous and scientific knowledge. *Dev Change* 26, 413–439.

Ahmad M, Zahir ZA, Asghar HN, Asghar M (2011) Inducing salt tolerance in mung bean through coinoculation with rhizobia and plant-growth-promoting rhizobacteria containing 1-aminocyclopropane-1-carboxylate deaminase. *Can J Microbiol* 57, 578–589.

Ahmad M, Khan I, Muhammad D, Mussarat M, Shafi MI (2019) Effect of phosphorus sources and their levels on spring maize. *Biological Sciences-PJSIR* 62, 8–14.

Ahmad N, Hussain S, Ali MA, Minhas A, Waheed W, Danish S, Fahad S, Ghafoor U, Baig, KS, Sultan H, Muhammad IH, Mohammad JA, Theodore DM (2022) Correlation of soil characteristics and citrus leaf nutrients contents in current scenario of layyah district. *Hortic* 8, 61. https://doi.org/10.3390/horticulturae8010061

Ahmad S, Kamran M, Ding R, Meng X, Wang H, Ahmad I, Fahad S, Han Q (2019) Exogenous melatonin confers drought stress by promoting plant growth, photosynthetic capacity and antioxidant defense system of maize seedlings. *PeerJ* 7, e7793. http://doi.org/10.7717/peerj.7793

Akbar H, Timothy JK, Jagadish T, Golam M, Apurbo KC, Muhammad F, Rajan B, Fahad S, Hasanuzzaman M (2020) Agricultural land degradation: processes and problems undermining future food security. In: Fahad S, Hasanuzzaman M, Alam M, Ullah H,Saeed M, Khan AK, Adnan M (Eds.) *Environment, Climate, Plant and Vegetation Growth.* Springer Publ Ltd, Springer Nature Switzerland, pp. 17–62. https://doi.org/10.1007/978-3-030-49732-3

Akram R, Turan V, Hammad HM, Ahmad S, Hussain S, Hasnain A, Maqbool MM, Rehmani MIA, Rasool A, Masood N, Mahmood F, Mubeen M, Sultana SR, Fahad S, Amanet K, Saleem M, Abbas Y, Akhtar HM, Waseem F, Murtaza R, Amin A, Zahoor SA, ul Din MS, Nasim W (2018a) Fate of organic and inorganic pollutants in paddy soils. In: Hashmi, MZ and Varma, A (Eds.), *Environmental Pollution of Paddy Soils, Soil Biology.* Springer Publ Ltd, Springer Nature Switzerland, pp. 197–214.

Akram R, Turan V, Wahid A, Ijaz M, Shahid MA, Kaleem S, Hafeez A, Maqbool MM, Chaudhary HJ, Munis, MFH, Mubeen M, Sadiq N, Murtaza R, Kazmi DH, Ali S, Khan N, Sultana SR, Fahad S, Amin A, Nasim

W (2018b) Paddy land pollutants and their role in climate change. In: Hashmi, MZ and Varma, A (Eds.), *Environmental Pollution of Paddy Soils, Soil Biology*. Springer Publ Ltd, Springer Nature Switzerland, pp. 113–124.

Alam S, Fatima A, Butt MS (2007) Sustainable development in Pakistan in the context of energy consumption demand and environmental degradation. *J Asian Econ* 18, 825–837.

Ali I, Ullah S, He L, Zhao Q, Iqbal A, Wei S, Shah T, Ali N, Bo Y, Adnan M (2020) Combined application of biochar and nitrogen fertilizer improves rice yield, microbial activity and N-metabolism in a pot experiment. *Peer J* 8, e10311.

Ali I, Adnan M, Ullah S, Zhao Q, Iqbal A, He L, Cheng F, Muhammad I, Ahmad S, Wei S (2021) Biochar combined with nitrogen fertilizer: a practical approach for increasing the biomass digestibility and yield of rice and promoting food and energy security. *Biofuels, Bioproducts & Biorefining* 16(5), 1304–1318. doi.org/10.1002/bbb.2334

Ali M, Ali A, Tahir M, Yaseen M (2012) Growth and yield response of hybrid maize through integrated phosphorus management. *Pak J Life Soc Sci* 10, 59–66.

Ali S, Hameed G, Muhammad A, Depeng W, Fahad S (2022) Comparative genetic evaluation of maize inbred lines at seedling and maturity stages under drought stress. *J Plant Growth Regul* 42, 989–1005. https://doi.org/10.1007/s00344-022-10608-2

Al-Zahrani HS, Alharby HF and Fahad S (2022) Antioxidative defense system, hormones, and metabolite accumulation in different plant parts of two contrasting rice cultivars as influenced by plant growth regulators under heat stress. *Front Plant Sci* 13, 911846. doi: 10.3389/fpls.2022.911846

Amanullah, Fahad S (Eds.) (2017) *Rice – Technology and Production*. IntechOpen Croatia 2017. http://dx.doi.org/10.5772/64480

Amanullah, Fahad S (Eds.) (2018a) *Corn – Production and Human Health in Changing Climate*. IntechOpen United Kingdom 2018. http://dx.doi.org/10.5772/intechopen.74074

Amanullah, Fahad S (Eds.) (2018b) *Nitrogen in Agriculture – Updates*. IntechOpen Croatia 2018. http://dx.doi.org/10.5772/65846

Amanullah, Muhammad I, Haider N, Shah K, Manzoor A, Asim M, Saif U, Izhar A, Fahad S, Adnan M et al. (2021) Integrated foliar nutrients application improve wheat (Triticum aestivum L.) productivity under calcareous soils in drylands. *Commun Soil Sci Plant Anal* 52(21), 2748–2766. https://Doi.Org/10.1080/00103624.2021.1956521

Amanullah, Shah K, Imran, Hamdan AK, Muhammad A, Abdel RA, Muhammad A, Fahad S, Azizullah S, Brajendra P (2020) Effects of climate change on irrigation water quality. In: Fahad S, Hasanuzzaman M, Alam M, Ullah H, Saeed M, Khan AK, Adnan M (Eds.) *Environment, Climate, Plant and Vegetation Growth*. Springer Publ Ltd, Springer Nature Switzerland, pp. 123–132. https://doi.org/10.1007/978-3-030-49732-3

Amir M, Muhammad A, Allah B, Sevgi Ç, Haroon ZK, Muhammad A, Emre A (2020) Bio fortification under climate change: The fight between quality and quantity. In: Fahad S, Hasanuzzaman M, Alam M, Ullah H, Saeed M, Khan AK, Adnan M (Eds.) *Environment, Climate, Plant and Vegetation Growth*. Springer Publ Ltd, Springer Nature Switzerland, pp. 173–228. https://doi.org/10.1007/978-3-030-49732-3

Amjad I, Muhammad H, Farooq S, Anwar H (2020) Role of plant bioactives in sustainable agriculture. In: Fahad S, Hasanuzzaman M, Alam M, Ullah H, Saeed M, Khan AK, Adnan M (Eds.) *Environment, Climate, Plant and Vegetation Growth*. Springer Publ Ltd, Springer Nature Switzerland, pp. 591–606. https://doi.org/10.1007/978-3-030-49732-3

Amjad SF, Mansoora N, Din IU, Khalid IR, Jatoi GH, Murtaza G, Yaseen S, Naz M, Danish S, Fahad S et al. (2021) Application of zinc fertilizer and mycorrhizal inoculation on physio-biochemical parameters of wheat grown under water-stressed environment. *Sustainability* 13, 11007. https://doi.org/10.3390/su13 1911007

Anam I, Huma G, Ali H, Muhammad K, Muhammad R, Aasma P, Muhammad SC, Noman W, Sana F, Sobia A, Fahad S (2021) Ameliorative mechanisms of turmeric-extracted curcumin on arsenic (As)-induced biochemical alterations, oxidative damage, and impaired organ functions in rats. *Environ Sci Pollut Res* 28(46), 66313–66326. https://doi.org/10.1007/s11356-021-15695-4

Arif M, Talha J, Muhammad R, Fahad S, Muhammad A, Amanullah, Kawsar A, Ishaq AM, Bushra K, Fahd R (2020) Biochar: a remedy for climate change. In: Fahad S, Hasanuzzaman M, Alam M, Ullah H, Saeed M, Khan AK, Adnan M (Eds.), *Environment, Climate, Plant and Vegetation Growth*. Springer Publ Ltd, Springer Nature Switzerland, pp. 151–172. https://doi.org/10.1007/978-3-030-49732-3

Arshad M, Shaharoona B, Mahmood T (2008) Inoculation with Pseudomonas spp. containing ACC-deaminase partially eliminates the effects of drought stress on growth, yield, and ripening of pea (Pisum sativum L.). *Pedosphere* 18, 611–620.

Ashfaq AR, Uzma Y, Niaz A, Muhammad AA, Fahad S, Haider S, Tayebeh Z, Subhan D, Süleyman T, Hesham AElE, Pramila T, Jamal MA, Sulaiman AA, Rahul D (2021) Toxicity of cadmium and nickel in the context of applied activated carbon biochar for improvement in soil fertility. *Saudi Society Agric Sci* 29(2), 743–750. https://doi.org/10.1016/j.sjbs.2021.09.035

Athar M, Masood IA, Sana S, Ahmed M, Xiukang W, Sajid F, Sher AK, Habib A, Faran M, Zafar H, Farhana G, Fahad S (2021) Bio-diesel production of sunflower through sulphur management in a semi-arid subtropical environment. *Environ Sci Pollution Res* 29(9), 13268–13278. https://doi.org/10.1007/s11356-021-16688-z

Atif B, Hesham A, Fahad S (2021) Biochar coupling with phosphorus fertilization modifies antioxidant activity, osmolyte accumulation and reactive oxygen species synthesis in the leaves and xylem sap of rice cultivars under high-temperature stress. *Physiol Mol Biol Plants* 27(9), 2083–2100. https://doi.org/10.1007/s12298-021-01062-7

Awais M, Tariq M, Ali A, Ali Q, Khan A, Tabassum B, Nasir IA, Husnain T (2017) Isolation, characterization and inter-relationship of phosphate solubilizing bacteria from the rhizosphere of sugarcane and rice. *Biocatal Agric Biotechnol* 11, 312–321.

Ayman EL Sabagh, Akbar Hossain, Celaleddin Barutçular, Muhammad Aamir Iqbal, Sohidul Islam M, Shah Fahad, Oksana Sytar, Fatih Çig, Ram Swaroop Meena, Murat Erman (2020) Consequences of salinity stress on the quality of crops and its mitigation strategies for sustainable crop production: an outlook of arid and semi-arid regions. In: Fahad S, Hasanuzzaman M, Alam M, Ullah H, Saeed M, Khan AK, Adnan M (Eds.), *Environment, Climate, Plant and Vegetation Growth*. Springer Publ Ltd, Springer Nature Switzerland, pp. 503–534. https://doi.org/10.1007/978-3-030-49732-3

Aziz K, Daniel KYT, Fazal M, Muhammad ZA, Farooq S, Fan W, Fahad S, Ruiyang Z (2017a) Nitrogen nutrition in cotton and control strategies for greenhouse gas emissions: a review. *Environ Sci Pollut Res* 24, 23471–23487. https://doi.org/10.1007/s11356-017-0131-y

Aziz K, Daniel KYT, Muhammad ZA, Honghai L, Shahbaz AT, Mir A, Fahad S (2017b) Nitrogen fertility and abiotic stresses management in cotton crop: a review. *Environ Sci Pollut Res* 24, 14551–14566. https://doi.org/10.1007/s11356-017-8920-x

Bado VB, Bationo A (2018) Integrated management of soil fertility and land resources in Sub-Saharan Africa: involving local communities. *Adv in Agronomy* 150, 1–33.

Barrera-Bassols N, Zinck JA (2003) Ethnopedology: a worldwide view on the soil knowledge of local people. *Geoderma* 111, 171–195.

Barrios E, Delve RJ, Bekunda M, Mowo J, Agunda J, Ramisch J, Trejo M, Thomas RJ (2006) Indicators of soil quality: a south–south development of a methodological guide for linking local and technical knowledge. *Geoderma* 135, 248–259.

Baseer M, Adnan M, Fazal M, Fahad S, Muhammad S, Fazli W, Muhammad A, Jr Amanullah, Depeng W, Saud S, Muhammad N, Muhammad Z, Fazli S, Beena S, Mian AR, Ishaq AM (2019) Substituting urea by organic wastes for improving maize yield in alkaline soil. *J Plant Nutrition* 42(19), 2423–2434. doi.org/10.1080/01904167.2019.1659344

Bayram AY, Seher Ö, Nazlican A (2020) Climate change forecasting and modeling for the year of 2050. In: Fahad S, Hasanuzzaman M, Alam M, Ullah H, Saeed M, Khan AK, Adnan M (Eds.), *Environment, Climate, Plant and Vegetation Growth*. Springer Publ Ltd, Springer Nature Switzerland, pp. 109–122. https://doi.org/10.1007/978-3-030-49732-3

Belachew T, Abera Y (2010) Assessment of soil fertility status with depth in wheat growing highlands of Southeast Ethiopia. *World J Agric Sci* 6, 525–531.

Belhadi D, De Lajudie P, Ramdani N, Le Roux C, Boulila F, Tisseyre P, Boulila A, Benguedouar A, Kaci Y, Laguerre G (2018) Vicia faba L. in the Bejaia region of Algeria is nodulated by Rhizobium leguminosarum sv. viciae, Rhizobium laguerreae and two new genospecies. *Syst Appl Microbiol* 41, 122–130.

Berazneva J, McBride L, Sheahan M, Güereña D (2018) Empirical assessment of subjective and objective soil fertility metrics in east Africa: implications for researchers and policy makers. *World Dev* 105, 367–382.

Berruti A, Lumini E, Balestrini R, Bianciotto V (2016) Arbuscular mycorrhizal fungi as natural biofertilizers: let's benefit from past successes. *Front Microbiol* 6, 1559.

Bukhari MA, Adnan NS, Fahad S, Javaid I, Fahim N, Abdul M, Mohammad SB (2021) Screening of wheat (*Triticum aestivum* L.) genotypes for drought tolerance using polyethylene glycol. *Arabian J Geosci* 14, 2808.

Cassán F, Diaz-Zorita M (2016) Azospirillum sp. in current agriculture: from the laboratory to the field. *Soil Biol and Biochem* 103, 117–130.

Chandra D, Sharma A (2017) Commercial microbial products: exploiting beneficial plant-microbe interaction. In: Dhananjaya Pratap Singh, Harikesh Bahadur Singh, Ratna Prabha (Eds.), *Plant–Microbe Interactions in Agro-Ecological Perspectives*. Springer, pp. 607–626.

Chang W, Qiujuan J, Evgenios A, Haitao L, Gezi L, Jingjing Z, Fahad S, Ying J (2021) Hormetic effects of zinc on growth and antioxidant defense system of wheat plants. *Sci Total Environ* 807, 150992. https://doi.org/10.1016/j.scitotenv.2021.150992

Chao W, Youjin S, Beibei Q, Fahad S (2022) Effects of asymmetric heat on grain quality during the panicle initiation stage in contrasting rice genotypes. *J Plant Growth Regul* 42, 630–636. https://doi.org/10.1007/s00344-022-10598-1

Chen J, Liu Y-Q, Yan X-W, Wei G-H, Zhang J-H, Fang L-C (2018) Rhizobium inoculation enhances copper tolerance by affecting copper uptake and regulating the ascorbate-glutathione cycle and phytochelatin biosynthesis-related gene expression in Medicago sativa seedlings. *Ecotoxicol and Environ Safety* 162, 312–323.

Chen Y, Guo Z, Dong L, Fu Z, Zheng Q, Zhang G, Qin L, Sun X, Shi Z, Fahad S, Xie F, Saud S (2021) Turf performance and physiological responses of native Poa species to summer stress in Northeast China. *Peer J* 9, e12252. http://doi.org/10.7717/peerj.12252

Chen Y, Rekha P, Arun A, Shen F, Lai W-A, Young CC (2006) Phosphate solubilizing bacteria from subtropical soil and their tricalcium phosphate solubilizing abilities. *Appl Soil Ecol* 34, 33–41.

Chukwuka K (2009) Soil fertility restoration techniques in sub-Saharan Africa using organic resources. *African J Agric Res* 4, 144–150.

Damir O, Mladen P, Božidar S, Srñan N (2011) Cultivation of the bacterium Azotobacter chroococcum for preparation of biofertilizers. *African J Biotechnol* 10, 3104–3111.

Danso S (1992) Twenty years of biological nitrogen fixation research. *Biol Nitrogen Fixation & Sustainability Trop Agriculture* 4, 3.

Deepranjan S, Ardith SO, Siva D, Sonam S, Shikha, Manoj P, Amitava R, Sayyed RZ, Abdul G, Mohammad JA, Subhan D, Fahad S, Rahul D (2021) Optimizing nutrient use efficiency, productivity, energetics, and economics of red cabbage following mineral fertilization and biopriming with compatible rhizosphere microbes. *Sci Rep* 11, 15680. https://doi.org/10.1038/s41598-021-95092-6

Depeng W, Fahad S, Saud S, Muhammad K, Aziz K, Mohammad NK, Hafiz MH, Wajid N (2018) Morphological acclimation to agronomic manipulation in leaf dispersion and orientation to promote "Ideotype" breeding: evidence from 3D visual modeling of "super" rice (*Oryza sativa* L.). *Plant Physiol Biochem* 135, 499–510. https://doi.org/10.1016/j.plaphy.2018.11.010

EL Sabagh A, Islam MS, Hossain A, Iqbal MA, Mubeen M, Waleed M, Reginato M, Battaglia M, Ahmed S, Rehman A, Arif M, Athar H-U-R, Ratnasekera D, Danish S, Raza MA, Rajendran K, Mushtaq M, Skalicky M, Brestic M, Soufan W, Fahad S, Pandey S, Kamran M, Datta R, Abdelhamid MT (2022) Phytohormones as growth regulators during abiotic stress tolerance in plants. *Front Agron* 4, 765068. doi: 10.3389/fagro.2022.765068

Emre B, Ömer SU, Martín LB, Andre D, Fahad S, Rahul D, Muhammad Z-ul-H, Ghulam SH, Subhan D (2021) Studying soil erosion by evaluating changes in physico-chemical properties of soils under different land-use types. *J Saudi Society Agricultural Sci* 20, 190–197.

Emrooz HBM, Maleki M, Rahmani A (2018) Azolla-derived hierarchical nanoporous carbons: from environmental concerns to industrial opportunities. *Journal of the Taiwan Institute of Chemical Engineers* 91, 281–290.

Fahad S, Bano A (2012) Effect of salicylic acid on physiological and biochemical characterization of maize grown in saline area. *Pak J Bot* 44, 1433–1438.

Fahad S, Chen Y, Saud S, Wang K, Xiong D, Chen C, Wu C, Shah F, Nie L, Huang J (2013) Ultraviolet radiation effect on photosynthetic pigments, biochemical attributes, antioxidant enzyme activity and hormonal contents of wheat. *J Food, Agri Environ* 11(3&4), 1635–1641.

Fahad S, Hussain S, Bano A, Saud S, Hassan S, Shan D, Khan FA, Khan F, Chen Y, Wu C, Tabassum MA, Chun MX, Afzal M, Jan A, Jan MT, Huang J (2014a) Potential role of phytohormones and plant growth-promoting rhizobacteria in abiotic stresses: consequences for changing environment. *Environ Sci Pollut Res* 22(7), 4907–4921. https://doi.org/10.1007/s11356-014-3754-2

Fahad S, Hussain S, Matloob A, Khan FA, Khaliq A, Saud S, Hassan S, Shan D, Khan F, Ullah N, Faiq M, Khan MR, Tareen AK, Khan A, Ullah A, Ullah N, Huang J (2014b) Phytohormones and plant responses to salinity stress: a review. *Plant Growth Regul* 75(2), 391–404. https://doi.org/10.1007/s10725-014-0013-y

Fahad S, Hussain S, Saud S, Tanveer M, Bajwa AA, Hassan S, Shah AN, Ullah A, Wu C, Khan FA, Shah F, Ullah S, Chen Y, Huang J (2015a) A biochar application protects rice pollen from high-temperature stress. *Plant Physiol Biochem* 96, 281–287.

Fahad S, Nie L, Chen Y, Wu C, Xiong D, Saud S, Hongyan L, Cui K, Huang J (2015b) Crop plant hormones and environmental stress. *Sustain Agric Rev* 15, 371–400.

Fahad S, Hussain S, Saud S, Hassan S, Chauhan BS, Khan F et al. (2016a) Responses of rapid viscoanalyzer profile and other rice grain qualities to exogenously applied plant growth regulators under high day and high night temperatures. *PLoS One* 11(7), e0159590. https://doi.org/10.1371/journal.pone.0159590

Fahad S, Hussain S, Saud S, Khan F, Hassan S, Jr A, Nasim W, Arif M, Wang F, Huang J (2016b) Exogenously applied plant growth regulators affect heat-stressed rice pollens. *J Agron Crop Sci* 202, 139–150.

Fahad S, Hussain S, Saud S, Hassan S, Ihsan Z, Shah AN, Wu C, Yousaf M, Nasim W, Alharby H, Alghabari F, Huang J (2016c) Exogenously applied plant growth regulators enhance the morphophysiological growth and yield of rice under high temperature. *Front Plant Sci* 7, 1250. https://doi.org/10.3389/fpls.2016. 01250

Fahad S, Hussain S, Saud S, Hassan S, Tanveer M, Ihsan MZ, Shah AN, Ullah A, Nasrullah KF, Ullah S, AlharbyH NW, Wu C, Huang J (2016d) A combined application of biochar and phosphorus alleviates heat-induced adversities on physiological, agronomical and quality attributes of rice. *Plant Physiol Biochem* 103, 191–198.

Fahad S, Bajwa AA, Nazir U, Anjum SA, Farooq A, Zohaib A, Sadia S, NasimW, Adkins S, Saud S, Ihsan MZ, Alharby H,Wu C,Wang D, Huang J (2017) Crop production under drought and heat stress: plant responses and management options. *Front Plant Sci* 8, 1147. https://doi.org/10.3389/fpls.2017.01147

Fahad S, Muhammad ZI, Abdul K, Ihsanullah D, Saud S, Saleh A, Wajid N, Muhammad A, Imtiaz AK, Chao W, Depeng W, Jianliang H (2018a) Consequences of high temperature under changing climate optima for rice pollen characteristics-concepts and perspectives. *Archives Agron Soil Sci* 64(11), 1473–1488. doi: 10.1080/03650340.2018.1443213

Fahad S, Abdul B, Adnan M (Eds.) (2018b) *Global Wheat Production*. IntechOpen United Kingdom 2018. http://dx.doi.org/10.5772/intechopen.72559

Fahad S, Rehman A, Shahzad B, Tanveer M, Saud S, Kamran M, Ihtisham M, Khan SU, Turan V, Rahman MHU (2019a) Rice responses and tolerance to metal/metalloid toxicity. In: Hasanuzzaman M, Fujita M, Nahar K, and Biswas JK (Eds.) *Advances in Rice Research for Abiotic Stress Tolerance*. Woodhead Publishing, UK, pp. 299–312.

Fahad S, Adnan M, Hassan S, Saud S, Hussain S, Wu C, Wang D, Hakeem KR, Alharby HF, Turan V, Khan MA, Huang J (2019b) Rice responses and tolerance to high temperature. In: Hasanuzzaman M, Fujita M, Nahar K, and Biswas, JK (Eds.) *Advances in Rice Research for Abiotic Stress Tolerance*. Woodhead Publishing Ltd, UK, pp. 201–224.

Fahad S, Hasanuzzaman M, Alam M, Ullah H, Saeed M, Ali Khan I, Adnan M (Eds.) (2020) *Environment, Climate, Plant and Vegetation Growth*. Springer Nature Switzerland AG 2020. doi: https://doi.org/10.1007/978-3-030-49732-3

Fahad S, Sönmez O, Saud S, Wang D, Wu C, Adnan M, Turan V (Eds.) (2021a) Plant growth regulators for climate-smart agriculture. First edition. *Footprints of Climate Variability on Plant Diversity*. CRC Press, Boca Raton.

Fahad S, Sonmez O, Saud S, Wang D, Wu C, Adnan M, Turan V (Eds.) (2021b) Climate change and plants: biodiversity, growth and interactions. First edition. *Footprints of Climate Variability on Plant Diversity*. CRC Press, Boca Raton.

Fahad S, Sonmez O, Saud S, Wang D, Wu C, Adnan M, Turan V (Eds.) (2021c) Developing climate resilient crops: improving global food security and safety. First edition. *Footprints of Climate Variability on Plant Diversity*. CRC Press, Boca Raton.

Fahad S, Sönmez O, Turan V, Adnan M, Saud S, Wu C, Wang D (Eds.) (2021d) Sustainable soil and land management and climate change. First edition. *Footprints of Climate Variability on Plant Diversity*. CRC Press, Boca Raton.

Fahad S, Sönmez O, Saud S, Wang D, Wu C, Adnan M, Arif M, Amanullah (Eds.) (2021e) Engineering tolerance in crop plants against abiotic stress. First edition. Footprints of Climate Variability on Plant Diversity. CRC Press, Boca Raton.

Fahad S, Saud S, Yajun C, Chao W, Depeng W (Eds.) (2021f) *Abiotic Stress in Plants*. IntechOpen United Kingdom 2021. http://dx.doi.org/10.5772/intechopen.91549

Fahad S, Adnan M, Saud S (Eds.) (2022a) Improvement of plant production in the era of climate change. First edition. *Footprints of Climate Variability on Plant Diversity*. CRC Press, Boca Raton.

Fahad S, Adnan M, Saud S, Nie L (Eds.) (2022b) Climate change and ecosystems: challenges to sustainable development. First edition. *Footprints of Climate Variability on Plant Diversity*. CRC Press, Boca Raton.

Fakhre A, Ayub K, Fahad S, Sarfraz N, Niaz A, Muhammad AA, Muhammad A, Khadim D, Saud S, Shah H, Muhammad ASR, Khalid N, Muhammad A, Rahul D, Subhan D (2021) Phosphate solubilizing bacteria optimize wheat yield in mineral phosphorus applied alkaline soil. *J Saudi Soc Agric Sci* 21(5), 339–348. https://doi.org/10.1016/j.jssas.2021.10.007

Farah R, Muhammad R, Muhammad SA, Tahira Y, Muhammad AA, Maryam A, Shafaqat A, Rashid M, Muhammad R, Qaiser H, Afia Z, Muhammad AA, Muhammad A, Fahad S (2020) Alternative and non-conventional soil and crop management strategies for increasing water use efficiency. In: Fahad S, Hasanuzzaman M, Alam M, Ullah H, Saeed M, Khan AK, Adnan M (Eds.), *Environment, Climate, Plant and Vegetation Growth*. Springer Publ Ltd, Springer Nature Switzerland, pp. 323–338. https://doi.org/10.1007/978-3-030-49732-3

Farhana G, Ishfaq A, Muhammad A, Dawood J, Fahad S, Xiuling L, Depeng W, Muhammad F, Muhammad F, Syed AS (2020) Use of crop growth model to simulate the impact of climate change on yield of various wheat cultivars under different agro-environmental conditions in Khyber Pakhtunkhwa, Pakistan. *Arabian J Geosci* 13, 112. https://doi.org/10.1007/s12517-020-5118-1

Farhat A, Hafiz MH, Wajid I, Aitazaz AF, Hafiz FB, Zahida Z, Fahad S, Wajid F, Artemi C (2020) A review of soil carbon dynamics resulting from agricultural practices. *J Environ Manage* 268, 110319.

Farhat UK, Adnan AK, Kai L, Xuexuan X, Muhammad A, Fahad S, Rafiq A, Mushtaq AK, Taufiq N, Faisal Z (2022) Influences of long-term crop cultivation and fertilizer management on soil aggregates stability and fertility in the Loess Plateau, Northern China. *J Soil Sci Plant Nutri* 22, 1446–1457. https://doi.org/10.1007/s42729-021-00744-1

Fazli W, Muhmmad S, Amjad A, Fahad S, Muhammad A, Muhammad N, Ishaq AM, Imtiaz AK, Mukhtar A, Muhammad S, Muhammad I, Rafi U, Haroon I, Muhammad A (2020) Plant–microbes interactions and functions in changing climate. In: Fahad S, Hasanuzzaman M, Alam M, Ullah H, Saeed M, Khan AK, Adnan M (Eds.), *Environment, Climate, Plant and Vegetation Growth*. Springer Publ Ltd, Springer Nature Switzerland, pp. 397–420. https://doi.org/10.1007/978-3-030-49732-3

Friedrich S, Platonova N, Karavaiko G, Stichel E, Glombitza F (1991) Chemical and microbiological solubilization of silicates. *Acta Biotechnologica* 11, 187–196.

Fukami J, Cerezini P, Hungria M (2018) Azospirillum: benefits that go far beyond biological nitrogen fixation. *Amb Express* 8, 1–12.

Funga A, Ojiewo CO, Turoop L, Mwangi GS (2016) Symbiotic effectiveness of elite rhizobia strains nodulating desi type chickpea (Cicer arietinum L.) varieties. *J Plant Sci* 4, 88–94.

Gadd GM (2010) Metals, minerals and microbes: geomicrobiology and bioremediation. *Microbiol* 156, 609–643.

Ghasem S, Morteza AS, Maryam T (2014) Effect of organic fertilizers on cucumber (Cucumis sativus) yield. *Int J Agric & Crop Sci* 7, 808.

Ghulam M, Muhammad AA, Donald LS, Sajid M, Muhammad FQ, Niaz A, Ateeq ur R, Shakeel A, Sajjad H, Muhammad A, Summia M, Aqib HAK, Fahad S, Rahul D, Mazhar I, Timothy DS (2021) Formalin fumigation and steaming of various composts differentially influence the nutrient release, growth and yield of muskmelon (*Cucumis melo* L.). *Sci Rep* 11, 21057.

Gonzalez AJ, Larraburu EE, Llorente BE (2015) Azospirillum brasilense increased salt tolerance of jojoba during in vitro rooting. *Ind Crops and Products* 76, 41–48.

Gopakumar L, Bernard NO, Donato V (2020) Soil microarthropods and nutrient cycling. In: Fahad S, Hasanuzzaman M, Alam M, Ullah H, Saeed M, Khan AK, Adnan M (Eds.), *Environment, Climate, Plant and Vegetation Growth*. Springer Publ Ltd, Springer Nature Switzerland, pp. 453–472. https://doi.org/10.1007/978-3-030-49732-3

Gouda S, Kerry RG, Das G, Paramithiotis S, Shin H-S, Patra JK (2018) Revitalization of plant growth promoting rhizobacteria for sustainable development in agriculture. *Microbiol Res* 206, 131–140.

Guofu L, Zhenjian B, Fahad S, Guowen C, Zhixin X, Hao G, Dandan L, Yulong L, Bing L, Guoxu J, Saud S (2021) Compositional and structural changes in soil microbial communities in response to straw mulching and plant revegetation in an abandoned artificial pasture in Northeast China. *Global Ecol and Conserv* 31(2021), e01871.

Gupta G, Parihar SS, Ahirwar NK, Snehi SK, Singh V (2015) Plant growth promoting rhizobacteria (PGPR): current and future prospects for development of sustainable agriculture. *J Microb Biochem Technol* 7, 96–102.

Guzman CD, Tilahun SA, Dagnew DC, Zegeye AD, Yitaferu B, Kay RW, Steenhuis TS (2018) Developing soil conservation strategies with technical and community knowledge in a degrading sub-humid mountainous landscape. *Land Degrad & Dev* 29, 749–764.

Habib ur R, Ashfaq A, Aftab W, Manzoor H, Fahd R, Wajid I, Md. Aminul I, Vakhtang S, Muhammad A, Asmat U, Abdul W, Syeda RS, Shah S, Shahbaz K, Fahad S, Manzoor H, Saddam H, Wajid N (2017) Application of CSM-CROPGRO-Cotton model for cultivars and optimum planting dates: evaluation in changing semi-arid climate. *Field Crops Res* 238, 139–152. http://dx.doi.org/10.1016/j.fcr.2017.07.007

Hafeez F, Ahmad T, Hameed S, Danso S, Malik K (1998) Comparison of direct and indirect methods of measuring nitrogen fixation in field grown chickpea genotypes. *Pak J Bot* 30, 199–208.

Hafeez M, Farman U, Muhammad MK, Xiaowei L, Zhijun Z, Sakhawat S, Muhammad I, Mohammed AA, Mandela F-G, Nicolas D, Muzammal R, Fahad S, Yaobin L (2021) Metabolic-based insecticide resistance mechanism and ecofriendly approaches for controlling of beet armyworm Spodoptera exigua: a review. *Environ Sci Pollution Res* 29, 1746–1762. https://doi.org/10.1007/s11356-021-16974-w

Hafiz MH, Wajid F, Farhat A, Fahad S, Shafqat S, Wajid N, Hafiz FB (2016) Maize plant nitrogen uptake dynamics at limited irrigation water and nitrogen. *Environ Sci Pollut Res* 24(3), 2549–2557. https://doi.org/10.1007/s11356-016-8031-0

Hafiz MH, Farhat A, Shafqat S, Fahad S, Artemi C, Wajid F, Chaves CB, Wajid N, Muhammad M, Hafiz FB (2018) Offsetting land degradation through nitrogen and water management during maize cultivation under arid conditions. *Land Degrad Dev* 29(5), 1–10. doi: 10.1002/ldr.2933

Hafiz MH, Muhammad A, Farhat A, Hafiz FB, Saeed AQ, Muhammad M, Fahad S, Muhammad A (2019) Environmental factors affecting the frequency of road traffic accidents: a case study of sub-urban area of Pakistan. *Environ Sci Pollut Res* 26, 11674–11685. https://doi.org/10.1007/s11356-019-04752-8

Hafiz MH, Farhat A, Ashfaq A, Hafiz FB, Wajid F, Carol Jo W, Fahad S, Gerrit H (2020a) Predicting kernel growth of maize under controlled water and nitrogen applications. *Int J Plant Prod* 14, 609–620. https://doi.org/10.1007/s42106-020-00110-8

Hafiz MH, Abdul K, Farhat A, Wajid F, Fahad S, Muhammad A, Ghulam MS, Wajid N, Muhammad M, Hafiz FB (2020b) Comparative effects of organic and inorganic fertilizers on soil organic carbon and wheat productivity under arid region. *Commun Soil Sci Plant Anal* 51, 1406–1422. doi: 10.1080/00103624.2020.1763385

Haider SA, Lalarukh I, Amjad SF, Mansoora N, Naz M, Naeem M, Bukhari SA, Shahbaz M, Ali SA, Marfo TD, Subhan D, Rahul D, Fahad S (2021) Drought stress alleviation by potassium-nitrate-containing chitosan/montmorillonite microparticles confers changes in Spinacia oleracea L. *Sustain* 13, 9903. https://doi.org/10.3390/su13179903

Hamza SM, Xiukang W, Sajjad A, Sadia Z, Muhammad N, Adnan M, Fahad S et al. (2021) Interactive effects of gibberellic acid and NPK on morpho-physio-biochemical traits and organic acid exudation pattern in coriander (Coriandrum sativum L.) grown in soil artificially spiked with boron. *Plant Physiol and Biochem* 167 (2021) 884–900.

Haoliang Y, Matthew TH, Ke L, Bin W, Puyu F, Fahad S, Holger M, Rui Y, De LL, Sotirios A, Isaiah H, Xiaohai T, Jianguo M, Yunbo Z, Meixue Z (2022) Crop traits enabling yield gains under more frequent extreme climatic events. *Sci Total Environ* 808(2022), 152170.

Hellriegel H, Wilfarth H (1888) Untersuchungen iiber die Stickstoffnahrung der Gramineen und Leguminosen. *Beilageheft zu der Ztschr.* Ver. Riibenzucker-Industrie Deutschen Reichs.

Hesham FA, Fahad S (2020) Melatonin application enhances biochar efficiency for drought tolerance in maize varieties: modifications in physio-biochemical machinery. *Agron J* 112(4), 1–22.

Huang Li-Y, Li Xiao-X, Zhang Yun-B, Fahad S, Wang F (2021) dep1 improves rice grain yield and nitrogen use efficiency simultaneously by enhancing nitrogen and dry matter translocation. *J Integrative Agri* 21(11), 3185–3198. doi: 10.1016/S2095-3119(21)63795-4

Hussain M, Mehboob I, Zahir Z, Naveed M, Asghar H (2009) Potential of Rhizobium spp. for improving growth and yield of rice (Oryza sativa L.). *Soil Environ* 28, 49–55

Hussain MA, Fahad S, Rahat S, Muhammad FJ, Muhammad M, Qasid A, Ali A, Husain A, Nooral A, Babatope SA, Changbao S, Liya G, Ibrar A, Zhanmei J, Juncai H (2020) Multifunctional role of brassinosteroid and its analogues in plants. *Plant Growth Regul* 92, 141–156. https://doi.org/10.1007/s10725-020-00647-8

Ibad U, Dost M, Maria M, Shadman K, Muhammad A, Fahad S, Muhammad I, Ishaq AM, Aizaz A, Muhammad HS, Muhammad S, Farhana G, Muhammad I, Muhammad ASR, Hafiz MH, Wajid N, Shah S, Jabar ZKK, Masood A, Naushad A, Rasheed Akbar M, Shah MK Jan B (2022) Comparative effects of biochar and NPK on wheat crops under different management systems. *Crop Pasture Sci* 74, 31–40. https://doi.org/10.1071/CP21146

Ibrar H, Muqarrab A, Adel MG, Khurram S, Omer F, Shahid I, Fahim N, Shakeel A, Viliam B, Marian B, Sami Al Obaid, Fahad S, Subhan D, Suleyman T, Hanife AKÇA, Rahul D (2021) Improvement in growth and yield attributes of cluster bean through optimization of sowing time and plant spacing under climate change Scenario. *Saudi J Bio Sci* 29(2), 781–792. https://doi.org/10.1016/j.sjbs.2021.11.018

Ibrar K, Aneela R, Khola Z, Urooba N, Sana B, Rabia S, Ishtiaq H, Mujaddad Ur Rehman, Salvatore M (2020) Microbes and environment: global warming reverting the frozen zombies. In: Fahad S, Hasanuzzaman M, Alam M, Ullah H, Saeed M, Khan AK, Adnan M (Eds.) *Environment, Climate, Plant and Vegetation Growth.* Springer Publ Ltd, Springer Nature Switzerland, pp. 607–634. https://doi.org/10.1007/978-3-030-49732-3

Ihsan MZ, Abdul K, Manzer HS, Liaqat A, Ritesh K, Hayssam MA, Amar M, Fahad S (2022) The response of Triticum aestivum treated with plant growth regulators to acute day/night temperature rise. *J Plant Growth Regul* 41, 2020–2033. https://doi.org/10.1007/s00344-022-10574-9

Ikram U, Khadim D, Muhammad T, Muhammad S, Fahad S (2021) Gibberellic acid and urease inhibitor optimize nitrogen uptake and yield of maize at varying nitrogen levels under changing climate. *Environ Sci Pollution Res* 29(5), 6568–6577. https://doi.org/10.1007/s11356-021-16049-w

Ilyas M, Mohammad N, Nadeem K, Ali H, Aamir HK, Kashif H, Fahad S, Aziz K, Abid U (2020) Drought tolerance strategies in plants: a mechanistic approach. *J Plant Growth Regul* 40, 926–944. https://doi.org/10.1007/s00344-020-10174-5

Ingram J, Dwyer J, Gaskell P, Mills J, de Wolf P (2018) Reconceptualising translation in agricultural innovation: a co-translation approach to bring research knowledge and practice closer together. *Land Use Policy* 70, 38–51.

Ingram J, Fry P, Mathieu A (2010) Revealing different understandings of soil held by scientists and farmers in the context of soil protection and management. *Land Use Policy* 27, 51–60.

Iqbal A, He L, Khan A, Wei S, Akhtar K, Ali I, Ullah S, Munsif F, Zhao Q, Jiang L (2019) Organic manure coupled with inorganic fertilizer: an approach for the sustainable production of rice by improving soil properties and nitrogen use efficiency. *Agronomy* 9, 651.

Iqbal A, He L, Ali I, Ullah S, Khan A, Khan A, Akhtar K, Wei S, Zhao Q, Zhang J (2020) Manure combined with chemical fertilizer increases rice productivity by improving soil health, post-anthesis biomass yield, and nitrogen metabolism. *Plos one* 15, e0238934.

Iqbal A, He L, Ali I, Ullah S, Khan A, Akhtar K, Wei S, Fahad S, Khan R, Jiang L (2021a) Co-incorporation of manure and inorganic fertilizer improves leaf physiological traits, rice production and soil functionality in a paddy field. *Sci Rep* 11, 1–16.

Iqbal A, He L, McBride SG, Ali I, Akhtar K, Khan R, Zaman M, We S, Guo Z, Jiang L (2021b) Manure applications combined with chemical fertilizer improves soil functionality, microbial biomass and rice production in a paddy field. *Research Square* 114(2), 1431–1446. doi: 10.21203/rs.3.rs-461485/v1.

Iqbal A, Khan A, Green SJ, Ali I, He L, Zeeshan M, Luo Y, Wu X, Wei S, Jiang L (2021c) Long-term straw mulching in a no-till field improves soil functionality and rice yield by increasing soil enzymatic activity and chemical properties in paddy soils. *J Plant Nutr & Soil Sci* 184, 622–634.

Iqbal B, Khan H, Saifullah, Khan I, Shah B, Naeem A, Ullah W, Khan N, Adnan M, Shah SRA, Junaid K, Ahmed N, Iqbal M (2016) Substrates evaluation for the quality, production and growth of oyster mushroom (*Pleurotusflorida*Cetto). *J Ento & Zool Studies* 4(3), 98–107.

Iqra M, Amna B, Shakeel I, Fatima K,Sehrish L, Hamza A, Fahad S (2020) Carbon cycle in response to global warming. In: Fahad S, Hasanuzzaman M, Alam M, Ullah H, Saeed M, Khan AK, Adnan M (Eds.), *Environment, Climate, Plant and Vegetation Growth.* Springer Publ Ltd, Springer Nature Switzerland, pp. 1–16. https://doi.org/10.1007/978-3-030-49732-3

Irfan M, Muhammad M, Muhammad JK, Khadim MD, Dost M, Ishaq AM, Waqas A, Fahad S, Saud S et al. (2021) Heavy metals immobilization and improvement in maize (Zea mays L.) growth amended with biochar and compost. *Sci Rep* 11, 18416.

Isfahani FM, Besharati H (2012) Effect of biofertilizers on yield and yield components of cucumber. *J Biol Earth Sci* 2, B83–92.

Itelima J, Bang W, Onyimba I, Sila M, Egbere O (2018) Bio-fertilizers as key player in enhancing soil fertility and crop productivity: a review. *Direct Res J Agric Food Sci* 6(3), 73–83.

Izhar Shafi M, Adnan M, Fahad S, Wahid F, Khan A, Yue Z, Danish S, Zafar-ul-Hye M, Brtnicky M, Datta R (2020) Application of single superphosphate with humic acid improves the growth, yield and phosphorus uptake of wheat (Triticum aestivum L.) in calcareous soil. *Agronomy* 10, 1224.

Jabborova D, Sulaymanov K, Sayyed RZ, Alotaibi SH, Enakiev Y, Azimov A, Jabbarov Z, Ansari MJ, Fahad S, Danish S et al. (2021) Mineral fertilizers improve the quality of turmeric and soil. *Sustain* 13, 9437. https://doi.org/10.3390/su13169437

Jan M, Muhammad Anwar-ul-Haq, Adnan NS, Muhammad Y, Javaid I, Xiuling L, Depeng W, Fahad S (2019) Modulation in growth, gas exchange, and antioxidant activities of salt-stressed rice (Oryza sativa L.) genotypes by zinc fertilization. *Arabian J Geosci* 12, 775. https://doi.org/10.1007/s12517-019-4939-2

Jilani MS, Bakar A, Waseem K, Kiran M (2009) Effect of different levels of NPK on the growth and yield of cucumber (Cucumis sativus) under the plastic tunnel. *J Agric Soc Sci* 5, 99–101.

Kamaran M, Wenwen C, Irshad A, Xiangping M, Xudong Z, Wennan S, Junzhi C, Shakeel A, Fahad S, Qingfang H, Tiening L (2017) Effect of paclobutrazol, a potential growth regulator on stalk mechanical strength, lignin accumulation and its relation with lodging resistance of maize. *Plant Growth Regul* 84, 317–332. https://doi.org/10.1007/ s10725-017-0342-8

Kanaujia S, Daniel M (2016) Integrated nutrient management for quality production and economics of cucumber on acid alfisol of Nagaland. *Annals Plant & Soil Res* 18, 375–380.

Karlen DL, Ditzler CA, Andrews SS (2003) Soil quality: why and how? *Geoderma* 114, 145–156.

Kaur P, Dhull SB, Sandhu KS, Salar RK, Purewal SS (2018a) Tulsi (Ocimum tenuiflorum) seeds: in vitro DNA damage protection, bioactive compounds and antioxidant potential. *J Food Meas & Charact* 12, 1530–1538.

Kaur P, Purewal SS (2019) Biofertilizers and their role in sustainable agriculture. In: Bhoopander Giri, Ram Prasad, Qiang-Sheng Wu, Ajit Varma (Eds.), *Biofertilizers for Sustainable Agriculture and Environment.* Springer, pp. 285–300

Kaur R, Kaur M, Purewal SS (2018b) Effect of incorporation of flaxseed to wheat rusks: antioxidant, nutritional, sensory characteristics, and in vitro DNA damage protection activity. *J Food Process Preserv* 42, e13585.

Khadim D, Fahad S, Jahangir MMR, Iqbal M, Syed SA, Shah AK, Ishaq AM, Rahul D et al. (2021a) Biochar and urease inhibitor mitigate NH_3 and N_2O emissions and improve wheat yield in a urea fertilized alkaline soil. *Sci Rep* 11, 17413.

Khadim D, Saif-ur-R, Fahad S, Syed SA, Shah AK et al. (2021b) Influence of variable biochar concentration on yield-scaled nitrous oxide emissions, wheat yield and nitrogen use efficiency. *Sci Rep* 11, 16774.

Khalid A, Arshad M, Shaharoona B, Mahmood T (2009) Plant growth promoting rhizobacteria and sustainable agriculture. In: Mohammad Saghir Khan, Almas Zaidi, Javed Musarrat (Eds.), *Microbial Strategies for Crop Improvement.* Springer, pp. 133–160.

Khan A, Ahmad D, Shah Hashmi H (2013) Review of Available Knowledge on Land Degradation in Pakistan. Aleppo, Syrian Arab Republic: International Center for Agricultural Research in the Dry Areas (ICARDA).

Khan A, Shah N, Muhammad, Khan MS, Ahmad MS, Farooq M, Adnan M, Jawad SM, Ullah H, Yousafzai AM (2016) Quantitative determination of lethal concentration LC50 of atrazine on biochemical parameters; Total protein and serum albumin of freshwater fish grass carp (*Ctenopharyngodon idella*). *Pol J Environ Stud* 25(4), 1–7.

Khan A, Yousafzai AM, Shah N, Adnan M, Jehan S, Jawad SM (2017) Scrutinizing impact of atrazine on stress biomarker: plasma glucose concentration of fresh water fish, grass carp (*Ctenopharyngodon idella*). *Pure Appl Biol* 6(4), 1319–1327.

Khan M, Ullah F, Zainub B, Khan M, Zeb A, Ahmad K, Arshad R (2017) Effects of poultry manure levels on growth and yield of cucumber cultivars. *Sci. Int. (Lahore)* 29, 1381–1386.

Khan MMH, Niaz A, Umber G, Muqarrab A, Muhammad AA, Muhammad I, Shabir H, Shah F, Vibhor A, Shams HA-H, Reham A, Syed MBA, Nadiyah MA, Ali TKZ, Subhan D, Rahul D (2021) Synchronization of Boron application methods and rates is environmentally friendly approach to improve quality attributes of *Mangifera indica* L. On sustainable basis. *Saudi J Bio Sci* 29(3), 1869–1880. https://doi.org/10.1016/j.sjbs.2021.10.036

Khatoon Z, Huang S, Rafique M, Fakhar A, Kamran MA, Santoyo G (2020) Unlocking the potential of plant growth-promoting rhizobacteria on soil health and the sustainability of agricultural systems. *J Environ Manage* 273, 111118.

Khatun M, Sarkar S, Era FM, Islam AKMM, Anwar MP, Fahad S, Datta R, Islam AKMA (2021) Drought stress in grain legumes: effects, tolerance mechanisms and management. *Agron* 11, 2374. https://doi.org/10.3390/agronomy11122374

Kumar M, Kathayat K, Singh SK, Singh L, Singh T (2018) Influence of bio-fertilizers application on growth, yield and quality attributes of cucumber (Cucumis sativus L.): a review. *Plant Arch* 18, 2329–2334.

Kumar R, Kumar R, Prakash O (2019) The impact of chemical fertilizers on our environment and ecosystem. In: *Research Trends in Environmental Sciences*, 2nd Edition, pp. 71–86. *Chief Editor*, pp. 35, 69.

Kumari P, Meena M, Gupta P, Dubey MK, Nath G, Upadhyay R (2018) Plant growth promoting rhizobacteria and their biopriming for growth promotion in mung bean (Vigna radiata (L.) R. Wilczek). *Biocatalysis & Agric Biotechnol* 16, 163–171.

Lal R (2018) Managing agricultural soils of Pakistan for food and climate. *Soil & Environ* 37, 1–10. https://doi.org/10.25252/se/18/61527

Llorente BE, Alasia MA, Larraburu EE (2016) Biofertilization with Azospirillum brasilense improves in vitro culture of Handroanthus ochraceus, a forestry, ornamental and medicinal plant. *New Biotechnol* 33, 32–40.

Luo Y, Iqbal A, He L, Zhao Q, Wei S, Ali I, Ullah S, Yan B, Jiang L (2020) Long-term no-tillage and straw retention management enhances soil bacterial community diversity and soil properties in Southern China. *Agronomy* 10, 1233.

Mahanta D, Rai R, Dhar S, Varghese E, Raja A, Purakayastha T (2018) Modification of root properties with phosphate solubilizing bacteria and arbuscular mycorrhiza to reduce rock phosphate application in soybean-wheat cropping system. *Ecol Eng* 111, 31–43.

Mahar A, Amjad A, Altaf HL, Fazli W, Ronghua L, Muhammad A, Fahad S, Muhammad A, Rafiullah, Imtiaz AK, Zengqiang Z (2020) Promising technologies for Cd-contaminated soils: drawbacks and possibilities. In: Fahad S, Hasanuzzaman M, Alam M, Ullah H, Saeed M, Khan AK, Adnan M (Eds.) *Environment, Climate, Plant and Vegetation Growth.* Springer Publ Ltd, Springer Nature Switzerland, pp. 63–92. https://doi.org/10.1007/978-3-030-49732-3

Mahmood Ul H, Tassaduq R, Chandni I, Adnan A, Muhammad A, Muhammad MA, Muhammad H-ur-R, Mehmood AN, Alam S, Fahad S (2021) Linking plants functioning to adaptive responses under heat stress conditions: a mechanistic review. *J Plant Growth Regul* 41, 2596–2613. https://doi.org/10.1007/s00344-021-10493-1

Mahmoud E, Abd EL-Kader N, Robin P, Akkal-Corfini N, Abd El-Rahman L (2009) Effects of different organic and inorganic fertilizers on cucumber yield and some soil properties. *World J Agric Sci* 5, 408–414.

Mairura FS, Mugendi DN, Mwanje J, Ramisch JJ, Mbugua P, Chianu JN (2007) Integrating scientific and farmers' evaluation of soil quality indicators in Central Kenya. *Geoderma* 139, 134–143.

Malik K, Mirza M, Hassan U, Mehnaz S, Rasul G, Haurat J, Bally R, Normand P (2002) The role of plant-associated beneficial bacteria in rice-wheat cropping system. *Biofertilisers in Action.* Rural Industries Research and Development Corporation, Canberra, pp. 73–83.

Malusà E, Pinzari F, Canfora L (2016) *Efficacy of biofertilizers: challenges to improve crop production, Microbial inoculants in sustainable agricultural productivity*. Springer, pp. 17–40.

Manzer HS, Saud A, Soumya M, Abdullah A. Al-A, Qasi DA, Bander MA, Al-M, Hayssam MA, Hazem MK, Fahad S, Vishnu DR, Om PN (2021) Molybdenum and hydrogen sulfide synergistically mitigate arsenic toxicity by modulating defense system, nitrogen and cysteine assimilation in faba bean (Vicia faba L.) seedlings. *Environ Pollut* 290, 117953.

Md Enamul H, Shoeb AZM, Mallik AH, Fahad S, Kamruzzaman MM, Akib J, Nayyer S, Mehedi AKM, Swati AS, Md Yeamin A, Most SS (2020) Measuring vulnerability to environmental hazards: qualitative to quantitative. In: Fahad S, Hasanuzzaman M, Alam M, Ullah H, Saeed M, Khan AK, Adnan M (Eds.), *Environment, Climate, Plant and Vegetation Growth*. Springer Publ Ltd, Springer Nature Switzerland, pp. 421–452. https://doi.org/10.1007/978-3-030-49732-3

Md Jakir H, Allah B (2020) Development and applications of transplastomic plants: A way towards eco-friendly agriculture. In: Fahad S, Hasanuzzaman M, Alam M, Ullah H, Saeed M, Khan AK, Adnan M (Eds.) *Environment, Climate, Plant and Vegetation Growth*. Springer Publ Ltd, Springer Nature Switzerland, pp. 285–322. https://doi.org/10.1007/978-3-030-49732-3

Mehboob I, Zahir ZA, Arshad M, Tanveer A, Azam F (2011) Growth promoting activities of different Rhizobium spp. in wheat. *Pak J Bot* 43, 1643–1650.

Mehmood K, Bao Y, Saifullah, Bibi S, Dahlawi S, Yaseen M, Abrar MM, Srivastava P, Fahad S, Faraj TK (2022) Contributions of open biomass burning and crop straw burning to air quality: current research paradigm and future outlooks. *Front Environ Sci* 10, 852492. doi: 10.3389/fenvs.2022.852492

Mishra D, Rajvir S, Mishra U, Kumar SS (2013) Role of bio-fertilizer in organic agriculture: a review. *Res J Recent Sci* 2, 39–41. ISSN 2277, 2502.

Mishra P, Dash D (2014) Rejuvenation of biofertilizer for sustainable agriculture and economic development. *Consilience* 11, 41–61. https://doi.org/10.7916/consilience.v0i11.4650

Mohammad I. Al-Wabel, Munir Ahmad, Adel RA Usman, Mutair Akanji, Muhammad Imran Rafique (2020a) Advances in pyrolytic technologies with improved carbon capture and storage to combat climate change. In: Fahad S, Hasanuzzaman M, Alam M, Ullah H, Saeed M, Khan AK, Adnan M (Eds.), *Environment, Climate, Plant and Vegetation Growth*. Springer Publ Ltd, Springer Nature Switzerland, pp. 535–576. https://doi.org/10.1007/978-3-030-49732-3

Mohammad I. Al-Wabel, Abdelazeem S, Munir A, Khalid E, Adel RAU (2020b) Extent of climate change in Saudi Arabia and its impacts on agriculture: a case study from Qassim Region. In: Fahad S, Hasanuzzaman M, Alam M, Ullah H, Saeed M, Khan AK, Adnan M (Eds.) *Environment, Climate, Plant and Vegetation Growth. Springer Publ Ltd*, Springer Nature Switzerland, pp. 635–658. https://doi.org/10.1007/978-3-030-49732-3

Mubeen M, Ashfaq A, Hafiz MH, Muhammad A, Hafiz UF, Mazhar S, Muhammad Sami ul Din, Asad A, Amjed A, Fahad S, Wajid N (2020) Evaluating the climate change impact on water use efficiency of cotton-wheat in semi-arid conditions using DSSAT model. *J Water Clim Change* 11(4), 1661–1675. doi/10.2166/wcc.2019.179/622035/jwc2019179.pdf

Muhammad I, Khadim D, Fahad S, Imran M, Saud A, Manzer HS, Shah S, Jabar ZKK, Shamsher A, Shah H, Taufiq N, Hafiz MH, Jan B, Wajid N (2022) Exploring the potential effect of *Achnatherum splendens* L.-derived biochar treated with phosphoric acid on bioavailability of cadmium and wheat growth in contaminated soil. *Environ Sci Pollut Res* 29(25), 37676–37684. https://doi.org/10.1007/s11356-021-17950-0.

Muhammad N, Muqarrab A, Khurram S, Fiaz A, Fahim N, Muhammad A, Shazia A, Omaima N, Sulaiman AA, Fahad S, Subhan D, Rahul D (2021) Kaolin and Jasmonic acid improved cotton productivity under water stress conditions. *J Saudi Society Agricultural Sci* 28(2021), 6606–6614. https://doi.org/10.1016/j.sjbs.2021.07.043

Muhammad Tahir ul Qamar, Amna F, Amna B, Barira Z, Xitong Z, Ling-Ling C (2020) Effectiveness of conventional crop improvement strategies vs. omics. In: Fahad S, Hasanuzzaman M, Alam M, Ullah H, Saeed M, Khan AK, Adnan M (Eds.) *Environment, Climate, Plant and Vegetation Growth*. Springer Publ Ltd, Springer Nature Switzerland, pp. 253–284. https://doi.org/10.1007/978-3-030-49732-3

Muhammad Z, Abdul MK, Abdul MS, Kenneth BM, Muhammad S, Shahen S, Ibadullah J, Fahad S (2019) Performance of Aeluropus lagopoides (mangrove grass) ecotypes, a potential turfgrass, under high saline conditions. *Environ Sci Pollut Res* 26(13), 13410–13412. https://doi.org/10.1007/s11356-019-04838-3

Murage EW, Karanja NK, Smithson PC, Woomer PL (2000) Diagnostic indicators of soil quality in productive and non-productive smallholders. *Agric Ecosyst Environ* 79(1), 1–8.

Muzammal R, Fahad S, Guanghui D, Xia C, Yang Y, Kailei T, Lijun L, Fei-Hu L, Gang D (2021) Evaluation of hemp (Cannabis sativa L.) as an industrial crop: a review. *Environ Sci Pollution Res* 28(38), 52832–52843. https://doi.org/10.1007/s11356-021-16264-5

Nadeem SM, Zahir ZA, Naveed M, Asghar HN, Arshad M (2010) Rhizobacteria capable of producing ACC-deaminase may mitigate salt stress in wheat. *Soil Sci Society Am J* 74, 533–542.

Narayanamma M, Chiranjeevis C, Ahmed R, Chaturvedi A (2010) Influence of integrated nutrient management on the yield, nutrient status and quality of cucumber (Cucumis sativus L.). *Vegetable Science* 37, 61–63.

Naveed M, Mehboob I, Shaker MA, Hussain MB, Farooq M (2015) Biofertilizers in Pakistan: initiatives and limitations. *Int J Agric Biol* 17, 411–420.

Niaz A, Abdullah E, Subhan D, Muhammad A, Fahad S, Khadim D, Suleyman T, Hanife A, Anis AS, Mohammad JA, Emre B, Ömer SU, Rahul D, Bernard RG (2022) Mitigation of lead (Pb) toxicity in rice cultivated with either ground water or wastewater by application of acidified carbon. *J Environ Manage* 307, 114521.

Noor M, Naveed ur R, Ajmal J, Fahad S, Muhammad A, Fazli W, Saud S, Hassan S (2020) Climate change and coastal plant lives. In: Fahad S, Hasanuzzaman M, Alam M, Ullah H, Saeed M, Khan AK, Adnan M (Eds.) *Environment, Climate, Plant and Vegetation Growth.* Springer Publ Ltd, Springer Nature Switzerland, pp. 93–108. doi.org/10.1007/978-3-030-49732-3

Okoli P, Nweke I (2015) Effect of poultry manure and mineral fertilizer on the growth performance and quality of cucumber fruits. *J Exp Biology and Agricultural Sciences* 3, 362–367.

Ortaş I, Rafique M, Ahmed İA (2017) Application of arbuscular mycorrhizal fungi into agriculture. *Arbuscular Mycorrhizas and Stress Tolerance of Plants.* Springer, pp. 305–327.

Ortas I, Rafique M, Çekiç F (2021) Do mycorrhizal fungi enable plants to cope with abiotic stresses by overcoming the detrimental effects of salinity and improving drought tolerance? *Symbiotic Soil Microorganisms.* Springer, pp. 391–428.

Qamar-uz Z, Zubair A, Muhammad Y, Muhammad ZI, Abdul K, Fahad S, Safder B, Ramzani PMA, Muhammad N (2017) Zinc biofortification in rice: leveraging agriculture to moderate hidden hunger in developing countries. *Arch Agron Soil Sci* 64, 147–161. https://doi.org/10.1080/03650340.2017.1338343

Qin ZH, Nasib ur Rahman, Ahmad A, Yu-pei Wang, Sakhawat S, Ehmet N, Wen-juan Shao, Muhammad I, Kun S, Rui L, Fazal S, Fahad S (2022) Range expansion decreases the reproductive fitness of *Gentiana officinalis* (*Gentianaceae*). *Sci Rep* 12, 2461. https://doi.org/10.1038/s41598-022-06406-1

Qureshi M, Shahzad H, Imran Z, Mushtaq M, Akhtar N, Ali M, Mujeeb F (2013) Potential of Rhizobium species to enhance growth and fodder yield of maize in the presence and absence of l-tryptophan. *J Anim Plant Sci* 23, 1448–1454.

Rafique M, Ortas I (2018) Nutrient uptake-modification of different plant species in Mediterranean climate by arbuscular mycorrhizal fungi. *Eur J Hortic Sci* 83, 65–71.

Rafique M, Sultan T, Ortas I, Chaudhary HJ (2017) Enhancement of maize plant growth with inoculation of phosphate-solubilizing bacteria and biochar amendment in soil. *Soil Sci & Plant Nutr* 63, 460–469.

Rahman I, Ali S, Alam M, Adnan M, Basir A, Ullah H, Malik MFA, Shah AS, Ibrahim M (2016) Effect of seed priming on germination performance and Yield of okra (*abelmoschus esculentus* L.). *Pakistan J Agric Res* 29(3), 253–262.

Rajesh KS, Fahad S, Pawan K, Prince C, Talha J, Dinesh J, Prabha S, Debanjana S, Prathibha MD, Bandana B, Akash H, Gupta NK, Rekha S, Devanshu D, Dalpat LS, Ke L, Matthew TH, Saud S, Adnan NS, Taufiq N (2022) Beneficial elements: new players in improving nutrient use efficiency and abiotic stress tolerance. *Plant Growth Regul* 100, 237–265. https://doi.org/10.1007/s10725-022-00843-8

Rashid M, Qaiser H, Khalid SK, Mohammad I. Al-Wabel, Zhang A, Muhammad A, Shahzada SI, Rukhsanda A, Ghulam AS, Shahzada MM, Sarosh A, Muhammad FQ (2020) Prospects of biochar in alkaline soils to mitigate climate change. In: Fahad S, Hasanuzzaman M, Alam M, Ullah H, Saeed M, Khan AK, Adnan M (Eds.) *Environment, Climate, Plant and Vegetation Growth.* Springer Publ Ltd, Springer Nature Switzerland, pp. 133–150. https://doi.org/10.1007/978-3-030-49732-3

Rehman M, Fahad S, Saleem MH, Hafeez M, Muhammad Habib ur Rahman, Liu F, Deng G (2020) Red light optimized physiological traits and enhanced the growth of ramie (Boehmeria nivea L.). *Photosynthetica* 58(4), 922–931.

Roger P-A, Ladha J (1992) Biological N_2 fixation in wetland rice fields: estimation and contribution to nitrogen balance. *Biol Nitrogen Fixation Sustainable Agric* 141(1/2), 41–55.

Ruíz-Sánchez M, Armada E, Muñoz Y, de Salamone IEG, Aroca R, Ruíz-Lozano JM, Azcón R (2011) Azospirillum and arbuscular mycorrhizal colonization enhance rice growth and physiological traits under well-watered and drought conditions. *J Plant Physiol* 168, 1031–1037.

Sadam M, Muhammad Tahir ul Qamar, Ghulam M, Muhammad SK, Faiz AJ (2020) Role of biotechnology in climate resilient agriculture. In: Fahad S, Hasanuzzaman M, Alam M, Ullah H, Saeed M, Khan AK, Adnan M (Eds.) *Environment, Climate, Plant and Vegetation Growth.* Springer Publ Ltd, Springer Nature Switzerland, pp. 339–366. https://doi.org/10.1007/978-3-030-49732-3

Saeed KS, Ahmed SA, Hassan IA, Ahmed PH (2015) Effect of bio-fertilizer and chemical fertilizer on growth and yield in cucumber (Cucumis sativus) in green house condition. *Pak J Biol Sci* 18, 129–134.

Safi UK, Ullah F, Mehmood S, Fahad S, Ahmad Rahi A, Althobaiti F, et al. (2021) Antimicrobial, antioxidant and cytotoxic properties of Chenopodium glaucum L. *PLoS One* 16(10), e0255502. https://doi.org/10.1371/journal. pone.0255502

Sahoo RK, Ansari MW, Pradhan M, Dangar TK, Mohanty S, Tuteja N (2014) Phenotypic and molecular characterization of native Azospirillum strains from rice fields to improve crop productivity. *Protoplasma* 251, 943–953.

Sahrish N, Shakeel A, Ghulam A, Zartash F, Sajjad H, Mukhtar A, Muhammad AK, Ahmad K, Fahad S, Wajid N, Sezai E, Carol Jo W, Gerrit H (2022) Modeling the impact of climate warming on potato phenology. *European J Agro N* 132, 126404.

Sahu PK, Singh DP, Prabha R, Meena KK, Abhilash P (2019) Connecting microbial capabilities with the soil and plant health: options for agricultural sustainability. *Ecol Indic* 105, 601–612.

Sajid H, Jie H, Jing H, Shakeel A, Satyabrata N, Sumera A, Awais S, Chunquan Z, Lianfeng Z, Xiaochuang C, Qianyu J, Junhua Z (2020) Rice production under climate change: adaptations and mitigating strategies. In: Fahad S, Hasanuzzaman M, Alam M, Ullah H, Saeed M, Khan AK, Adnan M (Eds.) *Environment, Climate, Plant and Vegetation Growth.* Springer Publ Ltd, Springer Nature Switzerland, pp. 659–686. https://doi.org/10.1007/978-3-030-49732-3

Sajjad H, Muhammad M, Ashfaq A, Waseem A, Hafiz MH, Mazhar A, Nasir M, Asad A, Hafiz UF, Syeda RS, Fahad S, Depeng W, Wajid N (2019) Using GIS tools to detect the land use/land cover changes during forty years in Lodhran district of Pakistan. *Environ Sci Pollut Res* 27, 39676–39692. https://doi.org/10.1007/s11356-019-06072-3

Sajjad H, Muhammad M, Ashfaq A, Fahad S, Wajid N, Hafiz MH, Ghulam MS, Behzad M, Muhammad T, Saima P (2021) Using space–time scan statistic for studying the effects of COVID-19 in Punjab, Pakistan: a guideline for policy measures in regional Agriculture. *Environ Sci Pollut Res* 30, 42495–42508. https://doi.org/10.1007/s11356-021-17433-2

Sajjad H, Muhammad M, Ashfaq A, Nasir M, Hafiz MH, Muhammad A, Muhammad I, Muhammad U, Hafiz UF, Fahad S, Wajid N, Hafiz MRJ, Mazhar A, Saeed AQ, Amjad F, Muhammad SK, Mirza W (2021) Satellite-based evaluation of temporal change in cultivated land in Southern Punjab (Multan region) through dynamics of vegetation and land surface temperature. *Open Geo Sci* 13, 1561–1577.

Salar RK, Purewal SS, Sandhu KS (2017) Bioactive profile, free-radical scavenging potential, DNA damage protection activity, and mycochemicals in Aspergillus awamori (MTCC 548) extracts: a novel report on filamentous fungi. *3 Biotech* 7, 1–9.

Saleem M, Arshad M, Hussain S, Bhatti AS (2007) Perspective of plant growth promoting rhizobacteria (PGPR) containing ACC deaminase in stress agriculture. *J Ind Microbiol & Biotechnol* 34, 635–648.

Saleem MH, Fahad S, Adnan M, Mohsin A, Muhammad SR, Muhammad K, Qurban A, Inas AH, Parashuram B, Mubassir A, Reem MH (2020a) Foliar application of gibberellic acid endorsed phytoextraction of copper and alleviates oxidative stress in jute (Corchorus capsularis L.) plant grown in highly copper-contaminated soil of China. *Environ Sci Pollution Res* 27, 37121–37133. https://doi.org/10.1007/s11356-020-09764-3

Saleem MH, Rehman M, Fahad S, Tung SA, Iqbal N, Hassan A, Ayub A, Wahid MA, Shaukat S, Liu L, Deng G (2020b) Leaf gas exchange, oxidative stress, and physiological attributes of rapeseed (Brassica napus L.) grown under different light-emitting diodes. *Photosynthetica* 58(3), 836–845.

Saleem MH, Fahad S, Shahid UK, Mairaj D, Abid U, Ayman ELS, Akbar H, Analía L, Lijun L (2020c) Copper-induced oxidative stress, initiation of antioxidants and phytoremediation potential of flax (Linum usitatissimum L.) seedlings grown under the mixing of two different soils of China. *Environ Sci Poll Res* 27, 5211–5221. https://doi.org/10.1007/s11356-019-07264-7

Saman S, Amna B, Bani A, Muhammad Tahir ul Qamar, Rana MA, Muhammad SK (2020) QTL mapping for abiotic stresses in cereals. In: Fahad S, Hasanuzzaman M, Alam M, Ullah H, Saeed M, Khan AK, Adnan M (Eds.) *Environment, Climate, Plant and Vegetation Growth.* Springer Publ Ltd, Springer Nature Switzerland, pp. 229–252. https://doi.org/10.1007/978-3-030-49732-3

Sana U, Shahid A, Yasir A, Farman UD, Syed IA, Mirza MFAB, Fahad S, Al-Misned F, Usman A, Xinle G, Ghulam N, Kunyuan W (2022) Bifenthrin induced toxicity in Ctenopharyngodon idella at an acute concentration: a multi-biomarkers based study. *J King Saud Uni Sci* 34 (2022) 101752.

Sarfaraz Q, Silva L, Drescher G, Zafar M, Severo F, Kokkonen A, Molin G, Shafi M, Shafique Q, Solaiman Z (2020) Characterization and carbon mineralization of biochars produced from different animal manures and plant residues. *Scientific Reports* 10, 1–9.

Saud S, Chen Y, Long B, Fahad S, Sadiq A (2013) The different impact on the growth of cool season turf grass under the various conditions on salinity and drought stress. *Int J Agric Sci Res* 3, 77–84.

Saud S, Li X, Chen Y, Zhang L, Fahad S, Hussain S, Sadiq A, Chen Y (2014) Silicon application increases drought tolerance of Kentucky bluegrass by improving plant water relations and morph physiological functions. *Sci World J* 2014, 1–10. https://doi.org/10.1155/2014/368694

Saud S, Chen Y, Fahad S, Hussain S, Na L, Xin L, Alhussien SA (2016) Silicate application increases the photosynthesis and its associated metabolic activities in Kentucky bluegrass under drought stress and post-drought recovery. *Environ Sci Pollut Res* 23(17), 17647–17655. https://doi.org/10.1007/s11356-016-6957-x

Saud S, Fahad S, Yajun C, Ihsan MZ, Hammad HM, Nasim W, Amanullah Jr, Arif M and Alharby H (2017) Effects of nitrogen supply on water stress and recovery mechanisms in Kentucky bluegrass plants. *Front Plant Sci* 8, 983. doi: 10.3389/fpls.2017.00983

Saud S, Fahad S, Cui G, Chen Y, Anwar S (2020) Determining nitrogen isotopes discrimination under drought stress on enzymatic activities, nitrogen isotope abundance and water contents of Kentucky bluegrass. *Sci Rep* 10, 6415. https://doi.org/10.1038/s41598-020-63548-w

Saud S, Fahad S, Hassan S (2022a) Developments in the investigation of nitrogen and oxygen stable isotopes in atmospheric nitrate. *Sustainable Chem Clim Action* 1, 100003.

Saud S, Li X, Jiang Z, Fahad S, Hassan S (2022b) Exploration of the phytohormone regulation of energy storage compound accumulation in microalgae. *Food Energy Secur* 2022, e418.

Senol C (2020) The effects of climate change on human behaviors. In: Fahad S, Hasanuzzaman M, Alam M, Ullah H, Saeed M, Khan AK, Adnan M (Eds.), *Environment, Climate, Plant and Vegetation Growth.* Springer Publ Ltd, Springer Nature Switzerland, pp. 577–590. https://doi.org/10.1007/978-3-030-49732-3

Sethi SK, Adhikary SP (2012) Azotobacter: a plant growth-promoting rhizobacteria used as biofertilizer. *Dyn Biochem, Process Biotechnol & Mol Biol* 6, 68–74.

Shafi MI, Adnan M, Fahad S, Fazli W, Ahsan K, Zhen Y, Subhan D, Zafar-ul-Hye M, Martin B, Rahul D (2020) Application of single superphosphate with humic acid improves the growth, yield and phosphorus uptake of wheat (Triticum aestivum L.) in *calcareous soil. Agron* 10, 1224. doi:10.3390/agronomy10091224

Shafi MI, Sharif M (2019) Soil extractable phosphorus contents as affected by phosphatic fertilizer sources applied with different levels of humic acid. *Sarhad J Agric* 35(4), 1084–1093.

Shah F, Lixiao N, Kehui C, Tariq S, Wei W, Chang C, Liyang Z, Farhan A, Fahad S, Huang J (2013) Rice grain yield and component responses to near 2°C of warming. *Field Crop Res* 157, 98–110.

Shah S, Shah H, Liangbing X, Xiaoyang S, Shahla A, Fahad S (2022) The physiological function and molecular mechanism of hydrogen sulfide resisting abiotic stress in plants. *Brazilian J Botany* 45, 563–572. https://doi.org/10.1007/s40415-022-00785-5

Shahzad S, Khalid A, Arshad M, Khalid M, Mehboob I (2008) Integrated use of plant growth promoting bacteria and P-enriched compost for improving growth, yield and nodulation of chickpea. *Pakistan J Bot* 40, 1441–1735.

Shakir MA, Bano A, Arshad M (2012) Short communciation: rhizosphere bacteria containing ACC-deaminase conferred drought tolerance in wheat grown under semi-arid climate. *Soil & Environ* 31(1), 108–112.

Sheng XF, Zhao F, He LY, Qiu G, Chen L (2008) Isolation and characterization of silicate mineral-solubilizing Bacillus globisporus Q12 from the surfaces of weathered feldspar. *Can J Microbiol* 54, 1064–1068.

Shirinbayan S, Khosravi H, Malakouti MJ (2019) Alleviation of drought stress in maize (Zea mays) by inoculation with Azotobacter strains isolated from semi-arid regions. *Appl Soil Ecol* 133, 138–145.

Shivprasad S, Page WJ (1989) Catechol formation and melanization by Na+-dependent Azotobacter chroococcum: a protective mechanism for aeroadaptation? *Applied and Environm Microbiol* 55, 1811–1817.

Sidra K, Javed I, Subhan D, Allah B, Syed IUSB, Fatma B, Khaled DA, Fahad S, Omaima N, Ali TKZ, Rahul D (2021) Physio-chemical characterization of indigenous agricultural waste materials for the development of potting media. *J Saudi Society Agricultural Sci* 28(12), 7491–7498. https://doi.org/10.1016/j.sjbs.2021.08.058

Singh JS, Pandey VC, Singh DP (2011) Efficient soil microorganisms: a new dimension for sustainable agriculture and environmental development. *Agric, Ecosyst & Environ* 140, 339–353.

Smith P, Calvin K, Nkem J, Campbell D, Cherubini F, Grassi G, Korotkov V, Le Hoang A, Lwasa S, McElwee P (2020) Which practices co-deliver food security, climate change mitigation and adaptation, and combat land degradation and desertification? *Global Change Biol* 26, 1532–1575.

Socolofsky M, Wyss O (1962) Resistance of the Azotobacter cyst. *J Bacteriol* 84, 119–124.

Solaiman ZM, Shafi MI, Beamont E, Anawar HM (2020) Poultry litter biochar increases mycorrhizal colonisation, soil fertility and cucumber yield in a fertigation system on sandy soil. *Agriculture* 10, 480.

Soma K, van den Burg SW, Hoefnagel EW, Stuiver M, van der Heide CM (2018) Social innovation–A future pathway for blue growth? *Marine Policy* 87, 363–370.

Spaepen S, Vanderleyden J (2015) Auxin signaling in Azospirillum brasilense: a proteome analysis. *Biol Nitrogen Fixation* 937–940.

Subhan D, Zafar-ul-Hye M, Fahad S, Saud S, Martin B, Tereza H, Rahul D (2020) Drought stress alleviation by ACC deaminase producing *Achromobacter xylosoxidans* and *Enterobacter cloacae*, with and without timber waste biochar in maize. *Sustain* 12(15), 6286 doi:10.3390/su12156286

Tariq M, Ahmad S, Fahad S, Abbas G, Hussain S, Fatima Z, Nasim W, Mubeen M, ur Rehman MH, Khan MA, Adnan M (2018) The impact of climate warming and crop management on phenology of sunflower-based cropping systems in Punjab, Pakistan. *Agri and Forest Met* 256, 270–282.

Ullah S, Liang H, Ali I, Zhao Q, Iqbal A, Wei S, Shah T, Yan B, Jiang L (2020) Biochar coupled with contrasting nitrogen sources mediated changes in carbon and nitrogen pools, microbial and enzymatic activity in paddy soil. *J Saudi Chem Society* 24, 835–849.

Umesha S, Singh PK, Singh RP (2018) Microbial biotechnology and sustainable agriculture. In: Ram Lakhan Singh, Sukanta Mondal (Eds,), *Biotechnology for Sustainable Agriculture* Elsevier, pp. 185–205.

Unsar Naeem-U, Muhammad R, Syed HMB, Asad S, Mirza AQ, Naeem I, Muhammad H ur R, Fahad S, Shafqat S (2020) Insect pests of cotton crop and management under climate change scenarios. In: Fahad S, Hasanuzzaman M, Alam M, Ullah H, Saeed M, Khan AK, Adnan M (Eds.) *Environment, Climate, Plant and Vegetation Growth.* Springer Publ Ltd, Springer Nature Switzerland, pp. 367–396. https://doi.org/10.1007/978-3-030-49732-3

Uosif M, Mostafa A, Elsaman R, Moustafa E-S (2014) Natural radioactivity levels and radiological hazards indices of chemical fertilizers commonly used in Upper Egypt. *J Radiation Res & Appl Sci* 7, 430–437.

Vanlauwe B, Kihara J, Chivenge P, Pypers P, Coe R, Six J (2011) Agronomic use efficiency of N fertilizer in maize-based systems in sub-Saharan Africa within the context of integrated soil fertility management. *Plant & Soil* 339, 35–50.

Vasanthi N, Saleena L, Raj SA (2018) Silica solubilization potential of certain bacterial species in the presence of different silicate minerals. *Silicon* 10, 267–275.

Wahid F, Fahad S, Subhan D, Adnan M, Zhen Y, Saud S, Manzer HS, Martin B, Tereza H, Rahul D (2020) Sustainable management with mycorrhizae and phosphate solubilizing bacteria for enhanced phosphorus uptake in calcareous soils. *Agri* 10(8), 334. doi:10.3390/agriculture10080334

Wahid F, Sharif M, Fahad S, Adnan M, Khan IA, Aksoy E, Ali A, Sultan T, Alam M, Saeed M (2019) Arbuscular mycorrhizal fungi improve the growth and phosphorus uptake of mung bean plants fertilized with composted rock phosphate fed dung in alkaline soil environment. *J Plant Nutr* 42, 1760–1769.

Wajid N, Ashfaq A, Asad A, Muhammad T, Muhammad A, Muhammad S, Khawar J, Ghulam MS, Syeda RS, Hafiz MH, Muhammad IAR, Muhammad ZH, Muhammad Habib ur R, Veysel T, Fahad S, Suad S, Aziz K, Shahzad A (2017) Radiation efficiency and nitrogen fertilizer impacts on sunflower crop in contrasting environments of Punjab. *Pakistan Environ Sci Pollut Res* 25, 1822–1836. https://doi.org/10.1007/s11356-017-0592-z

Wiqar A, Arbaz K, Muhammad Z, Ijaz A, Muhammad A, Fahad S (2022) Relative efficiency of biochar particles of different sizes for immobilising heavy metals and improving soil properties. *Crop Pasture Sci* 42(2), 112–120. https://doi.org/10.1071/CP20453.

Wu C, Kehui C, She T, Ganghua L, Shaohua W, Fahad S, Lixiao N, Jianliang H, Shaobing P, Yanfeng D (2020) Intensified pollination and fertilization ameliorate heat injury in rice (Oryza sativa L.) during the flowering stage. *Field Crops Res* 252, 107795.

Wu C, Tang S, Li G, Wang S, Fahad S, Ding Y (2019) Roles of phytohormone changes in the grain yield of rice plants exposed to heat: a review. *Peer J* 7, e7792. doi: 10.7717/peerj.7792

Wu K, Ali I, Xie H, Ullah S, Iqbal A, Wei S, He L, Huang Q, Wu X, Cheng F (2021) Impact of fertilization with reducing in nitrogen and phosphorous application on growth, yield and biomass accumulation of rice (Oryza sativa L.) under a dual cropping system. *PeerJ* 9, e11668.

Xie H, Wu K, Iqbal A, Ali I, He L, Ullah S, Wei S, Zhao Q, Wu X, Huang Q (2021) Synthetic nitrogen coupled with seaweed extract and microbial inoculants improves rice (Oryza sativa L.) production under a dual cropping system. *Italian J Agron* 6, 1800.

Xue B, Huang L, Li X, Lu J, Gao R, Kamran M, Fahad S (2022) Effect of clay mineralogy and soil organic carbon in aggregates under straw incorporation. *Agron* 12, 534. https://doi.org/10.3390/ agronomy12020534

Yageta Y, Osbahr H, Morimoto Y, Clark J (2019) Comparing farmers' qualitative evaluation of soil fertility with quantitative soil fertility indicators in Kitui County, Kenya. *Geoderma* 344, 153–163.

Yang H, Schroeder-Moreno M, Giri B, Hu S (2018) Arbuscular mycorrhizal fungi and their responses to nutrient enrichment. In: Bhoopander Giri, Ram Prasad, Ajit Varma (Eds.), *Root Biology*. Springer, pp. 429–449.

Yang R, Dai P, Wang B, Jin T, Liu K, Fahad S, Harrison MT, Man J, Shang J, Meinke H, Deli L, Xiaoyan W, Yunbo Z, Meixue Z, Yingbing T, Haoliang Y(2022) Over-optimistic projected future wheat yield potential in the North China Plain: the role of future climate extremes. *Agron* 12, 145. https://doi.org/10.3390/ agronomy12010145

Yang Z, Zhang Z, Zhang T, Fahad S, Cui K, Nie L, Peng S, Huang J (2017) The effect of season-long temperature increases on rice cultivars grown in the central and southern regions of China. *Front Plant Sci* 8, 1908. https://doi.org/10.3389/fpls.2017.01908

Yao Y, Zhang M, Tian Y, Zhao M, Zeng K, Zhang B, Zhao M, Yin B (2018) Azolla biofertilizer for improving low nitrogen use efficiency in an intensive rice cropping system. *Field Crops Res* 216, 158–164.

Yuan P, Li X, Ni M, Cao C, Jiang L, Iqbal A, Wang J (2022) Effects of straw return and feed addition on the environment and nitrogen use efficiency under different nitrogen application rates in the rice–crayfish system. *Plant and Soil* 475, 411–426.

Zafar-ul-Hye M, Muhammad N, Subhan D, Fahad S, Rahul D, Mazhar A, Ashfaq AR, Martin B, Jiˇrí H, Zahid HT, Muhammad N (2020a) Alleviation of cadmium adverse effects by improving nutrients uptake in bitter gourd through cadmium tolerant rhizobacteria. *Environ* 7(8), 54. doi:10.3390/environments7080054

Zafar-ul-Hye M, Muhammad T ahzeeb-ul-Hassan, Muhammad A, Fahad S, Martin B, Tereza D, Rahul D, Subhan D (2020b) Potential role of compost mixed biochar with rhizobacteria in mitigating lead toxicity in spinach. *Scientific Rep* 10, 12159. https://doi.org/10.1038/s41598-020-69183-9.

Zafar-ul-Hye M, Akbar MN, Iftikhar Y, Abbas M, Zahid A, Fahad S, Datta R, Ali M, Elgorban AM, Ansari MJ et al. (2021) Rhizobacteria inoculation and caffeic acid alleviated drought stress in lentil plants. *Sustain* 13, 9603. https://doi.org/10.3390/su13179603

Zahida Z, Hafiz FB, Zulfiqar AS, Ghulam MS, Fahad S, Muhammad RA, Hafiz MH, Wajid N, Muhammad S (2017) Effect of water management and silicon on germination, growth, phosphorus and arsenic uptake in rice. *Ecotoxicol Environ Saf* 144, 11–18.

Zahir SM, Zheng-HG, Ala Ud D, Amjad A, Ata Ur R, Kashif J, Shah F, Saud S, Adnan M, Fazli W, Saud A, Manzer HS, Shamsher A, Wajid N, Hafiz MH, Fahad S (2021) Synthesis of silver nanoparticles using Plantago lanceolata extract and assessing their antibacterial and antioxidant activities. *Sci Rep* 11, 20754.

Zahir ZA, Arshad M (2004) Perspectives in agriculture. *Adv Agron* 81, 97.

Zahir ZA, Asghar HN, Akhtar MJ, Arshad M (2005) Precursor (L-tryptophan)-inoculum (Azotobacter) interaction for improving yields and nitrogen uptake of maize. *J Plant Nutr* 28, 805–817.

Zaman I, Ali M, Shahzad K, Tahir MS, Matloob A, Ahmad W, Alamri S, Khurshid MR, Qureshi MM, Wasaya A, Khurram SB, Manzer HS, Fahad S, Rahul D (2021) Effect of plant spacings on growth, physiology, yield and fiber quality attributes of cotton genotypes under nitrogen fertilization. *Agron* 11, 2589. https:// doi.org/ 10.3390/agronomy11122589

Zia-ur-Rehman M (2020) Environment, climate change and biodiversity. In: Fahad S, Hasanuzzaman M, Alam M, Ullah H, Saeed M, Khan AK, Adnan M (Eds.) *Environment, Climate, Plant and Vegetation Growth*. Springer Publ Ltd, Springer Nature Switzerland, pp. 473–502. https://doi.org/10.1007/ 978-3-030-49732-3

6 Phosphate-mobilizing Mycorrhizal-based Bio-fertilizers

*Muhammad Jawad, Ayesha Jabeen, Ali Raza Gurmani, and Mazhar Rafique**
Department of Soil & Climate Sciences, The University of Haripur, Khyber Pakhtunkhwa, Pakistan
Correspondence: mazhar.rafique@uoh.edu.pk

CONTENTS

6.1 INTRODUCTION

Fertilizers are compounds that are either natural or man-made (Fuentes-Ramirez & Caballero-Mellado 2005). These compounds stimulate crop development, and improve soil nutrients and soil fertility when applied to soil and plants through irrigation water, or by fertigation (Benaffari et al. 2022). Important macronutrients (phosphorus, nitrogen, potassium, sulphur, calcium, and magnesium) as well as several micronutrients (boron, copper, zinc, iron, and molybdenum) are provided to plants through various sources of chemical and bio-fertilizers (Waqeel & Khan 2022; Fahad et al. 2016; Fahad et al. 2015a,b; Fahad et al. 2020; Fahad et al. 2021a,,cb,d,e,f; Fahad et al.

DOI: 10.1201/9781003286233-6

2022a,b). Some fertilizers are in high demand such as nitrogen-based urea (Singh 2018), calcium ammonium nitrate (CAN), ammonium sulphate (AS), phosphorus based di-ammonium phosphate (DAP), ammonium nitrate (AN), superphosphates, powdered rock phosphates, and potassium (potassium chloride/potash). Modern agriculture relies on synthetic chemical fertilizers, which results in creating environmental challenges such as the greenhouse effect, soil degradation, and air and water pollution (Chen et al. 2018; Adnan et al. 2018a,b; Adnan et al. 2020; Adnan et al. 2019; Atif et al. 2021). Furthermore, suitable agricultural practices are necessary for appropriate cost-effective food production to accommodate the increasing human population, reduced energy, and upcoming environmental challenges (Williams et al. 2016; Khan et al. 2016; Fakhre et al. 2021; Muhammad et al. 2022; Shafi et al. 2020; Wahid et al. 2020; Zahida et al. 2017).

Bio-fertilizers are implemented in farming as an alternative to traditional fertilizers (Barragán-Ocaña & del Carmen del-Valle-Rivera 2016). Bio-fertilizers include microorganisms such as bacteria, fungi, and algae that are natural, eco-friendly, and cost-effective, to sustain soil structure and biodiversity (Fakhar et al. 2020). Microbial bio-fertilizers improve plant development by enhancing the effective absorption or uptake of nutrients for plants and inhibiting phytopathogenicity, in addition to delivering nutrient enrichment to the soil (Amna et al. 2015; Rafique et al. 2017; Thomas & Singh 2019). Organic fertilizers primarily augment nutrients by fixing atmospheric nitrogen, solubilizing phosphorus, and manufacturing plant development chemicals (Rafique & Ortas 2018). Since the green revolution, chemical fertilizers have reduced soil health by making the soil ecosystem unsuitable for soil microbiota, which are essential for maintaining soil health and supplying crops with essential nutrients (John & Babu 2021).

Bio-fertilizers are known as a product that include one or more species of microbial organism (John & Babu 2021), that can mobilize soil nutrients and make them into an available form in soil. They include compost, phosphorus-solubilizing bacteria, nitrogen-fixing bacteria, and release of plant growth-promoting substances (Umesha et al. 2018). Thus, bio-fertilizers are inoculants of bacteria, algae, and fungi, or natural fertilizers that increase the accessibility of nutrients to plants (Kumar et al. 2017). Bio-fertilizers are gaining increased significance in agriculture, specifically with the escalating expense of conventional fertilizers and their possible detrimental effects on soil fertility (Igiehon & Babalola 2017; Ortaş et al. 2017).

6.2 CLASSIFICATION OF BIO-FERTILIZERS

Bio-fertilizers have been proven to help soil fertility in a variety of ways such as enhancing crop productivity and the provision of ecological benefits for maximizing crop yield (Kumar et al. 2017). Bio-fertilizers have several key functions or responsibilities in agriculture. In the development of bio-fertilizers, a diversity of microbes and their relationship with agricultural plants is of key importance (John & Babu 2021). Bio-fertilizers can be classified on the basis of their functions (Kaur & Purewal 2019) (Table 6.1).

i. ***Rhizobium***: This is a soil-dwelling microorganism that take over the roots of legumes, providing atmospheric nitrogen in a symbiotic manner. *Rhizobium* comes in a variety of shapes and physiologies, ranging from free-living to nodular bacteroid.
ii. ***Azotobacter***: The most common *Azotobacter* species is *A. chroococcum*. It may fix nitrogen (2–15 mg N_2 fixed/g of source of carbon) and is present in arable soils. These bacteria produce slime, which aids in soil clumping. Because of the presence of hostile bacteria and the scarcity of organic matter content in Asian soils, the population of *A. chroococcum* rarely reaches 10^5 per gram of soil (Mishra et al. 2013).

    Azospirillum lipoferum and *A. brasilense* are present in soil, rhizosphere, and root cortex intercellular spaces (called *Spirillum lipoferum* in the earlier literature) of gramineous plants

TABLE 6.1
Types of Bio-fertilizer

Group of Bio-fertilizers	Subgroup	Example
Bio-fertilizers for micronutrients	Silicate and zinc solubilizers	*Bacillus* sp.
Plant growth-promoting rhizobacteria (PGPR)	*Pseudomonas*	*Pseudomonas fluorescens*
P Mobilizing bio-fertilizer	Orchid mycorrhiza	*Rhizoctonia solani*
	mycorrhizae Ericoid	*Pezizellaericae*
	Ectomycorrhiza	*Amanita* sp., *Boletus* sp., *Pisolithus* sp., *Laccaria* sp.
	Arbuscular mycorrhizal fungi (AMF)	*Sclerocystis* sp., *Gigaspora* sp., *Glomussp.*, *Acaulospora* sp., *Scutellospora* sp.
Nitrogen fixing	Free living	*Rhodopseudomonas, Klebsiella, Nostoc, Chromatium, Bacillus polymyxa, Rhodospirillum, Desulfovibrio. Derxia, Aulosira, Clostridium, Cylindrospermum Anabaena, Azotobacter, Stigonema, Tolypothrix,*
	Associative	*Bacillus Herbaspirillum* spp., *Enterobacter, Klebsiella, Azoarcus* spp., *Azospirillum* spp., *A. brasilense, A. halopraeferens, Acetobacter diazotrophicus, A. lipoferum, A. irakense A. amazonense, Pseudomonas.*
	Symbiotic	*Anabaena azollae, Trichodesmium Rhizobia (Mesorhizobium Allorhizobium Rhizobium, Bradyrhizobium, Sinorhizobium, Azorhizobium)* and *Frankia.*
Phosphorus (microphos)	Phosphate solubilizing	*Burkholderia, Achromobacter, Microccocus, B. subtilis, B. megaterium var. phosphaticum, Achromobacter, Achromobacter, Agrobacterium, B. circulans, B. polymyxa, Pseudomonas striata, Penicillium* spp., *Aspergillus awamori, Erwinia*

Source: Thomas & Singh (2019)

as they have a symbiotic relationship. Apart from nitrogen fixation, *Azospirillum* inoculation has several additional benefits, including the production of growth-promoting chemicals such as indole acetic acid, mitigating drought stress, and providing disease resistance (Mishra et al. 2013).

iii. **Cyanobacteria**: Rice cultivation in India has incorporated cyanobacteria, which can be either free-living or symbiotic. In India, cyanobacteria have been advertised as a rice bio-fertilizer but they have not attracted the attention of farmers. Under ideal conditions, the advantages of algalization (a technique for mass cultivation of blue-green algae to be used as a bio-fertilizer in paddy fields) could be as high as 20–30 kg N/ha, albeit the labour-intensive process for producing blue-green algae bio-fertilizer is in itself a restriction (Thomas & Singh 2019).

iv. ***Azolla***: *Azolla* is a free-floating aquatic plant that fixes atmospheric nitrogen in collaboration with a nitrogen-fixing blue-green algae, *Anabaena azollae*. It can be utilized as an alternative to conventional nitrogen fertilizers. *Azolla* is a wetland rice bio-fertilizer that provides 40–60 kg N/ha to rice crop (Mishra et al. 2013).

v. **Phosphate-solubilizing microorganisms**: Many fungi and soil bacteria, including *Penicillium*, *Bacillus*, *Aspergillus*, *Pseudomonas*, and others, release organic acids that lower the soil pH in their surrounding area, allowing binding phosphates in soil to dissolve.

Increased wheat and potato yields were obtained by inoculating *Pseudomonas striata* peat-based and *Bacillus polymyxa* cultures (Mishra et al. 2013).

vi. **Arbuscular mycorrhiza fungi**: This is an obligatory intracellular fungus endosymbiont of the genera *Acaulospora*, *Glomus*, *Endogone*, *Sclerocysts*, and *Gigaspora*. The AMF have vesicles for storing arbuscles and nutrients to funnel these nutrients into the root system, and they oversee delivering nutrients from the soil to the root cortex cells. *Glomus* seems to be the most common genus, with many species identified in the soil (Van Der Heijden et al. 2006).

vii. **Plant growth-promoting rhizobacteria (PGPR):** These are bacteria that form colonies in plant roots or the soil's rhizosphere that are beneficial to plants. The PGPR inoculant boosts development by suppressing plant diseases (bio-protectants), improving nutrient acquisition (bio-fertilizers), or producing phytohormones (bio-stimulants). Growth regulators or phytohormones produced by *Pseudomonas* and *Bacillus* species encourage plants to develop more healthy roots, improving the absorptive superficial area of plant roots for nutrient absorption and water. The phytohormones synthesized by such PGPR are called bio-stimulants. The bio-stimulants include gibberellins, indole acetic acid, cytokinin, and inhibitors of ethylene synthesis (Mishra et al. 2013).

6.3 IMPORTANCE OF BIO-FERTILIZERS IN AGRICULTURE

Long-term viability of agriculture in changing environmental and climatic conditions is an emerging concern (Spicka et al. 2019). Based on the aforementioned data, it appears that long-term use of bio-fertilizers over chemical fertilizers is cost-effective, environmentally friendly, productively viable, and easily accessible to the farming community (Itelima et al. 2018). Hence, the use of bio-fertilizers is generally required for two reasons. First, enhanced fertilizer use increases agricultural output; secondly, a high amount of chemical fertilizer application destroys soil health and produces numerous environmental problems (Khan et al. 2016; Thomas & Singh 2019). Bio-fertilizers, commonly known as bio-inoculants, are a type of fertilizer where nutrients are supplied by a living organism (microbe) such as a bacterium for enhancing agricultural productivity in terms of N and P (Itelima et al. 2018). A bio-fertilizer is an organic substance that contains a large number of efficient microbes isolated from the plant's root or the rhizosphereic soil (Umesha et al. 2018). They reproduce quickly and form a dense community in the rhizosphere. As far as their application is concerned, they are used as a seed treatment, or soil application, or through seedling root dip. Bio-fertilizers fix atmospheric nitrogen, solubilize plant nutrients, and promote plant growth and phosphates, by synthesizing growth-promoting chemicals (Itelima et al. 2018). Bio-fertilizers are gaining attention in agriculture because they are non-toxic, eco-friendly, and non-hazardous (Itelima et al. 2018). *Azolla*, *Azospirillum*, P-solubilizing microbes, blue-green algae, *Azotobacter*, *Sinorhizobium*, and mycorrhizae, are examples of bio-fertilizers that promote crop yield (Mishra et al. 2013).

Bio-fertilizers have a lot of potential in terms of providing plant nutrients while reducing the use of chemical fertilizers (Kaur & Purewal 2019). The bio-inoculants are used as seed treatments or soil treatments to improve plant nutrient accessibility, which leads to increased crop growth and production (Chatterjee & Bandyopadhyay 2017; Ortas et al. 2017). Bio-fertilizers can also protect plants from soilborne diseases (Chatterjee & Bandyopadhyay 2017) and can help plants to increase crop yield by 10–25%. Plant pathogenic fungal activity is influenced by bio-fertilizer agents. *Rhizobium* culture applied to legume seeds controls seedborne fungi such as *Colletotrichum*, *Ascochyta*, and *Helminthosporium*. When rhizobia multiply on the seed and in the rhizosphere, they generate a poisonous toxin (Brahmaprakash et al. 2017).

Antibiotic compounds are produced by phosphate-solubilizing fungi such as *Aspergillus niger* and *Penicilla*, which kill plant pathogenic fungi (Devi et al. 2020). Plant pathogens are killed indirectly by creating healthy seedlings and phytoalexins (Pedras & Abdoli 2017). The use of mycorrhizae results in stronger root systems that are more resistant to root rotting, collar rot diseases, and soilborne pathogens. Bio-fertilizer application in the soil has been shown to encourage and augment the activity of saprophytic microorganisms in numerous studies (Naseer et al. 2016; Etesami et al. 2017). Under optimum agronomic and pest-free conditions, they can reduce the use of chemical fertilizers to no more than 40–50 kg N/ha (Asoegwu et al. 2020).

the benefits of employing bio-fertilizers include root development, improved mineral and water intake, vegetative growth, and nitrogen fixation (Ortaş et al. 2019; Singh 2018). Some bio-fertilizers (e.g., *Rhizobium* BGA, *Azotobacter* sp.) promote the formation of growth-promoting substances such as gibberellic acids, indole acetic acid (IAA), vitamin B complex, and other compounds (Mishra et al. 2013).

6.4 LIMITATIONS OF USING BIO-FERTILIZERS

1. As bio-fertilizers are living entities, they require careful handling during storage for lengthy periods of time (Itelima et al. 2018).
2. They have to be utilized before they expire or loose their activity (Vondeling et al. 2018).
3. They are less efficient if other microbes pollute the carrier medium or if an unwanted strain is present (Vondeling et al. 2018).
4. For bio-fertilizer organisms to flourish and function, the soil must have sufficient nutrients (Anand et al. 2016).
5. Other fertilizers can be supplemented with bio-fertilizers, but they cannot completely replace them (Panhwar et al. 2019).
6. If the soil is excessively warm or dry, bio-fertilizers lose their efficiency (Barati et al. 2017).
7. Inordinately acidic or alkaline soils can make it tough for helpful microbes to thrive. Furthermore, if the soil includes an abundance of the microbe's competition, they are much less effective (Martínez-Hidalgo et al. 2019).

6.5 CONTRIBUTION OF AMF IN NUTRIENT UPTAKE

Arbuscular mycorrhizal fungi (AMF) are a type of root-bound biotroph that help roughly 80% of plants by exchanging mutual benefits for plant growth and soil health improvement. In exchange for photosynthetic products, they assist the host with essential requirements of nutrients and water along with protection from pathogens. Consequently, AMF are an important organic component of soil whose absence can cause ecosystems to function inefficiently. With an eye to sustainable agriculture, the procedure of re-creating the usual level of AMF is a viable approach to current fertilization methods. The AMF association with maize plants results in osmolytes and aquaporin deposition, as well as antioxidant activities. PGPR and AMF are promising strategies for increasing P content at limited cost in plants, and their effects under low soil P supply are well understood (Zahoor et al. 2016). By modulating critical plant processes, these bacteria can also enhance the P content under varying stresses (Saia et al. 2015a, 2019a). This is especially true when both the plant and microorganisms meet their N requirements (Saia et al. 2020). Although these microorganisms can influence other plant strategies for receiving P by reducing direct absorption (and increasing plant needs for P from bacteria) or influencing root morphological traits, they cannot change the other plant mechanisms for acquiring P. Despite the importance of potassium to plants, less research has been done on the involvement of AMF in the accession of potassium by host plants. Potassium is abundant in soil, although it has low bioavailability to plants, and has been found in vesicles, hyphae, and spores. Phosphorus is a macronutrient that plants require for growth and evolution. Plants uptake P as orthophosphate (Pi) from the soil, but the amount of Pi in the soil solution rarely surpasses 10 M (Schachtman et al. 1998), and the diffusive movement of Pi in the soil solution is slow. When soil P levels are low, the symbiotic interaction between plants and AMF may be one strategy to improve P uptake. Plants deliver carbon to the AMF, and the fungus feeds P and other weakly accessible nutrients to the plant via an extensive hyphal network in this symbiotic interaction (Drew et al. 2003; Ortaş et al. 2016). AMF has also been shown to boost the activities of phosphatase enzymes and the solubilization of P (Tarafdar & Marschner 1994). However, an increase in P concentration in AMF-infected plants is not always associated with increased growth; it can also be associated with growth depression (Tran et al. 2019; Zhu et al. 2001), as AMF receive more C from the plants than they would otherwise contribute to plant growth, or AMF colonization may suppress direct P uptake by plant roots (Smith & Smith 2011; Zhu et al. 2001). Because of both environmental and genetic factors on colonization, plant susceptibility to AMF colonization is highly varied. Researchers have been studying the nutrient uptake processes in AM relationships (Ortaş & Rafique 2017). Plant genes are known to be stimulated by AM contact to make proteins that transport Pi. The AMF not only improve Pi absorption in plants, but also help to form the structure and shape of arbuscules and maintain symbiotic connections (Xie et al. 2013). It has been previously shown MtPT4 and LjPT4 Pi transporters are found in the root tips of *Medicago truncatula* and *Lotus japonica* plants not colonized by AMF because of AM interaction, and that *PT4* genes serve as elements of the Pi detecting machinery (Volpe et al. 2016). Plant transporter proteins can absorb and transmit inorganic nutrients from the peri-arbuscular apoplast to the cortical cell through the membrane mentioned earlier (Bapaume & Reinhardt 2012; Shah et al. 2016b). Ammonium transporters (enzymes that move ammonium from one place to another) produced by AMF have been found in the peri-arbuscular membrane of soybeans, indicating that they play a role in ammonium transfer to the cortex cell (Breuillin-Sessoms et al. 2015).

6.6 ARBUSCULAR MYCORRHIZAL FUNGI MECHANISM FOR PHOSPHORUS IN SOIL

There is an increasing demand to boost food production to fulfill the needs of the growing global population. This can be accomplished in one of two approaches: by expanding the cultivation area or by increasing the production per unit area. Previous methods are not achievable in several countries throughout the world owing to an assortment of limitations, such as a paucity of soil or water resources, drought, climate change, and soil salinization (Anukwonke et al. 2022). Alternatively, improving the nutritional characteristics of the soil is another approach to boost the yield per unit area. Phosphorus is necessary for energy generation, carbon metabolism, enzyme activation, energy transmission, nitrogen (N_2) fixation, and membrane development in plants (Lyu et al. 2022). Phosphorus also contributes to the formation of important biological components such as nucleic acids, ATP, and phospholipids (Park et al. 2022). Phosphorus scarcity is a critical limiting factor for crop development and productivity, affecting nearly half of all agricultural ecosystems worldwide (Etesami & Jeong 2021; Wu et al. 2022). To solve this issue, there has been a massive growth in the usage of phosphorus fertilizers all over the world. The increasing agricultural P demand has called into doubt the long-term viability of P mining for fertilizer manufacturure (Lun et al. 2021). Phosphorus fertilizers frequently result in the addition of an excessive amount of P to agricultural soils. Unfortunately, due to fixation and adsorption processes, more than 80% of P fertilizers applied to soil are lost in precipitated form (Bindraban et al. 2020) or converted into organic forms, which account for 40–80% of soil total P (Zhu et al. 2018), with phytates as the most familiar form (Menezes-Blackburn et al. 2018). As a result, the accessibility of this additional P to plants is restricted (almost 0.1% of the total P).

Plants can only absorb P in a limited range of soil conditions, such as HPO_4^{2-} and $H_2PO_4^-$, which have a pH of 6.5–7. Inorganic phosphate (Pi) is immobilized and mineralized into calcium phosphates when the pH of soil exceeds 7.0. At lower soil pH, soluble aluminium, manganese, iron, or similar hydrous oxides are usually absorbed by P (Etesami & Jeong 2021). Inorganic phosphate adsorbs to weathered silicates such as clay minerals at neutral pH, resulting in a 5–15% reduction in agricultural yield (Aziz et al. 2014; Hashizume et al. 2020).

The predicted improvement in plant development proficiency from the addition of chemical P fertilizers has peaked, hence more fertilization of chemical P is unlikely to enhance plant production (Yadav et al. 2022). According to the US Geological Survey, 22 million tons of P (3–4% of total P consumption) are removed yearly from natural sources (Etesami & Jeong 2021), as a result, natural P supplies are at risk from this reduction (Mnthambala et al. 2021). More effective use of P is required, which includes optimizing P collection and usage efficiency (Han et al. 2022).

Some plants can effectively accumulate and utilize P to uphold development and metabolism (Dissanayaka et al. 2021; Sarkar & Sadhukhan 2022). The following are some plant strategies for enhancing P uptake efficiency: (1) growth rearrangement among root types; (2) topsoil foraging; (3) soil investigation at a low metabolic rate; (4) root hair growth stimulation; (5) increased expression of high-affinity P carriers; (6) increased root-to-shoot ratio; (7) increased organic acids secretion (e.g., oxalate, malate, citrate,) from roots to the soil; and (8) improved phytase secretion or acid phosphatase (Wang et al. 2006).

Plants also have improved biotic relations with a variety of soil microbes that boost plant development. The most prevalent microorganisms are plant growth-promoting bacteria (PGPB) and AMF. PGPB and AMF, particularly the phosphate-solubilizing bacteria (PSB) subgroup, is recognized to assist plants in overcoming P deficiency (Khatoon et al. 2020; Rafique et al. 2017). AMF and PSB are part of important biogeochemical cycling mechanisms (Gouda et al. 2018; Pathak et al. 2017). Endomycorrhizal communities in which fungal hyphae enter root cell walls and contact the plasmalemma are known as AMF. They can be found in almost every plant habitat in the world (Wang et al. 2021). The formation of AMF has allowed plants to live and thrive in their natural settings for millions of years without the use of pesticides or fertilizers. AMF belong to the phylum *Glomeromycota* and the subphylum *Glomeromycotina*, which comprise 340 species (Bano & Uzair 2021; Reinhardt et al. 2021).

AMF are obligate symbionts that depend entirely on their plant hosts for organic carbon. This symbiosis is arguably the earliest kind of mycorrhiza, having originated 400–450 million years ago and involving a broad range of plants (Feijen et al. 2018). The symbiosis is typically mutualistic, relying on plant carbon exchange (4–20% of photosynthetically fixed carbon) and fungus-provided P (Ravi et al. 2021; Vijayakumar 2018; Watkinson 2016). This mycorrhizal symbiosis is thought to be capable of aiding more than 80% of the world's plant species (Vasar et al. 2021). The advantages of AMF in many plants (mainly crops) have been demonstrated (Chen et al. 2022). The AMF and bacterial species boost plant resilience to abiotic stressors, improve mineral absorption (especially of P), improve water relationships, and protect plants from soilborne diseases, all of which promote plant development (Fakhar et al. 2022; Ortas et al. 2021; Rafique et al. 2019; Sakthieaswari et al. 2022). AMF can support plants to acquire nutrients such as Mg, Cu, Zn, K, and N, especially when they are present in less soluble forms in soils (Bhantana et al. 2021; Saboor et al. 2021). These fungi enter the root cortical cell walls and create arbuscules, which resemble haustoria and mediate metabolite exchanges between the host cell and the fungi (Sakthieaswari et al. 2022). Increasing the root zone absorption area by 10–100%, AMF improve the plant's ability to use more soil resources. Because of the extraradical hyphae that enhance nutrient absorption and transport, mycorrhizal roots can reach a larger soil volume than non-mycorrhizal roots (Beltrano et al. 2013). AMF increase nutrient absorption by expanding the root's absorption area and secreting compounds such as glomalin, a glycoprotein produced by AMF hyphae and spores. Glomalin in the soil aids in the absorption of difficult-to-break-down nutrients such as Fe and P (Herath et al. 2021; Pal & Pandey 2014). Phosphorus is quickly absorbed from soil particles, and Pi-free zones form around the roots with ease. Mycorrhizal roots' extraradical hyphae reach beyond these P-depleted zones, absorbing bioavailable Pi that would otherwise be unavailable to plants.

Arbuscular mycorrhizal plant roots absorb P via two mechanisms. The first mechanism is shared by both non-arbuscular mycorrhizal and arbuscular mycorrhizal plants, and entails P absorption directly from the root epidermis and hairs. Phosphorus enters the root cortical cells in the second mechanism (intraradical mycelium) (González-González et al. 2020), and arbuscules or hyphal coils offer symbiotic interactions between fungal hyphae (cell-specific Pi transporter gene expression in mycorrhizal roots transports P from the interfacial apoplast) (Balestrini et al. 2007; Balestrini et al. 2021; Fochi et al. 2017). This is a fast P translocation that extends several centimetres. Recent molecular and physiological data show that the mycorrhizal pathway is active for P, regardless of plant growth responses (Smith & Smith 2011). The role of carriers has been carefully investigated, as has the translocation of Pi in fungi and the transfer of Pi to host plants (Chan et al. 2021; Ezawa & Saito 2018; Johri et al. 2015).

As previously stated, poor solubility of P in alkaline and acidic soils (e.g., less than 10 M) leads to very limited mobility (Seshadri et al. 2020). As a result, when roots absorb P, it takes a long time for it to be replaced in the bulk soil, leading to the formation of P-depleted zones, where all the available P has been rapidly removed from the environment around the roots, lowering P uptake by the root epidermal hairs (the first method of P absorption) (De Parseval et al. 2017; Smith et al. 2011). Plants should move beyond these deficient areas and demonstrate root activity in other parts of soil to increase their P acquisition. The success of this effort to absorb P (and other normally immobile soil nutrients) is defined by the root system surface area. The highly essential function of AMF hyphae is to enhance root surface area (depletion is reduced near small-diameter AMF hyphae) (Mei et al. 2014; Okiobe 2019). Furthermore, mycorrhizal plants can exhale organic acids like citrate and malate, which chelate Al^{3+} and Ca^{2+}, and dissolve calcium and aluminium phosphates (Seguel et al. 2017). With the help of AMF hyphae, plants gain improved access to orthophosphates and Pi in the soil solution by expanding their soil contact area (Rubin & Görres 2021). AMF hyphae allow the roots to immediately absorb released Pi because AMF roots lack a fungal coat, and they can potentially use both routes of nutrition absorption. The two nutrient absorption mechanisms in AMF symbiosis are thought to work in tandem (Xie et al. 2021). However, it is believed that AMF supply around 80% of P absorption in a mycorrhizal plant (Andrino et al. 2021). The AMF boosts legumes' ability to fix nitrogen and minimizes the quantity of inorganic N that percolates (Meena et al. 2018). Chlorophyll contains N_2, which is required for photosynthesis. Soil microorganisms such as AMF and PSB become more active when photosynthetic resources are transported to the roots.

The AMF can help P-deficient soils absorb more P by increasing the rate of P absorption (P input) per unit of AMF root. The enhanced P absorption rate with AMF is responsible for the improved efficiency with which hyphal surfaces absorb P from soil as compared to cylindrical root surfaces (Hammer et al. 2014). Absorbing Pi through fungal AMF hyphae, increasing the mycorrhizal hyphal network beyond the rhizosphere Pi transporters, which transfer Pi to internal fungal structures in root cortical cells, can be found up to 25 cm from the roots (Mbodj et al. 2018; Smith & Smith 2011). By successfully transporting P from soil to plant-based hyphae via appressoria and from extraradical mycelium to intraradical mycelium, the fungus can maintain low internal Pi levels (Preeti & Panwar 2013). Hyphae with a small diameter (2–20 m) allow the fungus to penetrate in small soil cores in search of phosphate and attain higher P inflow rates for a given surface area (Bitterlich et al. 2018); lowering the depletion zone around roots or hyphae (reduced rhizospheric Pi depletion) (Smith & Smith 2011). Phosphorus depletion surrounding the roots of *Capsicum annuum* L. or the hyphae of *Rhizophagus mossea* was just 0.06 cm in a study, meaning that only 7% of the soil P was available to the roots. The hyphae had complete access to soil since half the distance between close hyphae was barely 0.01 cm. In most cases, the high ability of hyphal surfaces to absorb P from soils may be sufficient to explain how AMF enhances accessible P absorption in soil (Sharif & Claassen 2011).

6.7 COMMERCIALLY AVAILABLE MYCORRHIZAL BIO-FERTILIZERS

6.7.1 Oregonism XL

Soluble root growth enhancer

Oregonism XL was particularly chosen for fruiting and blooming plants that can withstand intense fertilization. It improves nutrient availability and absorption, promotes strong root systems, and aids plant resistance to stress.

Ingredients

Oregonism XL is a proprietary mix of AMF, *Bacillus*, and *Pseudomonas* microorganisms along with humic acid. It works well in both soil and hydroponics. (www.planetnatural.com/product/roots-organics-oregonism-xl/)

6.7.2 Great White

The Great White mycorrhizae are a blend of carefully selected mycorrhizal fungus and beneficial bacteria that are well adapted to a wide range of soils, climates, and plants. These bacteria flourish in and on plant roots in nature, considerably enhancing plant growth and vitality.

Ingredients and features

- There are seven species of Endomycorrhizae and 11 species of Ectomycorrhizae
- There are 67,000 endo- and 1.5 billion ecto-propagules per pound.
- There are 19 species of bacteria and two species of *Trichoderma*.
- It is a biostimulant and vitamin package. (www.planetnatural.com/product/great-white-mycorrhizae/)

6.7.3 Myco Madness

Humboldt Nutrients' Myco Madness contains a biologically active powerhouse of nine mycorrhizal species, 15 helpful bacteria, and two *Trichoderma* species. This soluble powder, which is well suited to a variety of climates, soils, and plants, can assist plants in boosting nutrient and water absorption for improved plant performance. (www.planetnatural.com/product/myco-madness/)

6.7.4 Mycorrhizae (Soluble)

Mycorrhizae (Soluble) is a mixture of humic acid, vitamins, and 12 helpful bacteria (seven strains of endo- and five strains of ectomycorrhizae). It reproduces quickly once applied and acts to provide a favourable environment for the growth of seeds and transplants. It reduces transplant losses by reducing plant stress and boosting new root formation. It can be used in tank mixes containing organic fertilizers. Its addition aids in the preservation and expansion of beneficial bacteria and fungi in the soil. This increases biomass energy, which promotes soil tilth and nutrient absorption. Excessive application is not detrimental. (www.planetnatural.com/product/soluble-mycorrhizae/)

6.7.5 MycoStim

Organic Research Laboratories' MycoStim is a potent soil inoculant that mixes eight super-strains of mycorrhizae and two species of *Trichoderma* with amino acids, kelp extracts, and humic acid. These plant-friendly fungi promote root development, which provides:

- Increased growth and nutrient intake
- Resistance to transplant shock

- Vigorous plants
- Resistance to stress
- Increased plant quality, yield, and vigor.
- (www.planetnatural.com/product/mycostim-mycorrhizae/)

6.7.6 MYKOS

MYKOS (mycorrhizae) is a soil fungus that aids in the breakdown and transfer of nutrients to plant roots. MYKOS boosts the availability of both moisture and nutrients essential for plant growth, as well as connecting numerous microorganisms in healthy soil to host plants. MYKOS may change an average landscape into something with hugely improved performance with just one treatment.

(www.planetnatural.com/product/mykos-mycorrhizal-inoculant/)

6.7.7 Piranha

Pirhana is a 100% organic product. Advanced Nutrients' Piranha maximizes plant root mass by up to 700%. More roots guarantee that plants get the most nourishment, blooming, and harvest. It contains a unique blend of ecto- and endomycorrhizae to optimize the potential of vegetation.

- Helps roots to develop surface mass for maximum plant growth
- Increases the creation of essential oils
- Increases the production of plant aromatics
- Beneficial fungus increases root absorption of oxygen, nutrients, and water
- Increases the survival rate of clones and seedlings.
- (www.planetnatural.com/product/liquid-piranha/)

6.7.8 Plant Success (3-1-2)

This product improves soil and roots. Plant Prosperity Granular includes a variety of beneficial fungi (11 mycorrhizae and two *Trichoderma* species) that are well adapted to a wide range of soils, climates, and plants. Kelp meal, humus, vitamins, and amino acids are also used to increase mycorrhizal germination and efficacy.

- Contains both endo- and ectomycorrhizae
- Generates strong root systems and healthy plants
- Increases water and nutrient absorption
- Granular recipe is simple to use
- Contains a high-quality nutrition packet
- (www.planetnatural.com/product/plant-success-granular/)

6.7.9 Root Maximizer

Beneficial fungi may work their magic in any growing environment. Clonex Root Maximizer enables an endo-mycorrhizae network to promote optimal plant health in soil or indoor systems such as hydroponic or aeroponic setups.

- Mycorrhizal fungi, beneficial bacteria, and *Trichoderma* increase nutrition and water absorption
- Increases plant root surface area
- Protects plants from environmental stress and diseases

- Can be used dry or in a liquid solution
- Developed by Clonex, a pioneer in plant propagation resources for more than 30 years.
- (www.planetnatural.com/product/clonex-root-maximizer/)

6.7.10 Root Rally (0-3-0)

This product is totally organic. Age Old Root Rally contains endo- and ectomycorrhizae spores as well as vitamins, minerals, and nutrients. The inoculants encourage the growth of mycorrhizae in trees, shrubs, flowers, fruits, and vegetables. When applied to the plant's root system, this 100% organic combination reduces transplant shock, promotes root development, and increases water and nutrient intake. (www.planetnatural.com/product/root-rally/)

6.7.11 Rooters

This is a customized combination of eight mycorrhizal fungi chosen for a diverse range of plants, media, and habitats. Rooters of Earth Juice Mycorrhizae works with the plant's root system to increase the surface area of the root mass. It improves most plants' growth and vigor by increasing nutrient and water absorption. For use in soil and hydroponic systems. (www.planetnatural.com/product/rooters-mycorrhizae/)

6.7.12 SubCulture-M

This product increases root mass significantly to boost growth and yield. Subculture-M is a mycorrhizal root inoculant containing a diverse range of endo- and ectomycorrhizal fungi that colonize plant roots. These helpful fungi generate a thin network of fibrous strands that extend the plant's root system, increasing the root area and assisting plants in absorbing water and nutrients.

From germination or propagation through harvest, this product can be sued. The quantity of fertilizer used is reduced while still achieving exceptional growth and enormous harvests. When combined with Subculture-B, it creates a diversified microbial population that is favourable to plant health. (www.planetnatural.com/product/subculture-m-mycorrhizal-root-inoculant/)

6.7.13 White Widow

The mycorrhizal fungus in Humboldt Nutrients' White Widow can assist plant roots to have better access to water and nutrients in the soil. When applying this mycorrhizal powder, it is normal to see huge, bright white roots come out of the bottom of pots barely a week after transplanting. (www.planetnatural.com/product/white-widow/)

6.7.14 ZHO

This is an inoculant for the rhizosphere. Botanicare ZHO is a patented combination of mycorrhizae and *Trichoderma* fungus that establishes a natural microbial system in and around plant roots, significantly boosting plant and root growth, vigor, and production organically. (www.planetnatural.com/product/zho-root-inoculant/)

REFERENCES

Adnan M, Zahir S, Fahad S, Arif M, Mukhtar A, Imtiaz AK, Ishaq AM, Abdul B, Hidayat U, Muhammad A, Inayat-Ur R, Saud S, Muhammad ZI, Yousaf J, Amanullah, Hafiz MH, Wajid N (2018a)

Phosphate-solubilizing bacteria nullify the antagonistic effect of soil calcification on bioavailability of phosphorus in alkaline soils. *Sci Rep* 8, 4339. https://doi.org/10.1038/s41598-018-22653-7

Adnan M, Shah Z, Sharif M, Rahman H (2018b) Liming induces carbon dioxide (CO_2) emission in PSB inoculated alkaline soil supplemented with different phosphorus sources. *Environ Sci Poll Res* 25(10), 9501–9509.

Adnan M, Fahad S, Khan IA, Saeed M, Ihsan MZ, Saud S, Riaz M, Wang D, Wu C (2019) Integration of poultry manure and phosphate solubilizing bacteria improved availability of Ca bound P in calcareous soils. *3 Biotech* 9(10), 368.

Adnan M, Fahad S, Muhammad Z, Shahen S, Ishaq AM, Subhan D, Zafar-ul-Hye M, Martin LB, Raja MMN, Beena S, Saud S, Imran A, Zhen Y, Martin B, Jiri H, Rahul D (2020) Coupling phosphate-solubilizing bacteria with phosphorus supplements improve maize phosphorus acquisition and growth under lime induced salinity stress. *Plants* 9(7), 900. doi: 10.3390/plants9070900

Amna, Ali N, Masood S, Mukhtar T, Kamran MA, Rafique M, Munis MFH, Chaudhary HJ (2015) Differential effects of cadmium and chromium on growth, photosynthetic activity, and metal uptake of Linum usitatissimum in association with Glomus intraradices. *Environ Monit & Assess* 187, 311.

Anand K, Kumari B, Mallick M (2016) Phosphate solubilizing microbes: an effective and alternative approach as biofertilizers. *Int J Pharm Sci* 8, 37–40.

Andrino A, Guggenberger G, Sauheitl L, Burkart S, Boy J (2021) Carbon investment into mobilization of mineral and organic phosphorus by arbuscular mycorrhiza. *Biol & Fertil Soils* 57, 47–64.

Anukwonke CC, Tambe EB, Nwafor DC, Malik KT (2022) *Climate Change and Interconnected Risks to Sustainable Development, Climate Change*. Springer, pp. 71–86.

Asoegwu CR, Awuchi CG, Nelson K, Orji CG, Nwosu OU, Egbufor UC, Awuchi CG (2020) A review on the role of biofertilizers in reducing soil pollution and increasing soil nutrients. *Himalayan J Agric* 1, 34–38.

Atif B, Hesham A, Fahad S (2021) Biochar coupling with phosphorus fertilization modifies antioxidant activity, osmolyte accumulation and reactive oxygen species synthesis in the leaves and xylem sap of rice cultivars under high-temperature stress. *Physiol Mol Biol Plants* 27(9), 2083–2100. https://doi.org/10.1007/s12298-021-01062-7

Aziz T, Finnegan PM, Lambers H, Jost R (2014) Organ-specific phosphorus-allocation patterns and transcript profiles linked to phosphorus efficiency in two contrasting wheat genotypes. *Plant, Cell & Environ* 37, 943–960.

Balestrini R, Gómez-Ariza J, Lanfranco L, Bonfante P (2007) Laser microdissection reveals that transcripts for five plant and one fungal phosphate transporter genes are contemporaneously present in arbusculated cells. *Mol Plant–Microbe Interact* 20, 1055–1062.

Balestrini R, Perotto S, Fiorilli V (2021) Laser microdissection as a tool to study fungal gene expression in mycorrhizal endosymbioses. *Italian J Mycol* 50, 1–9.

Bano SA, Uzair B (2021) Arbuscular mycorrhizal fungi (AMF) for improved plant health and production. In: Manoj Kaushal, Ram Prasad (Eds.), *Microbial Biotechnology in Crop Protection*. Springer, pp. 147–169.

Bapaume L, Reinhardt D (2012) How membranes shape plant symbioses: signaling and transport in nodulation and arbuscular mycorrhiza. *Front Plant Sci* 3, 223.

Barati MR, Aghbashlo M, Ghanavati H, Tabatabaei M, Sharifi M, Javadirad G, Dadak A, Soufiyan MM (2017) Comprehensive exergy analysis of a gas engine-equipped anaerobic digestion plant producing electricity and biofertilizer from organic fraction of municipal solid waste. *Energy Convers & Manage* 151, 753–763.

Barragán-Ocaña A, del Carmen del-Valle-Rivera M (2016) Rural development and environmental protection through the use of biofertilizers in agriculture: an alternative for underdeveloped countries? *Technol Society* 46, 90–99.

Beltrano J, Ruscitti M, Arango MC, Ronco M (2013) Effects of arbuscular mycorrhiza inoculation on plant growth, biological and physiological parameters and mineral nutrition in pepper grown under different salinity and P levels. *J Soil Sci & Plant Nutr* 13, 123–141.

Benaffari W, Boutasknit A, Anli M, Ait-El-Mokhtar M, Ait-Rahou Y, Ben-Laouane R, Ben Ahmed H, Mitsui T, Baslam M, Meddich A (2022) The native arbuscular mycorrhizal fungi and vermicompost-based organic amendments enhance soil fertility, growth performance, and the drought stress tolerance of quinoa. *Plants* 11, 393.

Bhantana P, Malla R, Vista SP, Rana MS, Mohamed G, Joshi BD, Shah S, Khadka D, Prasad G, Timsina KP (2021) Use of arbuscular mycorrhizal fungi (AMF) and zinc fertilizers in an adaptation of plant from drought and heat stress. *Biomed J Sci & Tech Res* 8, 30357–30373.

Bindraban PS, Dimkpa CO, Pandey R (2020) Exploring phosphorus fertilizers and fertilization strategies for improved human and environmental health. *Biol & Fertil Soils* 56, 299–317.

Bitterlich M, Sandmann M, Graefe J (2018) Arbuscular mycorrhiza alleviates restrictions to substrate water flow and delays transpiration limitation to stronger drought in tomato. *Front Plant Sci* 9, 154.

Brahmaprakash G, Sahu PK, Lavanya G, Nair SS, Gangaraddi VK, Gupta A (2017) Microbial functions of the rhizosphere. In: Dhananjaya Pratap Singh, Harikesh Bahadur Singh, Ratna Prabha (Eds.), *Plant–Microbe Interactions in Agro-Ecological Perspectives*. Springer, pp. 177–210.

Breuillin-Sessoms F, Floss DS, Gomez SK, Pumplin N, Ding Y, Levesque-Tremblay V, Noar RD, Daniels DA, Bravo A, Eaglesham JB (2015) Suppression of arbuscule degeneration in Medicago truncatula phosphate transporter4 mutants is dependent on the ammonium transporter 2 family protein AMT2; 3. *Plant Cell* 27, 1352–1366.

Chan C, Liao Y-Y, Chiou T-J (2021) The impact of phosphorus on plant immunity. *Plant & Cell Physiol* 62, 582–589.

Chatterjee R, Bandyopadhyay S (2017) Effect of boron, molybdenum and biofertilizers on growth and yield of cowpea (Vigna unguiculata L. Walp.) in acid soil of eastern Himalayan region. *J Saudi Society of Agric Sci* 16, 332–336.

Chen K, Kleijn D, Scheper J, Fijen TP (2022) Additive and synergistic effects of arbuscular mycorrhizal fungi, insect pollination and nutrient availability in a perennial fruit crop. *Agric, Ecosyst & Environ* 325, 107742.

Chen X, Ma L, Ma W, Wu Z, Cui Z, Hou Y, Zhang F (2018) What has caused the use of fertilizers to skyrocket in China? *Nutrient Cycl Agroecosyst* 110, 241–255.

De Parseval H, Barot S, Gignoux J, Lata J-C, Raynaud X (2017) Modelling facilitation or competition within a root system: importance of the overlap of root depletion and accumulation zones. *Plant & Soil* 419, 97–111.

Devi R, Kaur T, Kour D, Rana KL, Yadav A, Yadav AN (2020) Beneficial fungal communities from different habitats and their roles in plant growth promotion and soil health. *Microbial Biosyst* 5, 21–47.

Dissanayaka D, Ghahremani M, Siebers M, Wasaki J, Plaxton WC (2021) Recent insights into the metabolic adaptations of phosphorus-deprived plants. *J Exp Botany* 72, 199–223.

Drew E, Murray R, Smith S, Jakobsen I (2003) Beyond the rhizosphere: growth and function of arbuscular mycorrhizal external hyphae in sands of varying pore sizes. *Plant & Soil* 251, 105–114.

Etesami H, Emami S, Alikhani HA (2017) Potassium solubilizing bacteria (KSB): mechanisms, promotion of plant growth, and future prospects A review. *J Soil Sci & Plant Nutr* 17, 897–911.

Etesami H, Jeong BR (2021) Contribution of arbuscular mycorrhizal fungi, phosphate–solubilizing bacteria, and silicon to P uptake by plant: a review. *Front Plant Sci* 12, 1355.

Ezawa T, Saito K (2018) How do arbuscular mycorrhizal fungi handle phosphate? New insight into fine-tuning of phosphate metabolism. *New Phytol* 220, 1116–1121.

Fahad S, Hussain S, Saud S, Tanveer M, Bajwa AA, Hassan S, Shah AN, Ullah A,Wu C, Khan FA, Shah F, Ullah S, Chen Y, Huang J (2015a) A biochar application protects rice pollen from high-temperature stress. *Plant Physiol Biochem* 96, 281–287.

Fahad S, Nie L, Chen Y, Wu C, Xiong D, Saud S, Hongyan L, Cui K, Huang J (2015b) Crop plant hormones and environmental stress. *Sustain Agric Rev* 15, 371–400.

Fahad S, Hussain S, Saud S, Hassan S, Tanveer M, Ihsan MZ, Shah AN, Ullah A, Nasrullah KF, Ullah S, Alharby HNW, Wu C, Huang J (2016) A combined application of biochar and phosphorus alleviates heat-induced adversities on physiological, agronomical and quality attributes of rice. *Plant Physiol Biochem* 103, 191–198.

Fahad S, Hasanuzzaman M, Alam M, Ullah H, Saeed M, Ali Khan I, Adnan M (Eds.) (2020) *Environment, Climate, Plant and Vegetation Growth*. Springer Nature Switzerland AG 2020. doi: https://doi.org/10.1007/978-3-030-49732-3

Fahad S, Sönmez O, Saud S, Wang D, Wu C, Adnan M, Turan V (Eds.) (2021a) Plant growth regulators for climate-smart agriculture. First edition. *Footprints of Climate Variability on Plant Diversity*. CRC Press, Boca Raton.

Fahad S, Sonmez O, Saud S, Wang D, Wu C, Adnan M, Turan V (Eds.) (2021b) Climate change and plants: biodiversity, growth and interactions. First edition. *Footprints of Climate Variability on Plant Diversity*. CRC Press, Boca Raton.

Fahad S, Sonmez O, Saud S, Wang D, Wu C, Adnan M, Turan V (Eds.) (2021c) Developing climate resilient crops: improving global food security and safety. First edition. *Footprints of Climate Variability on Plant Diversity*. CRC Press, Boca Raton.

Fahad S, Sönmez O, Turan V, Adnan M, Saud S, Wu C, Wang D (Eds.) (2021d) Sustainable soil and land management and climate change. First edition. *Footprints of Climate Variability on Plant Diversity*. CRC Press, Boca Raton.

Fahad S, Sönmez O, Saud S, Wang D, Wu C, Adnan M, Arif M, Amanullah (Eds.) (2021e) Engineering tolerance in crop plants against abiotic stress. First edition. *Footprints of Climate Variability on Plant Diversity*. CRC Press, Boca Raton.

Fahad S, Saud S, Yajun C, Chao W, Depeng W (Eds.) (2021f) *Abiotic Stress in Plants*. IntechOpen United Kingdom 2021. http://dx.doi.org/10.5772/intechopen.91549

Fahad S, Adnan M, Saud S (Eds.) (2022a) Improvement of plant production in the era of climate change. First edition. *Footprints of Climate Variability on Plant Diversity*. CRC Press, Boca Raton.

Fahad S, Adnan M, Saud S, Nie L (Eds.) (2022b) Climate change and ecosystems: challenges to sustainable development. First edition. *Footprints of Climate Variability on Plant Diversity*. CRC Press, Boca Raton.

Fakhre A, Ayub K, Fahad S, Sarfraz N, Niaz A, Muhammad AA, Muhammad A, Khadim D, Saud S, Shah H, Muhammad ASR, Khalid N, Muhammad A, Rahul D, Subhan D (2021) Phosphate solubilizing bacteria optimize wheat yield in mineral phosphorus applied alkaline soil. *J Saudi Soc Agric Sci* 21(5), 339–348. https://doi.org/10.1016/j.jssas.2021.10.007

Fakhar A, Gul B, Gurmani AR, Khan SM, Ali S, Sultan T, Chaudhary HJ, Rafique M, Rizwan M (2020) Heavy metal remediation and resistance mechanism of Aeromonas, Bacillus, and Pseudomonas: a review. *Critical Reviews in Environmental Science and Technology*, 1–48.

Fakhar A, Gul B, Gurmani AR, Khan SM, Ali S, Sultan T, Chaudhary HJ, Rafique M, Rizwan M (2022) Heavy metal remediation and resistance mechanism of Aeromonas, Bacillus, and Pseudomonas: a review. *Critical Reviews in Environmental Science and Technology* 52, 1868–1914.

Feijen FA, Vos RA, Nuytinck J, Merckx VS (2018) Evolutionary dynamics of mycorrhizal symbiosis in land plant diversification. *Sci Rep* 8, 1–7.

Fochi V, Falla N, Girlanda M, Perotto S, Balestrini R (2017) Cell-specific expression of plant nutrient transporter genes in orchid mycorrhizae. *Plant Sci* 263, 39–45.

Fuentes-Ramirez LE, Caballero-Mellado J (2005) Bacterial biofertilizers. In: Siddiqui ZA (Ed.), *PGPR: Biocontrol and Biofertilization*. Springer-Verlag, Heidelberg, Berlin. pp. Springer. 143–172. doi:10.1007/1-4020-4152-7_5

González-González MF, Ocampo-Alvarez H, Santacruz-Ruvalcaba F, Sánchez-Hernández CV, Casarrubias-Castillo K, Becerril-Espinosa A, Castañeda-Nava JJ, Hernández-Herrera RM (2020) Physiological, ecological, and biochemical implications in tomato plants of two plant biostimulants: Arbuscular mycorrhizal fungi and seaweed extract. *Front Plant Sci* 11, 999.

Gouda S, Kerry RG, Das G, Paramithiotis S, Shin H-S, Patra JK (2018) Revitalization of plant growth promoting rhizobacteria for sustainable development in agriculture. *Microbiol Res* 206, 131–140.

Hammer EC, Balogh-Brunstad Z, Jakobsen I, Olsson PA, Stipp SL, Rillig MC (2014) A mycorrhizal fungus grows on biochar and captures phosphorus from its surfaces. *Soil Biol & Biochem* 77, 252–260.

Han Y, Hong W, Xiong C, Lambers H, Sun Y, Xu Z, Schulze WX, Cheng L (2022) Combining analyses of metabolite profiles and phosphorus fractions to explore high phosphorus-utilization efficiency in maize. *J Exp Botany* 73(12), 4184–4203. doi: 10.1093/jxb/erac117

Hashizume M, Yoshida M, Demura M, Watanabe MM (2020) Culture study on utilization of phosphite by green microalgae. *J Appl Phycol* 32, 889–899.

Herath B, Madushan K, Lakmali J, Yapa P (2021) Arbuscular mycorrhizal fungi as a potential tool for bioremediation of heavy metals in contaminated soil. *World J Adv Res & Rev* 10, 217–228.

Igiehon NO, Babalola OO (2017) Biofertilizers and sustainable agriculture: exploring arbuscular mycorrhizal fungi. *Appl Microbiol & Biotechnol* 101, 4871–4881.

Itelima J, Bang W, Onyimba I, Sila M, Egbere O (2018) Bio-fertilizers as key player in enhancing soil fertility and crop productivity: a review. *Dir Res J Agric & Food Sci* 6(3), 73–83.

John DA, Babu GR (2021) Lessons from the aftermaths of green revolution on food system and health. *Front Sustainable Food Syst* 5, 644559.

Johri AK, Oelmueller R, Dua M, Yadav V, Kumar M, Tuteja N, Varma A, Bonfante P, Persson BL, Stroud RM (2015) Fungal association and utilization of phosphate by plants: success, limitations, and future prospects. *Front Microbiol* 6, 984.

Kaur P, Purewal SS (2019) Biofertilizers and their role in sustainable agriculture. In: Bhoopander Giri, Ram Prasad, Qiang-Sheng Wu, Ajit Varma (Eds.), *Biofertilizers for Sustainable Agriculture and Environment*. Springer, pp. 285–300.

Khan A, Shah N, Gul A, Sahar NU, Ismail A, Muhammad, Aziz F, Farooq M, Adnan M, Rizwan M (2016) Comparative study of toxicological impinge of glyphosate and atrazine (herbicide) on stress biomarkers: blood biochemical and hematological parameters of the freshwater common carp (*Cyprinus carpio*). *Pol J Environ Stud* 25(5), 1993–1999.

Khatoon Z, Huang S, Rafique M, Fakhar A, Kamran MA, Santoyo G (2020) Unlocking the potential of plant growth-promoting rhizobacteria on soil health and the sustainability of agricultural systems. *J Environ Manage* 273, 111118.

Koshila Ravi R, Balachandar M, Yuvarani S, Anaswara S, Pavithra L, Muthukumar T (2021) Arbuscular mycorrhiza in sustainable plant nitrogen nutrition: mechanisms and impact. In: Cristina Cruz, Kanchan Vishwakarma, Devendra Kumar Choudhary, Ajit Varma (Eds.), *Soil Nitrogen Ecology*. Springer, pp. 407–436.

Kumar R, Kumawat N, Sahu YK (2017) Role of biofertilizers in agriculture. *Pop Kheti* 5, 63–66.

Lun F, Sardans J, Sun D, Xiao X, Liu M, Li Z, Wang C, Hu Q, Tang J, Ciais P (2021) Influences of international agricultural trade on the global phosphorus cycle and its associated issues. *Global Environ Change* 69, 102282.

Lyu X, Sun C, Zhang J, Wang C, Zhao S, Ma C, Li S, Li H, Gong Z, Yan C (2022) Integrated proteomics and metabolomics analysis of nitrogen system regulation on soybean plant nodulation and nitrogen fixation. *Int J Mol Sci* 23, 2545.

Martínez-Hidalgo P, Maymon M, Pule-Meulenberg F, Hirsch AM (2019) Engineering root microbiomes for healthier crops and soils using beneficial, environmentally safe bacteria. *Can J Microbiol* 65, 91–104.

Mbodj D, Effa-Effa B, Kane A, Manneh B, Gantet P, Laplaze L, Diedhiou A, Grondin A (2018) Arbuscular mycorrhizal symbiosis in rice: establishment, environmental control and impact on plant growth and resistance to abiotic stresses. *Rhizosphere* 8, 12–26.

Meena RS, Vijayakumar V, Yadav GS, Mitran T (2018) Response and interaction of Bradyrhizobium japonicum and arbuscular mycorrhizal fungi in the soybean rhizosphere. *Plant Growth Regul* 84, 207–223.

Mei C, Lara-Chavez A, Lowman S, Flinn B (2014) The use of endophytes and mycorrhizae in switchgrass biomass production. In: Hong Luo, Yanqi Wu, Chittaranjan Kole (Eds.), *Compendium of Bioenergy Plants: Switchgrass*. CRC, pp. 67–108.

Menezes-Blackburn D, Giles C, Darch T, George TS, Blackwell M, Stutter M, Shand C, Lumsdon D, Cooper P, Wendler R (2018) Opportunities for mobilizing recalcitrant phosphorus from agricultural soils: a review. *Plant & Soil* 427, 5–16.

Mishra D, Rajvir S, Mishra U, Kumar SS (2013) Role of bio-fertilizer in organic agriculture: a review. *Res J Recent Sci* 2, 39–41. ISSN 2277, 2502.

Mnthambala F, Tilley E, Tyrrel S, Sakrabani R (2021) Phosphorus flow analysis for Malawi: identifying potential sources of renewable phosphorus recovery. *Resour, Conserv & Recycl* 173, 105744.

Muhammad I, Khadim D, Fahad S, Imran M, Saud A, Manzer HS, Shah S, Jabar ZKK, Shamsher A, Shah H, Taufiq N, Hafiz MH, Jan B, Wajid N (2022) Exploring the potential effect of *Achnatherum splendens* L.-derived biochar treated with phosphoric acid on bioavailability of cadmium and wheat growth in contaminated soil. *Environ Sci Pollut Res* 29(25), 37676–37684. https://doi.org/10.1007/s11356-021-17950-0

Naseer F, Baig J, Din SU, Nafees MA, Alam A, Hameed N, Adnan M, Arshad M, Khan MA, Romman M, Mian IA, Shah SRA (2016) Impact of water quality on distribution of macro-invertebrate in Jutialnallah, Gilgit-Baltistan, Pakistan. *Int J Biosci* 9(6), 451–459.

Okiobe ST (2019) Potential effects of arbuscular mycorrhiza fungi on denitrification potential activity and nitrous oxide (N_2O). *Emissions from a Fertile Agricultural Soil*. Freie Universitaet Berlin (Germany).

Ortaş I, Rafique M (2017) The mechanisms of nutrient uptake by arbuscular mycorrhizae, mycorrhiza-nutrient uptake. In: Ajit Varma, Ram Prasad, Narendra Tuteja (Eds.), *Biocontrol, Ecorestoration*. Springer, pp. 1–19.

Ortaş I, Rafique M, Ahmed İA (2017) Application of arbuscular mycorrhizal fungi into agriculture. *Arbuscular Mycorrhizas and Stress Tolerance of Plants*. Springer, pp. 305–327.

Ortas İ, Rafique M, Akpinar C, Kacar YA (2017) Growth media and mycorrhizal species effect on acclimatization and nutrient uptake of banana plantlets. *Scientia Hortic* 217, 55–60.

Ortas I, Rafique M, Çekiç F (2021) Do mycorrhizal fungi enable plants to cope with abiotic stresses by overcoming the detrimental effects of salinity and improving drought tolerance? In: Neeraj Shrivastava, Shubhangi Mahajan, Ajit Varma (Eds.), *Symbiotic Soil Microorganisms*. Springer, pp. 391–428.

Ortaş I, Rafique M, Iqbal MT (2019) Mycorrhizae resource allocation in root development and root morphology. In: Ajit Varma, Swati Tripathi, Ram Prasad (Eds.), *Plant Microbe Interface*. Springer, pp. 1–26.

Ortaş I, Razzaghi S, Rafique M (2016) Arbuscular mycorrhizae: effect of rhizosphere and relation with carbon nutrition. In: Devendra K. Choudhary, Ajit Varma, Narendra Tuteja (Eds.), *Plant–Microbe Interaction: An Approach to Sustainable Agriculture*. Springer, pp. 125–152.

Pal A, Pandey S (2014) Role of glomalin in improving soil fertility. *Int J Plant & Soil Sci* 3, 112–129.

Panhwar QA, Ali A, Naher UA, Memon MY (2019) Fertilizer management strategies for enhancing nutrient use efficiency and sustainable wheat production. In: Sarath Chandran, Unni M.R., Sabu Thomas (Eds.), *Organic Farming*. Elsevier, pp. 17–39.

Park Y, Solhtalab M, Thongsomboon W, Aristilde L (2022) Strategies of organic phosphorus recycling by soil bacteria: acquisition, metabolism, and regulation. *Environ Microbiol Rep* 14(1), 3–24.

Pathak D, Lone R, Koul K (2017) Arbuscular mycorrhizal fungi (AMF) and plant growth-promoting rhizobacteria (PGPR) association in potato (Solanum tuberosum L.): a brief review. In: Vivek Kumar, Manoj Kumar , Shivesh Sharma , Ram Prasad (Eds.), *Probiotics & Plant Health*. Springer, pp.401–420.

Pedras MSC, Abdoli A (2017) Pathogen inactivation of cruciferous phytoalexins: detoxification reactions, enzymes and inhibitors. *RSC Advances* 7, 23633–23646.

Preeti SK, Panwar J (2013) Mycorrhiza – its potential use for augmenting soil fertility and crop productivity. *Physiology of Nutrition and Environmental Stresses on Crop Productivity*, 111.

Rafique M, Ortas I (2018) Nutrient uptake-modification of different plant species in Mediterranean climate by arbuscular mycorrhizal fungi. Eur *J Hortic Sci* 83, 65–71.

Rafique M, Ortas I, Rizwan M, Sultan T, Chaudhary HJ, Işik M, Aydin O (2019) Effects of Rhizophagus clarus and biochar on growth, photosynthesis, nutrients, and cadmium (Cd) concentration of maize (Zea mays) grown in Cd-spiked soil. *Environ Sci & Pollut Res* 26, 20689–20700.

Rafique M, Sultan T, Ortas I, Chaudhary HJ (2017) Enhancement of maize plant growth with inoculation of phosphate-solubilizing bacteria and biochar amendment in soil. *Soil Sci & Plant Nutr* 63, 460–469.

Reinhardt D, Roux C, Corradi N, Di Pietro A (2021) Lineage-specific genes and cryptic sex: parallels and differences between arbuscular mycorrhizal fungi and fungal pathogens. *Trends Plant Sci* 26, 111–123.

Rubin JA, Görres JH (2021) Potential for mycorrhizae-assisted phytoremediation of phosphorus for improved water quality. *Int J Environ Res & Public Health* 18, 7.

Saboor A, Ali MA, Hussain S, El Enshasy HA, Hussain S, Ahmed N, Gafur A, Sayyed R, Fahad S, Danish S (2021) Zinc nutrition and arbuscular mycorrhizal symbiosis effects on maize (Zea mays L.) growth and productivity. *Saudi J Biol Sci* 28, 6339–6351.

Saia SM, Nelson N, Huseth AS, Griegerc K, Reich BJ (2020) Transitioning machine learning from theory to practice. *Nat Resour Manage* 435, 109257.

Sakthieaswari P, Kannan A, Baby S (2022) Role of mycorrhizosphere as a biostimulant and its impact on plant growth, nutrient uptake and stress management. In: Vijai Gupta (Ed.), *New and Future Developments in Microbial Biotechnology and Bioengineering*. Elsevier, pp. 319–336.

Sarkar AK, Sadhukhan S (2022) Imperative role of trehalose metabolism and trehalose-6-phosphate signalling on salt stress responses in plants. *Physiologia Plantarum* 174(1), e13647.

Schachtman DP, Reid RJ, Ayling SM (1998) Phosphorus uptake by plants: from soil to cell. *Plant Physiol* 116, 447–453.

Seguel A, Cornejo P, Ramos A, Von Baer E, Cumming J, Borie F (2017) Phosphorus acquisition by three wheat cultivars contrasting in aluminium tolerance growing in an aluminium-rich volcanic soil. *Crop & Pasture Sci* 68, 305–316.

Seshadri P, Seames W, Sisk MD, Bowman F, Benson S (2020) Mobility of semi-volatile trace elements from the fly ash generated by the combustion of a sub-bituminous coal – The effects of the combustion temperature. *Energy & Fuels* 34, 15411–15423.

Shafi MI, Adnan M, Fahad S, Fazli W, Ahsan K, Zhen Y, Subhan D, Zafar-ul-Hye M, Martin B, Rahul D (2020) Application of single superphosphate with humic acid improves the growth, yield and phosphorus uptake of wheat (Triticum aestivum L.) in c*alcareous soil. Agron* (10), 1224. doi:10.3390/agronomy10091224

Shah B, Khan IA, Khan A, Din MMU, Adnan M, Junaid K, Shah SRA, Zaman M, Ahmad N, Akbar R, Fayaz W, Rahman IU (2016b) Determination of physio-morphic basis of resistance in different maize cultivars against insect pests. *J Ento & Zool Studies* 4(1), 317–321.

Sharif M, Claassen N (2011) Action mechanisms of arbuscular mycorrhizal fungi in phosphorus uptake by Capsicum annuum L. *Pedosphere* 21, 502–511.

Singh B (2018) Are nitrogen fertilizers deleterious to soil health? *Agron* 8, 48.

Smith SE, Smith FA (2011) Roles of arbuscular mycorrhizas in plant nutrition and growth: new paradigms from cellular to ecosystem scales. *Annu Rev Plant Biol* 62, 227–250.

Smith SE, Jakobsen I, Grønlund M, Smith FA (2011) Roles of arbuscular mycorrhizas in plant phosphorus nutrition: interactions between pathways of phosphorus uptake in arbuscular mycorrhizal roots have important implications for understanding and manipulating plant phosphorus acquisition. *Plant Physiol* 156, 1050–1057.

Spicka J, Hlavsa T, Soukupova K, Stolbova M (2019) Approaches to estimation the farm-level economic viability and sustainability in agriculture: a literature review. *Agric Econ* 65, 289–297.

Tarafdar J, Marschner H (1994) Phosphatase activity in the rhizosphere and hyphosphere of VA mycorrhizal wheat supplied with inorganic and organic phosphorus. *Soil Biol & Biochem* 26, 387–395.

Thomas L, Singh I (2019) Microbial biofertilizers: types and applications. In: Bhoopander Giri, Ram Prasad, Qiang-Sheng Wu, Ajit Varma (Eds.), *Biofertilizers for Sustainable Agriculture and Environment.* Springer, pp. 1–19.

Tran BT, Cavagnaro TR, Watts-Williams SJ (2019) Arbuscular mycorrhizal fungal inoculation and soil zinc fertilisation affect the productivity and the bioavailability of zinc and iron in durum wheat. *Mycorrhiza* 29, 445–457.

Umesha S, Singh PK, Singh RP (2018) Microbial biotechnology and sustainable agriculture. In: Ram Lakhan Singh, Sukanta Mondal (Eds.), *Biotechnology for Sustainable Agriculture.* Elsevier, pp. 185–205.

Van Der Heijden MG, Streitwolf-Engel R, Riedl R, Siegrist S, Neudecker A, Ineichen K, Boller T, Wiemken A, Sanders IR (2006) The mycorrhizal contribution to plant productivity, plant nutrition and soil structure in experimental grassland. *New Phytologist* 172, 739–752.

Vasar M, Davison J, Sepp S-K, Öpik M, Moora M, Koorem K, Meng Y, Oja J, Akhmetzhanova AA, Al-Quraishy S (2021) Arbuscular mycorrhizal fungal communities in the soils of desert habitats. *Microorg* 9, 229.

Vijayakumar V (2018) *Symbiotic Tripartism in the Model Plant Family of Legumes and Soil Sustainability, Legumes for Soil Health and Sustainable Management.* Springer, pp. 173–203.

Volpe V, Giovannetti M, Sun XG, Fiorilli V, Bonfante P (2016) The phosphate transporters LjPT4 and MtPT4 mediate early root responses to phosphate status in non mycorrhizal roots. Plant, *Cell & Environ* 39, 660–671.

Vondeling GT, Cao Q, Postma MJ, Rozenbaum MH (2018) The impact of patent expiry on drug prices: a systematic literature review. *Appl Health Econ & Health Policy* 16, 653–660.

Wahid F, Fahad S, Subhan D, Adnan M,, Zhen Y, Saud S, Manzer HS, Martin B, Tereza H, Rahul D (2020) Sustainable management with mycorrhizae and phosphate solubilizing bacteria for enhanced phosphorus uptake in *calcareous soils. Agri* 10(8), 334. doi:10.3390/agriculture10080334

Wang H, Inukai Y, Yamauchi A (2006) Root development and nutrient uptake. *Critical Rev Plant Sci* 25, 279–301.

Wang Y, Li Y, Li S, Rosendahl S (2021) Ignored diversity of arbuscular mycorrhizal fungi in co-occurring mycotrophic and non-mycotrophic plants. *Mycorrhiza* 31, 93–102.

Waqeel J, Khan ST (2022) Microbial biofertilizers and micronutrients bioavailability: approaches to deal with zinc deficiencies. In: Shams Tabrez, Khan, Abdul Malik (Eds.), *Microbial Biofertilizers and Micronutrient Availability.* Springer, pp. 239–297.

Watkinson SC (2016) Mutualistic symbiosis between fungi and autotrophs. *The Fungi.* Elsevier, pp. 205–243.

Williams MR, King KW, Ford W, Fausey NR (2016) Edge-of-field research to quantify the impacts of agricultural practices on water quality in Ohio. *J Soil & Water Conserv* 71, 9A–12A.

Wu J, Zhao H, Wang X (2022) Soil microbes influence nitrogen limitation on plant biomass in alpine steppe in North Tibet. *Plant & Soil* 474, 395–409.

Xie K, Ren Y, Chen A, Yang C, Zheng Q, Chen J, Wang D, Li Y, Hu S, Xu G (2021) Plant nitrogen nutrition: the roles of arbuscular mycorrhizal fungi. *J Plant Physiol*, 153591.

Xie X, Huang W, Liu F, Tang N, Liu Y, Lin H, Zhao B (2013) Functional analysis of the novel mycorrhiza-specific phosphate transporter A s PT 1 and PHT 1 family from A stragalus sinicus during the arbuscular mycorrhizal symbiosis. *New Phytol* 198, 836–852.

Yadav AK, Gurnule GG, Gour NI, There U, Choudhary VC (2022) Micronutrients and fertilizers for improving and maintaining crop value: a review. *Int J Environ, Agric & Biotechnol* 7, 1.

Zahida Z, Hafiz FB, Zulfiqar AS, Ghulam MS, Fahad S, Muhammad RA, Hafiz MH,Wajid N, Muhammad S (2017) Effect of water management and silicon on germination, growth, phosphorus and arsenic uptake in rice. *Ecotoxicol Environ Saf* 144, 11–18.

Zahoor M, Afzal M, Ali M, Mohammad W, Khan N, Adnan M, Ali A, Saeed M (2016) Effect of organic waste and NPK fertilizer on potato yield and soil fertility. *Pure Appl Biol* 5(3), 439–445.

Zhu J, Li M, Whelan M (2018) Phosphorus activators contribute to legacy phosphorus availability in agricultural soils: a review. *Sci Total Environ* 612, 522–537.

Zhu Y, Christie P, Laidlaw AS (2001) Uptake of Zn by arbuscular mycorrhizal white clover from Zn-contaminated soil. *Chemosphere* 42, 193–199.

7 Quality Standards for Production and Marketing of Biofertilizers

Adeel Ahmad,[1] Amir Aziz,[2] Muhammad Irfan,[3] Mukkram Ali Tahir,[2] Faiza Rawish,[4] and Noman Younas[1]*
[1] Institute of Soil and Environmental Sciences, University of Agriculture, Faisalabad, Pakistan
[2] Department of Soil and Environmental Sciences, College of Agriculture, University of Sargodha, Sargodha, Pakistan
[3] School of Environmental Science and Engineering, Tianjin University, Tianjin, PR China
[4] Superior College of Commerce Sat Sira Chowk, Mandi Bahauddin, Punjab, Pakistan
* Correspondence: Adeelgujjar046@yahoo.com

CONTENTS

7.1 BACKGROUND TO BIOFERTILIZERS

In recent decades, the term "biofertilizer" has been defined in a variety of ways, owing to an improved understanding of rhizosphere–microorganism–plant interactions. Soil nutrient utilization is improved with the use of biofertilizers such as mycorrhizal fungi and plant growth-promoting rhizobacteria (PGPR), which were first proposed by Okon and Labandera-Gonzalez (1994). Vessey (2003) describes the term "biofertilizer" as "a substance which contains living microorganisms". When biofertilizer is applied to seeds, plant surfaces, or soil, it colonizes the rhizosphere of the plant or the interior of the plant and promotes growth by increasing the supply or availability of primary nutrients to the host plant. Biological fertilizer was shortened to biofertilizer, which implies the use

DOI: 10.1201/9781003286233-7

of living organisms. Biofertilizer is "a product that contains living microorganisms that exert direct or indirect beneficial effects on plant growth and crop yield through different mechanisms", as later defined by Fuentes-Ramirez and Caballero-Mellado (2005). Biological control of plant pathogens was included in the expanded definition. But microorganisms that promote plant growth by controlling harmful organisms, such as biofungicides, bioinsecticides, bionematocides, or other products with similar activities favouring plant health, are generally defined as biopesticides, not biofertilizers (Siddiqui and Mahmood 1999; Vessey 2003).

A new term has been proposed to describe a microorganism's primary mechanism of action in promoting plant growth: biofertilizers. Rhizoremediators are those that degrade organic pollutants that can harm plant growth. In contrast, bioenhancers or phytostimulators are those that produce phytohormones to boost plant growth (Somers et al. 2004; Ali et al. 2018). A species' primary mechanism of action may be used to characterize its classification. However, according to Kloepper (1993) and Vessey (2003), even a single microorganism in the field will often reveal multiple mechanisms of action, making further sub-classification a purely theoretical endeavour.

Descriptions that could lead to incorrect classifications are also a factor in the need for a legal definition of biofertilizers. Gianinazzi and Vosatka (2004) classified mycorrhizal fungi as "plant natural parts" rather than microorganisms, which could save time and money by avoiding lengthy registration processes. However, it also opens the market to fraudulent products.

7.2 TYPES OF BIOFERTILIZERS

Biofertilizers are classified into four categories:

- **Nitrogen biofertilizer (N-BF)**
  - *Rhizobium* – symbiotic BF with legumes of all varieties
  - *Azotobacter* – non-symbiotic BF for cereals, vegetables, and horticultural crops.
  - *Azospirillum* – the associative BF for grains such as millets (blue-green algae, azolla, *Gluconoacetobacter diazotrophicus* is also N-BF, but is yet to be included in FCO)
- **Phosphorus biofertilizers (P-BF)**
  - P-solubilizes – PSB (*Bacillus*, *Pseudomonas* etc., for all crops)
  - P-mobilizers – mycorrhizae (*Glomus*, *Gigaspora* etc. for all crops)
- **Potash biofertilizers – K-BF** (microbes like *B. mucilaginosus* and *F. aurantia* for all crops).
- **Zinc solubilizers – Z-BF** (*Bacillus* microbes are capable of zinc solubilization).

If you're looking for a biofertilizer, you'll typically find it in either solid form (like peat and other types of lignite) or liquid form (like broths with polyvinyl pyrrolidone or biosurfactants). Single or multiple strains of microorganisms can be used to produce biofertilizers, which can also be frozen, polyacrylamide-encased, or granular. It has yet to be developed with the best production technology and packaging.

7.3 PRODUCTION PROCESSES FOR BIOFERTILIZERS

Biofertilizer production involves a wide range of considerations, including the microbes' growth profiles, the types and ideal conditions of organisms, and the formulation of inoculum. The formulation of the inoculant, the method of application, and the storage of the product are all crucial to the success of a biological product. Making biofertilizer involves five major steps. These include the isolation and selection of target microbes, the selection of method and carrier material, the selection of propagation method, prototype testing, and large-scale testing. Figure 7.1 elaborates the steps in biofertilizer production.

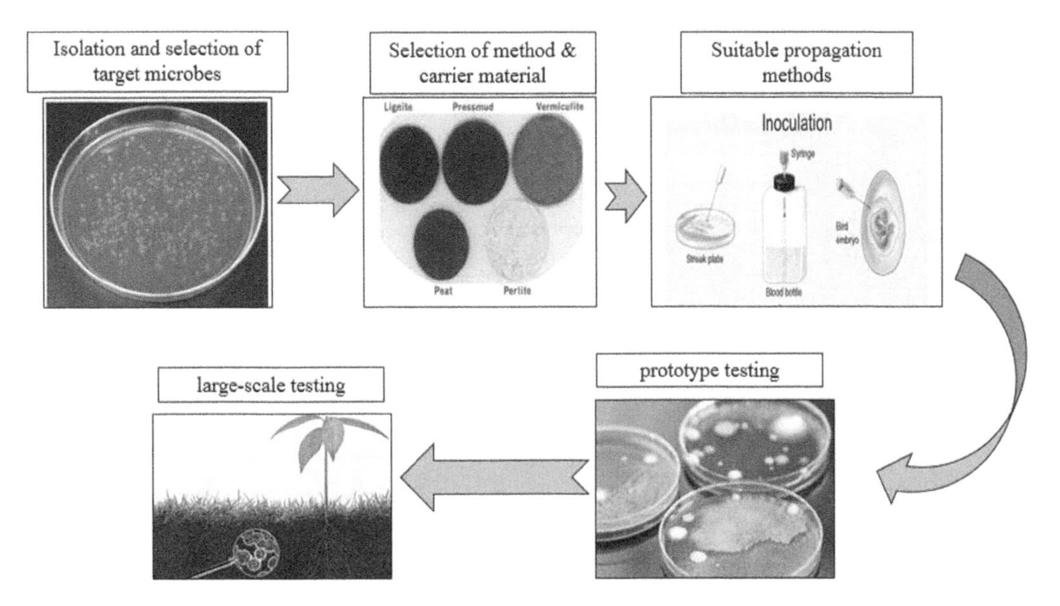

FIGURE 7.1 Step-by-step production process for biofertilizers.

- **Production flow**
    1. The materials can be composted and fermented. Composting and fermentation are required steps in the production of a quality organic fertilizer.
    2. Crushing and blending. Hammer crusher, chain crusher, and semi-wet material crusher are some of the machines used for crushing. Single-shaft continuous mixer, double-shaft horizontal mixer, and so on are some of the mixing machines available.
    3. Organic fertilizer production relies heavily on the granulating process. Today, the granulation industry employs various techniques such as double-roller extrusion granulation, disc, and rotary drum.
    4. Granulating produces a large number of particles, but the strength and water content of these particles are lower than expected. This is why drying is so important.
    5. A conveyor belt transfers the materials from the drying machine to the cooling machine, where they cool. Using this machine, the particle strengths can be improved even further, and the water content can be reduced.
    6. After cooling, a powdery substance is left behind. The fine powders and large particles can be screened out by this machine. The fine powder is then re-involved in the granulation process.
    7. Afterwards, a protective film is applied to the finished products, insulating the dust particles from the ambient air.
    8. Packaging. The automatic packing machine takes care of the packaging of qualified products.

When compared to chemical fertilizers, the production of biofertilizers is more cost-effective and simpler. However, the production of biofertilizers must take into account a number of factors, including field applications, microbial strains, formulation type, and carrier materials (Malusa et al. 2012; Mohammadi & Sohrabi 2012; Arif et al. 2015). Standardizing the commercial production of biofertilizers necessitates the consideration of six key steps (Figure 7.2). In the first step, microbes that may be beneficial to plant growth are isolated, identified, and functionally characterized. Common

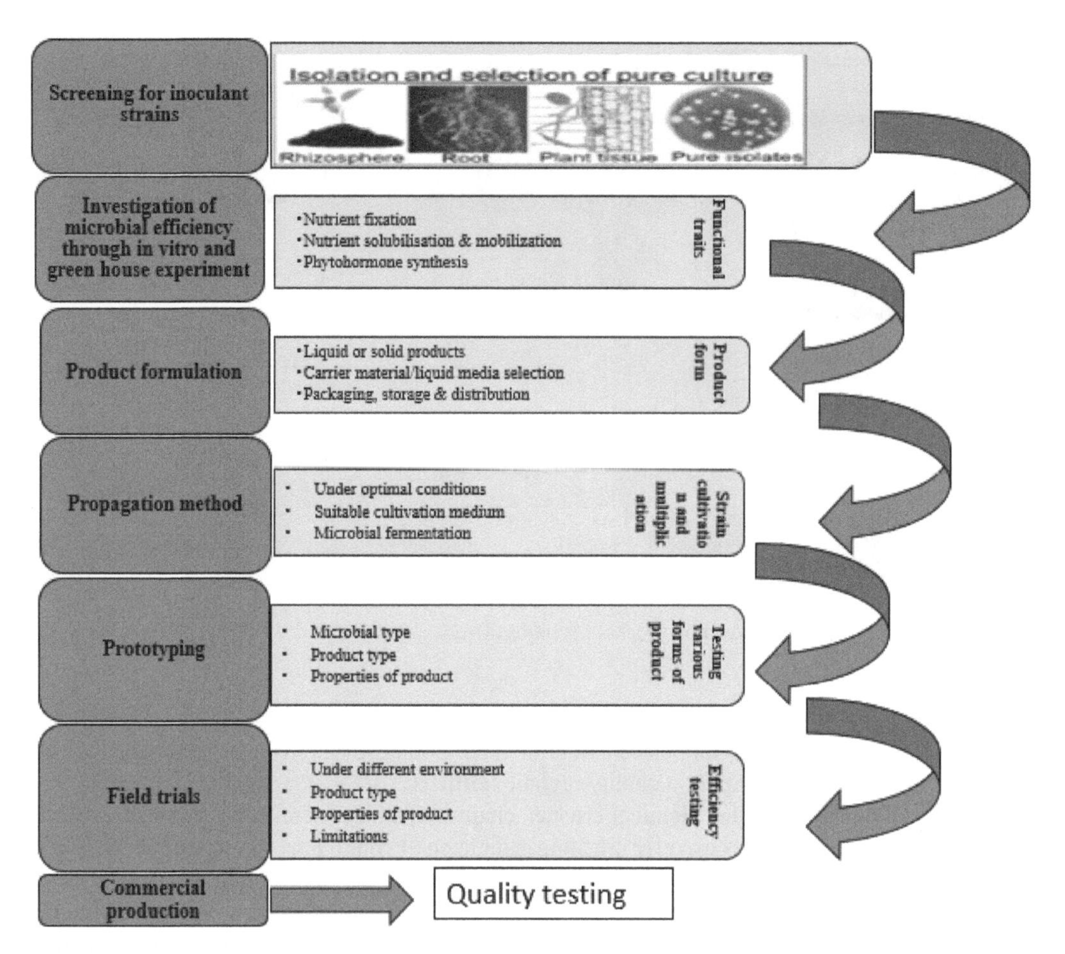

FIGURE 7.2 Standardization process for commercial biofertilizer production. In order to produce biofertilizer, microbes must be present in the rhizosphere and plant tissues, as well as the surrounding environment. A fermenter is used to cultivate and multiply candidate microbes in the presence of optimal growth factors. After that, the effectiveness of the product is put to the test in the field. Biofertilizers need to be tested for quality at each stage of standardization and formulated with appropriate carriers in order to produce a high-quality product.

sources of microbial strains are the plant tissues (flowers, leaves, stems, roots, and seeds), bulk soil, or rhizosphere (Thomas & Singh 2019; Gupta et al. 2014). Microbial strains are characterized using a variety of laboratory techniques, such as qualitative testing and differential culture media. Because functional characteristics can be checked quickly and cheaply, they are an inoculant's main way to make sure it's good.

When selecting a biofertilizer, it's important to keep in mind how it will be used in the field, such as for nitrogen fixation, nutrient solubilization, or phytohormone production. In vitro testing: selected strains are then subjected to additional processes (Figure 7.2), including growth on specific media and potency quantification (Deaker et al. 2011). In greenhouse experiments, the strains are also tested as part of the selection process before going into the field for field trials and applications (Bhattacharjee & Dey 2014; Murphy et al. 2018). Analysis of plant growth-promoting mechanisms and pathways is also part of this step. Genomic, metabolomics, proteomic, and transcriptomic techniques, along with culture-based methods, have made it possible to compare the ability of inoculant strains to promote plant growth (Bilqees et al. 2019; Krishnamoorthy et al. 2020; Luziatelli et al. 2020; Kumari &

Thakur 2018; Terra et al. 2020). Using this information, biofertilizer formulations can be developed that can thrive in a wide range of agricultural ecosystems. Molybdenum, for example, could be used to make N-fixing inoculants if the soil doesn't have enough of it (Deaker et al. 2011).

In the third step, to determine the final product form, it is crucial to select the right formulation materials, which include liquid or carrier-based, such as granules, powder, or slurry. The carrier is essential in order to keep the microbes in a viable state and in the proper quantity. Propagation methods must be chosen for the cultivation and multiplication of the selected strains in a laboratory under ideal conditions to preserve their inherent properties for effective field performance in the fourth step. Monitoring the microbial growth profile under various conditions helps to determine the best conditions for microbial growth. Typically, strains are multiplied using a conventional system of fermenters (Suthar et al. 2017; Khan et al. 2016). For biofertilizer production, solid-state and submerged fermentation are two of the most common methods. The fifth step is prototyping, which entails testing a variety of different products. This step ensures that the most efficient product is chosen for use in the field. Before setting up a standard commercial production process (Figure 7.2), the products must be tested in the field on a large scale to find out how well and how well they work in different ecological areas and conditions (step six) (Mohammadi & Sohrabi 2012; Bhattacharjee & Dey 2014).

Because AM fungi are obligate symbionts, their recovery and multiplication strategies differ from those of other microbes, such as bacteria and fungi. Using monogenic culture, AM fungi can be multiplied in a controlled environment where they are grown on a host plant (Parnell et al. 2016). A pot-house is used for commercial production because of the controlled conditions. This method can be used to monitor the spores and mycelia of inoculum attached to the roots of host plants. Obtaining high levels of synergy between the host organism and the AM fungi inoculum requires careful attention to soil nutrient conditions. There should be good affinity for AM fungi colonization and rapid growth of the host plant with significant root development (Treseder 2013). A contaminant-free biofertilizer can be made in an in vitro monogenic culture (Mukhongo et al. 2016) if certain rules are followed.

7.4 BIOFERTILIZER PRODUCTION PROTOCOLS

a. **Safety measures**
   - No smoking
   - No eating
   - No drinking
   - Good washing facilities
   - Good disposal facilities
   - Responsibilities

The laboratory supervisor/manager must ensure that:

1. The lab is a safe place to work.
2. All of the equipment is in working order.
3. It is important that all technicians be aware of the risks.
4. All technicians are properly trained to carry out their duties.
5. Be responsible for chemicals.

Technicians are responsible for:

1. Maintaining the cleanliness of the benches.
2. Keep the floors neat and tidy.
3. Glassware cleaning and proper storage.

4. Chemicals should be stored in the proper locations (cupboards, stores).
5. The upkeep of machinery.
6. Notifying the supervisor of any problems or potential problems.

b. Laboratory requirements

A microbiology lab must have the following items as a bare minimum for effective operation:

- **Equipment**
    1. Low, high, and oil-immersion lenses, as well as micrometres and haemocytometres, are included in a compound biological microscope for microscopic examination, sizing, and counting of vegetative and spore/budding/cyst specimens
    2. An autoclave is used to sterilize media,
    3. Hot-air oven for glassware sterilization.
    4. Incubator for cell growth (temperature range of 25–40°C).
    5. Colony counter (cfu) for counting colonies in an agar plate.
    6. Agar media melting by using a water bath.
    7. Refrigerator in order to store strains.
    8. pH meter for measurement of media acidity and alkalinity.
    9. Weighing balance for weighing of media ingredients.
    10. Bunsen burner/spirit lamp.

- **Glassware and other necessary materials**
    1. Petri dishes for storing molten media.
    2. Conical flasks ranging in size from 50 ml to 1000 ml.
    3. Culture tubes with caps or test tubes without rims (Biju bottles).
    4. Funnels for pouring media.
    5. Pipettes of various sizes, from capillary to 10 ml.
    6. Cylinders to take measurements.
    7. 50–1000 ml beakers.
    8. Glass slides and cover slips.
    9. Bent glass rods (spreaders).
    10. Needles and loops for inoculation.
    11. Dropping stain and reagent bottles.
    12. Saucepans for boiling media.
    13. Wire baskets for Petri dishes and tubes.
    14. Filtering papers.
    15. Wax markers and pencils.
    16. Steel pipette and Petri dish storage container.
    17. Cheese cloth and cotton wool.
    18. Forceps and a scalpel.
    19. Disinfectant-filled bottles.
    20. Cleansing agent in powder form.
    21. Necessary chemicals.

7.5 PREVENTION OF CONTAMINATION

In agricultural systems, the use of biofertilizers instead of chemical fertilizers holds great promise. However, technical reliability and accessibility for farmers are two major challenges that must be overcome. Aside from the potential impact on human and ecosystem health, it is important to consider the long-term consequences. For these reasons, we believe that all inoculant research and development should be evaluated using a "One Health" approach that considers the health of plants, animals, people, and ecosystems as a whole (Van Bruggen et al. 2019). Because we don't yet have a complete picture of how a product interacts with the soil and other microorganisms, we can't yet determine how dangerous it is. A variety of inoculants have been studied in an attempt to close knowledge gaps and develop guidelines for evaluating the risks associated with each one. An accurate assessment of the risks posed by the large-scale introduction of microorganisms into agricultural systems depends on an understanding of the mechanisms underlying successful PGP.

7.6 WAYS TO IMPROVE PRODUCTION SYSTEMS

In addition to selecting specific microorganisms or functions, the success of biofertilizers is dependent on developing new formulations to ensure the survival of inoculated strains. In this chapter, we reviewed a variety of techniques for improving bioformulations, including the use of biofilm-producing strains, microencapsulation with alginate, and fluidized bed dryer (FBD) processes. In an ideal world, new technologies would focus on carriers and additives that are both inexpensive and simple to use, but also capable of storing and applying a greater number of viable cells. In addition, a "One Health" approach should be used to evaluate the biosafety of microbes that are inoculated. In order to ensure the safety of the product, proper screening tests (e.g., toxicity and pathogenicity testing) must be carried out. Research into ecological interactions and how plants shape their microbiomes in agricultural systems is still vital. Soil microorganisms play a critical role in many biogeochemical processes, and these processes could be negatively impacted by climate change. Due to the genetic diversity and complexity of soil and plant microorganisms, not all fields will benefit from a single formulation. However, designing biofertilizers for every single field is impossible and unrealistic. Thus, we agree with Bell et al. (2019) that the design, refinement, and validation of microbial products should be oriented toward optimal and suboptimal environmental ranges (i.e., agricultural practices, crop, climate, and soil type). Public and private funding are essential to filling critical knowledge gaps in this field. This effort needs to be supported by regulatory agencies and policymakers who support biofertilizers and sustainable practices.

7.7 LIMITATIONS IN MARKETING

Before they can have the desired effect on plant growth and fitness, introduced microbes have to deal with a number of big issues (Figure 7.3), like how they interact with plants, how long they last, and how symbiotic or parasitic they are (e.g., PGP performance, symbiotic/parasitic behaviour). During the lab-to-field transition, a microbial strain that performs well in the lab may perform poorly in greenhouse and field trials (Uddin et al. 2016; Keswani et al. 2019; Parnell et al. 2016). It is difficult to predict the outcome of an inoculation because we only take into account and control for a small number of variables, often without taking into account the complex interactions between them (Busby et al. 2017; Moutia et al. 2010; Sasaki et al. 2010). Plant-related factors (e.g., concentration, viability) and edaphic/environmental factors (e.g., additives) are among the most important factors influencing inoculation success (Malusà et al. 2016).

7.7.1 Edaphic and Environmental Conditions

Both edaphic and environmental factors play a significant role in the variability and low reproducibility of biofertilizer results in field trials. Initially, biofertilizer testing is done in an aseptic environment, which allows for an unbiased evaluation of the microorganism under study. There are many uncontrolled biotic and abiotic factors that can interfere with inoculation success when testing in growth chambers, greenhouses, and even in the field (Bacilio et al. 2017; Nkebiwe et al.

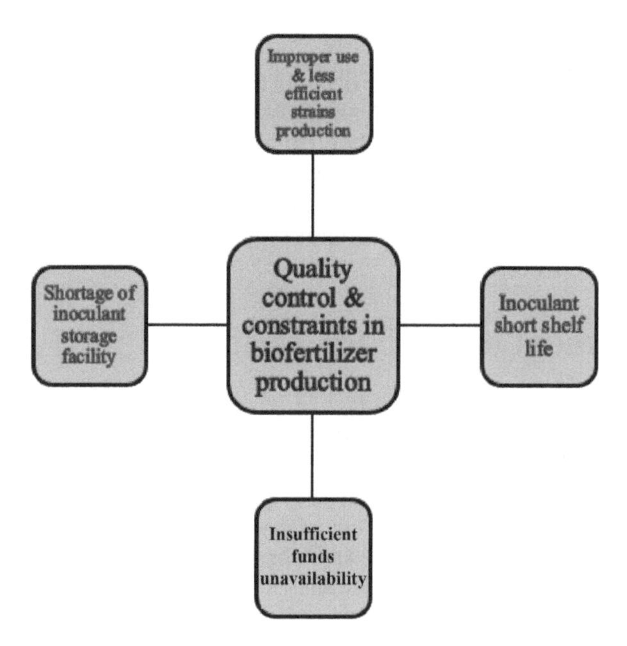

FIGURE 7.3 Constraints in successful utilization of biofertilizers.

2016). An important consideration in inoculation outcomes is whether or not there are any antagonistic microorganisms in the resident microbiome that can act as competitors, predators, or other antagonists. Climate and edaphic factors can also affect the efficiency and yield of biofertilization on crop nutrient use (Al-Zahrani et al. 2022; Rajesh et al. 2022; Anam et al. 2021; Deepranjan et al. 2021; Haider et al. 2021; Amjad et al. 2021; Sajjad et al. 2021a,b; Fakhre et al. 2021; Khatun et al. 2021; Ibrar et al. 2021; Bukhari et al. 2021; Haoliang et al. 2021; Sana et al. 2022; Abid et al. 2021; Zaman et al. 2021; Rehana et al. 2021; Yang et al. 2022; Ahmad et al. 2022; Shah et al. 2022). Schütz et al. (2018) conducted a meta-analysis and found that biofertilizers perform better in dry climates, with high levels of soil P (especially for N_2-fixers), and, for AMF, lower soil organic matter content and neutral or slightly acidic pH. Higher growth and yield responses to biofertilization can be expected when the plant is able to fully benefit from the interaction with the introduced microbe. According to a number of studies, there is a negative correlation between biofertilization and soil nutrient content, both for AMF colonization (Jansa et al. 2009a,b; Kaeppler et al. 2000; Ullah et al. 2016) and rhizobia nodulation with soil N.

Additional inoculation success variability can be caused by edaphic properties that are particularly sensitive to agricultural practices. Inoculation of microbes with chemical fertilizers, organic amendments, pesticides, and tillage have all been shown to interact so far (Da Costa et al. 2014; Ozturk et al. 2003; Jansa et al. 2009a,b), but there is still more research to be done (Manandhar et al. 2017; Mulas et al. 2015; Miller et al. 1995; Muhammad et al. 2022; Wiqar et al. 2022; Farhat et al. 2022; Niaz et al. 2022; Ihsan et al. 2022; Chao et al. 2022, Qin et al. 2022; Xue et al. 2022; Ali et al. 2022; Mehmood et al. 2022; El Sabagh et al. 2022; Ibad et al. 2022). Many different things can affect how well biofertilization works. Crop rotations and cover crops, for example, can change the microbial communities in the soil (Buysens et al. 2016).

7.7.2 Plant-related Factors

The results of biofertilization can vary depending on the type of crop that is being grown. Genetic markers (such as quantitative trait loci or QTLs) have been identified in some studies, but the underlying plant factors have not yet been fully understood (Remans et al. 2008; Deepranjan et al. 2021; Haider et al. 2021; Huang Li et al. 2021; Ikram et al. 2021; Jabborova et al. 2021; Khadim et al. 2021a,b; Manzer et al. 2021; Muzammal et al. 2021; Abdul et al. 2021a,b; Ashfaq et al. 2021; Amjad et al. 2021; Atif et al. 2021; Athar et al. 2021; Adnan et al. 2018a,b; Adnan et al. 2019; Akram et al. 2018a,b; Aziz et al. 2017a,b; Chang et al. 2021; Chen et al. 2021; Emre et al. 2021). Through rhizodeposits and root architecture changes, plants are widely accepted to have an impact on the rhizosphere environment (Saleem et al. 2018; Somers et al. 2004; Habib et al. 2017; Hafiz et al. 2016; Hafiz et al. 2019; Ghulam et al. 2021; Guofu et al. 2021; Hafeez et al. 2021; Khan et al. 2021; Kamaran et al. 2017; Muhmmad et al. 2019; Safi et al. 2021; Sajjad et al. 2019; Saud et al. 2013; Saud et al. 2014; Saud et al. 2017; Saud et al. 2016; Shah et al. 2013; Saud et al. 2020; Saud et al. 2022a,b; Qamar et al. 2017; Hamza et al. 2021; Irfan et al. 2021;Wajid et al. 2017; Yang et al. 2017; Zahida et al. 2017; Depeng et al. 2018; Hussain et al. 2020; Hafiz et al. 2020 a,b). Plant roots secrete secondary metabolites, such as antibiotics, flavonoids, and hormones, which are recognized and interacted with by PGP microbes in their numerous rhizodeposits (Bais et al. 2004). Rhizobia–legume symbioses, in which the compatibility of microorganisms with host plants is critical to colonization success, can have a high degree of specificity in this signalling process (Thilakarathna and Raizada 2017; Hirsch et al. 2003). Arbuscular mycorrhiza may not require as much host specificity as previously thought (Koyama et al. 2017), but the genotype of the plant can still influence colonization and PGP (Hoeksema et al. 2010; Yao et al. 2001; Linderman and Davis 2004). AMF–host plant compatibility has been explained by differences in root architecture, aerial architecture, P utilization, and uptake efficiency (Declerck et al. 1995; Kaeppler et al. 2000; Yao et al. 2001). Also, PGPRs such as *Azotobacter* sp. and *Pseudomonas* sp. were found to have a high degree of specificity for plant genotypes, as were endophytes (Anbi et al. 2020; Yoon et al. 2016; Vujanovic and Germida 2017; Mezei et al. 1997). Most plant breeding programmes do not take into account interactions between plants and microbes, resulting in a wide range of biofertilization outcomes. Because of this, the impact of genotypes isn't just based on how old and healthy the plants are, but also how the environment is and when the plants were inoculated (Dennis et al. 2010; Shafi et al. 2020; Wahid et al. 2020; Subhan et al. 2020; Zafar-ul-Hye et al. 2020a,b; Zafar et al. 2021; Adnan et al. 2020; Ilyas et al. 2020; Saleem et al. 2020a,b,c; Rehman 2020; Farhat et al. 2020; Wu et al. 2020; Mubeen et al. 2020; Farhana 2020; Jan et al. 2020; Wu et al. 2019; Ahmad et al. 2019; Baseer et al. 2019; Hafiz et al. 2018; Tariq et al. 2018).

7.7.3 Inoculant-related Factors

Selecting a microbial genotype that is compatible with the plant host genotype is just as important as determining PGP functions (Vargas et al. 2012; Ehinger et al. 2014; Linderman and Davis 2004). Among the many processes influenced by bacterial characteristics are inoculation, establishment, colonization, and persistence. Genetic and PGP traits, rather than ecological traits, are typically the focus of inoculant development. However, inoculation success is ultimately determined by ecological traits (Hart et al. 2018; Kaminsky et al. 2019). Osmotic tolerance and psychotolerance, for example, are two traits that can be used to select strains that are better suited to the environment in which they are grown (Garca et al. 2017; Rawat et al. 2019). Another option is to isolate native strains that have been proven to improve biofertilization performance (Ahmed et al. 2013; Maltz and Treseder 2015; Melchiorre et al. 2011). When developing an inoculant, there is a trade-off between the ability to establish and survive as well as the PGP traits (Kaminsky et al. 2019; Parnell et al. 2016). An important question arises as a result of this trade-off: should we pursue highly targeted biofertilizers or broad-spectrum versatile ones, given the high degree of microbial strain specificity to environment and genotype (Tosi et al. 2020; Parnell et al. 2016; Bell et al. 2019)? A mixed inoculant with a wide range of ecological adaptability would be the preferred method of achieving this goal (i.e., functionally redundant strains that encompass a wider range of environmental adaptation).

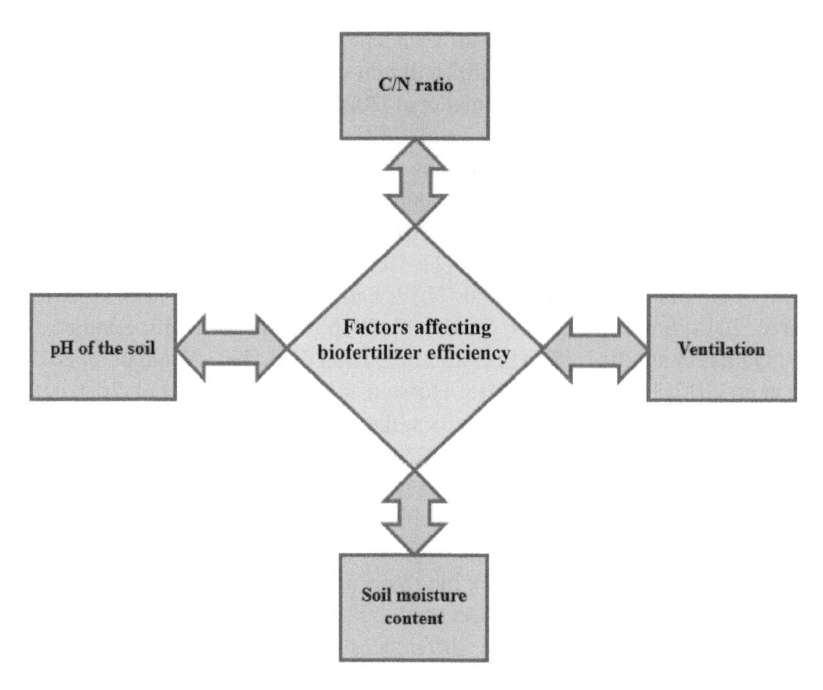

FIGURE 7.4 Factors governing the efficiency of biofertilizers.

If a mixed inoculant interferes with other parts of the inoculant, there is a lot of room for this inoculant to work, even if there are some challenges (Ballesteros-Almanza et al. 2010; Xavier and Germida 2003; Remans et al. 2008). Major factors limiting the efficiency of biofertilizers are described in Figure 7.4.

For the microbes to be able to withstand and protect themselves from environmental constraints, it is important to know how the biofertilizer is formulated and delivered. For a plant to respond effectively, the inoculant formulation must support microbial growth while protecting a sufficient number of viable cells (Bashan et al. 2014; Herrmann and Lesueur 2013). It's not uncommon for the strain to undergo genetic or physiological changes as a result of scaling up and commercialization, and there's also the risk of viability loss (especially due to desiccation) and contamination (Greffe and Michiels 2020; Parnell et al. 2016; Glick 2020). There has been an increase in the variety of biofertilizer formulations as a result of the growth of the industry, including the development of new carriers (solids, liquids, and slurries), additives (such as nutrients and stimulants), and preservatives (such as adhesives and surfactants) to improve the inoculants' physical properties (Preininger et al. 2018; Bashan et al. 2014). In the distribution, storage, and application of microbes, these additives allow them to withstand the fluctuating and suboptimal conditions. Some think that adding biofertilizer carriers and other additives is a last-minute thing, but they can be very important for obtaining good results (Lee et al. 2016; Gomez et al. 1997; Herrmann and Lesueur 2013; Gopakumar et al. 2020; Zia-ur-Rehman 2020; Ayman et al. 2020; Mohammad I. Al-Wabel et al. 2020a,b; Senol 2020; Amjad et al. 2020; Ibrar et al. 2020; Sajid et al. 2020; Muhammad et al. 2021; Sidra et al. 2021; Zahir et al. 2021; Sahrish et al. 2022).

Similar to delivery methods (such as seed or soil), timing and frequency of application can be crucial for inoculation success (Parnell et al. 2016). It has been found that soil applications of a granular *Rhizobium* inoculant on *Pisum sativa* resulted in higher PGP than seed applications of a liquid formulation inoculant (Clayton et al. 2004). Endophytic and phyllosphere microorganisms could also be delivered via foliar and flower applications, which were considered safer and more effective (Preininger et al. 2018; Mitter et al. 2016; Iqra et al. 2020; Akbar et al. 2020; Mahar et al.

2020; Noor et al. 2020; Bayram et al. 2020; Amanullah, Fahad S 2018a,b; Amanullah, Fahad S 2017; Amanullah et al. 2020; Amanullah et al. 2021; Rashid et al. 2020; Arif et al. 2020; Amir et al. 2020; Saman et al. 2020; Muhammad Tahir et al. 2020; Md Jakirand Allah 2020; Mahmood et al. 2021; Farah et al. 2020; Sadam et al. 2020; Unsar et al. 2020; Fazli et al. 2020; Md. Enamul et al. 2020). Consideration should also be given to other agricultural practices, such as fertilization and transplant (Pii et al. 2019; Fahad and Bano 2012; Fahad et al. 2017; Fahad et al. 2013; Fahad et al. 2014a,b; Fahad et al. 2016a,b,c,d; Fahad et al. 2015a,b; Fahad et al. 2018a,b; Fahad et al. 2019a,b; Fahad et al. 2020; Fahad et al. 2021a,b,c,d,e,f; Fahad et al. 2022a,b; Hesham and Fahad 2020; Sajjad et al. 2021a,b), timing and frequency of applications, as well as plant growth stages (Sohn et al. 2003; Bashan 1986; Fallik et al. 1988; Linderman & Davis 2004). After the microbe is applied, it needs time to establish itself before it can provide any benefit to a plant. This time period is critical.

7.8 STRATEGIES TO IMPROVE THE MARKETING PROCESS

On the other hand, there are a number of practical considerations inherent in biofertilizer development that should not be overlooked in light of the technical constraints we've already discussed in this chapter. Farmers and manufacturers alike have expressed concerns about the ease of use and accessibility of inoculants in comparison to chemical fertilizers. Accessibility to biofertilizers is influenced by a number of factors, including price/benefit ratios, versatility, robustness and reproducibility, shelf life and storage requirements, user-friendliness (in terms of handling and application), adaptability to agricultural practices and machinery, and biosecurity (Bell et al. 2019; Bashan et al. 2014; Parnell et al. 2016). If biofertilizers are to have a long-term positive impact on the environment, farmers need to be educated about the long-term benefits of using biofertilizers and encouraged to use them. More rarely discussed are intellectual property rights and patent development, which may help transfer technology between the academic and industrial sectors, but may also limit researchers' ability to freely share ideas and materials (Glick 2020). Proper regulations, including standard and universal testing protocols and guidelines for risk evaluation, are also required for the commercialization of new products (Timmusk et al. 2017; Sessitsch et al. 2019; Kaminsky et al. 2019). For the most part, there aren't any standard procedures for evaluating how well microbes are inoculated and how to monitor them after they've been used in the field.

7.9 CONCLUSIONS AND RECOMMENDATIONS

Solubilization of nutrients, increased absorption of nutrients, or decomposition of organic residues are some of the ways in which biofertilizers can increase crop productivity. Reducing chemical fertilizer use by substituting with biofertilizers will save money while also reducing pollution caused by excessive chemical fertilizer use. A low-input farming system can be achieved with the use of biological and organic fertilizers. This helps ensure the long-term sustainability of farms.

Sustainable agriculture relies heavily on the use of biofertilizers. Biofertilizers are still being produced in small quantities, with little public awareness of their existence, and few people are using them. To combat food insecurity and soil depletion, biofertilizers should be used more frequently. An effective extension programme, public–private partnerships, and adequate training can help increase the use of biofertilizers. Cooperative efforts between government agencies and biofertilizer technology companies should be encouraged. These strategies will help high-quality biofertilizers to be manufactured that are widely used and valued.

In addition, research should focus on how to increase the efficiency of biofertilizers in the field. These new findings could help us better understand how beneficial microbes promote plant growth, as well as how to better utilize soil and plant microbiomes in general. Plant-associated microbes have been established as having key roles in the growth and development of plants. As a result, improving plant productivity and health by modifying these functions is clearly an area of research that merits further investigation. If researchers want to make new biofertilizers, they need to find and make inoculant strains that can work well in a wide range of field conditions and with a wide range of plant species.

Developing agencies should work together to define the quality of biofertilizers and develop specific standards for different biofertilizers. Biofertilizers and their quality testing methods should also be re-evaluated in the light of changing market demands and feedback from farmers. Development agencies should take the initiative to train various production units, state government extension staff, and farmers about quality control in biofertilizers.

REFERENCES

Abdul S, Muhammad AA, Shabir H, Hesham A El E, Sajjad H, Niaz A, Abdul G, Sayyed RZ, Fahad S, Subhan D, Rahul D (2021a) Zinc nutrition and arbuscular mycorrhizal symbiosis effects on maize (Zea mays L.) growth and productivity. *J Saudi Soc Agri Sci* 28(11):6339–6351. https://doi.org/10.1016/j.sjbs.2021.06.096

Abdul S, Muhammad AA, Subhan D, Niaz A, Fahad S, Rahul D, Mohammad JA, Omaima N, Muhammad Habib ur R, Bernard RG (2021b) Effect of arbuscular mycorrhizal fungi on the physiological functioning of maize under zinc-deficient soils. *Sci Rep* 11:18468.

Abid M, Khalid N, Qasim A, Saud A, Manzer HS, Chao W, Depeng W, Shah S, Jan B, Subhan D, Rahul D, Hafiz MH, Wajid N, Muhammad M, Farooq S, Fahad S (2021) Exploring the potential of moringa leaf extract as bio stimulant for improving yield and quality of black cumin oil. *Sci Rep* 11:24217. https://doi.org/10.1038/s41598-021-03617-w

Adnan M, Zahir S, Fahad S, Arif M, Mukhtar A, Imtiaz AK, Ishaq AM, Abdul B, Hidayat U, Muhammad A, Inayat-Ur R, Saud S, Muhammad ZI, Yousaf J, Amanullah, Hafiz MH, Wajid N (2018a) Phosphate-solubilizing bacteria nullify the antagonistic effect of soil calcification on bioavailability of phosphorus in alkaline soils. *Sci Rep* 8:4339. https://doi.org/10.1038/s41598-018-22653-7

Adnan M, Shah Z, Sharif M, Rahman H (2018b) Liming induces carbon dioxide (CO_2) emission in PSB inoculated alkaline soil supplemented with different phosphorus sources. *Environ Sci Poll Res.* 25(10):9501–9509.

Adnan M, Fahad S, Khan IA, Saeed M, Ihsan MZ, Saud S, Riaz M, Wang D, Wu C (2019) Integration of poultry manure and phosphate solubilizing bacteria improved availability of Ca bound P in calcareous soils. *Biotech* 9(10):368.

Adnan M, Fahad S, Muhammad Z, Shahen S, Ishaq AM, Subhan D, Zafar-ul-Hye M, Martin LB, Raja MMN, Beena S, Saud S, Imran A, Zhen Y, Martin B, Jiri H, Rahul D (2020) Coupling phosphate-solubilizing bacteria with phosphorus supplements improves maize phosphorus acquisition and growth under lime induced salinity stress. *Plants* 9(7):900. doi: 10.3390/plants9070900

Ahmad N, Hussain S, Ali MA, Minhas A, Waheed W, Danish S, Fahad S, Ghafoor U, Baig, KS, Sultan H, Muhammad IH, Mohammad JA, Theodore DM (2022) Correlation of soil characteristics and citrus leaf nutrients contents in current scenario of Layyah District. *Hortic* 8: 61. https://doi.org/10.3390/horticulturae8010061

Ahmad S, Kamran M, Ding R, Meng X, Wang H, Ahmad I, Fahad S, Han Q (2019) Exogenous melatonin confers drought stress by promoting plant growth, photosynthetic capacity and antioxidant defense system of maize seedlings. *PeerJ* 7:e7793. http://doi.org/10.7717/peerj.7793

Ahmed AH, Darwish ES, AH, Hamoda SAF, Alobaidy MG (2013) Effect of putrescine and humic acid on growth, yield and chemical composition of cotton plants grown under saline soil conditions. *Am-Eurasian J Agric Environ Sci* 13: 479–497.

Akbar H, Timothy JK, Jagadish T, Golam M, Apurbo KC, Muhammad F, Rajan B, Fahad S, Hasanuzzaman M (2020) Agricultural land degradation: processes and problems undermining future food security. In: Fahad S, Hasanuzzaman M, Alam M, Ullah H,Saeed M, Khan AK, Adnan M (Eds.) *Environment, Climate, Plant and Vegetation Growth*. Springer Publ Ltd, Springer Nature Switzerland, pp. 17–62. https://doi.org/10.1007/978-3-030-49732-3

Akram R, Turan V, Hammad HM, Ahmad S, Hussain S, Hasnain A, Maqbool MM, Rehmani MIA, Rasool A, Masood N, Mahmood F, Mubeen M, Sultana SR, Fahad S, Amanet K, Saleem M, Abbas Y, Akhtar HM, Waseem F, Murtaza R, Amin A, Zahoor SA, ul Din MS, Nasim W (2018a) Fate of organic and inorganic pollutants in paddy soils. In: Hashmi MZ and Varma A (Eds.), *Environmental Pollution of Paddy Soils, Soil Biology*. Springer International Publishing AG, Switzerland, pp. 197–214.

Akram R, Turan V, Wahid A, Ijaz M, Shahid MA, Kaleem S, Hafeez A, Maqbool MM, Chaudhary HJ, Munis, MFH, Mubeen M, Sadiq N, Murtaza R, Kazmi DH, Ali S, Khan N, Sultana SR, Fahad S, Amin A, Nasim W (2018b) Paddy land pollutants and their role in climate change. In: Hashmi MZ and Varma A (Eds.) *Environmental Pollution of Paddy Soils, Soil Biology*. Springer International Publishing AG, Switzerland, pp. 113–124.

Ali S, Hameed G, Muhammad A, Depeng W, Fahad S (2022) Comparative genetic evaluation of maize inbred lines at seedling and maturity stages under drought stress. *J Plant Growth Regul* 42:989–1005. https://doi.org/10.1007/s00344-022-10608-2

Ali S, Xua S, Ahmad I, Jia Q, Ma X, Ullah H, Alam M, Adnan M, Daur I, Ren X, Cai T, Zhang J, Jia Z (2018) Tillage and deficit irrigation strategies to improve winter wheat production through regulating root development under simulated rainfall conditions. *Agric Water Manag* 209:44–54.

Al-Zahrani HS, Alharby HF, Fahad S (2022) Antioxidative defense system, hormones, and metabolite accumulation in different plant parts of two contrasting rice cultivars as influenced by plant growth regulators under heat stress. *Front Plant Sci* 13:911846. doi: 10.3389/fpls.2022.911846

Amanullah, Fahad S (Eds.) (2017) *Rice – Technology and Production*. IntechOpen Croatia 2017. http://dx.doi.org/10.5772/64480

Amanullah, Fahad S (Eds.) (2018a) *Corn – Production and Human Health in Changing Climate*. IntechOpen United Kingdom 2018. http://dx.doi.org/10.5772/intechopen.74074

Amanullah, Fahad S (Eds.) (2018b) Nitrogen in Agriculture – Updates. IntechOpen Croatia 2018. http://dx.doi.org/10.5772/65846

Amanullah, Muhammad I, Haider N, Shah K, Manzoor A, Asim M, Saif U, Izhar A, Fahad S, Adnan M et al. (2021) Integrated foliar nutrients application improves wheat (Triticum aestivum L.) productivity under calcareous soils in drylands. *Commun Soil Sci Plant Analysis* 52(21), 2748–2766. https://Doi.Org/10.1080/00103624.2021.1956521

Amanullah, Shah K, Imran, Hamdan AK, Muhammad A, Abdel RA, Muhammad A, Fahad S, Azizullah S, Brajendra P (2020) Effects of climate change on irrigation water quality. In: Fahad S, Hasanuzzaman M, Alam M, Ullah H,Saeed M, Khan AK, Adnan M (Eds.) *Environment, Climate, Plant and Vegetation Growth*. Springer Publ Ltd, Springer Nature Switzerland, pp. *123–132*. https://doi.org/10.1007/978-3-030-49732-3

Amir M, Muhammad A, Allah B, Sevgi Ç, Haroon ZK, Muhammad A, Emre A (2020) Biofortification under climate change: the fight between quality and quantity. In: Fahad S, Hasanuzzaman M, Alam M, Ullah H, Saeed M, Khan AK, Adnan M (Eds.) *Environment, Climate, Plant and Vegetation Growth*. Springer Publ Ltd, Springer Nature Switzerland, pp. 173–228. https://doi.org/10.1007/978-3-030-49732-3

Amjad I, Muhammad H, Farooq S, Anwar H (2020) Role of plant bioactives in sustainable agriculture. In: Fahad S, Hasanuzzaman M, Alam M, Ullah H, Saeed M, Khan AK, Adnan M (Eds.) *Environment, Climate, Plant and Vegetation Growth*. Springer Publ Ltd, Springer Nature Switzerland, pp. 591–606. https://doi.org/10.1007/978-3-030-49732-3

Amjad SF, Mansoora N, Din IU, Khalid IR, Jatoi GH, Murtaza G, Yaseen S, Naz M, Danish S, Fahad S et al. (2021) Application of zinc fertilizer and mycorrhizal inoculation on physio-biochemical parameters of wheat grown under water-stressed environment. *Sustainability* 13:11007. https://doi.org/10.3390/su131911007

Anam I, Huma G, Ali H, Muhammad K, Muhammad R, Aasma P, Muhammad SC, Noman W, Sana F, Sobia A, Fahad S (2021) Ameliorative mechanisms of turmeric-extracted curcumin on arsenic (As)-induced biochemical alterations, oxidative damage, and impaired organ functions in rats. *Environ Sci Pollut Res* 28(46): 66313–66326. https://doi.org/10.1007/s11356-021-15695-4

Anbi AA, Mirshekari B, Eivazi A, Yarnia M, Behrouzyar EK (2020) PGPRs affected photosynthetic capacity and nutrient uptake in different Salvia species. *J Plant Nutr* 43(1):108–121.

Arif M, Shah T, Ilyas M, Ahmad W, Mian AA, Jadoon MA, Adnan M (2015) Effect of organic manures and their levels on weeds density and maize yield. *Pak J Weed Sci Res* 21(4):517–522.

Arif M, Talha J, Muhammad R, Fahad S, Muhammad A, Amanullah, Kawsar A, Ishaq AM, Bushra K, Fahd R (2020) Biochar; a remedy for climate change. In: Fahad S, Hasanuzzaman M, Alam M, Ullah H,Saeed

M, Khan AK, Adnan M (Eds.) *Environment, Climate, Plant and Vegetation Growth.* Springer Publ Ltd, Springer Nature Switzerland, pp. 151–172. https://doi.org/10.1007/978-3-030-49732-3

Ashfaq AR, Uzma Y, Niaz A, Muhammad AA, Fahad S, Haider S, Tayebeh Z, Subhan D, Süleyman T, Hesham AElE, Pramila T, Jamal MA, Sulaiman AA, Rahul D (2021) Toxicity of cadmium and nickel in the context of applied activated carbon biochar for improvement in soil fertility. *Saudi Society Agricultural Sci* 29(2):743–750. https://doi.org/10.1016/j.sjbs.2021.09.035

Athar M, Masood IA, Sana S, Ahmed M, Xiukang W, Sajid F, Sher AK, Habib A, Faran M, Zafar H, Farhana G, Fahad S (2021) Bio-diesel production of sunflower through sulphur management in a semi-arid subtropical environment. *Environ Sci Pollution Res* 29(9):13268–13278. https://doi.org/10.1007/s11356-021-16688-z

Atif B, Hesham A, Fahad S (2021) Biochar coupling with phosphorus fertilization modifies antioxidant activity, osmolyte accumulation and reactive oxygen species synthesis in the leaves and xylem sap of rice cultivars under high-temperature stress. *Physiol Mol Biol Plants* 27(9):2083–2100. https://doi.org/10.1007/s12298-021-01062-7

Ayman EL Sabagh, Akbar Hossain, Celaleddin Barutçular, Muhammad Aamir Iqbal, Sohidul Islam M, Shah Fahad, Oksana Sytar, Fatih Çig, Ram Swaroop Meena, and Murat Erman (2020) Consequences of salinity stress on the quality of crops and its mitigation strategies for sustainable crop production: an outlook of arid and semi-arid regions. In: Fahad S, Hasanuzzaman M, Alam M, Ullah H, Saeed M, Khan AK, Adnan M (Eds.) *Environment, Climate, Plant and Vegetation Growth.* Springer Publ Ltd, Springer Nature Switzerland, pp. 503–534. https://doi.org/10.1007/978-3-030-49732-3

Aziz K, Daniel KYT, Fazal M,Muhammad ZA, Farooq S, Fan W, Fahad S, Ruiyang Z (2017a) Nitrogen nutrition in cotton and control strategies for greenhouse gas emissions: a review. *Environ Sci Pollut Res* 24:23471–23487. https://doi.org/10.1007/s11356-017-0131-y

Aziz K, Daniel KYT, Muhammad ZA, Honghai L, Shahbaz AT, Mir A, Fahad S (2017b) Nitrogen fertility and abiotic stresses management in cotton crop: a review. *Environ Sci Pollut Res* 24:14551–14566. https://doi.org/10.1007/s11356-017-8920-x

Bacilio M, Moreno M, Lopez-Aguilar DR, Bashan Y (2017) Scaling from the growth chamber to the greenhouse to the field: Demonstration of diminishing effects of mitigation of salinity in peppers inoculated with plant growth-promoting bacterium and humic acids. *Appl Soil Ecol* 119:327–338.

Bais HP, Fall R, Vivanco JM (2004) Biocontrol of Bacillus subtilis against infection of Arabidopsis roots by Pseudomonas syringae is facilitated by biofilm formation and surfactin production. *Plant Physiol* 134(1):307–319.

Ballesteros-Almanza L, Altamirano-Hernandez J, Pena-Cabriales JJ, Santoyo G, Sanchez-Yanez JM, Valencia-Cantero E, Farias-Rodriguez R (2010) Effect of co-inoculation with mycorrhiza and rhizobia on the nodule trehalose content of different bean genotypes. *Open Microbiol J* 4:83.

Baseer M, Adnan M, Fazal M, Fahad S, Muhammad S, Fazli W, Muhammad A, Jr. Amanullah, Depeng W, Saud S, Muhammad N, Muhammad Z, Fazli S, Beena S, Mian AR, Ishaq AM (2019) Substituting urea by organic wastes for improving maize yield in alkaline soil. *J Plant Nutr* 42(19):2423–2434. doi.org/10.1080/01904167.2019.1659344

Bashan Y (1986) Alginate beads as synthetic inoculant carriers for slow release of bacteria that affect plant growth. *Appl & Environ Microbiol* 51(5):1089–1098.

Bashan Y, de-Bashan LE, Prabhu SR, Hernandez JP (2014) Advances in plant growth-promoting bacterial inoculant technology: formulations and practical perspectives (1998–2013). *Plant & Soil* 378(1):1–33.

Bayram AY, Seher Ö, Nazlican A (2020) Climate change forecasting and modeling for the year of 2050. In: Fahad S, Hasanuzzaman M, Alam M, Ullah H,Saeed M, Khan AK, Adnan M (Eds.) *Environment, Climate, Plant and Vegetation Growth.* Springer Publ Ltd, Springer Nature Switzerland, pp. 109–122. https://doi.org/10.1007/978-3-030-49732-3

Bell A, Chetty R, Jaravel X, Petkova N, Van-Reenen J (2019) Do tax cuts produce more Einsteins? The impacts of financial incentives versus exposure to innovation on the supply of inventors. *J Eur Econ Assoc* 17(3):651–677.

Bhattacharjee R, Dey U (2014) Biofertilizer, a way towards organic agriculture: a review. *African J Microbiol Res* 8(24):2332–2343.

Bilqees R, Romman M, Subhan M, Jan S, Parvez R, Adnan M, Ikram M, Iqbal S (2019) Cardiac function profiling and antidiabetic effect of Vigna radiata in Alloxan monohydrate induced diabetic rabbits. *Pure Appl Biol* 9(1):390–395.

Bukhari MA, Adnan NS, Fahad S, Javaid I, Fahim N, Abdul M, Mohammad SB (2021) Screening of wheat (*Triticum aestivum* L.) genotypes for drought tolerance using polyethylene glycol. *Arabian J Geosci* 14:2808.

Busby PE, Soman C, Wagner MR, Friesen ML, Kremer J, Bennett A, Dangl JL (2017) Research priorities for harnessing plant microbiomes in sustainable agriculture. *PLoS Biology* 15(3):2001793.

Buysens C, César V, Ferrais F, de-Boulois HD, Declerck S (2016) Inoculation of Medicago sativa cover crop with Rhizophagus irregularis and Trichoderma harzianum increases the yield of subsequently-grown potato under low nutrient conditions. *Appl Soil Ecol* 105:137–143.

Chang W, Qiujuan J, Evgenios A, Haitao L, Gezi L, Jingjing Z, Fahad S, Ying J (2021) Hormetic effects of zinc on growth and antioxidant defense system of wheat plants. *Sci Total Environ* 807, 150992. https://doi.org/10.1016/j.scitotenv.2021.150992

Chao W, Youjin S, Beibei Q, Fahad S (2022) Effects of asymmetric heat on grain quality during the panicle initiation stage in contrasting rice genotypes. *J Plant Growth Regul* 42, 630–636. https://doi.org/10.1007/s00344-022-10598-1

Chen Y, Guo Z, Dong L, Fu Z, Zheng Q, Zhang G, Qin L, Sun X, Shi Z, Fahad S, Xie F, Saud S (2021) Turf performance and physiological responses of native Poa species to summer stress in Northeast China. *Peer J* 9:e12252. http://doi.org/10.7717/peerj.12252

Clayton GW, Rice WA, Lupwayi NZ, Johnston AM, Lafond GP, Grant CA, Walley F (2004) Inoculant formulation and fertilizer nitrogen effects on field pea: nodulation, N_2 fixation and nitrogen partitioning. *Can J Plant Sci* 84(1): 79–88.

Da-Costa-Rocha I, Bonnlaender B, Sievers H, Pischel I, Heinrich M (2014) Hibiscus sabdariffa L. – A phytochemical and pharmacological review. *Food Chem* 165:424–443.

Deaker R, Kecskés ML, Rose MT, Amprayn K, Ganisan K, Tran TKC, Kennedy IR (2011) *Practical Methods for the Quality Control of Inoculant Biofertilisers*. Australian Centre for International Agricultural Research (ACIAR).

Declerck S, Plenchette C, Strullu DG (1995) Mycorrhizal dependency of banana (Musa acuminata, AAA group) cultivar. *Plant & Soil* 176(1):183–187.

Deepranjan S, Ardith SO, Siva D, Sonam S, Shikha, Manoj P, Amitava R, Sayyed RZ, Abdul G, Mohammad JA, Subhan D, Fahad S, Rahul D (2021) Optimizing nutrient use efficiency, productivity, energetics, and economics of red cabbage following mineral fertilization and biopriming with compatible rhizosphere microbes. *Sci Rep* 11:15680. https://doi.org/10.1038/s41598-021-95092-6

Dennis JH, Lopez RG, Behe BK, Hall CR, Yue C, Campbell BL (2010) Sustainable production practices adopted by greenhouse and nursery plant growers. *HortSci* 45(8):1232–1237.

Depeng W, Fahad S, Saud S, Muhammad K, Aziz K, Mohammad NK, Hafiz MH, Wajid N (2018) Morphological acclimation to agronomic manipulation in leaf dispersion and orientation to promote "Ideotype" breeding: evidence from 3D visual modeling of "super" rice (*Oryza sativa* L.). *Plant Physiol Biochem* 135:499–510. https://doi.org/10.1016/j.plaphy.2018.11.010

Ehinger BV, Fischer P, Gert AL, Kaufhold L, Weber F, Pipa G, König P (2014) Kinesthetic and vestibular information modulate alpha activity during spatial navigation: a mobile EEG study. *Front Human Neurosci* 8:71.

EL Sabagh A, Islam MS, Hossain A, Iqbal MA, Mubeen M, Waleed M, Reginato M, Battaglia M, Ahmed S, Rehman A, Arif M, Athar H-U-R, Ratnasekera D, Danish S, Raza MA, Rajendran K, Mushtaq M, Skalicky M, Brestic M, Soufan W, Fahad S, Pandey S, Kamran M, Datta R, Abdelhamid MT (2022) Phytohormones as growth regulators during abiotic stress tolerance in plants. *Front Agron* 4:765068. doi: 10.3389/fagro.2022.765068

Emre B, Ömer SU, Martín LB, Andre D, Fahad S, Rahul D, Muhammad Z-ul-H, Ghulam SH, Subhan D (2021) Studying soil erosion by evaluating changes in physico-chemical properties of soils under different land-use types. *J Saudi Society Agric Sci* 20:190–197.

Fahad S, Bano A (2012) Effect of salicylic acid on physiological and biochemical characterization of maize grown in saline area. *Pak J Bot* 44:1433–1438.

Fahad S, Chen Y, Saud S, Wang K, Xiong D, Chen C, Wu C, Shah F, Nie L, Huang J (2013) Ultraviolet radiation effect on photosynthetic pigments, biochemical attributes, antioxidant enzyme activity and hormonal contents of wheat. *J Food, Agri Environ* 11(3&4):1635–1641.

Fahad S, Hussain S, Bano A, Saud S, Hassan S, Shan D, Khan FA, Khan F, Chen Y, Wu C, Tabassum MA, Chun MX, Afzal M, Jan A, Jan MT, Huang J (2014a) Potential role of phytohormones and plant growth-promoting rhizobacteria in abiotic stresses: consequences for changing environment. *Environ Sci Pollut Res* 22(7):4907–4921. https://doi.org/10.1007/s11356-014-3754-2

Fahad S, Hussain S, Matloob A, Khan FA, Khaliq A, Saud S, Hassan S, Shan D, Khan F, Ullah N, Faiq M, Khan MR, Tareen AK, Khan A, Ullah A, Ullah N, Huang J (2014b) Phytohormones and plant responses to salinity stress: a review. *Plant Growth Regul* 75(2):391–404. https://doi.org/10.1007/s10725-014-0013-y

Fahad S, Hussain S, Saud S, Tanveer M, Bajwa AA, Hassan S, Shah AN, Ullah A,Wu C, Khan FA, Shah F, Ullah S, Chen Y, Huang J (2015a) A biochar application protects rice pollen from high-temperature stress. *Plant Physiol Biochem* 96:281–287.

Fahad S, Nie L, Chen Y, Wu C, Xiong D, Saud S, Hongyan L, Cui K, Huang J (2015b) Crop plant hormones and environmental stress. *Sustain Agric Rev* 15:371–400.

Fahad S, Hussain S, Saud S, Hassan S, Chauhan BS, Khan F et al. (2016a) Responses of rapid viscoanalyzer profile and other rice grain qualities to exogenously applied plant growth regulators under high day and high night temperatures. *PLoS One* 11(7):e0159590. https://doi.org/10.1371/journal.pone.0159590

Fahad S, Hussain S, Saud S, Khan F, Hassan S, Jr A, Nasim W, Arif M, Wang F, Huang J (2016b) Exogenously applied plant growth regulators affect heat-stressed rice pollens. *J Agron Crop Sci* 202:139–150.

Fahad S, Hussain S, Saud S, Hassan S, Ihsan Z, Shah AN, Wu C, Yousaf M, Nasim W, Alharby H, Alghabari F, Huang J (2016c) Exogenously applied plant growth regulators enhance the morphophysiological growth and yield of rice under high temperature. *Front Plant Sci* 7:1250. https://doi.org/10.3389/fpls.2016.01250

Fahad S, Hussain S, Saud S, Hassan S, Tanveer M, Ihsan MZ, Shah AN, Ullah A, Nasrullah KF, Ullah S, AlharbyH NW, Wu C, Huang J (2016d) A combined application of biochar and phosphorus alleviates heat-induced adversities on physiological, agronomical and quality attributes of rice. *Plant Physiol Biochem* 103:191–198.

Fahad S, Bajwa AA, Nazir U, Anjum SA, Farooq A, Zohaib A, Sadia S, Nasim W, Adkins S, Saud S, Ihsan MZ, Alharby H, Wu C, Wang D, Huang J (2017) Crop production under drought and heat stress: plant responses and management options. *Front Plant Sci* 8:1147. https://doi.org/10.3389/fpls.2017.01147

Fahad S, Hasanuzzaman M, Alam M, Ullah H, Saeed M, Ali Khan I, Adnan M (Eds.) (2020) *Environment, Climate, Plant and Vegetation Growth*. Springer Nature Switzerland AG 2020. doi: https://doi.org/10.1007/978-3-030-49732-3

Fahad S, Muhammad ZI, Abdul K, Ihsanullah D, Saud S, Saleh A, Wajid N, Muhammad A, Imtiaz AK, Chao W, Depeng W, Jianliang H (2018a) Consequences of high temperature under changing climate optima for rice pollen characteristics-concepts and perspectives. *Arch Agron Soil Sci* 64(11):1473–1488 doi: 10.1080/03650340.2018.1443213

Fahad S, Abdul B, Adnan M (Eds.) (2018b) *Global Wheat Production*. IntechOpen United Kingdom 2018. http://dx.doi.org/10.5772/intechopen.72559

Fahad S, Rehman A, Shahzad B, Tanveer M, Saud S, Kamran M, Ihtisham M, Khan SU, Turan V, Rahman MHU (2019a) Rice responses and tolerance to metal/metalloid toxicity. In: Hasanuzzaman M, Fujita M, Nahar K, Biswas JK (Eds.) *Advances in Rice Research for Abiotic Stress Tolerance*. Woodhead Publishing Ltd, UK, pp. 299–312.

Fahad S, Adnan M, Hassan S, Saud S, Hussain S, Wu C, Wang D, Hakeem KR, Alharby HF, Turan V, Khan MA, Huang J (2019b) Rice responses and tolerance to high temperature. In: Hasanuzzaman M, Fujita M, Nahar K, and Biswas, JK (Eds.) *Advances in Rice Research for Abiotic Stress Tolerance*. Woodhead Publishing Ltd, UK. pp. 201–224.

Fahad S, Sönmez O, Saud S, Wang D, Wu C, Adnan M, Turan V (Eds.) (2021a) Plant growth regulators for climate-smart agriculture. First edition. *Footprints of Climate Variability on Plant Diversity*. CRC Press, Boca Raton.

Fahad S, Sonmez O, Saud S, Wang D, Wu C, Adnan M, Turan V (Eds.) (2021b) Climate change and plants: biodiversity, growth and interactions. First edition. *Footprints of Climate Variability on Plant Diversity*. CRC Press, Boca Raton.

Fahad S, Sonmez O, Saud S, Wang D, Wu C, Adnan M, Turan V (Eds.) (2021c) Developing climate resilient crops: improving global food security and safety. First edition. *Footprints of Climate Variability on Plant Diversity*. CRC Press, Boca Raton.

Fahad S, Sönmez O, Turan V, Adnan M, Saud S, Wu C, Wang D (Eds.) (2021d) Sustainable soil and land management and climate change. First edition. *Footprints of Climate Variability on Plant Diversity*. CRC Press, Boca Raton.

Fahad S, Sönmez O, Saud S, Wang D, Wu C, Adnan M, Arif M, Amanullah (Eds.) (2021e) Engineering tolerance in crop plants against abiotic stress, First edition. *Footprints of Climate Variability on Plant Diversity*. CRC Press, Boca Raton.

Fahad S, Saud S, Yajun C, Chao W, Depeng W (Eds.) (2021f) Abiotic Stress in Plants. IntechOpen United Kingdom 2021. http://dx.doi.org/10.5772/intechopen.91549

Fahad S, Adnan M, Saud S (Eds.) (2022a) Improvement of plant production in the era of climate change. First edition. *Footprints of Climate Variability on Plant Diversity*. CRC Press, Boca Raton.

Fahad S, Adnan M, Saud S, Nie L (Eds.) (2022b) Climate change and ecosystems: challenges to sustainable development. First edition. *Footprints of Climate Variability on Plant Diversity*. CRC Press, Boca Raton.

Fakhre A, Ayub K, Fahad S, Sarfraz N, Niaz A, Muhammad AA, Muhammad A, Khadim D, Saud S, Shah H, Muhammad ASR, Khalid N, Muhammad A, Rahul D, Subhan D (2021) Phosphate solubilizing bacteria optimize wheat yield in mineral phosphorus applied alkaline soil. *J Saudi Soc Agric Sci* 21(5):339–348. https://doi.org/10.1016/j.jssas.2021.10.007

Fallik E, Okon Y, Epstein E, Goldman A, Fischer M (1989) Identification and quantification of IAA and IBA in Azospirillum brasilense-inoculated maize roots. *Soil Biol & Biochem* 21(1):147–153.

Farah R, Muhammad R, Muhammad SA, Tahira Y, Muhammad AA, Maryam A, Shafaqat A, Rashid M, Muhammad R, Qaiser H, Afia Z, Muhammad AA, Muhammad A, Fahad S (2020) Alternative and non-conventional soil and crop management strategies for increasing water use efficiency. In: Fahad S, Hasanuzzaman M, Alam M, Ullah H, Saeed M, Khan AK, Adnan M (Eds.) *Environment, Climate, Plant and Vegetation Growth*. Springer Publ Ltd, Springer Nature Switzerland, pp. 323–338. https://doi.org/10.1007/978-3-030-49732-3

Farhana G, Ishfaq A, Muhammad A, Dawood J, Fahad S, Xiuling L, Depeng W, Muhammad F, Muhammad F, Syed AS (2020) Use of crop growth model to simulate the impact of climate change on yield of various wheat cultivars under different agro-environmental conditions in Khyber Pakhtunkhwa, Pakistan. *Arabian J Geosci* 13:112. https://doi.org/10.1007/s12517-020-5118-1

Farhat A, Hafiz MH, Wajid I, Aitazaz AF, Hafiz FB, Zahida Z, Fahad S, Wajid F, Artemi C (2020) A review of soil carbon dynamics resulting from agricultural practices. *J Environ Manage* 268(2020): 110319.

Farhat UK, Adnan AK, Kai L, Xuexuan X, Muhammad A, Fahad S, Rafiq A, Mushtaq AK, Taufiq N, Faisal Z (2022) Influences of long-term crop cultivation and fertilizer management on soil aggregates stability and fertility in the Loess Plateau, Northern China. *J Soil Sci Plant Nutri* 22:1446–1457. https://doi.org/10.1007/s42729-021-00744-1

Fazli W, Muhmmad S, Amjad A, Fahad S, Muhammad A, Muhammad N, Ishaq AM, Imtiaz AK, Mukhtar A, Muhammad S, Muhammad I, Rafi U, Haroon I, Muhammad A (2020) Plant–microbes interactions and functions in changing climate. In: Fahad S, Hasanuzzaman M, Alam M, Ullah H, Saeed M, Khan AK, Adnan M (Eds.) *Environment, Climate, Plant and Vegetation Growth*. Springer Publ Ltd, Springer Nature Switzerland, pp. 397–420. https://doi.org/10.1007/978-3-030-49732-3

Fuentes-Ramirez LE, Caballero-Mellado J (2005) Bacterial biofertilizers. In: Siddiqui ZA (Ed.), *PGPR: Biocontrol and Biofertilization*. Springer, pp. 143–172. doi:10.1007/1-4020-4152-7_5

Garca P, Dublán-García O, Gómez-Oliván LM, Baeza-Jiménez R, López-Martínez LX (2017) In vitro antioxidant and bioactive properties of corn (Zea mays L.). *Arch. Latinoam. Nutr* 67:300–308.

Ghulam M, Muhammad AA, Donald LS, Sajid M, Muhammad FQ, Niaz A, Ateeq ur R, Shakeel A, Sajjad H, Muhammad A, Summia M, Aqib HAK, Fahad S, Rahul D, Mazhar I, Timothy DS (2021) Formalin fumigation and steaming of various composts differentially influence the nutrient release, growth and yield of muskmelon (*Cucumis melo* L.). *Sci Rep* 11:21057.

Gianinazzi S, Vosátka M (2004) Inoculum of arbuscular mycorrhizal fungi for production systems: science meets business. *Canadian J Botany* 82(8):1264–1271.

Glick BR (2020) Introduction to plant growth-promoting bacteria. *Beneficial Plant–Bacterial Interactions* 1–37.

Gomez DE, Alonso DF, Yoshiji H, Thorgeirsson UP (1997) Tissue inhibitors of metalloproteinases: structure, regulation and biological functions. *European Journal of Cell Biology* 74(2):111–122.

Gopakumar L, Bernard NO, Donato V (2020) Soil microarthropods and nutrient cycling. In: Fahad S, Hasanuzzaman M, Alam M, Ullah H, Saeed M, Khan AK, Adnan M (Eds.) *Environment, Climate, Plant and Vegetation Growth*. Springer Publ Ltd, Springer Nature Switzerland, pp. 453–472. https://doi.org/10.1007/978-3-030-49732-3

Greffe VRG, Michiels J (2020) Desiccation-induced cell damage in bacteria and the relevance for inoculant production. *Applied Microbiology and Biotechnology* 104(9):3757–3770.

Guofu L, Zhenjian B, Fahad S, Guowen C, Zhixin X, Hao G, Dandan L, Yulong L, Bing L, Guoxu J, Saud S (2021) Compositional and structural changes in soil microbial communities in response to straw mulching and plant revegetation in an abandoned artificial pasture in Northeast China. *Glob Ecol Conserv* 31(2021):e01871.

Gupta S, Meena MK, Datta S (2014) Isolation, characterization of plant growth promoting bacteria from the plant Chlorophytum borivilianum and in-vitro screening for activity of nitrogen fixation, phosphate solubilization and IAA production. *Int J Curr Microbiol Appl Sci* 3(7):1082–1090.

Habib ur R, Ashfaq A, Aftab W, Manzoor H, Fahd R, Wajid I, Md Aminul I, Vakhtang S, Muhammad A, Asmat U, Abdul W, Syeda RS, Shah S, Shahbaz K, Fahad S, Manzoor H, Saddam H, Wajid N (2017) Application of CSM-CROPGRO-cotton model for cultivars and optimum planting dates: evaluation in changing semi-arid climate. *Field Crops Res* 238:139–152. http://dx.doi.org/10.1016/j.fcr.2017.07.007

Hafeez M, Farman U, Muhammad MK, Xiaowei L, Zhijun Z, Sakhawat S, Muhammad I, Mohammed AA, Mandela F-G, Nicolas D, Muzammal R, Fahad S, Yaobin L (2021) Metabolic-based insecticide resistance mechanism and ecofriendly approaches for controlling of beet armyworm Spodoptera exigua: a review. *Environ Sci Pollution Res* 29:1746–1762. https://doi.org/10.1007/s11356-021-16974-w

Hafiz MH, Wajid F, Farhat A, Fahad S, Shafqat S, Wajid N, Hafiz FB (2016) Maize plant nitrogen uptake dynamics at limited irrigation water and nitrogen. *Environ Sci Pollut Res* 24(3):2549–2557. https://doi.org/10.1007/s11356-016-8031-0

Hafiz MH, Farhat A, Shafqat S, Fahad S, Artemi C, Wajid F, Chaves CB, Wajid N, Muhammad M, Hafiz FB (2018) Offsetting land degradation through nitrogen and water management during maize cultivation under arid conditions. *Land Degrad Dev* 29(5), 1–10. doi: 10.1002/ldr.2933

Hafiz MH, Muhammad A, Farhat A, Hafiz FB, Saeed AQ, Muhammad M, Fahad S, Muhammad A (2019) Environmental factors affecting the frequency of road traffic accidents: a case study of sub-urban area of Pakistan. *Environ Sci Pollut Res* 26:11674–11685. https://doi.org/10.1007/s11356-019-04752-8

Hafiz MH, Farhat A, Ashfaq A, Hafiz FB, Wajid F, Carol Jo W, Fahad S, Gerrit H (2020a) Predicting kernel growth of maize under controlled water and nitrogen applications. *Int J Plant Prod* 14, 609–620. https://doi.org/10.1007/s42106-020-00110-8

Hafiz MH, Abdul K, Farhat A, Wajid F, Fahad S, Muhammad A, Ghulam MS, Wajid N, Muhammad M, Hafiz FB (2020b) Comparative effects of organic and inorganic fertilizers on soil organic carbon and wheat productivity under arid region. *Commun Soil Sci Plant Anal* 51, 1406–1422. doi: 10.1080/00103624.2020.1763385

Haider SA, Lalarukh I, Amjad SF, Mansoora N, Naz M, Naeem M, Bukhari SA, Shahbaz M, Ali SA, Marfo TD, Subhan D, Rahul D, Fahad S (2021) Drought stress alleviation by potassium-nitrate-containing chitosan/montmorillonite microparticles confers changes in Spinacia oleracea L. *Sustain* 13:9903. https://doi.org/10.3390/su13179903

Hamza SM, Xiukang W, Sajjad A, Sadia Z, Muhammad N, Adnan M, Fahad S et al. (2021) Interactive effects of gibberellic acid and NPK on morpho-physio-biochemical traits and organic acid exudation pattern in coriander (Coriandrum sativum L.) grown in soil artificially spiked with boron. *Plant Physiol & Biochem* 167(2021):884–900.

Haoliang Y, Matthew TH, Ke L, Bin W, Puyu F, Fahad S, Holger M, Rui Y, De LL, Sotirios A, Isaiah H, Xiaohai T, Jianguo M, Yunbo Z, Meixue Z (2022) Crop traits enabling yield gains under more frequent extreme climatic events. *Sci Total Environ* 808(2022):152170.

Hart AG, Carpenter WS, Hlustik-Smith E, Reed M, Goodenough AE (2018) Testing the potential of Twitter mining methods for data acquisition: evaluating novel opportunities for ecological research in multiple taxa. *Methods Ecol & Evol* 9(11):2194–2205.

Herrmann L, Lesueur D (2013) Challenges of formulation and quality of biofertilizers for successful inoculation. *Appl Microbiol & Biotechnol* 97(20):8859–8873.

Hesham FA and Fahad S (2020) Melatonin application enhances biochar efficiency for drought tolerance in maize varieties: modifications in physio-biochemical machinery. *Agron J* 112(4):1–22.

Hirsch LR, Stafford RJ, Bankson JA, Sershen SR, Rivera B, Price RE, West JL (2003) Nanoshell-mediated near-infrared thermal therapy of tumors under magnetic resonance guidance. *Proceedings of the National Academy of Sciences* 100(23):13549–13554.

Hoeksema JD, Chaudhary VB, Gehring CA, Johnson NC, Karst J, Koide RT, Umbanhowar J (2010) A meta-analysis of context-dependency in plant response to inoculation with mycorrhizal fungi. *Ecol Lett* 13(3):394–407.

Huang Li-Y, Li Xiao-X, Zhang Yun-B, Fahad S, Wang F (2021) dep1 improves rice grain yield and nitrogen use efficiency simultaneously by enhancing nitrogen and dry matter translocation. *J Integrative Agri* 21(11):3185–3198. doi: 10.1016/S2095-3119(21)63795-4

Hussain MA, Fahad S, Rahat S, Muhammad FJ, Muhammad M, Qasid A, Ali A, Husain A, Nooral A, Babatope SA, Changbao S, Liya G, Ibrar A, Zhanmei J, Juncai H (2020) Multifunctional role of brassinosteroid and its analogues in plants. *Plant Growth Regul* 92:141–156. https://doi.org/10.1007/s10725-020-00647-8

Ibad U, Dost M, Maria M, Shadman K, Muhammad A, Fahad S, Muhammad I, Ishaq AM, Aizaz A, Muhammad HS, Muhammad S, Farhana G, Muhammad I, Muhammad ASR, Hafiz MH, Wajid N, Shah S, Jabar ZKK, Masood A, Naushad A, Rasheed Akbar M, Shah MK Jan B (2022) Comparative effects of biochar and NPK on wheat crops under different management systems. *Crop Pasture Sci* 74:31–40 https://doi.org/10.1071/CP21146

Ibrar H, Muqarrab A, Adel MG, Khurram S, Omer F, Shahid I, Fahim N, Shakeel A, Viliam B, Marian B, Sami Al Obaid, Fahad S, Subhan D, Suleyman T, Hanife AKÇA, Rahul D (2021) Improvement in growth and yield attributes of cluster bean through optimization of sowing time and plant spacing under climate change Scenario. *Saudi J Bio Sci* 29(2):781–792. https://doi.org/10.1016/j.sjbs.2021.11.018

Ibrar K, Aneela R, Khola Z, Urooba N, Sana B, Rabia S, Ishtiaq H, Mujaddad Ur Rehman, Salvatore M (2020) Microbes and environment: global warming reverting the frozen zombies. In: Fahad S, Hasanuzzaman M, Alam M, Ullah H, Saeed M, Khan AK, Adnan M (Eds.) *Environment, Climate, Plant and Vegetation Growth*. Springer Publ Ltd, Springer Nature Switzerland, pp. 607–634. https://doi.org/10.1007/978-3-030-49732-3

Ihsan MZ, Abdul K, Manzer HS, Liaqat A, Ritesh K, Hayssam MA, Amar M, Fahad S (2022) The response of Triticum aestivum treated with plant growth regulators to acute day/night temperature rise. *J Plant Growth Regul* 41:2020–2033. https://doi.org/10.1007/s00344-022-10574-9

Ikram U, Khadim D, Muhammad T, Muhammad S, Fahad S (2021) Gibberellic acid and urease inhibitor optimize nitrogen uptake and yield of maize at varying nitrogen levels under changing climate. *Environ Sci Pollution Res* 29(5):6568–6577 https://doi.org/10.1007/s11356-021-16049-w

Ilyas M, Mohammad N, Nadeem K, Ali H, Aamir HK, Kashif H, Fahad S, Aziz K, Abid U (2020) Drought tolerance strategies in plants: a mechanistic approach. *J Plant Growth Regulation* 40:926–944 https://doi.org/10.1007/s00344-020-10174-5

Iqra M, Amna B, Shakeel I, Fatima K,Sehrish L, Hamza A, Fahad S (2020) Carbon cycle in response to global warming. In: Fahad S, Hasanuzzaman M, Alam M, Ullah H,Saeed M, Khan AK, Adnan M (Eds.) *Environment, Climate, Plant and Vegetation Growth*. Springer Publ Ltd, Springer Nature Switzerland, pp. 1–16. https://doi.org/10.1007/978-3-030-49732-3

Irfan M, Muhammad M, Muhammad JK, Khadim MD, Dost M, Ishaq AM, Waqas A, Fahad S, Saud S et al. (2021) Heavy metals immobilization and improvement in maize (Zea mays L.) growth amended with biochar and compost. *Sci Rep* 11:18416.

Jabborova D, Sulaymanov K, Sayyed RZ, Alotaibi SH, Enakiev Y, Azimov A, Jabbarov Z, Ansari MJ, Fahad S, Danish S et al. (2021) Mineral fertilizers improve the quality of turmeric and soil. *Sustain* 13: 9437. https://doi.org/10.3390/su13169437

Jan M, Muhammad Anwar-ul-Haq, Adnan NS, Muhammad Y, Javaid I, Xiuling L, Depeng W, Fahad S (2019) Modulation in growth, gas exchange, and antioxidant activities of salt-stressed rice (Oryza sativa L.) genotypes by zinc fertilization. *Arabian J Geosci* 12:775. https://doi.org/10.1007/s12517-019-4939-2

Jansa J, Oberholzer HR, Egli S (2009a) Environmental determinants of the arbuscular mycorrhizal fungal infectivity of Swiss agricultural soils. *Eur J Soil Biol* 45(5–6):400–408.

Jansa SA, Giarla TC, Lim BK (2009b) The phylogenetic position of the rodent genus Typhlomys and the geographic origin of Muroidea. *J Mammal* 90(5):1083–1094.

Kaeppler SM, Kaeppler HF, Rhee Y (2000) Epigenetic aspects of somaclonal variation in plants. *Plant Gene Silencing* 43(2–3):179–188.

Kamaran M, Wenwen C, Irshad A, Xiangping M, Xudong Z, Wennan S, Junzhi C, Shakeel A, Fahad S, Qingfang H, Tiening L (2017) Effect of paclobutrazol, a potential growth regulator on stalk mechanical strength, lignin accumulation and its relation with lodging resistance of maize. *Plant Growth Regul* 84:317–332. https://doi.org/10.1007/ s10725-017-0342-8

Kaminsky LM, Trexler RV, Malik RJ, Hockett KL, Bell TH (2019) The inherent conflicts in developing soil microbial inoculants. *Trends Biotechnol* 37(2):140–151.

Keswani B, Mohapatra AG, Mohanty A, Khanna A, Rodrigues JJ, Gupta D, De-Albuquerque VHC (2019) Adapting weather conditions based IoT enabled smart irrigation technique in precision agriculture mechanisms. *Neural Comput & Appl* 31(1):277–292.

Khadim D, Fahad S, Jahangir MMR, Iqbal M, Syed SA, Shah AK, Ishaq AM, Rahul D et al. (2021a) Biochar and urease inhibitor mitigate NH_3 and N_2O emissions and improve wheat yield in a urea fertilized alkaline soil. *Sci Rep* 11:17413.

Khadim D, Saif-ur-R, Fahad S, Syed SA, Shah AK et al. (2021b) Influence of variable biochar concentration on yield-scaled nitrous oxide emissions, wheat yield and nitrogen use efficiency. *Sci Rep* 11:16774.

Khan IA, Shah B, Khan A, Zaman M, M. Din MMU, Junaid K, Shah SRA, Adnan M, Ahmad N, Akbar R, Fayaz W, Rahman IU (2016) Study on the population densities of grass hopper and armyworm on different maize cultivars at Peshawar. *J Ento & Zool Studies* 4(1):28–331.

Khan MMH, Niaz A, Umber G, Muqarrab A, Muhammad AA, Muhammad I, Shabir H, Shah F, Vibhor A, Shams HA-H, Reham A, Syed MBA, Nadiyah MA, Ali TKZ, Subhan D, Rahul D (2021) Synchronization of boron application methods and rates is environmentally friendly approach to improve quality attributes of *Mangifera indica* L. on sustainable basis. *Saudi J Bio Sci* 29(3):1869–1880 https://doi.org/10.1016/ j.sjbs.2021.10.036

Khatun M, Sarkar S, Era FM, Islam AKMM, Anwar MP, Fahad S, Datta R, Islam AKMA (2021) Drought stress in grain legumes: effects, tolerance mechanisms and management. *Agron* 11: 2374. https://doi.org/10.3390/agronomy11122374

Kloepper JW (1993) Host specificity in microbe–microbe interactions. *Bioscience* 46(6): 406–409.

Koyama A, Pietrangelo O, Sanderson L, Antunes PM (2017) An empirical investigation of the possibility of adaptability of arbuscular mycorrhizal fungi to new hosts. *Mycorrhiza* 27(6): 553–563.

Krishnamoorthy A, Agarwal T, Kotamreddy JNR, Bhattacharya R, Mitra A, Maiti TK, Maiti MK (2020) Impact of seed-transmitted endophytic bacteria on intra-and inter-cultivar plant growth promotion modulated by certain sets of metabolites in rice crop. *Microbiol Res* 241:126582.

Kumari M, Thakur IS (2018) Biochemical and proteomic characterization of Paenibacillus sp. ISTP10 for its role in plant growth promotion and in rhizostabilization of cadmium. *Bioresour Technol Rep* 3:59–66.

Lee K, Kang H (2016) Emerging roles of RNA-binding proteins in plant growth, development, and stress responses. *Mol & Cells* 39(3):179.

Linderman RG, Davis EA (2004) Varied response of marigold (Tagetes spp.) genotypes to inoculation with different arbuscular mycorrhizal fungi. *Scientia Hortic* 99(1):67–78.

Luziatelli F, Ficca AG, Bonini P, Muleo R, Gatti L, Meneghini M, Ruzzi M (2020) A genetic and metabolomic perspective on the production of indole-3-acetic acid by Pantoea agglomerans and use of their metabolites as biostimulants in plant nurseries. *Front Microbiol* 11, 1475.

Mahar A, Amjad A, Altaf HL, Fazli W, Ronghua L, Muhammad A, Fahad S, Muhammad A, Rafiullah, Imtiaz AK, Zengqiang Z (2020) Promising technologies for Cd-contaminated soils: drawbacks and possibilities. In: Fahad S, Hasanuzzaman M, Alam M, Ullah H,Saeed M, Khan AK, Adnan M (Eds.) *Environment, Climate, Plant and Vegetation Growth*. Springer Publ Ltd, Springer Nature Switzerland, pp. 63–92. https://doi.org/10.1007/978-3-030-49732-3

Mahmood Ul H, Tassaduq R, Chandni I, Adnan A, Muhammad A, Muhammad MA, Muhammad H-ur-R, Mehmood AN, Alam S, Fahad S (2021) Linking plants functioning to adaptive responses under heat stress conditions: a mechanistic review. *J Plant Growth Regul* 41:2596–2613. https://doi.org/10.1007/ s00344-021-10493-1

Maltz MR, Treseder KK (2015) Sources of inocula influence mycorrhizal colonization of plants in restoration projects: a meta-analysis. *Restor Ecol* 23(5):625–634.

Malusá E, Sas-Paszt L, Ciesielska J (2012) Technologies for beneficial microorganisms inocula used as biofertilizers. *Sci World J* 2012:491206.

Malusà MG, Resentini A, Garzanti E (2016) Hydraulic sorting and mineral fertility bias in detrital geochronology. *Gondwana Res* 31:1–19.

Manandhar S, Tuladhar R, Prajapati K, Singh A, Varma A (2017) Effect of Azotobacter chroococcum and Piriformospora indica on Oryza sativa in presence of vermicompost. In: Ajit Varma, Ram Prasad, Narendra Tuteja (Eds.), *Mycorrhiza-Nutrient Uptake, Biocontrol, Ecorestoration*. Springer, pp. 327–339.

Manzer HS, Saud A, Soumya M, Abdullah A, Al-A, Qasi DA, Bander MA, Al-M, Hayssam MA, Hazem MK, Fahad S, Vishnu DR, Om PN (2021) Molybdenum and hydrogen sulfide synergistically mitigate arsenic toxicity by modulating defense system, nitrogen and cysteine assimilation in faba bean (*Vicia faba* L.) seedlings. *Environ Pollut* 290:117953.

Md Enamul H, Shoeb AZM, Mallik AH, Fahad S, Kamruzzaman MM, Akib J, Nayyer S, Mehedi AKM, Swati AS, Md Yeamin A, Most SS (2020) Measuring vulnerability to environmental hazards: qualitative to quantitative. In: Fahad S, Hasanuzzaman M, Alam M, Ullah H, Saeed M, Khan AK, Adnan M (Eds.), *Environment, Climate, Plant and Vegetation Growth*. Springer Publ Ltd, Springer Nature Switzerland, pp. 421–452. https://doi.org/10.1007/978-3-030-49732-3

Md Jakir H, Allah B (2020) Development and applications of transplastomic plants: a way towards eco-friendly agriculture. In: Fahad S, Hasanuzzaman M, Alam M, Ullah H, Saeed M, Khan AK, Adnan M (Eds.) *Environment, Climate, Plant and Vegetation Growth*. Springer Publ Ltd, Springer Nature Switzerland, pp. 285–322. https://doi.org/10.1007/978-3-030-49732-3

Mehmood K, Bao Y, Saifullah, Bibi S, Dahlawi S, Yaseen M, Abrar MM, Srivastava P, Fahad S, Faraj TK (2022) Contributions of open biomass burning and crop straw burning to air quality: current research paradigm and future outlooks. *Front Environ Sci* 10:852492. doi: 10.3389/fenvs.2022.852492

Melchiorre M, De-Luca MJ, Gonzalez Anta G, Suarez P, Lopez C, Lascano R, Racca RW (2011) Evaluation of bradyrhizobia strains isolated from field-grown soybean plants in Argentina as improved inoculants. *Biol & Fertil Soils* 47(1):81–89.

Mezei S, Popović M, Kovačev L, Mrkovački N, Nagl N, Malenčić D (1997) Effect of Azotobacter strains on sugar beet callus proliferation and nitrogen metabolism enzymes. *Biologia Plant* 40(2):277–283.

Miller M, Dick RP (1995) Thermal stability and activities of soil enzymes as influenced by crop rotations. *Soil Biol & Biochem* 27(9):1161–1166.

Mitter B, Pfaffenbichler N, Sessitsch A (2016) Plant–microbe partnerships in 2020. *Microbial Biotechnol* 9(5):635–640.

Mohammad I. Al-Wabel, Munir Ahmad, Adel RA Usman, Mutair Akanji, and Muhammad Imran Rafique (2020a) Advances in pyrolytic technologies with improved carbon capture and storage to combat climate change. In: Fahad S, Hasanuzzaman M, Alam M, Ullah H, Saeed M, Khan AK, Adnan M (Eds.) *Environment, Climate, Plant and Vegetation Growth*. Springer Publ Ltd, Springer Nature Switzerland, pp. 535–576. https://doi.org/10.1007/978-3-030-49732-3

Mohammad I. Al-Wabel, Abdelazeem S, Munir A, Khalid E, Adel RAU (2020b) Extent of climate change in Saudi Arabia and its impacts on agriculture: a case study from Qassim Region. In: Fahad S, Hasanuzzaman M, Alam M, Ullah H, Saeed M, Khan AK, Adnan M (Eds.) *Environment, Climate, Plant and Vegetation Growth*. Springer Publ Ltd, Springer Nature Switzerland, pp. 635–658. https://doi.org/10.1007/978-3-030-49732-3

Mohammadi K, Sohrabi Y (2012) Bacterial biofertilizers for sustainable crop production: a review. *ARPN J Agric Biol Sci* 7(5):307–316.

Moutia JFY, Saumtally S, Spaepen S, Vanderleyden J (2010) Plant growth promotion by Azospirillum sp. in sugarcane is influenced by genotype and drought stress. *Plant & Soil* 337(1):233–242.

Mubeen M, Ashfaq A, Hafiz MH, Muhammad A, Hafiz UF, Mazhar S, Muhammad Sami ul Din, Asad A, Amjed A, Fahad S, Wajid N (2020) Evaluating the climate change impact on water use efficiency of cotton-wheat in semi-arid conditions using DSSAT model. *J Water Climate Change* 11(4):1661–1675. doi/10.2166/wcc.2019.179/622035/jwc2019179.pdf

Muhammad I, Khadim D, Fahad S, Imran M, Saud A, Manzer HS, Shah S, Jabar ZKK, Shamsher A, Shah H, Taufiq N, Hafiz MH, Jan B, Wajid N (2022) Exploring the potential effect of *Achnatherum splendens* L.-derived biochar treated with phosphoric acid on bioavailability of cadmium and wheat growth in contaminated soil. *Environ Sci Pollut Res* 29(25):37676–37684. https://doi.org/10.1007/s11356-021-17950-0.

Muhammad N, Muqarrab A, Khurram S, Fiaz A, Fahim N, Muhammad A, Shazia A, Omaima N, Sulaiman AA, Fahad S, Subhan D, Rahul D (2021) Kaolin and jasmonic acid improved cotton productivity under water stress conditions. *J Saudi Society Agric Sci* 28(2021):6606–6614. https://doi.org/10.1016/j.sjbs.2021.07.043

Muhammad Tahir ul Qamar, Amna F, Amna B, Barira Z, Xitong Z, Ling-Ling C (2020) Effectiveness of conventional crop improvement strategies vs. omics. In: Fahad S, Hasanuzzaman M, Alam M, Ullah H, Saeed M, Khan AK, Adnan M (Eds.) *Environment, Climate, Plant and Vegetation Growth*. Springer Publ Ltd, Springer Nature Switzerland, pp. 253–284. https://doi.org/10.1007/978-3-030-49732-3

Muhammad Z, Abdul MK, Abdul MS, Kenneth BM, Muhammad S, Shahen S, Ibadullah J, Fahad S (2019) Performance of Aeluropus lagopoides (mangrove grass) ecotypes, a potential turfgrass, under high saline conditions. *Environ Sci Pollut Res* 26(13):13410–13412. https://doi.org/10.1007/s11356-019-04838-3

Mukhongo RW, Tumuhairwe JB, Ebanyat P, Abdel-Gadir AH, Thuita M, Masso C (2016) Production and use of arbuscular mycorrhizal fungi inoculum in Sub-Saharan Africa: challenges and ways of improving. *Int J Soil Sci* 11(3):108–122.

Mulas D, Seco V, Casquero PA, Velázquez E, González-Andrés F (2015) Inoculation with indigenous rhizobium strains increases yields of common bean (Phaseolus vulgaris L.) in northern Spain, although its efficiency is affected by the tillage system. *Symbiosis* 67(1):113–124.

Murphy BR, Doohan FM, Hodkinson TR (2018) From concept to commerce: developing a successful fungal endophyte inoculant for agricultural crops. *J Fungi* 4(1):24.

Muzammal R, Fahad S, Guanghui D, Xia C, Yang Y, Kailei T, Lijun L, Fei-Hu L, Gang D (2021) Evaluation of hemp (Cannabis sativa L.) as an industrial crop: a review. *Environ Sci Pollution Res* 28(38):52832–52843. https://doi.org/10.1007/s11356-021-16264-5

Niaz A, Abdullah E, Subhan D, Muhammad A, Fahad S, Khadim D, Suleyman T, Hanife A, Anis AS, Mohammad JA, Emre B, Ömer SU, Rahul D, Bernard RG (2022) Mitigation of lead (Pb) toxicity in rice cultivated with either ground water or wastewater by application of acidified carbon. *J Environ Manage* 307:114521.

Nkebiwe PM, Weinmann M, Bar-Tal A, Müller T (2016) Fertilizer placement to improve crop nutrient acquisition and yield: a review and meta-analysis. *Field Crops Res* 196:389–401.

Noor M, Naveed ur R, Ajmal J, Fahad S, Muhammad A, Fazli W, Saud S, Hassan S (2020) Climate change and coastal plant lives. In: Fahad S, Hasanuzzaman M, Alam M, Ullah H,Saeed M, Khan AK, Adnan M (Eds.) *Environment, Climate, Plant and Vegetation Growth*. Springer Publ Ltd, Springer Nature Switzerland, pp. 93–108. https://doi.org/10.1007/978-3-030-49732-3

Okon Y, Labandera-Gonzalez CA (1994) Agronomic applications of Azospirillum. In: de Bruijn (Ed.), *Improving Plant Productivity with Rhizosphere Bacteria*. Wiley, pp. 274–278.

Ozturk A, Caglar O, Sahin F (2003) Yield response of wheat and barley to inoculation of plant growth promoting rhizobacteria at various levels of nitrogen fertilization. *J Plant Nutrition & Soil Sci* 166(2):262–266.

Parnell JJ, Berka R, Young HA, Sturino JM, Kang Y, Barnhart DM, DiLeo MV (2016) From the lab to the farm: an industrial perspective of plant beneficial microorganisms. *Front Plant Sci* 7:1110.

Pii Y, Aldrighetti A, Valentinuzzi F, Mimmo T, Cesco S (2019) Azospirillum brasilense inoculation counteracts the induction of nitrate uptake in maize plants. *J Exp Botany* 70(4):1313–1324.

Preininger C, Sauer U, Bejarano A, Berninger T (2018) Concepts and applications of foliar spray for microbial inoculants. *Appl Microbiol & Biotechnol* 102(17):7265–7282.

Qamar-uz Z, Zubair A, Muhammad Y, Muhammad ZI, Abdul K, Fahad S, Safder B, Ramzani PMA, Muhammad N (2017) Zinc biofortification in rice: leveraging agriculture to moderate hidden hunger in developing countries. *Arch Agron Soil Sci* 64:147–161. https://doi.org/10.1080/03650340.2017.1338343

Qin ZH, Nasib ur Rahman, Ahmad A, Yu-pei Wang, Sakhawat S, Ehmet N, Wen-juan Shao, Muhammad I, Kun S, Rui L, Fazal S, Fahad S (2022) Range expansion decreases the reproductive fitness of *Gentiana officinalis* (*Gentianaceae*). *Sci Rep* 12:2461 https://doi.org/10.1038/s41598-022-06406-1

Rajesh KS, Fahad S, Pawan K, Prince C, Talha J, Dinesh J, Prabha S, Debanjana S, Prathibha MD, Bandana B, Akash H, Gupta NK, Rekha S, Devanshu D, Dalpat LS, Ke L, Matthew TH, Saud S, Adnan NS, Taufiq N (2022) Beneficial elements: new players in improving nutrient use efficiency and abiotic stress tolerance. *Plant Growth Regul* 100:237–265. https://doi.org/10.1007/s10725-022-00843-8

Rashid M, Qaiser H, Khalid SK, Mohammad I. Al-Wabel, Zhang A, Muhammad A, Shahzada SI, Rukhsanda A, Ghulam AS, Shahzada MM, Sarosh A, Muhammad FQ (2020) Prospects of biochar in alkaline soils to mitigate climate change. In: Fahad S, Hasanuzzaman M, Alam M, Ullah H,Saeed M, Khan AK, Adnan M (Eds.) *Environment, Climate, Plant and Vegetation Growth*. Springer Publ Ltd, Springer Nature Switzerland, pp. 133–150. https://doi.org/10.1007/978-3-030-49732-3

Rawat M, Arunachalam K, Arunachalam A, Alatalo J, Pandey R (2019) Associations of plant functional diversity with carbon accumulation in a temperate forest ecosystem in the Indian Himalayas. *Ecol Indicators* 98:861–868.

Rehana S, Asma Z, Shakil A, Anis AS, Rana KI, Shabir H, Subhan D, Umber G, Fahad S, Jiri K, Sami Al Obaid, Mohammad JA, Rahul D (2021) Proteomic changes in various plant tissues associated with chromium stress in sunflower. *Saudi J Bio Sci* 29(4):2604–2612. https://doi.org/10.1016/j.sjbs.2021.12.042

Rehman M, Fahad S, Saleem MH, Hafeez M, Muhammad Habib ur Rahman, Liu F, Deng G (2020) Red light optimized physiological traits and enhanced the growth of ramie (Boehmeria nivea L.). *Photosynthetica* 58(4):922–931.

Remans R, Ramaekers L, Schelkens S, Hernandez G, Garcia A, Reyes JL, Vanderleyden J (2008) Effect of Rhizobium–Azospirillum coinoculation on nitrogen fixation and yield of two contrasting Phaseolus vulgaris L. genotypes cultivated across different environments in Cuba. *Plant & Soil* 312(1):25–37.

Sadam M, Muhammad Tahir ul Qamar, Ghulam M, Muhammad SK, Faiz AJ (2020) Role of biotechnology in climate resilient agriculture. In: Fahad S, Hasanuzzaman M, Alam M, Ullah H, Saeed M, Khan AK, Adnan M (Eds.) *Environment, Climate, Plant and Vegetation Growth.* Springer Publ Ltd, Springer Nature Switzerland, pp. 339–366. https://doi.org/10.1007/978-3-030-49732-3

Safi UK, Ullah F, Mehmood S, Fahad S, Ahmad Rahi A, Althobaiti F et al. (2021) Antimicrobial, antioxidant and cytotoxic properties of Chenopodium glaucum L. *PLoS One* 16(10):e0255502. https://doi.org/10.1371/journal. pone.0255502

Sahrish N, Shakeel A, Ghulam A, Zartash F, Sajjad H, Mukhtar A, Muhammad AK, Ahmad K, Fahad S, Wajid N, Sezai E, Carol Jo W, Gerrit H (2022) Modeling the impact of climate warming on potato phenology. *Eur J Agro N* 132:126404.

Sajid H, Jie H, Jing H, Shakeel A, Satyabrata N, Sumera A, Awais S, Chunquan Z, Lianfeng Z, Xiaochuang C, Qianyu J, Junhua Z (2020) Rice production under climate change: Adaptations and mitigating strategies. In: Fahad S, Hasanuzzaman M, Alam M, Ullah H, Saeed M, Khan AK, Adnan M (Eds.), *Environment, Climate, Plant and Vegetation Growth.* Springer Publ Ltd, Springer Nature Switzerland, pp. 659–686. https://doi.org/10.1007/978-3-030-49732-3

Sajjad H, Muhammad M, Ashfaq A, Waseem A, Hafiz MH, Mazhar A, Nasir M, Asad A, Hafiz UF, Syeda RS, Fahad S, Depeng W, Wajid N (2019) Using GIS tools to detect the land use/land cover changes during forty years in Lodhran district of Pakistan. *Environ Sci Pollut Res* 27:39676–39692. https://doi.org/10.1007/s11356-019-06072-3

Sajjad H, Muhammad M, Ashfaq A, Fahad S, Wajid N, Hafiz MH, Ghulam MS, Behzad M, Muhammad T, Saima P (2021) Using space–time scan statistic for studying the effects of COVID-19 in Punjab, Pakistan: a guideline for policy measures in regional agriculture. *Environ Sci Pollut Res* 30:42495–42508. https://doi.org/10.1007/s11356-021-17433-2

Sajjad H, Muhammad M, Ashfaq A, Nasir M, Hafiz MH, Muhammad A, Muhammad I, Muhammad U, Hafiz UF, Fahad S, Wajid N, Hafiz MRJ, Mazhar A, Saeed AQ, Amjad F, Muhammad SK, Mirza W (2021) Satellite-based evaluation of temporal change in cultivated land in Southern Punjab (Multan region) through dynamics of vegetation and land surface temperature. *Open Geo Sci* 13:1561–1577.

Saleem M, Law AD, Sahib MR, Pervaiz ZH, Zhang Q (2018) Impact of root system architecture on rhizosphere and root microbiome. *Rhizosphere* 6:47–51.

Saleem MH, Fahad S, Adnan M, Mohsin A, Muhammad SR, Muhammad K, Qurban A, Inas AH, Parashuram B, Mubassir A, Reem MH (2020a) Foliar application of gibberellic acid endorsed phytoextraction of copper and alleviates oxidative stress in jute (Corchorus capsularis L.) plant grown in highly copper-contaminated soil of China. *Environ Sci Pollution Res* 27:37121–37133. https://doi.org/10.1007/s11 356-020-09764-3

Saleem MH, Rehman M, Fahad S, Tung SA, Iqbal N, Hassan A, Ayub A, Wahid MA, Shaukat S, Liu L, Deng G (2020b) Leaf gas exchange, oxidative stress, and physiological attributes of rapeseed (Brassica napus L.) grown under different light-emitting diodes. *Photosynthetica* 58(3):836–845.

Saleem MH, Fahad S, Shahid UK, Mairaj D, Abid U, Ayman ELS, Akbar H, Analía L, Lijun L (2020c) Copper-induced oxidative stress, initiation of antioxidants and phytoremediation potential of flax (Linum usitatissimum L.) seedlings grown under the mixing of two different soils of China. *Environ Sci Poll Res* 27:5211–5221. https://doi.org/10.1007/s11356-019-07264-7

Saman S, Amna B, Bani A, Muhammad Tahir ul Qamar, Rana MA, Muhammad SK (2020) QTL mapping for abiotic stresses in cereals. In: Fahad S, Hasanuzzaman M, Alam M, Ullah H, Saeed M, Khan AK, Adnan M (Eds.) *Environment, Climate, Plant and Vegetation Growth.* Springer Publ Ltd, Springer Nature Switzerland, pp. 229–252. https://doi.org/10.1007/978-3-030-49732-3

Sana U, Shahid A, Yasir A, Farman UD, Syed IA, Mirza MFAB, Fahad S, Al-Misned F, Usman A, Xinle G, Ghulam N, Kunyuan W (2022) Bifenthrin induced toxicity in Ctenopharyngodon idella at an acute concentration: a multi-biomarkers based study. *J King Saud Uni – Sci* 34(2022):101752.

Sasaki T, Mori IC, Furuichi T, Munemasa S, Toyooka K, Matsuoka K, Yamamoto Y (2010) Closing plant stomata requires a homolog of an aluminum-activated malate transporter. *Plant & Cell Physiol* 51(3):354–365.

Saud S, Chen Y, Long B, Fahad S, Sadiq A (2013) The different impact on the growth of cool season turf grass under the various conditions on salinity and drought stress. *Int J Agric Sci Res* 3:77–84.

Saud S, Li X, Chen Y, Zhang L, Fahad S, Hussain S, Sadiq A, Chen Y (2014) Silicon application increases drought tolerance of Kentucky bluegrass by improving plant water relations and morph physiological functions. *Sci World J* 2014:1–10. https://doi.org/10.1155/2014/ 368694

Saud S, Chen Y, Fahad S, Hussain S, Na L, Xin L, Alhussien SA (2016) Silicate application increases the photosynthesis and its associated metabolic activities in Kentucky bluegrass under drought stress and post-drought recovery. *Environ Sci Pollut Res* 23(17):17647–17655. https://doi.org/10.1007/s11356-016-6957-x

Saud S, Fahad S, Yajun C, Ihsan MZ, Hammad HM, Nasim W, Amanullah Jr, Arif M and Alharby H (2017) Effects of nitrogen supply on water stress and recovery mechanisms in Kentucky bluegrass plants. *Front Plant Sci* 8:983. doi: 10.3389/fpls.2017.00983

Saud S, Fahad S, Cui G, Chen Y, Anwar S (2020) Determining nitrogen isotopes discrimination under drought stress on enzymatic activities, nitrogen isotope abundance and water contents of Kentucky bluegrass. *Sci Rep* 10:6415. https://doi.org/10.1038/s41598-020-63548-w

Saud S, Fahad S, Hassan S (2022a) Developments in the investigation of nitrogen and oxygen stable isotopes in atmospheric nitrate. *Sustainable Chem Clim Action* 1:100003.

Saud S, Li X, Jiang Z, Fahad S, Hassan S (2022b) Exploration of the phytohormone regulation of energy storage compound accumulation in microalgae. *Food Energy Secur* 2022:e418.

Schütz L, Gattinger A, Meier M, Müller A, Boller T, Mäder P, Mathimaran N (2018) Improving crop yield and nutrient use efficiency via biofertilization – a global meta-analysis. *Front Plant Sci* 8, 2204.

Senol C (2020) The effects of climate change on human behaviors. In: Fahad S, Hasanuzzaman M, Alam M, Ullah H, Saeed M, Khan AK, Adnan M (Eds.) *Environment, Climate, Plant and Vegetation Growth.* Springer Publ Ltd, Springer Nature Switzerland, pp. 577–590. https://doi.org/10.1007/978-3-030-49732-3

Sessitsch A, Pfaffenbichler N, Mitter B (2019) Microbiome applications from lab to field: facing complexity. *Trends Plant Sci* 24(3):194–198.

Shafi MI, Adnan M, Fahad S, Fazli W, Ahsan K, Zhen Y, Subhan D, Zafar-ul-Hye M, Martin B, Rahul D (2020) Application of single superphosphate with humic acid improves the growth, yield and phosphorus uptake of wheat (Triticum aestivum L.) in calcareous soil. *Agron* (10):1224. doi:10.3390/agronomy10091224

Shah F, Lixiao N, Kehui C, Tariq S, Wei W, Chang C, Liyang Z, Farhan A, Fahad S, Huang J (2013) Rice grain yield and component responses to near 2°C of warming. *Field Crop Res* 157:98–110.

Shah S, Shah H, Liangbing X, Xiaoyang S, Shahla A, Fahad S (2022) The physiological function and molecular mechanism of hydrogen sulfide resisting abiotic stress in plants. *Brazilian J Botany* 45:563–572 https://doi.org/10.1007/s40415-022-00785-5

Siddiqui ZA, Mahmood I (1999) Role of bacteria in the management of plant parasitic nematodes: a review. *Bioresour Technol* 69(2):167–179.

Sidra K, Javed I, Subhan D, Allah B, Syed IUSB, Fatma B, Khaled DA, Fahad S, Omaima N, Ali TKZ, Rahul D (2021) Physio-chemical characterization of indigenous agricultural waste materials for the development of potting media. *J Saudi Society Agricultural Sci* 28(12):7491–7498. https://doi.org/10.1016/j.sjbs.2021.08.058

Sohn BK, Kim KY, Chung SJ, Kim WS, Park SM, Kang JG, Lee JH (2003) Effect of the different timing of AMF inoculation on plant growth and flower quality of chrysanthemum. *Scientia Hortic* 98(2):173–183.

Somers E, Vanderleyden J, Srinivasan M (2004) Rhizosphere bacterial signalling: a love parade beneath our feet. *Critical Rev Microbiol* 30(4):205–240.

Subhan D, Zafar-ul-Hye M, Fahad S, Saud S, Martin B, Tereza H, Rahul D (2020) Drought stress alleviation by ACC deaminase producing *Achromobacter xylosoxidans* and *Enterobacter cloacae*, with and without timber waste biochar in maize. *Sustain* 12(15):6286. doi:10.3390/su12156286

Suthar H, Hingurao K, Vaghashiya J, Parmar J (2017) Fermentation: a process for biofertilizer production. In: Deepak G Panpatte, Yogeshvari K Jhala, Rajababu V Vyas, Harsha N Shelat (Eds.), *Microorg Green Revolution* 229–252.

Tariq M, Ahmad S, Fahad S, Abbas G, Hussain S, Fatima Z, Nasim W, Mubeen M, ur Rehman MH, Khan MA, Adnan M (2018) The impact of climate warming and crop management on phenology of sunflower-based cropping systems in Punjab, Pakistan. *Agri and Forest Met* 256:270–282.

Terra LA, de-Soares CP, Meneses CH, Tadra Sfeir MZ, de-Souza EM, Silveira V, Schwab S (2020) Transcriptome and proteome profiles of the diazotroph Nitrospirillum amazonense strain CBAmC in response to the sugarcane apoplast fluid. *Plant & Soil* 451(1):145–168.

Thilakarathna MS, Raizada MN (2017) A meta-analysis of the effectiveness of diverse rhizobia inoculants on soybean traits under field conditions. *Soil Biol & Biochem* 105:177–196.

Thomas L, Singh I (2019) Microbial biofertilizers: types and applications. *Biofertilizers for Sustainable Agriculture and Environment* 1–19.

Timmusk S, Behers L, Muthoni J, Muraya A, Aronsson AC (2017) Perspectives and challenges of microbial application for crop improvement. *Front Plant Sci* 8:49.

Tosi M, Mitter EK, Gaiero J, Dunfield K (2020) It takes three to tango: the importance of microbes, host plant, and soil management to elucidate manipulation strategies for the plant microbiome. *Can J Microbiol* 66(7):413–433.

Treseder KK (2013) The extent of mycorrhizal colonization of roots and its influence on plant growth and phosphorus content. *Plant & Soil* 371(1):1.

Uddin MN, Ali M, Muhammad, Farooq M, Ahmad N, Jamil J, Kalsoom, Adnan M, Shah N, Khan A (2016) Characterizing microbial populations in petroleum-contaminated soils of Swat District, Pakistan. *Pol J Environ Stud* 2(4):11691–1697.

Ullah H, Khan WU, Alam M, Khalil IH, Adhikari KH, Shahwar D, Jamal Y, Jan I, Adnan M (2016) Assessment of G × E interaction and heritability for simplification of selection in spring wheat genotypes. *Can J Plant Sci* 96(3):1021–1025.

Unsar Naeem-U, Muhammad R, Syed HMB, Asad S, Mirza AQ, Naeem I, Muhammad H ur R, Fahad S, Shafqat S (2020) Insect pests of cotton crop and management under climate change scenarios. In: Fahad S, Hasanuzzaman M, Alam M, Ullah H, Saeed M, Khan AK, Adnan M (Eds.) *Environment, Climate, Plant and Vegetation Growth*. Springer Publ Ltd, Springer Nature Switzerland, pp. 367–396. https://doi.org/10.1007/978-3-030-49732-3

Van-Bruggen AH, Goss EM, Havelaar A, Van-Diepeningen AD, Finckh MR, Morris Jr JG (2019) One Health – Cycling of diverse microbial communities as a connecting force for soil, plant, animal, human and ecosystem health. *Sci Total Environ* 664:927–937.

Vargas F, González Z, Sánchez R, Jiménez L, Rodríguez A (2012) Cellulosic pulps of cereal straws as raw material for the manufacture of ecological packaging. *BioResour* 7(3):4161–4170.

Vessey JK (2003) Plant growth promoting rhizobacteria as biofertilizers. *Plant & Soil* 255(2):571–586.

Vujanovic V, Germida JJ (2017) Seed endosymbiosis: a vital relationship in providing prenatal care to plants. *Can J Plant Sci* 97(6):972–981.

Wahid F, Fahad S, Subhan D, Adnan M, Zhen Y, Saud S, Manzer HS, Martin B, Tereza H, Rahul D (2020) Sustainable management with mycorrhizae and phosphate solubilizing bacteria for enhanced phosphorus uptake in calcareous soils. *Agri* 10(8):334. doi:10.3390/agriculture10080334

Wajid N, Ashfaq A, Asad A, Muhammad T, Muhammad A, Muhammad S, Khawar J, Ghulam MS, Syeda RS, Hafiz MH, Muhammad IAR, Muhammad ZH, Muhammad Habib ur R, Veysel T, Fahad S, Suad S, Aziz K, Shahzad A (2017) Radiation efficiency and nitrogen fertilizer impacts on sunflower crop in contrasting environments of Punjab. *Pakistan Environ Sci Pollut Res* 25:1822–1836. https://doi.org/10.1007/s11356-017-0592-z

Wiqar A, Arbaz K, Muhammad Z, Ijaz A, Muhammad A, Fahad S (2022) Relative efficiency of biochar particles of different sizes for immobilising heavy metals and improving soil properties. *Crop Pasture Sci* 42(2):112–120 https://doi.org/10.1071/CP20453.

Wu C, Kehui C, She T, Ganghua L, Shaohua W, Fahad S, Lixiao N,Jianliang H, Shaobing P, Yanfeng D (2020) Intensified pollination and fertilization ameliorate heat injury in rice (Oryza sativa L.) during the flowering stage. *Field Crops Res* 252:107795.

Wu C, Tang S, Li G, Wang S, Fahad S, Ding Y (2019) Roles of phytohormone changes in the grain yield of rice plants exposed to heat: a review. *PeerJ* 7:e7792. doi 10.7717/peerj.7792

Xavier LJ, Germida JJ (2003) Bacteria associated with Glomus clarum spores influence mycorrhizal activity. *Soil Biol & Biochem* 35(3):471–478.

Xue B, Huang L, Li X, Lu J, Gao R, Kamran M, Fahad S (2022) Effect of clay mineralogy and soil organic carbon in aggregates under straw incorporation. *Agron* 12:534. https://doi.org/10.3390/ agronomy12020534

Yang R, Dai P, Wang B, Jin T, Liu K, Fahad S, Harrison MT, Man J, Shang J, Meinke H, Deli L, Xiaoyan W, Yunbo Z, Meixue Z, Yingbing T, Haoliang Y (2022) Over-optimistic projected future wheat yield potential in the North China Plain: the role of future climate extremes. *Agron* 12:145. https://doi.org/10.3390/ agronomy12010145

Yang Z, Zhang Z, Zhang T, Fahad S, Cui K, Nie L, Peng S, Huang J (2017) The effect of season-long temperature increases on rice cultivars grown in the central and southern regions of China. *Front Plant Sci* 8:1908. https://doi.org/10.3389/fpls.2017.01908

Yao C, Moreshet S, Aloni B (2001) Water relations and hydraulic control of stomatal behaviour in bell pepper plant in partial soil drying. *Plant, Cell & Environ* 24(2):227–235.

Yoon V, Tian G, Vessey JK, Macfie SM, Dangi OP, Kumer AK, Tian L (2016) Colonization efficiency of different sorghum genotypes by Gluconacetobacter diazotrophicus. *Plant & Soil* 398(1):243–256.

Zafar-ul-Hye M, Muhammad N, Subhan D, Fahad S, Rahul D, Mazhar A, Ashfaq AR, Martin B, Jiˇrí H, Zahid HT, Muhammad N (2020a) Alleviation of cadmium adverse effects by improving nutrients uptake in bitter gourd through cadmium tolerant rhizobacteria. *Environ* 7(8), 54. doi:10.3390/environments7080054

Zafar-ul-Hye M, Muhammad T ahzeeb-ul-Hassan, Muhammad A, Fahad S, Martin B, Tereza D, Rahul D, Subhan D (2020b) Potential role of compost mixed biochar with rhizobacteria in mitigating lead toxicity in spinach. *Scientific Rep* 10:12159. https://doi.org/10.1038/s41598-020-69183-9.

Zafar-ul-Hye M, Akbar MN, Iftikhar Y, Abbas M, Zahid A, Fahad S, Datta R, Ali M, Elgorban AM, Ansari MJ et al. (2021) Rhizobacteria inoculation and caffeic acid alleviated drought stress in lentil plants. *Sustain* 13:9603. https://doi.org/10.3390/su13179603

Zahida Z, Hafiz FB, Zulfiqar AS, Ghulam MS, Fahad S, Muhammad RA, Hafiz MH,Wajid N, Muhammad S (2017) Effect of water management and silicon on germination, growth, phosphorus and arsenic uptake in rice. *Ecotoxicol Environ Saf* 144:11–18.

Zahir SM, Zheng-HG, Ala Ud D, Amjad A, Ata Ur R, Kashif J, Shah F, Saud S, Adnan M, Fazli W, Saud A, Manzer HS, Shamsher A, Wajid N, Hafiz MH, Fahad S (2021) Synthesis of silver nanoparticles using Plantago lanceolata extract and assessing their antibacterial and antioxidant activities. *Sci Rep* 11:20754.

Zaman I, Ali M, Shahzad K, Tahir, MS, Matloob A, Ahmad W, Alamri S, Khurshid MR, Qureshi MM, Wasaya A, Khurram SB, Manzer HS, Fahad S, Rahul D (2021) Effect of plant spacings on growth, physiology, yield and fiber quality attributes of cotton genotypes under nitrogen fertilization. *Agron* 11:2589. https:// doi.org/ 10.3390/agronomy11122589

Zia-ur-Rehman M (2020) Environment, climate change and biodiversity. In: Fahad S, Hasanuzzaman M, Alam M, Ullah H, Saeed M, Khan AK, Adnan M (Eds.), *Environment, Climate, Plant and Vegetation Growth*. Springer Publ Ltd, Springer Nature Switzerland, pp. 473–502. https://doi.org/10.1007/ 978-3-030-49732-3

8 Limitations of Using Biofertilizers as an Alternative to Chemical Fertilizers

Muhammad Haroon,[1] Fazli Wahid,[1] Rafi Ullah,[1] Muhammad Adnan,[1] Mukhtar Alam,[1] Hidayat Ullah,[1] Muhammad Saeed,[1] Muhammad Saeed,[2] Shah Fahad,[3]* Muhammad Romman,[4] Nazeer Ahmed,[1] Taufiq Nawaz,[5] Anas Iqbal,[6] Zia Ur Rehman,[1] Ayman El Sabagh,[7] Shah Saud,[8] and Shah Hassan[9]*

[1] Department of Agriculture, University of Swabi, Swabi, Pakistan
[2] Department of Weed Science and botany, University of Agriculture, Peshawar, Pakistan
[3] Department of Agriculture, Abdul Wali Khan University Mardan, Khyber Pakhtunkhwa, Pakistan
[4] Department of Botany, University of Chitral, Khyber Pakhtunkhwa, Pakistan
[5] Department of Food Science and Technology, The University of Agriculture, Peshawar, Pakistan
[6] College of Agriculture, Guangxi University, Nanning, China
[7] Department of Agronomy, Faculty of Agriculture, University of Kafrelsheikh, Kafr El-Sheikh, Egypt
[8] College of Life Science, Linyi University, Linyi, Shandong, China
[9] Department of Agricultural Extension Education and Communication, The University of Agriculture, Peshawar, Pakistan
* Correspondence: madnan@uoswabi.edu.pk; shah_fahad80@yahoo.com

CONTENTS

DOI: 10.1201/9781003286233-8

8.1 INTRODUCTION

The agriculture sector plays an important role in improving the standard of living in a country. Hence, it is necessary to ensure food security and access for the global population. The agriculture sector has experienced a massive change (Ajmal et al. 2016). Today, the agriculture sector not only improves food supply but also improves the living standard of all. To fulfil this demand, farmers have used an excessive amount of agrochemicals to enhance crop growth and productivity (Aktar et al. 2009; Santos et al. 2012). Agrochemicals are synthetic products that contain essential elements, e.g. nitrogen, potassium, and phosphorus. Nowadays, to increase soil fertility, farmers mostly use synthetic fertilizers, which cause numerous environmental issues, e.g. health problems, air, soil, and water pollution, climate change, loss of beneficial biodiversity, etc. In long term, they may destroy soil fertility (Bhardwaj et al. 2014; Khan et al. 2016). According to Chun-Li (2014), agrochemical applications increase soil acidity and result in environmental pollution, with plants also becoming susceptible to many diseases (Khosro and Yousef 2012). Researchers have developed a new approach in the formulation of biofertilizers, which keep the soil rich in all nutrients and protect the soil from pollution.

Biofertilizers are the most essential part of integrated nutrients management that play a vital role in productivity and soil sustainability when added to soil. Biofertilizers do not supply nutrients directly to the plant, they use strains of efficient beneficial microbes added to soil, seeds, and compost to accelerate the process and to enhance nutrient availability to plants that help in plant growth (Yadav and Sarkar 2019). Biofertilizers include nitrogen-fixing bacteria (*Rhizobium*, *Azospirillum*, cyanobacteria, *Clostridium*, *azotobacter*, and *Bacillus* polymyxin), phosphate-solubilizing bacteria (*Aspergillus*, *Bacillus*, *Fusarium*, etc.), potassium-solubilizing bacteria (*Bacillus edaphicus*, *Bacillus mucilagenosus*), sulphur-solubilizing bacteria (chemolithotrophs, *Thiobacillus denitrificans*, etc.) and arbuscular mycorrhiza (Gautam et al. 2021). These microbes are used in different combinations for the preparation of biofertilizer formulations for enhancing crop productivity. Biofertilizers are an eco-friendly approach that help to enhance soil fertility, sustainability, and plant growth and productivity (Mohapatra et al. 2013). The above-mentioned aspects aid farmer incomes through a noticeable reduction in the cost of agrochemicals (Fundases 2005). Biofertilizer application can be a feasible option with many social, economic, and environmental benefits (Carvajal-Munoz and Carmona-García 2012). However, biofertilizer implementation needs more practice, studies of environmental variables, assets, and time in research (Vanegas 2003; Fresco 2003). In order to achieve sustainability, it is necessary to follow an implementation plan, business plan, project plan, and production plan, and to implement sustainable technology toward the minimization of environmental impacts and improved subsequent profits for growers.

8.2 ADVANTAGES OF BIOFERTILIZERS

The advantages of biofertilizer use include the following:

1. Enhances crop yield and soil health.
2. Safe to the environment and health.

3. Provides basic nutrients, e.g. NPK to plants.
4. Maintains soil's natural fertility and microbial biodiversity.
5. Helps to control plant and soil pathogens.
6. Helps to increase resistance against harsh environmental conditions.
7. Easy to apply, cost-effective, and affordable to farmers.
8. Increases plant growth and productivity.

8.3 SOURCES OF BIOFERTILIZERS

8.3.1 Nitrogen-based Biofertilizer

Nitrogen is an important crop nutrient. However, it is deficient in most soils. To meet crop demand, nitrogen-based biofertilizer is a new approach that is used worldwide. It contains beneficial microbes such as *Azotobactor*, *Azospirillum*, *Actinobacteria*, and *Rhizobium* (El-Sirafi et al. 2005). It helps to lower the requirement for nitrogen fertilizer by up to 35% (Favilli et al. 1987).

8.3.2 Phosphorus-based Biofertilizer

Phosphorus is an important plant growth-limiting factor. Its deficiency may disturb plant growth and other biological process (Azziz et al. 2012; Tak et al. 2012; Ilyas et al. 2016). Phosphorus-solubilizing microbes help to improve the soil nutrition status and also increase its availability to plants through biological processes. Phosphorus biofertilizer is a possible way to increase the availability of phosphorus to plants. In an ecosystem many beneficial microbes are involved in releasing phosphorus, these are called phosphorus-solubilizing microorganisms (Bhattacharyya and Jha 2012). Bacteria and fungi have been found to be effective in phosphorus solubilization, with bacteria contributing up to 50%; while the fungal potential is very low, i.e. 0.1 to 0.5% (Alam et al. 2002; Chen et al. 2006). Phosphate-solubilizing bacteria are a group of plant growth promotors reported in many crops including rice, tomato, etc. (Rodriguez and Fraga 1999; Hafeez et al. 2004; Hilda and Fraga 2000).

8.3.3 Compost-based Biofertilizer

Compost biofertilizers are those containing different microorganisms that have the potential to convert organic waste material from unavailable to available forms to enrich soil with important elements through biological activities (Vessey 2003; Ebrahimi et al. 2010). The application of compost improves soil porosity, fertility, cation exchange capacity, water retention capacity, and microbial activities (Ahmad et al. 2008; Fiorentino and Fagnano 2011).

8.4 CROP RESPONSE TO BIOFERTILIZER

Biofertilizer is a sustainable tool used to enhance soil health and crop yield (Igiehon and Babalola 2017; Bhattacharrya et al. 2020). The incorporation of biofertilizer is a crop management practice that helps to improve yield-attributing characteristics and soil fertility. Biofertilizers impart better health to soil and plants. *Azolla* is a common biofertilizer used in paddy fields as it is quickly decomposed in soil and enables efficient nitrogen availability to rice crop. *Azospirillum lipoferum* produces chemicals which stimulate plant growth and enhance productivity (Yadav and Gautam 2019). Likewise, Park et al. (2015) reported that PGPR has a significant effect on crop aerial growth, i.e., number of leaves, plant height, shoot biomass, and stem width. Also, Adiprasetyo et al. (2014) observed that the application of multi-microbial biofertilizer was able to increase the growth-related parameters of oil palm plants. Yang et al. (2014) detected increased water use efficiency in arbuscular mycorrhiza-treated seedlings. Some beneficial microorganisms (*Bacillus megaterium* and *Frateuria aurantia*) have been reported as effective phosphorus- and potassium-mobilizing bacteria in crop cultivation (Subhashini 2014; Ghaffari et al. 2018). It has also been reported that adding biofertilizer can efficiently increase *Angelica dahurica* growth and yield (Doan and Cao 2020).

8.5 RESPONSE OF SOIL FERTILITY TO BIOFERTILIZER

Soil fertility is an important function of soil that provides the essential nutrients to soil. Crop growth is mainly dependent on soil fertility status. As decreasing of soil fertility occurrs continuously, to overcome this problem, it is necessary to boost the soil condition and fertility. Among the soil management practices, biofertilizer is a good approach to manage soil nutrients. Moreover, biofertilizers are an ecofriendly approach that helps to increase the soil properties, i.e. macro- and micronutrients, increase soil pH, ion exchange capacity, and biological properties (Kariada et al. 2003). In addition, microbial activities help in the availability and recovery of nutrients, hence they increase soil quality (Upadhyay et al. 2012a; Arafat et al. 2016; Yadav and Sarkar 2019). Biofertilizers have the ability to improve soil fertility by delaying the nitrification process for a long time.

8.6 LIMITATIONS OF BIOFERTILIZERS

Biofertilizers are nutrient-based organic fertilizers that are mostly used as a supplement to synthetic fertilizers. However, they need improvement to replace synthetic fertilizers totally. There are many limitations of biofertilizer use that need studies and research to find a better solution for the future.

8.6.1 Agri-climatic Conditions

Many factors affect agriculture worldwide. The use of synthetic fertilizers hardens the soil, decreases soil productivity, increases salinity, with irregularity in soil nutrients and loss of water-holding capacity (Savci 2012). To stop the use of synthetic fertilizers, biofertilizers have emerged as alternative and effective fertilizers. Biofertilzers can be a good option for basic inputs of nutrients for sustainable production with more ecofriendly approaches (Al-Taey 2018). However, the performances of biofertilizers are inconsistent due to many factors. Among these factors, the most limiting are the agro-environmental conditions (Mathenge et al. 2019). Agroclimatic conditions are the essential factor that influences the performance of a biofertilizer. Results have been reported that agroclimatic conditions greatly affect the soil pH, and reduce phosphorus and the nodulation process (Rengel 2002). Likewise, the maximum phosphorus availability was reported to be very low in mineral soil (Wolf 1999). The reaction of phosphate with other elements like iron, calcium, and aluminium may reduce its availability to plants (Gyaneshwar et al. 2002; Ali et al. 2013). Although the application of biofertilizers may increase the input by up to 35% in saline and acidic soil, it is less effective in neutral and calcareous soil (Hashem 2001). A recent report has shown that may beneficial microbes such as *Bacillus*, *Azotobacter*, *Enterobacter*, and *Paenibacillus* help in minimizing the agrochemical toxicity (Shahid et al. 2019).

However, its efficacy depends on whether the strain can survive and also sufficient knowledge about the soil pH for each type of biofertilizer is needed.

Agrochemicals are widely used in agriculture for managing different pests and disease. The high use of agrochemicals damages the agroecosystem in terms of soil microbial biodiversity. The massive use of fertilizers also disturbs microbial activity and biodiversity (Stolte 2016). Likewise herbicides also cause everlasting changes to the microflora of soil, also affecting soil health and crop growth by inhibiting the activity of nitrogen-fixing bacteria (Sachin 2009) and interfering with ammonification (Reinhardt et al. 2008).

8.6.2 Inconsistent Nature

Over the past few years, significant success has been recorded for biofertilizers which offer a successful and natural option in the agriculture sector. Biofertilizers are gaining increased awareness each day with their role in improving plant health, productivity, and soil fertility. However, they are not widely accepted due to difficulty in reproducing their useful outcomes on natural ecosystem swhere there are variations in the environmental conditions.

Extreme environmental conditions often cause a loss of their effectiveness. Highly alkaline and acidic soils also affect the growth of useful microbes, which has a great influence if the soil contains excessive natural enemies. It also affects its efficiency in soil, which depends on the particular strain availability, growing medium, and specific environment.

8.6.3 Formulation and Storage Problems

Formulation is an important step in maintaining microorganism viability, to provide a long shelf life and ensure that microorganisms establish themselves (Soumare et al. 2019). Formulation is a major step that provides optimal conditions to enhance the activity and persistence of microorganisms in soil to obtain maximum benefits for the host plants (McQuilken et al. 1998). A good bioformulation should be non-toxic, inexpensive, easily available, with good adhesion to seeds and have a good pH to be recommended as a carriers (Hassan and Bano 2016). A short shelf life and inadequate formulation discourage sellers and farmers because they cannot store the product for a long time. A problem for the production of biofertilizers is the low demand in the market and slow adoption rate. Thus it is important to improve the formulation and shelf life of locally formulated biofertilizers in various condition to ensure their significance, activity, persistence, and product viability over a period time.

8.6.4 Nutrient Availability Constraints

Plants are continuously exposed to many microorganisms and interact with plants differently (Shah et al. 2017; Glick and Gamalaro 2021). These plant–microbe interactions are important and have been used to improve crop yield and production. These activities help to improve soil physical properties, structure, access to nutrients, and resistance against pathogens (Yadav et al. 2011). Beneficial bacteria provide important nutrients to improve crop growth (Basu et al. 2021). Any nutrient deficiencies may lead to malfunction and growth imbalance in plant growth. Kumar et al. (2019) reported that zinc deficiency in wheat crop causes yellowing of leaves and stunted growth. Many beneficial microbes are involved in fixing zinc availability to plants (Sindhu et al. 2019a). Nutrient deficiencies such as phosphorus and molybdenum negatively affect the growth and population of rhizobia (Giller 2001). Many environmental factors are involved that affect nutrient availability, composition, and microbial diversity (Hartmann et al. 2017; Ahmad et al. 2016). Biofertilizers must be kept at room temperature or cold storage away from sunlight and heat (Mishra and Dadluck 2010). Abiotic stress can change the rhizosphere and may affect the potential, survival, and diversity of these beneficial microbes (Etesami 2020). For any biofertilizer product to produce the desired effects on crop growth and production, it must be well formulated and a suitable source of all the essential nutrients to support plant growth.

8.6.5 Extension and Market Constraints

Biofertilizers are an emerging industry and market growth is extremely important. However, biofertilizers face many challenges due high market competition. Therefore, it is necessary to develop the right marketing strategy to maintain and increase the demand for biofertilizers. Therefore, the growth prospects of the biofertilizer market and its product availability are equally important (Giazinnazi and Vosatka 2004; Tariq et al. 2013). A poorly developed marketing strategy will certainly have a negative impact with limited involvement of customers and the private sector. A proper marketing plan is necessary to get consumer satisfaction. Consumers can gain awareness of biofertilizers through active marketing plans, research, and the availability of products (Banayo et al. 2012).

8.6.6 Farmer Awareness

Challenges common to biofertilizers include inadequate biofertilizer awareness among farmers. It is necessary to explore the potential of biofertilizers for crops and the environment. It is need of the day that farmers should be aware of their sources and bioinoculants. Inadequate awareness may lead to low adoption of biofertilizers by farmers. Therefore, it is necessary to educate farmers on the societal and ecological benefits of biofertilizers and to improve the application of biofertilizers by growers. Inadequate information and poor awareness may lead to low development and poor acceptance of biofertilizers among growers.

8.7 FUTURE PROSPECTS FOR BIOFERTILIZERS

Growing crops sustainably allows growers to protect the soil and increase crop productivity in a way that is less harmful to the environment. Sustainable crop management approaches help to minimize the impact on the natural habitat, improve soil health, and protect water resources. This can be achieved only by implementing sustainable management methods in agriculture. Biofertilizer application is the most important part of sustainable agriculture and is a good option as an alternative to chemical fertilizers. However, there are many challenges in the preparation and development of biofertilizers, and additional research is needed to support the promotion of biofertilizer use. In addition, public and private sectors, and commercial investors have the potential to improve the formulation and sale of biofertilizer substitutes to replace synthetic fertilizers. Thus, biofertilizer development faces many challenges that pose many threats.

New interest in biofertilizers to increase productivity and reduce health, environmental and commercial pressures for agrochemicals have caused significant advances in the research into biofertilizers. Using modern approaches, the efficacy of biofertilizers can be improved. Moreover, advances in production technology lead to a resuction in the cost of products. However, there remain many challenges that need to be solved to reflect the full potential for commercial biofertilizer production.

It is very important to develop integrated nutrient programs that can offer a good option and economic choices to favour the use of biofertilizers. Synthetic fertilizers should be used at a minimal level to reduce environmental pollution and support the use of biofertilizers. Different approaches should be developed to protect soil fertility, and to avoid health and environmental risks. Biofertilizer efficiency can be enhanced by using different approaches such as advances in technology and formulation.

REFERENCES

Adiprasetyo T, Purnomo B, Handajaningsih M, Hidayat H (2014) The usage of BIOM3G-Biofertilizer to improve and support sustainability of land system of independent oil palm smallholders. *Int J Adv Sci Eng Inf Technol* 4:345–348.

Ahmad R, Shehzad SM, Khalid A, Arshad M, Mahmood MH (2008) Growth and yield response of wheat and maize to nitrogen and L tryptophan enriched compost. *Pak J Bot* 39(2):541–549.

Ahmad W, Hassan G, Ali M, Khan N, Ishaq M, Afridi K, Shah IA, Adnan M, Tajudin R, Ullah L, Rehman AU, Hussain B (2016) Estimates of heritability, genetic advance and correlation in F3 populations of wheat. *Pure Appl Biol* 5(4):1142–1150.

Ajmal M, Ali HI, Saeed R, Akhtar A, Tahir M, Mehboob MZ, Ayub A (2016) Biofertilizer as an alternative for chemical fertilizers. *J Agric Allied Sci* 7(1):1–7.
Aktar MW, Sengupta D, Chowdhury A (2009) Impact of pesticides use in agriculture: their benefits and hazards. *Interdiscip Toxicol* 2:1–12.
Alam S, Khalil S, Ayub N, Rashid M (2002) In vitro solubilization of inorganic phosphate by phosphate solubilizing microorganism (PSM) from maize rhizosphere. *Int J Agric Biol* 4:454–458.
Ali K, Arif M, Khan Z, Tariq M, Waqas M, Gul B, Bibi S, Zia-ud-Din, Ali M, Shafi B, Adnan M (2013) Effect of cutting on productivity and associated weeds of canola. *Pak J Weed Sci Res* 19(4):393–401.
Al-Taey DKA, Majid ZZ (2018) Study effect of kinetin, bio-fertilizers and organic matter application in lettuce under salt stress. *J Glob Pharma Technol* 10:148–164.
Arafat Y, Shafi M, Khan MA, Adnan M, Basir A, Rahman IU, Arshad M, Khan A, Saleem N, Romman M, Rahman Z, Shah JA (2016) Yield response of wheat cultivars to zinc application rates and methods. *Pure Appl Biol* 5(4):1260–1270.
Azziz G, Bajsa N, Haghjou, T, Taulé, C, Valverde A, Igual, J, Arias A (2012) Abundance, diversity and prospecting of culturable phosphate solubilizing bacteria on soils under crop–pasture rotations in a no-tillage regime in Uruguay. *Appl Soil Ecol* 61:320–326.
Banayo NPM, Cruz PCS, Aguilar EA, Badayos RB, Haefele SM (2012) Evaluation of biofertilizers in irrigated rice: effects on grain yield at different fertilizer rates. *Agric* 2:73–86.
Basu A, Prasad P, Das SN, Kalam S, Sayyed RZ, Reddy MS, El-Enshasy H (2021). Plant growth promoting rhizobacteria (PGPR) as green bioinoculants: recent developments, constraints, and prospects. *Sustainability* 13:1140.
Bhardwaj D, Ansari MW, Sahoo RK, Tuteja N (2014) Biofertilizers function as key player in sustainable agriculture by improving soil fertility, plant tolerance and crop productivity. *Microbial Cell Factories* 13:66.
Bhattacharrya C, Roy R, Tribedi P, Ghosh A, Ghosh A (2020) Biofertilizers as substitute to commercial agrochemicals. *Pesticide and Chemical Fertilizer* 263–290.
Bhattacharyya PN, Jha DK (2012) Plant growth-promoting rhizobacteria (PGPR): emergence in agriculture. *World J Microbiol Biotechnol* 28:1327–1350.
Carvajal-Munoz JS, Carmona-García CE (2012) Benefits and limitations of biofertilization in agricultural practices. *Livest Res Rural Dev* 24(3), 1–8.
Chen YP, Rekha PD, Arun-shen AB, Lai WA, Young CC (2006) Phosphate solubilizing bacteria from subtropical soil and their tricalcium phosphate solubilizing abilities. *Appl Soil Ecol* 34:33–41.
Chun-Li W (2014) Present situation and future perspective of biofertilizer for environmentally friendly agriculture. Annal Reports, 1–5. Daily Trust News Paper, 9th September 2016, pp. 11.
Doan HTT, Cao V (2020) Effects of additional bio-fertilizer on the growth and yield of Angelica dahurica. *Asian J Plant Sci* 19:63–67.
Ebrahimi A, Moaveni P, Aliabadi Farahani H (2010) Effects of planting dates and compost on mucilage variations in borage (Borago officinalis L.) under different chemical fertilization systems. *Int J Biotechnol Mol Biol Res* 1(5):58–61.
El-Sirafy HJ, Woodard, El-Noriar EM (2005) Contribution of biofertilizers and fertilizer nitrogen to nutrient uptake and yield of Egyptian winter wheat. *J Plant nutr* 24(4):587–599.
Etesami H (2020) Plant–microbe interactions in plants and stress tolerance. In: *Plant Life Under Changing Environment*. Elsevier BV: Amsterdam, The Netherlands, pp. 355–396.
Favilli F, Balloni W, Cappellini A, Granchi L, Savoini G (1987) Esperienze pluriennali di batterizzazione campo con *Azospirillum* spp. di coltore cereasicole. [Longterm field experiments' spreading *Azospirillum* spp. on cereal crops.] *Annals of Microbiol* 37:169–181.
Fiorentino N, Fagnano M (2011) Soil fertilization with composted solid waste: short term effects on lettuce production and mineral N availability. *Geophysic Res Abs* 13:10520.
Fresco L (2003) Fertilize the plant, not the soil. *UN Chronicle* 40(3):62.
Fundases Y, Asocolflores (2005) Biofertilizantes como alternativa nutricional en ornamentales. Revista Virtual Pro. www.revistavirtualpro.com/revista/index.php? ed=2006-10-01&pag=11
Gautam K, Sirohi C, Singh NR, Thakur Y, Jatav SS, Rana K, Chitara M, Meena RP, Singh AK, Parihar M (2021) Microbial biofertilizer: types, applications, and current challenges for sustainable agricultural production. *Advs Bio-Inoculants* 1:3–19.

Ghaffari H, Gholizadeh A, Biabani A, Fallah A, Mohammadian M (2018) Plant growth promoting rhizobacteria (PGPR) application with different nitrogen fertilizer levels in rice (*Oryza sativa* L.). *Pertanika J Trop Agric Sci* 41(2):715–728.

Gianinnazi S, Vosatka M (2004) Inoculum of arbuscular mycorrhizal fungi for production systems: Science meets business. *Canadian J Bot* 82(8):1264–1271.

Giller KE (2001) *Nitrogen Fixation in Tropical Cropping Systems.* CABI Publishing, Wallingford, UK.

Glick BR, Gamalaro E (2021) Recent developments in the study of plant microbiomes. *Microorganisms* 9:153.

Gyaneshwar P, Naresh KG, Parekh LJ, Poole PS (2002) Role of soil microorganisms in improving P nutrition of plants. *Plant & Soil* 245:83–93.

Hafeez FY, Safdar ME, Chaudhry AU, Malik KA (2004) Rhizobial inoculation improves seedling emergence, nutrient uptake, and growth of cotton. *Aust J Exp Agric* 446:617–622.

Hartmann M, Brunner I, Hagedorn F, Bardgett RD, Stierli B, Herzog C, Chen X, Zingg A, Graf-Pannatier E, Rigling, A, Frey B (2017) A decade of irrigation transforms the soil microbiome of a semi-arid pine forest. *Mol Ecol* 26:1190–1206.

Hashem MA (2001) Problems and prospects of cyanobacterial biofertilizer for rice cultivation. *Aust J Plant Physiol* 28(9):881–888.

Hassan TU, Bano A (2016) Biofertilizer: a novel formulation for improving wheat growth, physiology and yield. *Pak J Bot* 48(6):2233–2241.

Hilda R, Fraga R (2000) Phosphate solubilizing bacteria and their role in plant growth promotion. *Biotech Adv* 17:319–359.

Igiehon NO, Babalola OO (2017) Biofertilizers and sustainable agriculture: exploring arbuscular mycorrhizal fungi. *Appl Microbiol Biotechnol* 101:4871–4881.

Ilyas H, Alam SS, Khan I, Shah B, Naeem A, Khan N, Ullah W, Adnan M, Shah SRA, Junaid K, Ahmed N, Iqbal M (2016) Medicinal plants rhizosphere exploration for the presence of potential biocontrol fungi. *J Ento & Zool Studies* 4(3):108–113.

Khan A, Yousafzai AM, Shah N, Muhammad, Ahmad MS, Farooq M, Aziz F, Adnan M, Rizwan M, Jawad SM (2016) Enzymatic profile aactivity of grass carp (*Ctenopharyngodon idella*) after exposure to the pollutant named atrazine (herbicide). *Pol J Environ Stud* 25(5):2001–2006.

Khosro M, Yousef S (2012) Bacterial bio-fertilizers for sustainable crop production: A review. *ARPN J Agric Biol Sci* 7(5):237–308.

Kumar A, Dewangan S, Lewate P, Bahadar I, Parajapati S (2019) Zinc solubilizing bacteria: a boon for sustainable agriculture. In: Sayyed RZ, Arora NK, Reddy MS (Eds.), *Plant Growth Promoting Rhizobacteria for Sustainable Stress Management.* Springer, pp: 139–155.

Mathenge C, Thuita M, Masso C, Gweyi-Onyango J, Vanlauwe B (2019) Variability of soybean response to rhizobia inoculant, vermicompost, and a legume-specific fertilizer blend in Siaya County of Kenya. *Soil Tillage Res* 194:104290.

McQuilken MP, Halmer P, Rhodes DJ (1998) Application of microorganisms to seeds. In: Burges HD (ed) *Formulation of Microbial Biopesticides.* Springer, Berlin, pp. 255–285.

Mishra BK, Dadhick SK (2010) *Methodology of nitrogen bio-fertilizer production unpublished B.Sc. Thesis,* Department of Molecular and Biotechnology, RCA Udaipur, pp. 4–16.

Mohapatra SK, Munsi PS, Mohapatra PN (2013) Effect of integrated nutrient management on growth, yield and economics of broccoli (Brassica oleracea var. italica Plenck). *Vegetable Science* 40(1):69–72.

Park YS, Park K, Kloepper JW, Ryu CM (2015) Plant growth-promoting Rhizobacteria stimulate vegetative growth and asexual reproduction of Kalanchoe daigremontiana. *Plant Pathol J* 31:310–315.

Reinhardt EL, Ramos PL, Manfio GP, Barbosa HR, Pavan C, Filho CAM (2008) Molecular characterization of nitrogen-fixing bacteria isolated from Brazilian agricultural plants at also Paulo state. *Braz J Microbiol* 39:414–422.

Rengel Z (2002) Breeding for better symbiosis. *Plant and Soil* 245:147–162.

Rodriguez H, Fraga R (1999) Phosphate solubilizing bacteria and their role in plant growth promotion. *Biotechnol Adv* 17:319–339.

Sachin DN (2009) Effect of *Azotobacter chroococcum* (PGPR) on the growth of bamboo (*Bambusa bamboo*) and maize (*Zea mays*) plants. *Biofront* 1:24–31.

Santos VB, Araujo SF, Leite LF, Nunes LA, Melo JW (2012) Soil microbial biomass and organic matter fractions during transition from conventional to organic farming systems. *Geoderma* 170:227–231.

Savci S (2012) An agricultural pollutant: chemical fertilizer Int. *J Environ Sci Dev* 3:73.

Shah N, Tanzeela, Khan A, Khisroon M, Adnan M, Jawad SM (2017) Seroprevalence and risk factors of toxoplasmosis in pregnant women of district Swabi. *Pure Appl Biol* 6(4):1306–1313.
Shahid M, Zaidi A, Ehtram A, Khan MS (2019) In vitro investigation to explore the toxicity of different groups of pesticides for an agronomically important rhizosphere isolate Azotobacter vinelandii. *Pest Biochem Physiol* 157:33–44.
Sindhu SS, Sharma R, Sindhu S, Phour M (2019a) Plant nutrient management through inoculation of zinc solubilizing bacteria for sustainable agriculture. In: Giri B, Prasad R, Wu QS, Verma A (Eds.) *Biofertilizers for Sustainable Agriculture and Environment*, Springer Nature Springer, Pte Ltd. pp. 173–201.
Soumare A, Boubekri K, Lyamlouli K, Hafidi M, Ouhdouch Y, Kouisni L (2019) From isolation of phosphate solubilizing microbes to their formulation and use as biofertilizers: Status and needs. *Front Bioengin Biotechnol* 7:425.
Stolte J, Tesfai M, Keizer J (2016) *Soil Threats in Europe: Status, Methods, Drivers and Effects on Ecosystem Services*. European Commission DG Joint Research Centre; Brussels, Belgium.
Subhashini DV (2014) Growth promotion and increased potassium uptake of tobacco by potassium-mobilizing bacterium *Frateuria aurantia* grown at different potassium levels in vertisols. *Commun Soil Sci Plan Anal* 46(2):210–220.
Tak HI, Ahmad F, Babalola OO, Inam A (2012) Growth, photosynthesis and yield of chickpea as influenced by urban wastewater and different levels of phosphorus. *Int J Plant Res* 2:6–13.
Tariq M, Khan Z, Arif M, Ali K, Waqas M, Naveed K, Ali M, Khan MA, Shafi B, Adnan M (2013) Effect of nitrogen application timings on the seed yield of Brassica cultivars and associated weeds. *Pak J Weed Sci Res* 19(4):493–502.
Upadhyay SK, Singh JS, Saxena AK, Singh DP (2012) Impact of PGPR inoculation on growth and antioxidant status of wheat under saline conditions. *Plant Biol* 14(4):605–611.
Vanegas R (2003) La transición: la búsqueda del cambio hacia sistemas sustentables de producción agropecuaria. Foro Agrario "Sistemas Agrícolas Sustentables". *Quito: Universidad Central del Ecuador*, pp. 18–19.
Vessey JK (2003) Plant growth promoting rhizobacteria as biofertilizers. *Plant Soil* 255:571–586.
Wolf B (Ed.) (1999). The Fertile Triangle: the interrelationship of air, water, and nutrients. In: *Maximizing Soil Productivity*, The Haworth Press Inc.
Yadav J, Verma JP, Tiwari KN (2011) Plant growth promoting activities of fungi and their effect on chickpea plant growth. *Asian J Biol Sci* 4(3): 291–299.
Yadav KK, Sarkar S (2019) Biofertilizers, impact on soil fertility and crop productivity under sustainable agriculture. *Environ Ecol* 37(1):89–93.
Yang Y, Tang M, Sulpice R, Chen H, Tian S, Ban Y (2014) Arbuscular mycorrhizal fungi alter fractal dimension characteristics of *Robinia pseudoacacia* L. seedlings through regulating plant growth, leaf water status, photosynthesis, and nutrient concentration under drought stress. *J Plant Growth Regulion* 33: 612–625.

9 Phosphorus-solubilizing Bio-fertilizers

Kinza Iqbal,[1] *Sajid Masood,*[1] *Qaiser Hussain,*[1*]
Rabia Khalid,[1] *Khalid Saifullah Khan,*[1]
Muhammad Akmal,[1] *Shahzada Sohail Ijaz,*[1]
Muhammad Jamil,[2] *Muhammad Azeem,*[1] *and Servat Jehan*[1]
[1] Institute of Soil & Environmental Sciences, PMAS-Arid Agriculture University, Rawalpindi, Pakistan
[2] Soil and Water Testing Laboratory, Sahiwal, Pakistan
Correspondence: qaiser.hussain@uaar.edu.pk; qaiseruaf@gmail.com

CONTENTS

9.1 INTRODUCTION

Due to its low soil availability, phosphorus (P) is an important macronutrient for optimal plant growth, especially in tropical environments (Santana et al. 2016). Plant phosphorus concentrations range from 0.2% to 0.8% (Sharma et al. 2013), and it is, after nitrogen, the second most important plant nutrient (Aziz et al. 2017a,b). Because of the formation of insoluble phosphorus complexes in soil, phosphorus availability in soil–plant systems is a problem. Insoluble forms of P, they if get precipitated, are unavailable to plants (Rengel and Marschner 2005; Adnan et al. 2018a,b; Adnan et al. 2020; Adnan et al. 2019; Atif et al. 2021; Fahad et al. 2016; Fahad et al. 2015a,b; Fahad et al. 2020; Fahad et al. 2021a,b,c,d,e,f; Fahad et al. 2022a,b). This means that, in most cases, the phosphorus concentration in the soil is 0.05% (w/w), with only 0.1% of that being available to plants (Zhou et al. 1992). Phosphorus fixation is a term used to describe the process of removing readily available phosphate from soil solutions and storing it in the solid soil phase (Barber 1995; Fakhre et al. 2021; Muhammad et al. 2022; Shafi et al. 2020; Wahid et al. 2020; Zahida et al. 2017).

DOI: 10.1201/9781003286233-9

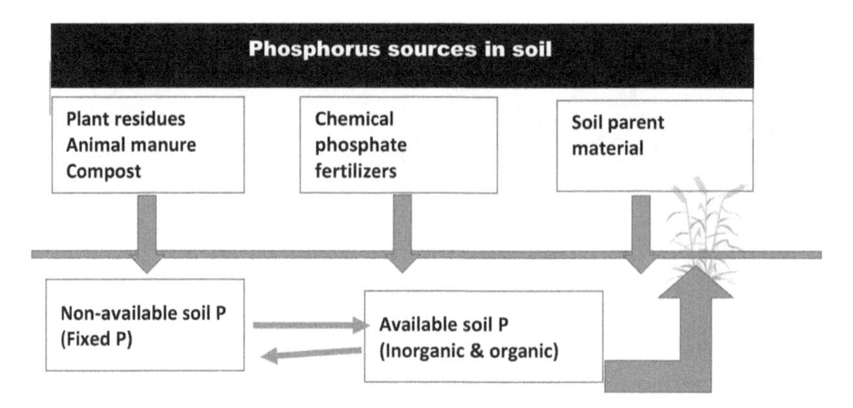

FIGURE 9.1 A schematic diagram of P sources and forms in soil.

Phosphorus-solubilizing microbes (PSMs) not only promote phosphorus cycling by mineralizing organic compounds, but also dissolve insoluble phosphorus complexes and increase phosphorus bioaccumulation in soil–plant systems. It is possible for PSM to convert soil P into plant-available P (mainly PO_4^{3-}, HPO_4^{2-}, and $H_2PO_4^-$) through exudations, root acidification, and enhanced enzymatic metal chelation (Billah et al. 2019). Because of this, it is critical to investigate management strategies that can increase agricultural yields, reduce pollution caused by phosphorus loss in soil, and improve the efficiency of phosphorus fertilizers. Numerous soil and rhizosphere microorganisms have been found to improve phosphorus availability and plant phosphorus uptake (Bhattacharyya and Jha 2012; Ahmad et al. 2017). It is even possible for these PSMs to mineralize organic phosphorus, and dissolve inorganic phosphorus, including in the presence of salt (Zhu et al. 2011).

As a highly active element, phosphorus does not exist in its elemental state in the soil. It's possible to find soil solutions that contain both organic phosphorus (Po) and inorganic phosphorus (Pi) (Walpola and Yoon 2012). Sedimentary circulation of P is possible because, unlike nitrogen, it is not exchanged with any form of P in the atmosphere and there is no major atmospheric P stock available for biological use. As a result, phosphorus deficiency leads to a reduction of plant growth and yield (Rodriguez and Fraga 1999). Most of this (95–99%) is in the form of insoluble phosphate, despite the fact that soil test results are usually much higher (Pradhan and Sukla 2005). There is a wide range of soluble phosphorus concentrations within soil solutions, ranging from ppb to 01 mg/L in highly fertilized soils (Khan, M.S. et al. 2009b). Organisms and plant residues that have not yet decomposed, as well as organophosphorus compounds, make up the bulk of the soil's organophosphorus compounds. Organophosphorus mineralization and immobilization can be affected by soil microbial activity. Cations such as Ca^{2+} immobilize P to form $Ca_3(PO_4)^2$ (calcium phosphate) in normal and calcareous soils, while in acidic soils, Al^{3+} and Fe^{3+} immobilize P to produce AlPO and FePO (Satyaprakash et al. 2017). A variety of P forms can be taken up by plant cells, but the majority of phosphorus is taken up as phosphate anions, primarily HPO_4^{2-} or $H_2PO_4^-$, depending on the pH level of the soil (Mahidi et al. 2011). As a result, PSM's strategy could mobilize the P and transform it into soluble P and plant-required P forms. Figure 9.1 depicts soil P transformations and soil–plant P availability in a comprehensive way.

9.2 PHOSPHORUS-SOLUBILIZING MICROORGANISMS (PSMS)

Phosphorus-solubilizing microorganisms (PSMs) play important roles in the soil phosphorus cycle by mineralizing organic phosphorus, dissolving inorganic phosphorus minerals, and storing vast quantities of phosphorus in biomass (Liang et al. 2020). Many microorganisms have the ability to lyse and mineralize, including bacteria, fungi, actinomycetes, and algae. Plant growth-promoting

TABLE 9.1
PSM Types

Bacteria	*Alcaligenes* sp., *Achromobacter* sp., *Aerobactor aerogenes*, *Actinomadura oligospora*, *Azospirillum brasilense*, *Agrobacterium* sp., *B. mycoides*, *Bacillus circulans*, *B. fusiformis*, *Bacillus* sp., *B. megaterium*, *B. pumils*, *B. cereus*, *B. polymyxa*, *B. coagulans*, *B. chitinolyticus*, *Bradyrhizobium* sp., *B. subtilis*, *Brevibacterium* sp., *Pseudomonas* sp., *P. striata*, *P. putida*, *Enterobacter asburiae*, *Escherichia intermedia*, *Nitrobacter* sp., *T. thioxidans*, *Thiobacillus ferroxidans*, *Rhizobium meliloti*, *Xanthomonas* sp.
Fungi	*Aspergillus awamori*, *A. tereus*, *A. wentii*, *A. niger*, *A. nidulans*, *A. flavus*, *A. foetidus*, *Alternaria teneius*, *Achrothcium* sp. *Fusarium oxysporum*, *P. lilacinium*, *Penicillium digitatum*, *P. funicolosum*, *P. balaji*, *Cladosprium* sp., *Cephalosporium* sp., etc
Actinomycetes	*Actinomyces*, *Streptomyces*
Cyanobacteria	*Anabena* sp., *Calothrix braunii*, *Nostoc* sp., *Scytonema* sp.,
VAM	*Glomus fasciculatum*

bacteria (PGPB) and other microorganisms have been shown to increase P availability to plants via solubilization and mineralization (Adnan et al. 2018c). Many fungi have also been shown to increase the availability of phosphorus in soil–plant systems. To dissolve inorganic phosphorus, *Actinomyces*, *Micromonospora*, and *Streptomyces* can produce acids such as gluconic, lactic, citric, 2-ketogluconic, tartaric, and oxalic. It has also been discovered that algae such as cyanobacteria are capable of dissolving phosphorus (Sharma et al. 2013). PSMs that live in alkaline, acidic, or organic environments frequently form metal complexes (Bashan et al. 2013a,b) (Table 9.1).

9.3 MECHANISM AND PROCESSES OF INORGANIC P-SOLUBILIZATION BY PSMS

PSMs primarily solubilize P by releasing complex or inorganic solvent compounds (organic acid anions and siderophores), CO_2, protons, hydroxyl ions, and the release of P during substrate degradation by extracellular enzymes (Sharma et al. 2013). The direct oxidation pathway, which produces organic acids in the periplasmic space, results in lower soil pH and increases P availability (Seshachala and Tallapragada 2012). Lower pH (less than 7) of microbial cells and the adjacent environment occurs when H^+ replaces Ca^{2+} during organic acid excretion, leading to the release of P ions (Goldstein 1994). Illmer and Schinner (1995) proposed a H^+ acidification mechanism. They described that H^+ excretion and NH_4^+ uptake are related to the release of H^+, which increases P solubilization. Krishnaraj et al. (1998) suggest that protons ejected from cells are the key to P solubilization. When H^+ is released into the rhizosphere in response to cation uptake or with the help of H^+ ATP, organic acids are produced and inorganic P is solubilized (Rodriguez and Fraga 1999; Ali et al. 2018). Phosphorus is solubilized by microorganisms without the formation of organic acids via the protons released during the absorption of NH^{4+} (Sharma et al. 2013). Gram-negative bacteria, for example, can aid in the solubilization of inorganic P by oxidizing gluconic acid (Goldstein 2000). Pyrroloquinoline quinone (PQQ) functions as a redox cofactor in glucose dehydrogenase (GDH), resulting in phosphate solubilization (Rodriguez et al. 2006). In addition, hydrogen sulfide (H_2S) is formed, which reacts with ferric phosphate to form ferrous sulphate and phosphate (Swaby and Sperber 1958).

Organic acid excretion by PSMs, which then chelates Fe and Al ions to release P, can make moderately labile Pi (partly from Fe/Al–P and sesquioxide surface) available for use by soil organisms (Baumann et al. 2018). Carboxyl groups found in organic acids have the ability to bind phosphorus through either the substitution of cations or the competition for phosphorus adsorption sites. This

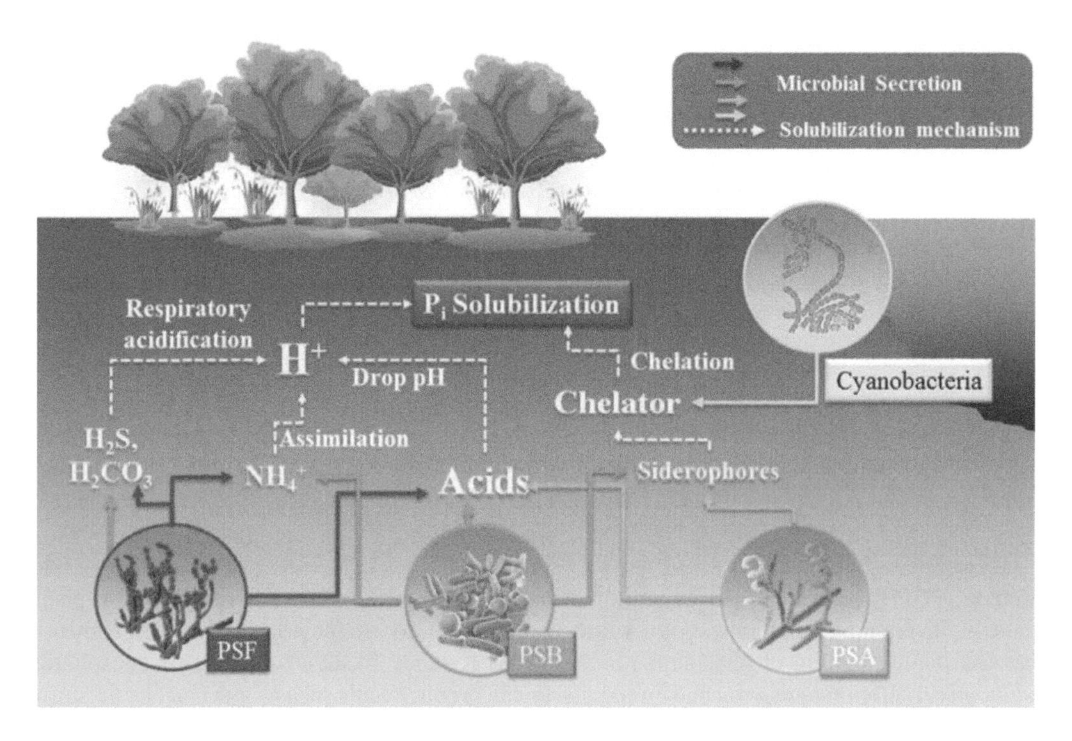

FIGURE 9.2 Phosphorus solubilization mechanisms/processes by PSMs in soil.

results in an increase in the soil's capacity to take up PO_4^{3-} and the solubilization of phosphorus. Phosphorus-solubilizing bacteria (PSBs) are known to be rich in exopolysaccharides (EPSs), which are necessary polymers made up of carbohydrates and play an important part in the process of Pi solubilization (Sharma et al. 2013). EPS has the potential to cause a disruption in the homeostasis of organic acids or H^+ that are involved in the process of Pi solubilization by keeping free P in the medium. This could result in an increased release of P from Pi minerals (Yi et al. 2008). Both the *Streptomyces* species and the *Micromonospora* species have been demonstrated in the past to be capable of dissolving phosphate rock while simultaneously producing calcium chelators or siderophores (Hamdali et al. 2008; Yandigeri et al. 2011). In addition to the mechanisms that have already been described, the production of chelators (such as calcium, aluminium, and iron chelators) by PSB, phosphorus-solubilizing actinomycetes (PSA), and blue-green algae (BGA) is another way that phosphorus can be released from Pi crystals (Figure 9.2).

9.3.1 Non-specific Acid Phosphatases (NSAPs)

An enzyme known as non-specific acid phosphatase (NSAPs) is responsible for dephosphorylation of phosphoester or phosphoanhydride bonds present in organic matter. Phosphomonoesterases, also known as phosphatases, are the most common and have been the most researched of all the different types of phosphatase enzymes generated by PSMs (Nannipieri et al. 2011). Depending on the optimum pH at which they operate, these enzymes are classified as either acidic or alkaline phosphomonoesterases. Depending on environmental conditions, both types can be generated by PSMs (Jorquera et al. 2008; Basir et al. 2016). The presence of alkaline phosphatases is more prevalent in neutral and alkaline soils, while acid phosphatases are more prevalent in acidic soils (Renella et al. 2006).

Although plant roots can produce acid phosphatase, alkaline phosphatase is not very often produced in significant amounts, suggesting that this could be a potential niche for PSMs (Criquet et al. 2004). It is also difficult to distinguish between PSM-produced and root-produced phosphatases, but some evidence proposes that microbial-derived phosphatases have a greater affinity for Po compounds than phosphatases of plant origin (Tarafdar et al. 2001). At present, it is unclear whether or not there is a link between phosphatase activity, PSM that was injected into the soil, and future Po mineralization (Chen et al. 2003).

9.3.2 Phytases

The consequent breakdown of the phytate leads to liberation of the element phosphorus. This is the primary source of phosphate and the most common form of phosphorus stored in pollen and plant seeds, and phytate is an essential component of the organic phosphorus that is found in soil (Richardson 1994). When fed phytate, *Arabidopsis* plants that had been genetically transformed with the phytase (phyA) gene obtained from *Aspergillus niger* showed significant improvements in P growth and nutrition. This was the case regardless of the fact that plants have a limited capacity to obtain phosphorus directly from phytate (Richardson 2001). Because of this, plant nutritional P increased to the point where plant growth and P content were comparable to control plants that had been given inorganic P. This occurred because it led to an increase in plant nutritional P. This is due to an increase in the amount of nutrition provided by phosphorus.

9.3.3 PSM-derived P Desorption from Clay Minerals

Adsorption reactions speed up the removal of soluble P forms from soil solutions, allowing for their sorption in the solid phase. Volcanic ash and highly weathered tropical soils are rich in aluminium hydroxides and amorphous iron and are particularly vulnerable to these chemical reactions (Osorio and Habte 2012). Compared to their variable charge counterparts, permanently charged minerals have a greater adsorption capacity and binding energy for phosphorus. Iron hydroxides and aluminium hydroxides are the most significant minerals with a variable charge (He and Zu 1998). Most of the phosphorus in clay minerals in fertilized soil is bound to aluminium hydroxides, whereas iron-bound phosphorus is the most common form in unfertilized soils (Eriksson et al. 2015). Hydroxides of Fe–P (Al–P) are formed because of the close relationship between iron (Al) and orthophosphate (P). The high adsorption capacity of these hydroxides causes phosphate to be deposited in the soil (Wu et al. 2019). Orthophosphate can be adsorbed by soil organic matter or immobilized by it, resulting in relatively stable Po molecules and lowering the concentration of dissolved phosphate in the soil (Qiao et al. 2017).

The P desorption ability of PSM depends on the clay mineral surfaces, saturation of Pi adsorbent sites, and adsorption capacity. This is the case despite the fact that allophane has the greatest ability to desorb P. Moreover, since it causes higher levels of absorbed phosphorus in a wide variety of soils, inoculation with *Mortiella* sp. is able to significantly increase the phosphorus solution of the soil in situ by desorption. It can be observed that the ability of PSM to increase soluble Pi in soil is proportional to the amount of P taken up by soil and clay minerals, while the extent of Pi desorption by PSM is inversely proportional to the Pi-absorptive capacity of soil and clay minerals. This is something to keep in mind (Osrio and Habte 2012). Desorption of phosphorus from PSM usually occurs in conjunction with a decrease in pH. The desorption of P is mainly caused by the increased solubility of Fe and Al due to the potential complexation of Fe and Al with low-molecular-weight organic acids (Hoberg et al. 2005; Iqbal et al. 2016). During the process of phosphorus mineral (Pi) solubilization, plant roots and PSM may be responsible for releasing a variety of low-molecular-weight organic acids. Some of these organic acids, such as citrate, malate, and oxalate are well known for their ability to increase Pi availability in soil through desorption (Dodd and Sharopely 2015).

Pi can be displaced from adsorption sites by organic acids and anions due to ligand exchange caused by microbial activity and temporary blockage of Pi adsorption sites (Andrade et al. 2013). Depending on their dissociation characteristics and the carboxylic groups they contain, organic acids can have a variety of different negative charges that help promote Pi desorption (Hoberg et al. 2005).

9.3.4 Siderphore Production

Almost all microorganisms produce siderophores, which are complexing agents with a strong affinity for iron. Solubilizers of iron from mineral or organic molecules are provided by siderophores in the case of iron deficiency. About 500 known siderophores are used by a wide range of microbes and plants, while some are used only by microbial species and the strains that produce them (Crowley 2007). Aside from one report (Hamdali et al. 2008a; Zahoor et al. 2016) that siderophores are released by PSM, the production of siderophores has not been consistently linked to P-solubilization. Mineral dissolution dominates the P-solubilizing role of organic acid anions, so the potential role of siderophores in increasing P availability should be obvious.

9.3.5 Solubilization via Exopolysaccharides (EPS)

Polysaccharides have been shown to have a significant impact on phosphorus solubilization by microbes (Yi et al. 2008). Microbial exopolysaccharides (EPS) are polymers secreted by bacteria and fungi that are known as microorganism exopolysaccharides. Some of these polysaccharides are homopolysaccharides, while others are heteropolysaccharides, with a variety of organic and inorganic substituents in between. Their structure and composition can differ greatly from one another (Sutherland 2001). Yi et al. (2008) reported that tricalcium phosphate (TCP) was solubilized using four different bacterial strains. Among the strains tested were *Artrobacter* sp. (ArHy-505), *Enterobacter* sp. (EnHy-401 and EnHy-402), and *Azotobacter* sp. (AzHy-510). Bacteria with high phosphate solubility were able to produce large amounts of exopolysaccharides (EPS). More research is needed to better understand the relationship between EPS synthesis and phosphate solubilization.

9.4 FACTORS RESPONSIBLE FOR P-SOLUBILIZATION

The ability of PSM to modify organic and insoluble inorganic phosphorus is dependent on the nutrient content of the soil as well as the organism's physiological and growth state. Compared to PSMs residing in temperate soil conditions, PSMs from extreme soil conditions, such as heavy-metal-contaminated soils, high nutrient-deficient soils, saline soils, or soils subject to extreme temperatures, are more likely to dissolve phosphate (Zhu et al. 2011).

9.4.1 Temperature

Temperature has been discussed in relation to phosphorus solubilization by bacteria. Varsha (2002) discovered that 28°C was the optimum temperature to use for maximum microbial phosphorus solubilization, whereas White et al. (1997) found that 20–25°C was the best temperature to use. Others have reported that 30°C is the optimal temperature for solubilizing P. Nautiyal et al. (2000) found maximum P solubilization in desert soil at 45°C, whereas Johri et al. (1999) discovered P solubilization in the same medium at 10°C.

9.4.2 Soil pH

The main mechanism responsible for phosphorus solubilization in soil is the reduction of soil pH caused by the release of protons or microbial synthesis of organic acids (Pradhan and Sukla 2005).

In alkaline soils, calcium ions react with phosphate to form calcium phosphates, which are insoluble in soil and include rock phosphate. After that, the calcium phosphates can precipitate (fluorapatite and francolite). As the pH of the soil decreases, so does its soluble capacity. PSMs produce organic acids, which lower soil pH and make more phosphorus available to plants (Satyaprakash et al. 2017). Furthermore, aliphatic acids are more effective than phenolic, citric, and fumaric acids in solubilizing phosphate, while tri- and dicarboxylic acids outperform monobasic and aromatic acids (Mahidi et al. 2011; Rehman et al. 2016). Oxalic, citric, acetic, gluconic, lactic, 2-ketogluconic, fumaric, adipic, malic, tartaric, succinic, malonic, glutaric, butyric, propionic, and glyoxalic acids are the most common organic acids that can solubilize phosphates (Kumar et al. 2018).

9.4.3 Chelation

PSMs are responsible for the dissolution of insoluble soil phosphates because they produce both organic and inorganic acids. They accomplish this by chelating cations and engaging in competition with phosphate in the soil for adsorption sites (Khan, A. et al. 2009). The hydroxyl and carboxylic groups that are present in the acids are responsible for the chelation of the cations that are linked to the phosphate, as well as the transformation of those cations into forms that are soluble. These acids are able to serve as attachment sites for the insoluble oxides of aluminium and iron because, following their reaction with the insoluble oxides of those elements, the resulting compound is stabilized. 2-Ketogluconic acid is one of these, and it is a calcium chelator that is exceptionally effective (Walpola and Yoon 2012). It has been observed that inorganic acids such as sulfuric nitric acid and carbonic acid are produced as a result of the observations that have been made. As a result of a reaction involving nitric and sulphuric acids, calcium phosphate will have a greater degree of solubility (Khan et al. 2007).

9.4.4 Mineralization

Another method for releasing phosphorus from soil is mineralization. PSMs convert organic phosphate into usable form through the mineralization process. This transformation occurs in the soil and consumes a variety of organic phosphorus compounds, including phosphonates, phospholipids, nucleic acids, sugar phosphates, polyphosphates, and phytic acid (Khan, A. et al. 2009). Organic phosphorus mineralization and immobilization in agricultural soil is a critical step in the phosphorus cycling process. PSMs are responsible for mineralizing organic phosphorus in the soil by producing phosphatases such as phytase (Aseri et al. 2009). These phosphatases hydrolyse organic phosphate compounds, releasing inorganic phosphorus for use by plants. Both acid and alkaline phosphatases use organic phosphate as a substrate in the chemical reaction to convert it to phosphate.

Extracellular enzymes such as phosphoesterases, phosphodiesterases, phytases, and phospholipases are produced in the soil by *Bacillus* and *Streptomyces* spp. Bacteria can mineralize complex organic phosphates thanks to these enzymes (Walpola and Yoon 2012). The most efficient method of mineralizing organic phosphate is to use a phosphorus-solubilizing bio-fertilizer composed of mixed cultures of PSM (*Bacillus*, *Streptomyces*, and *Pseudomonas*). According to Dodor and Tabatabai (2003), some PSMs can produce siderophores, which hydrolyse soil organic phosphorus and make it more accessible.

REFERENCES

Adnan M, Zahir S, Fahad S, Arif M, Mukhtar A, Imtiaz AK, Ishaq AM, Abdul B, Hidayat U, Muhammad A, Inayat-Ur R, Saud S, Muhammad ZI, Yousaf J, Amanullah, Hafiz MH, Wajid N (2018a)

Phosphate-solubilizing bacteria nullify the antagonistic effect of soil calcification on bioavailability of phosphorus in alkaline soils. *Sci Rep* 8:4339. https://doi.org/10.1038/s41598-018-22653-7

Adnan M, Shah Z, Sharif M, Rahman H (2018b) Liming induces carbon dioxide (CO_2) emission in PSB inoculated alkaline soil supplemented with different phosphorus sources. *Environ Sci Poll Res* 25(10):9501–9509.

Adnan M, Shah Z, Sharif M, Rahman H (2018c) Liming induces carbon dioxide (CO_2) emission in PSB inoculated alkaline soil supplemented with different phosphorus sources. *Environ Sci Pollut Res* 25:9501–9509.

Adnan M, Fahad S, Khan IA, Saeed M, Ihsan MZ, Saud S, Riaz M, Wang D, Wu C (2019) Integration of poultry manure and phosphate solubilizing bacteria improved availability of Ca bound P in calcareous soils. *3 Biotech* 9(10):368.

Adnan M, Fahad S, Muhammad Z, Shahen S, Ishaq AM, Subhan D, Zafar-ul-Hye M, Martin LB, Raja MMN, Beena S, Saud S, Imran A, Zhen Y, Martin B, Jiri H, Rahul D (2020) Coupling phosphate-solubilizing bacteria with phosphorus supplements improves maize phosphorus acquisition and growth under lime induced salinity stress. *Plants* 9(7):900. doi: 10.3390/plants9070900

Ahmad I, Jadoon SA, Said A, Adnan M, Mohammad F, Munsif F (2017) Response of sunflower varieties to NPK fertilization. *Pure Appl Biol* 6(1):272–277.

Ali S, Xua S, Ahmad I, Jia Q, Ma X, Ullah H, Alam M, Adnan M, Daur I, Ren X, Cai T, Zhang J, Jia Z (2018) Tillage and deficit irrigation strategies to improve winter wheat production through regulating root development under simulated rainfall conditions. *Agric Water Manag* 209:44–54.

Andrade FV, Mendonça ES, Silva IR (2013) Organic acids and diffusive flux of organic and inorganic phosphorus in sandy-loam and clayey latosols. *Commun Soil Sci Plant* 44:1211–1223.

Aseri GK, Jain N, Tarafdar JC (2009) Hydrolysis of organicphosphate forms by phosphatases and phytase producingfungi of arid and semi-arid soils of India. *Am-Eurasian J Agric & Environ Sci* 5:564–570.

Atif B, Hesham A, Fahad S (2021) Biochar coupling with phosphorus fertilization modifies antioxidant activity, osmolyte accumulation and reactive oxygen species synthesis in the leaves and xylem sap of rice cultivars under high-temperature stress. *Physiol Mol Biol Plants* 27(9):2083-2100. https://doi.org/10.1007/s12298-021-01062-7

Aziz K, Daniel KYT, Fazal M,Muhammad ZA, Farooq S, FanW, Fahad S, Ruiyang Z (2017a) Nitrogen nutrition in cotton and control strategies for greenhouse gas emissions: a review. *Environ Sci Pollut Res* 24:23471–23487. doi.org/10.1007/s11356-017-0131-y

Aziz K, Daniel KYT, Muhammad ZA, Honghai L, Shahbaz AT, Mir A, Fahad S (2017b) Nitrogen fertility and abiotic stresses management in cotton crop: a review. *Environ Sci Pollut Res* 24:14551–14566. doi.org/10.1007/s11356-017-8920-

Barber SA (1995) *Soil nutrient bioavailability. A Mechanistic Approach.* Wiley, New York.

Bashan Y, Kamnev AA, de Bashan LE (2013a) A proposal for isolating and testing phosphate-solubilizing bacteria that enhance plant growth. *Biol Fertil Soils* 49:1–2.

Bashan Y, Kamnev AA, de Bashan LE (2013b) Tricalcium phosphate is inappropriate as a universal selection factor for isolating and testing phosphate-solubilizing bacteria that enhance plant growth: a proposal for an alternative procedure. *Biol Fertil Soils* 49:465–479.

Basir A, Jan MT, Alam M, Shah AS, Afridi K, Adnan M, Ali K, Mian IA (2016) Impacts of tillage, stubble management and nitrogen on wheat production and soil properties. *Can J Soil Sci* 97(2):133–140.

Baumann K, Jung P, Samolov E, Lehnert LW, Budel B, Karsten U, Bendix J, Achilles S, Schermer M, Matus F (2018) Biological soil crusts along a climatic gradient in Chile: richness and imprints of phototrophic microorganisms in phosphorusbiogeochemical cycling. *Soil Biol Biochem* 127:286–300.

Bhattacharyya PN, Jha DK (2012) Plant growth-promoting rhizobacteria (PGPR): emergence in agriculture. *World J Microbiol Biotechnol* 28:1327–1350. doi: 10.1007/s11274-011-0979-9

Billah M, Khan M, Bano A, Hassan TU, Munir A, Gurmani AR (2019) Phosphorus and phosphate solubilizing bacteria: keys for sustainable agriculture. *Geomicrobiol J* 36:904–916.

Chen CR, Condron LM, Davis MR, Sherlock RR (2003) Seasonal changes in soil phosphorus and associated microbial properties under adjacent grassland and forest in New Zealand. *Forest Ecol Manag* 117:539–557.

Criquet S, Ferre E, Farner EM, Le Petit J (2004) Annual dynamics of phosphatase activities in an evergreen oak litter—influence of biotic and abiotic factors. *Soil Biol Biochem* 36:1111–1118.

Crowley DE (2007) Microbial siderophores in the plant rhizosphere. In: Barton LL, Abadia J (Eds.) *Iron Nutrition in Plants and Rhizospheric Microorganisms*. Springer, Dordrecht, pp. 169–198.

Dodd RJ, Sharpley AN (2015) Recognizing the role of soil organic phosphorus in soil fertility and water quality. *Resour Conserv Recy* 105:282–293.

Dodor DE, Tabatabai MA (2003) Effect of cropping systems on phosphatases in soils. *J Plant Nutr & Soil Sci* 166(1):7–13.

Eriksson AK, Gustafsson JP, Hesterberg D (2015) Phosphorus speciation of clay fractions from long-term fertility experiments in Sweden. *Geoderma* 241–242:68–74.

Fahad S, Hussain S, Saud S, Tanveer M, Bajwa AA, Hassan S, Shah AN, Ullah A,Wu C, Khan FA, Shah F, Ullah S, Chen Y, Huang J (2015a) A biochar application protects rice pollen from high-temperature stress. *Plant Physiol Biochem* 96:281–287.

Fahad S, Nie L, Chen Y, Wu C, Xiong D, Saud S, Hongyan L, Cui K, Huang J (2015b) Crop plant hormones and environmental stress. *Sustain Agric Rev* 15:371–400.

Fahad S, Hussain S, Saud S, Hassan S, Tanveer M, Ihsan MZ, Shah AN, Ullah A, Nasrullah KF, Ullah S, AlharbyH NW, Wu C, Huang J (2016) A combined application of biochar and phosphorus alleviates heat-induced adversities on physiological, agronomical and quality attributes of rice. *Plant Physiol Biochem* 103:191–198.

Fahad S, Hasanuzzaman M, Alam M, Ullah H, Saeed M, Ali Khan I, Adnan M (Eds.) (2020) *Environment, Climate, Plant and Vegetation Growth*. Springer Nature Switzerland AG 2020. doi: https://doi.org/10.1007/978-3-030-49732-3

Fahad S, Sönmez O, Saud S, Wang D, Wu C, Adnan M, Turan V (Eds.) (2021a) Plant growth regulators for climate-smart agriculture. First edition. *Footprints of Climate Variability on Plant Diversity*. CRC Press, Boca Raton.

Fahad S, Sonmez O, Saud S, Wang D, Wu C, Adnan M, Turan V (Eds.) (2021b) Climate change and plants: biodiversity, growth and interactions. First edition. *Footprints of Climate Variability on Plant Diversity*. CRC Press, Boca Raton.

Fahad S, Sonmez O, Saud S, Wang D, Wu C, Adnan M, Turan V (Eds.) (2021c) Developing climate resilient crops: improving global food security and safety. First edition. *Footprints of Climate Variability on Plant Diversity*. CRC Press, Boca Raton.

Fahad S, Sönmez O, Turan V, Adnan M, Saud S, Wu C Wang D (Eds.) (2021d) Sustainable soil and land management and climate change. First edition. *Footprints of Climate Variability on Plant Diversity*. CRC Press, Boca Raton.

Fahad S, Sönmez O, Saud S, Wang D, Wu C, Adnan M, Arif M, Amanullah (Eds.) (2021e) Engineering tolerance in crop plants against abiotic stress. First edition. *Footprints of Climate Variability on Plant Diversity*. CRC Press, Boca Raton.

Fahad S, Saud S, Yajun C, Chao W, Depeng W (Eds.) (2021f) Abiotic stress in plants. *IntechOpen United Kingdom* 2021. http://dx.doi.org/10.5772/intechopen.91549

Fahad S, Adnan M, Saud S (Eds.) (2022a) Improvement of plant production in the era of climate change. First edition. *Footprints of Climate Variability on Plant Diversity*. CRC Press, Boca Raton.

Fahad S, Adnan M, Saud S, Nie L (Eds.) (2022b) Climate change and ecosystems: challenges to sustainable development. First edition. *Footprints of Climate Variability on Plant Diversity*. CRC Press, Boca Raton.

Fakhre A, Ayub K, Fahad S, Sarfraz N, Niaz A, Muhammad AA, Muhammad A, Khadim D, Saud S, Shah H, Muhammad ASR, Khalid N, Muhammad A, Rahul D, Subhan D (2021) Phosphate solubilizing bacteria optimize wheat yield in mineral phosphorus applied alkaline soil. *J Saudi Soc Agric Sci* 21(5): 339–348. https://doi.org/10.1016/j.jssas.2021.10.007

Goldstein AH (1994) Involvement of the quinoprotein glucose dehydrogenase in the solubilization of exogenous phosphates by gram-negative bacteria. In: TorrianiGorini A, Yagil E, Silver S (Eds.) *Phosphate in Microorganisms: Cellular and Molecular Biology*, ASM Press,Washington, DC. pp. 197–203.

Goldstein AH (2000) Bioprocessing of rock phosphate ore: essential technical considerations for the development of a successful commercial technology. In: Proceedings of the 4th International Fertilizer Association Technical Conference (Paris: IFA), 220.

Gross A, Lin Y, Weber PK, Pett-Ridge J, Silver WL (2020) The role of soil redox conditions in microbial phosphorus cycling in humid tropical forests. *Ecol* 101:e02928.

Hamdali H, Bouizgarne B, Hafidi M, Lebrihi A, Virolle MJ, Ouhdouch Y (2008) Screening for rock phosphate solubilizing *Actinomycetes* from Moroccan phosphate mines. *Appl Soil Ecol* 38:12–19.

He Z, Zhu J (1998) Microbial utilization and transformation of phosphate adsorbed by variable charge minerals. *Soil Biol Biochem* 30:917–923.

Hoberg E, Marschner P, Lieberei R (2005) Organic acid exudation and pH changes by *Gordonia* sp. and *Pseudomonas fluorescens* grown with P adsorbed to goethite. *Microbiol Res* 160:177–187.

Illmer P, Schinner F (1995). Solubilization of inorganic calcium phosphates—solubilization mechanisms. *Soil Biol Biochem* 27:257–263. doi: 10.1016/0038-0717(94)00190-C

Iqbal B, Khan H, Saifullah, Khan I, Shah B, Naeem A, Ullah W, Khan N, Adnan M, Shah SRA, Junaid K, Ahmed N, Iqbal M (2016) Substrates evaluation for the quality, production and growth of oyster mushroom (*Pleurotus florida* Cetto). *J Ento & Zool Studies* 4(3):98–107.

Johri JK, Surange S, Nautiyal CS (1999) Occurrence of salt, pH and temperature tolerant phosphate solubilizing bacteria in alkaline soils. *Curr Microbiol* 39:89–93.

Jorquera MA, Hernandez MT, Rengel Z, Marschner P, Mora MD (2008) Isolation of culturable phosphor bacteria with both phytate-mineralization and phosphate-solubilization activity from the rhizosphere of plants grown in a volcanic soil. *Biol Fertil Soils* 44:1025–1034.

Khan A, Jilani V, Akhtar MS, Naqvi SMS, Rasheed M (2009) Phosphorus solubilizing bacteria: occurrence, mechanisms and their role in crop production. *J Agric & Biol Sci* 1:48–58.

Khan MS, Zaidi A, Wani PA (2007) Role of phosphate solubilizing microorganisms in sustainable agriculture—a review. *Agron Sustainable Dev* 27(1):29–43.

Khan MS, Zaidi A, Wani PA (2009) Role of phosphate solubilizing microorganisms in sustainable agriculture. In: Lictfouse E et al. (Eds.) *Sustainable Agriculture*. Springer. p. 552. doi: 10.1007/978-90-481-2666-8_34

Krishnaraj PU, Khanuja SPS, Sadashivam KV (1998) Mineral phosphate solubilization (MPS) and mps genes-components in eco-friendly P fertilization. Abstracts of Indo US Workshop on Application of Biotechnology for Clean Environment and Energy, National Institute of Advanced Studies, Bangalore, p. 27.

Kumar A, Kumar A, Patel H (2018) Role of microbes in phosphorus availability and acquisition by plants. *Int J Curr Microbiol & Appl Sci* 7(5):1344–1347.

Liang JL, Liu J, Jia P, Yang TT, Zeng QW, Zhang SC, Liao B, Shu WS, Li JT (2020) Novel phosphate-solubilizing bacteria enhance soil phosphorus cycling following ecological restoration of land degraded by mining. *ISME J* 14:1600–1613.

Mahidi SS, Hassan GI, Hussain A, Faisul-Ur-Rasool (2011) Phosphorus availability issue-its fixation and role of phosphate solubilizing bacteria in phosphate solubilization-case study. *Res J Agric Sci* 2:174–179.

Muhammad I, Khadim D, Fahad S, Imran M, Saud A, Manzer HS, Shah S, Jabar ZKK, Shamsher A, Shah H, Taufiq N, Hafiz MH, Jan B, Wajid N (2022) Exploring the potential effect of *Achnatherum splendens* L.–derived biochar treated with phosphoric acid on bioavailability of cadmium and wheat growth in contaminated soil. *Environ Sci Pollut Res* 29(25):37676–37684. https://doi.org/10.1007/s11356-021-17950-0.

Nannipieri P, Giagnoni L, Landi L, Renella G (2011) Role of phosphatase enzymes in soil. In: Bunemann E, Oberson A, Frossard E (Eds.) *Phosphorus in Action: Biological Processes in Soil Phosphorus Cycling. Soil Biology*, 26. Springer, Heidelberg, pp. 251–244.

Nautiyal CS, Bhadauria S, Kumar P, Lal H, Mondal R, Verma D (2000) Stress induced phosphate solubilization in bacteria isolated from alkaline soils. *FEMS Microbiol Lett* 182:291–296.

Osorio NW, Habte M (2012) Phosphate desorption from the surface of soil mineral particles by a phosphate-solubilizing fungus. *Biol Fert Soils* 49:481–486.

Pradhan N, Sukla LB (2005) Solubilization of inorganic phosphate by fungi isolated from agriculture soil. *African J Biotechnol* 5:850–854.

Qiao Z, Hong J, Li L, Liu C (2017) Effect of phosphobacterias on nutrient, enzyme activities and phosphorus adsorption–desorption characteristics in a reclaimed soil. *J Soil Water Conserv* 31:166–171.

Rehman I, Ali S, Rahman I, Adnan M, Ullah H, Basir A, Malik FA, Shah AS, Ibrahim M, Arshad M (2016) Effect of pre-storage seed priming on biochemical changes in okra seed. *Pure Appl Biol* 5(1):165–171.

Renella G, Egamberdiyeva D, Landi L, Mench M, Nannipieri P (2006) Microbial activity and hydrolase activities during decomposition of root exudates released by an artificial root surface in Cd-contaminated soils. *Soil Biol Biochem* 38:702–708.

Rengel Z, Marschner P (2005) Nutrient availability and management in the rhizosphere: exploiting genotypic differences. *New Phytol* 168:305–312.

Richardson AE (1994) Soil microorganisms and phosphorus availability. In: Pankhurst CE, Doubeand BM, Gupta VVSR (Eds.) *Soil Biota: Management in Sustainable Farming Systems*. CSIRO, Victoria, Australia, pp. 50–62.

Richardson AE (2001) Prospects for using soil microorganisms to improve the acquisition of phosphorus by plants. *Aust J Plant Physiol* 28:897–906.

Rodríguez H, Fraga R (1999) Phosphate solubilizing bacteria and their role in plant growth promotion. *Biotechnol Adv* 17:319–339. doi: 10.1016/S0734-9750(99)00014-2

Rodriguez H, Fraga R, Gonzalez T, Bashan Y (2006) Genetics of phosphate solubilization and its potential applications for improving plant growth-promoting bacteria. *Plant Soil* 287:15–21.

Santana EB, Marques ELS, Dias JCT (2016) Effects of phosphate-solubilizing bacteria, native microorganisms and rock dust on *Jatropha curcas* L. growth. *Genetics and Molecular Research* 15(4):gmr.15048729.

Satyaprakash IM, Nikitha T, Reddi EUB, Sadhana B, Vani SS (2017) A review on phosphorous and phosphate solubilising bacteria and their role in plant nutrition. *Int J Curr Microbiol & Appl Sci* 6:2133–2144.

Seshachala U, Tallapragada P (2012) Phosphate solubilizers from the rhizosphere of *Piper nigrum* L. in Karnataka, India. *Chil J Agric Res* 72:397–403.

Shafi MI, Adnan M, Fahad S, Fazli W, Ahsan K, Zhen Y, Subhan D, Zafar-ul-Hye M, Martin B, Rahul D (2020) Application of single superphosphate with humic acid improves the growth, yield and phosphorus uptake of wheat (Triticum aestivum L.) in calcareous soil. *Agron* 10:1224. doi:10.3390/agronomy10091224

Sharma SB, Sayyed RZ, Trivedi MH, Gobi TA (2013) Phosphate solubilizing microbes: sustainable approach for managing phosphorusdeficiency in agricultural soils. *Springer Plus* 2:587–600. doi: 10.1186/2193-1801-2-587

Sutherland IW (2001) Biofilm exopolysaccharides: a strong and sticky framework. *Microbiol* 147:3–9.

Swaby R, Sperber JI (1958) Phosphate dissolving microorganisms in the rhizosphere of legume, nutrition of legumes; Proc. Univ. Nottingham 5th Easter Sch. Agril. Sci. (CSIRO Adelaide). *Soils & Fert* 22, 286(1959):289–294.

Tarafdar JC, Yadav RS, Meena SC (2001) Comparative efficiency of acid phosphatase originated from plant and fungal sources. *J Plant Nutr Soil Sci* 164:279–282.

Varsha NHH (2002) *Aspergillus aculeatus* as a rock phosphate solubilizer. *Soil Biol Biochem* 32:559–565.

Wahid F, Fahad S, Subhan D, Adnan M,, Zhen Y, Saud S, Manzer HS, Martin B, Tereza H, Rahul D (2020) Sustainable management with mycorrhizae and phosphate solubilizing bacteria for enhanced phosphorus uptake in calcareous soils. *Agri* 10(8):334. doi:10.3390/agriculture10080334

Walpola BC, Yoon M (2012) Prospectus of phosphate solubilizing microorganisms and phosphorus availability inagricultural soils: a review. *African J Microbiol Res* 6:6600–6605.

White C, Sayer JA, Gadd GM (1997) Microbial solubilization and immobilization of toxic metals: key biogeochemical processes for treatment of contamination. *FEMS Microbiol Rev* 20:503–516.

Wu SJ, Zhao YP, Chen YY, Dong XM, Wang MY, Wang GX (2019) Sulfur cycling in freshwater sediments: a cryptic driving force of iron deposition and phosphorus mobilization. *Sci Total Environ* 657:1294–1303.

Yandigeri MS, Yadav AK, Srinivasan R, Kashyap S, Pabbi S (2011) Studies on mineral phosphate solubilization by *cyanobacteria* Westiellopsis and Anabaena. *Microbiol* 80:558–565.

Yi Y, Huang W, Ge Y (2008) Exopolysaccharide: a novel important factor in the microbial dissolution of tricalcium phosphate. *World J Microbiol Biotechnol* 24:1059–1065.

Zahida Z, Hafiz FB, Zulfiqar AS, Ghulam MS, Fahad S, Muhammad RA, Hafiz MH, Wajid N, Muhammad S (2017) Effect of water management and silicon on germination, growth, phosphorus and arsenic uptake in rice. *Ecotoxicol Environ Saf* 144:11–18.

Zahoor M, Khan N, Ali M, Saeed M, Ullah Z, Adnan M, Ahmad B (2016) Integrated effect of organic waste and NPK fertilizers on nutrients uptake in potato crop and soil fertility. *Pure Appl Biol* 5(3):601–607.

Zhou K, Binkley D, Doxtader KG (1992) A new method for estimating gross phosphorus mineralization and immobilization rates in soils. *Plant Soil* 147:243–250.

Zhu F, Qu L, Hong X, Sun X (2011) Isolation and characterization of a phosphate solubilizing halophilic bacterium *Kushneriasp*. YCWA18 from Daqiao Saltern on the coast of yellow sea of China. *Evid Based Complement Alternat Med* 2011:615032. doi: 10.1155/2011/615032

10 Potential Applications of Algae-based Bio-fertilizers

Hafiz Muhammad Rashad Javeed,[1] Mazhar Ali,[1] Rafi Qamar,[2] Fahim Nawaz,[3] Humaira Yasmin,[4] Koushik Chakraborty,[5] Zainul Abideen,[6] Muhammad Zahid Ihsan,[7] and Muhammad Adnan Bukhari[8]*

[1] Department of Environmental Sciences, COMSATS University Islamabad, Punjab, Pakistan
[2] Department of Agronomy, College of Agriculture, University of Sargodha, Sargodha, Punjab, Pakistan
[3] Department of Agronomy, Muhammad Nawaz Shareef University of Agriculture, Multan, Punjab, Pakistan
[4] Department of Biosciences, COMSATS University Islamabad, Islamabad, Pakistan
[5] Department of Crop Physiology & Biochemistry, ICAR-National Rice Research Institute, Bidyadharpur, Cuttack, Odisha, India
[6] Dr. Muhammad Ajmal Khan Institute of Sustainable Halophyte Utilization, University of Karachi, Sindh, Pakistan
[7] The Cholistan Institute of Desert Studies, Faculty of Agriculture and Environment Sciences, The Islamia University of Bahawalpur, Bahawalpur, Punjab, Pakistan
[8] Department of Agronomy, Faculty of Agriculture and Environment, The Islamia University of Bahawalpur, Bahawalpur, Punjab, Pakistan
Correspondence: rashadjaveed@cuivehari.edu.pk

CONTENTS

DOI: 10.1201/9781003286233-10

10.1 ALTERNATIVE FERTILIZER SOURCES: A NEED OF THE DAY

The main challenge in meeting rising energy demands and avoiding the negative environmental impact of industrial waste is to overcome fossil fuel mismanagement and the global industrial revolution. Industrial waste, in addition to agricultural waste, is the primary source of water contamination (Hussain et al. 2021). Water scarcity or pollution, as a severe environmental concern, has drawn a large number of people to seek solutions. In the last 120 years, global power generation has been steadily expanding, accounting for more than 39% of global electricity generation. Even though burning coal in nuclear reactors emits CO_2, the most environmentally harmful greenhouse gas, and a major contributor to observed global climate change (GCC), numerous regional and global efforts to reduce coal use for electricity generation and CO_2 emissions have been made in the last 30 years (Iyovo et al. 2010). Other than hydroelectric, renewable energies increased from 0.7 percent in 1980 to 6.8 percent in 2015. Replacement of coal power plants has taken place with widely available renewable energy sources (Zou et al. 2021). At the same time, as fossil fuel supplies become depleted, the need for renewable energy is growing. In comparison to other physicochemical treatment approaches, microalgae are the most appealing biological agents for dealing with both energy emergencies and wastewater treatment (Zhao and Magoulès 2012). Synthetic fertilizer, pesticide, and other chemical manufacturers and users have proliferated to fulfil the growing need for food, polluting the planet and posing a serious threat to all living things. Moreover, as a result of intensive farming operations and climate change, agricultural land is losing its fertility (Besser and Hamed 2021). Pathogens, cyanobacteria, microbes, and other organisms are gaining popularity as long-term substitutes for synthetic chemicals due to their capacity to increase soil fertility, fix atmospheric nitrogen for plant availability, manufacture plant anabolic steroids and biotoxins, and more (Lee and Ryu 2021).

10.2 INTRODUCTION OF ALGAE

Algae are a varied collection of microorganisms that can perform photosynthesis by absorbing energy from the sun. Algae are widely employed in agriculture as bio-fertilizers and soil stabilizers (Park and Kim 2018). Algae, especially seaweeds, are employed as fertilizers, resulting in less nitrogen and phosphorus discharge than when cattle dung is used. As a result, the quality of water flowing into rivers and seas improves. These organisms are grown and utilized as human food supplements all over the world. They can also generate clean and carbon-neutral food, and can be produced on abandoned sites and arid desert regions with little freshwater requirement. Iodine may be found in abundance in seaweed (Arshad et al. 2016; Kulikova et al. 2018). The amount of iodine in milk is determined by the diet of the cow that produced it. According to Fuzhou Wonderful Biological Technology, feeding seaweeds to milk calves can improve the amount of iodine in milk. Algae feed additives also boost the egg-laying rate of hens (Bhuyar et al. 2018).

10.3 POTENTIAL APPLICATIONS OF ALGAE

It is unknown if algae can be employed in large-scale farming. Although large-scale algal production has been optimized for developing treatments, there is no evidence that algae can be used for large-scale agricultural cultivation. Algal determinants that enhance plant development and immunity have been identified and categorized as secretory products. Plants that were inoculated with cell wall components like glucans grew faster and their defensive mechanisms were triggered (Adnan et al. 2018; Kumar et al. 2020). When algae are cultivated on liquid medium, their released products may be gathered in enormous quantities (Lin et al. 2020). These defensive hormones greatly boost plant defence systems when given exogenously. Items released by algae in liquid culture have a lot of promise for use in the field. In general, large-scale algae production, usually by heterotrophic culture, is done to collect algal cells. The non-cellular components are regarded as waste products that must be detoxified (Reddy et al. 2019). The potential uses of cell-free extracts will substantially expand if they can be utilized for plants. However, before applying algae to crop plants, a number of issues must be addressed. First and foremost, the potential risks of cell-free algal extracts must be analysed and eliminated. Algae frequently emit dangerous compounds during cultivation. As a result, quality monitoring of algal liquid culture is crucial. Secondly, the development of algae determinants must be improved for a complex programme. Third, consideration should be given to the preparation of cell-free extracts (Reddy et al. 2019). An immersion or drip irrigation method could be used to give the cell-free extract. It is, however, crucial to obtain a significant volume of extracts. As a result, the extract should be evaporated and purified using chemical and physical methods, with the completed substance demonstrating great efficacy and being employed in agricultural (Arun et al. 2020). Furthermore, determinants are granulated similarly to other agricultural goods like fertilizers and agrochemicals. Finally, in the near future, a specialized process for isolating beneficial algae for plant health enhancement must be created. The extract should be evaporated and purified using chemical and physical methods, with the completed substance demonstrating great efficacy and being employed in agriculture (Das et al. 2019). Despite the fact that microorganisms have been employed to increase production fitness, a new study shows that phytoplankton may also improve crop output and function as biocontrol agents against illnesses by directly blocking pathogen development and activating plant immune systems. As a result, cyanobacteria represent a unique biochemical supply that may be used as bio-fertilizers and plant protectants, implying that microalgae are a naturally occurring reactive ecosystem constituent (Win et al. 2018).

10.4 APPLICATION OF ALGAE-BASED BIO-FERTILIZERS IN AGRICULTURE

Fertilizers are used extensively in agricultural production operations to boost crop output, however the widespread use of synthetic and pesticide fertilizers presently available could pose a severe environmental threat (Khan et al. 2016; Godlewska et al. 2019). As a result, fertilizer research is concentrating on employing microorganisms as a more environmentally friendly method for sustainable agriculture. Microorganisms that may coexist with higher plants are included in the microbiological toolbox (Vishwakarma et al. 2018).

Bio-stimulants are inorganic substances produced by diverse microorganisms that help seeds, plants, and soil bacterial consortia thrive by supplying vital micronutrients, phosphorus, copper, as well as other nutrients (Zahra et al. 2020; Saud et al. 2022b). Bio-fertilizers provide contemporary agriculture with renewable and environmentally friendly solutions, in specific, that may assist to enable productivity expansion. Fertilizers, on either hand, face a slew of issues, such as a short shelf life (3–4 weeks) and stringent storage requirements (Stella et al. 2019).

Different species have been shown to be useful as bio-fertilizers and soil conditioners in recent research. Chrysophyta are the major groups of microalgae. The only photosynthetic prokaryotes capable of producing oxygen are cyanobacteria (Stella et al. 2019).

There are over 1 million kinds of both unicellular microalgae and more sophisticated eukaryotic eukaryotes. Microalgae are divided into 150 genera and roughly 2000 species. Eukaryotic microalgae and bacterial cyanobacteria are both classified as algae (Schaap et al. 2012).

10.5 INDUSTRIAL APPLICATIONS OF ALGAE

The capacity of algae to flourish with low water resources and to use locations that would otherwise be unavailable to agriculture provides economic advantages. Microalgae are commonly assumed to have a large effect on important ecosystem processes as they can recover extra nutrients and recycle water for reuse and may be created in wastewater and agricultural runoff (Priyadarshani and Rath 2012). They can also help to reduce greenhouse gas emissions by capturing nitrous oxide from industrial sources. The most basic photosynthetic plants on the Earth are microalgae. These bacteria can make sustenance out of inorganic substances and can be found in large numbers in liquid environments (Hussain et al. 2016; Zahra et al. 2020). Microbes have various unique traits that make them a desirable bio-fertilizer supplier for enhancing soil physio-chemical properties, including moisture capacity, rapid generation time, ability to fix atmospheric N_2, and resistance to hostile settings. Microbes can also produce biologically active compounds that improve soil chemical characteristics, facilitate nutrient movement from the ground to crops, and cause soil aggregation (Wijffels et al. 2013).

10.6 APPLICATION OF ALGAL BIO-FERTILIZER

Recent studies, both in fundamental and practical research, have demonstrated that algal bio-fertilizers have good effects on plants, crop output, soil microorganisms, fruit nutritional value, and seed germination. Researchers from academia and industry have created three fundamental models for employing algae in crops, motivated by the environmental and economic benefits (Altaf et al. 2016; Das et al. 2019). To commence with, biofuel is employed as a slow-release fertilizer that may be digested by soil bacteria and which delivers nutrients to plants on a continuous basis. Second, live algae cells could be used to control soil moisture, improve soil fertility, and regulate the microbial community. Finally, as a solvent plant, an algae extract containing amino acids and minerals could be applied to the surfaces of plant leaves. Gradual bio-fertilizers, ammonia microalgae, and aqueous plant all have distinct manufacturing techniques and action mechanisms, but they all have the potential to accelerate plant growth and improve soil conditions (Iqbal et al. 2021).

10.7 SLOW-RELEASE BIO-FERTILIZERS

Microbial activity accelerates biomass degradation, which extends the life of algal bio-fertilizers. As a consequence, an algal fertilizer with a delayed release performe better than an ordinary chemical fertilizer. Slow-release algal fertilizer is made by combining it with an algal culture (Ramli 2019). Dehydration is used to evaporate the water of microalgae biomass and keep the bio-fertilizer from decaying too quickly during storage. When paired with algae growth and biomass collection, this technique can be successful, resulting in a high algae spore cost. Gradual algal fertilizer is used in agriculture for a variety of reasons, including stimulating plant growth and improving soil health (Ali et al. 2016; Chen et al. 2020). To begin with, plants may ingest algal fertilizer nutrients and minerals, phosphate, hydrocarbons, and micronutrients. According to Coppens et al. (2016), applying algal bio-fertilizer to tomato plants boosted leaf elongation and fresh and dry weight, as well as increasing the accumulation of sugar, sucrose, and antioxidants in tomatoes. Several more researchers have confirmed the beneficial benefits of algal bio-fertilizer on plant development (Yao et al. 2013).

Second, the nutrients produced by algal fertilizers have the potential to boost soil fertility. Microalgae have been shown to enhance the amount of nitrogen, phosphorus, and potash accessible in topsoil, supplying more nutrition to plants. Furthermore, the amounts of ammonium, phosphate, and potash that remained in algal biomass-fertilized soil at harvest were greater than the amounts of soil quality fertilized by other treatments. As a result, long-term use of an algal bio-fertilizer may aid in soil health preservation (Cole et al. 2016).

10.8 CYANOBACTERIA NITROGEN FIXATION

Cyanobacteria that fix nitrogen are beneficial to agriculture in five ways (Cole et al. 2016). Agricultural crops have been reported to obtain up to 20–30 kg nitrogen ha^{-1} using cyanobacteria fertilizer. As a consequence, the amount of chemical nitrogen fertilizer used in farming could be lowered (Agawin et al. 2007). As a conclusion, broad adoption of nitrogen-fixing cyanobacteria could pave the way for long-term agricultural viability. In practice, cyanobacteria can be employed as a bacterial biofilm or as a microbial inoculant. To begin with, it is the most straightforward technique for microbiological inoculating of soil with cyanobacteria organisms. Photosynthetic bacteria have been shown in the past to improve soil quality and aid plant growth. In certain circumstances, microbial inoculants may struggle to survive in soil, and their PGP powers are reliant on their ability to thrive in this environment. Second, the cyanobacteria biofilm can be created before organisms are inoculated into the soil. Photosynthetic bacteria can be coupled with other microbes to improve the efficacy of the biofilm in soil protection in the real world (Meeks 1998).

10.9 LIQUID BIO-FERTILIZERS

Aqueous plants are algae extracts that are filled with nutritional requirements for plant development. When algal extracts are sprayed om leaves, nutrients may reach plants or crops through leaf pores. A liquid bio-fertilizer is more efficient than a slow-release fertilizer in terms of plant and crop use. As a response, eco-sustainable farming is increasingly turning to liquid bio-fertilizers made from microalgae (Maheswari and Elakkiya 2014). High-pressure gas treatment, bead grinding, and ultrasonic therapy are examples of physical techniques that alter the structure of algal cells by physical hits or shocks (Dey 2021). It is worth noting, however, that certain physical approaches consume a lot of energy and require specialized equipment. Hydrolytic enzymes involve the process of using enzymes like cellulase, mannanase, xylanase, or pectinase to break down particular components in cell walls. Enzymatic hydrolysis, as opposed to physical and chemical methods, may take place in a much softer micro-environment, decreasing the damage caused by wall breaking of value-added biological composites (Sivamurugan et al. 2018). In some cases, the aforementioned procedures for breaking down algal cell walls can be combined to ensure the most effective efficiency (Sivamurugan et al. 2018).

10.10 ALGAE-BASED BIO-FERTILIZERS

Cyanobacteria provide the most remarkable potential because they can thrive in the midst of highly concentrated pollutants in a range of pollution flows that are poisonous to living organisms. For agricultural production to become more effective and efficient, this is crucial (Mahapatra et al. 2018).

10.10.1 Soil Fertility

Despite the lack of information on how algal biomass development from wastewater treatment influences soil nitrogen kinetics, it has potential as a soil amendment. In studies concentrating on indigenous *Anabaena* species, this strain's ability to enhance soil fertility while reducing soil density was established (Bhardwaj et al. 2014).

The study of microalgae and microbial communities in bio-fertilizer applications is another fascinating research topic. In fact, it may be more efficient than employing individual microbes, not just for contaminant detoxification and nutrient removal from wastewaters, but also for maximizing the utilization of available nitrogen, phosphorus, and K in the soil. If algae and bacteria could minimize pollution, consortium engineering would be a triumph. Moreover, data suggest that microalgal partnerships have a lot of potential for soil amendment on marginal lands, which might help with the conversion of the soil in these areas (Singh et al. 2020).

10.10.2 Nitrogen Fixation

The capacity of phytoplankton to fix ammonium is one of the reasons they are used as bio-fertilizers. Microbes transform inorganic nitrogen (N_2) from the air into organic nitrogen that higher plants may utilize. Microalgae have been used to boost rice output in India and Chile. In Chile, local cyanobacterial strains have been demonstrated to increase the efficiency of nitrogen absorption in rice paddies (Bhat et al. 2015).

Inoculants were employed in another investigation to produce colonies in chickpea production. Furthermore, when utilized as a bio-stimulant, bacterial association (21 species including proteobacteria, bacteriodetes, chlorophyta, and others) was demonstrated to get a higher potential for nitrification (10,294 nmol ethylene/g dry weight/h) (Mohammadi et al. 2010).

10.10.3 Production of Plant Growth Stimulants

Several algal compounds have been found to boost root development, either actively or passively, or vegetation symbiotic involving soil microbes, hence enhancing bioavailability. To test this theory, plant growth-stimulating chemicals were employed to cultivate *Lupinus termis* utilizing cyanobacteria and bacteria (Win et al. 2018).

10.10.4 Bio-pesticidal Substances

In vitro studies have shown that cyanobacteria extracts and exudates impede hatching and cause immobilization and fatality in adolescent plant parasite microbes, suggesting that microalgae might be employed as nematocidal biocontrol agents (Kalra and Khanuja 2007). The antifungal and antibacterial characteristics of culture filtrate were studied when it showed hydrolytic action against phytopathogens (Morales-Borrell et al. 2020). A most commercially significant fungal disease is *Fusarium* sp., although other fungal diseases have been managed under the same conditions. According to studies, the biocidal capabilities of algae have revealed significant promise for creating novel pest control systems. Further study is required to establish their frequency and economic effect (Abbey et al. 2019).

10.10.5 Algal Toleration and Mutation

10.10.5.1 Tolerance to Extreme Environmental Conditions

Microalgae can resist a number of pressures in the ecosystem. Ningthoujam et al. observed that *Anabaena variabilis* could survive 100 lg/mL malathion in terms of pesticide tolerance. Jha et al. demonstrated that Mn and sodium may negatively affect cyanobacteria, which is an example of salinity tolerance ability (Na, –30.19 percent) (Shehata and Whitton 1982).

Sinha and Hader looked at cyanobacteria's UV-B photosensitizing process, which might be effective in bio-stimulants for field crops. The ability of active microorganisms to adapt to their environment is a major practical issue with commercial soil amendment (Lee et al. 2017).

10.10.5.2 Mutation of Algae for Better Bio-fertilizers

As a result, novel methods for creating cyanobacterial mutants are appropriate for investigating cyanobacteria's potential (Win et al. 2018). To establish a gene transfer mechanism in *Oscillatoria*

MKU 277, Ravindran et al. produced the plasmid pRL489 and electroporated it into the cyanobacterium. As a result of this research, the methods for producing more potent and viable bio-fertilizer strains have improved (Shah et al. 2016c; Chakraborty and Akhtar 2021). For nitrate shortage, this mutant is employed in paddy fields as a soil amendment alongside commercial nitrate fertilizer without impacting N_2-fixation. Singh et al. also improved an *A. variabilis* mutant that was grown in a weedkiller environment. Advancements in synthetic biology have produced a new study area in algal biotech which may provide a stronger solution to these challenges (Irisarri 2006).

10.10.5.3 Large-scale Algal Growth

Biomass production for farming production, particularly from waste streams, has become a profitable industry. Several research groups have linked the generation of microalgae biomass with the bioremediation of industrial effluent. It is reported that using cyanobacteria will help only to improve soil quality but it will not reduce the use of pesticides (Gupta et al. 2015). Its potential to increase soil's lateral rigidity, availability of nutrients, and crop growth has been further exploited under a water-constrained environment, which is critical for extended crop growth. Indigenous algae strains are superior for fertilizer applications, according to the findings of previous investigations (Schreiber et al. 2017).

Algae's role in removing toxins from various ecosystems is presently undervalued. Several studies have emphasized the importance of economically viable algal production. Algal biomass can be utilized for grazing, bioenergy fuel, or organic fertilizers after harvesting (Chisti 2016).

On hydrocarbon discharge, municipal wastewater, slaughter house effluent, municipal wastewater, residential sewerage, industrial discharges, farmed effluent, and milk raw sewage, phytoplankton varieties were effectively produced. Barminski et al. used tank techniques, dump approach, outdoor process, and nurseries combining algal methods of production to produce medium compositions for continuous development of N-fixing cyanobacteria (Lian et al. 2018).

10.11 FORMULATION OF ALGAL BIO-FERTILIZERS

A commercially feasible algal bio-fertilizer composition has been created and tested. Particulate inoculants, for example, were used by Dubey et al. for longer-term strain inoculation in soil. The algal population in the soil was 10–70 times higher in the injected regions and after 4 months. Mishra et al. supplied technologies to growers after they received a soil-based bacterial strain, allowing them to produce bio-fertilizer themselves with minimal additional imports. Tripathi et al. used a technology known as fly-ash (FA) to improve rice plant growth rate and yield by combining cyanobacteria with nitrogen fertilizers. As a result of this method, the demand for nitrogen fertilizers has decreased (Saif et al. 2021).

Using transparent containers, dry blending, and agricultural residues as a carrier can extend the shelf life of macroalgal soil amendments. The use of phytoplankton as a carrier boosted inoculation absorption and longevity. To reduce the risk of methanotrophic illness after preservation, it was observed that tobacco waste outperformed its competitors. Moreover, because akinetes may be dried for at least several weeks, it was suggested that they be used as a bio-fertilizer after drying. The effects of temperature and pH on algal growth have been studied (Prasanna et al. 2020). Wet blue-green algae (BGA) can be air dried or baked for 24 hours at 35–40°C in the dark to preserve cyanobacteria. This method is simple to use and ensures a high germination rate. Grzesik et al. studied foliar bio-fertilization by algae on willow monocultures and discovered that it improved yield potential (Prasanna et al. 2015). Foliar applications of aqueous extracts of *Acutodesmus dimorphus*, as well as the formulation of the nutrient broth, have been proven to have a positive impact on germination and seedling growth in studies. *A. dimorphus* cellular extract and dry biomass were shown to be effective as a bio-stimulant and bio-fertilizer in Roma seedlings, promoting faster germination, plant growth, and flower production. The carrier-based formulation is better for

N_2-fixing fertilizers and soil preparation, whereas the intervening factors composition is superior for germinating boosting impacts (Holajjer et al. 2013).

10.12 CHALLENGES AND MEASURES FOR COMMERCIALIZATION

Finding an appropriate carrier for each algae, as well as soil and climatic conditions, and biotic and abiotic stresses in the field, are all major obstacles to commercial use of microalgae as a bio-fertilizer. Because of their habitat and growth needs, cyanobacteria-based organic fertilizers have been frequently employed in improving paddy field productivity (Zou et al. 2021).

According to various studies, macroalgal bio-fertilizers are limited in places where pesticide wastes, herbicide by-products, toxic substances (including aluminium and copper), or highly saline soils are prevalent. Algal development, cellular photosynthesis, and nitrogen fixation could all be hampered by these chemicals. All four cyanobacterial cultures have been utilized as tools for researching intracellular metabolic processes involved in the creation of medicinal and economically useful compounds for a long time. Sharma et al. suggested integrating multidisciplinary methodologies with a multiproduct process (biorefinery) strategy in order to get the most out of cyanobacteria (Sampathkumar et al. 2019).

10.12.1 New Pollutant Factors Discovered in Wastewater

Microalgae have been shown to be able to absorb nutrients from wastewater, including ammonia, phosphate, and hydrocarbons. Algal blooms, on the other hand, are not a good strategy for effluent treatment since they might bring more polluting components into the aqueous phase. Algal development, for example, can raise the acidity of effluent, therefore acidic effluent cannot be released immediately (Molinuevo-Salces et al. 2019). After algal cultivation, the pH of wastewater reached 9.32 in Khan et al (2019)'s research. *Spirulina* sp., a kind of cyanobacteria commonly used as a bio-fertilizer feedstock, elevated the pH of the culture medium to 10.60, according to Cardoso et al. (2021). Second, a variety of prokaryotic cyanobacteria may emit poisonous components, resulting in ecological disasters in waterbodies (Win et al. 2018). Finally, because most technologies cannot guarantee 100% system throughput, some live algae may persist in the water phase after wastewater treatment. Living creatures will penetrate streams as a result of effluent discharged, promoting algae growth and disturbing ecological equilibrium. The functions of algae in treating wastewater should be thoroughly researched due to the different pollution elements supplied by phytoplankton, notably cyanobacteria. These new contamination components, especially bacterial toxins and buffering capacity, may be more detrimental to the ecosystem than eutrophic wastewater, in our opinion (Win et al. 2018).

10.12.2 High Cost of Algal Biomass

One of the most significant barriers to commercializing algal biomass is the excessive cost of manufacturing. Choosing algal varieties, developing plankton, collecting algae, drying biomass, and preparing fertilizer are all steps in the manufacturing of algal bio-fertilizers, which all result in excessive organic matter and algae bio-fertilizer prices (Sharma et al. 2021). Owing to the reduced economics of conventional agriculture, utilizing algal bio-fertilizer to replace chemical fertilizer in large-scale agricultural output is not the best option for farmers in underdeveloped countries (Iqbal et al. 2021).

10.12.3 Contamination of Algal Biomass

Toxic pollutants in sewage, especially heavy metals, may contaminate algal biomass, resulting in toxic pollutants building up in the soil and crops. Furthermore, algae with hydroxyls and carboxyl on their cell surfaces are charged negatively and function well in the adsorption of heavy metals. As a result, algae cultivation entails solubilizing and accumulating heavy metals while also devouring minerals from growth materials (Han et al. 2020). Constant irrigation with wastewater may result in heavy metal accumulation in the soil and, as a result, increased heavy metal absorption by crops and plants. As a result, soil irrigation with sewage is strictly prohibited in many countries, as it has the potential to cause huge ecological disasters and pollution. If heavy metal ions in sewage are not properly cleaned, employing algae for soil fertilization is another method of irrigating fields with wastewater, since algae may convey dangerous compounds into the surface (Tripathi et al. 2008).

10.12.4 Water Consumption and Water Loss

Algae cultivation is a water-intensive process that takes a huge amount of water and produces effluent containing leftover nutrients and/or other pollutants (Zou et al. 2021).

10.12.5 Potential Threats of Cyanobacteria in the Environment

Microcystins have been seen to accumulate in salad lettuce and clover, posing a threat to food safety. Secondly, live cyanobacteria used as bio-fertilizers can spread to neighboring bodies of water, generating an algal bloom and biological invasions. This problem poses a serious threat to the modern agricultural environment, which frequently involves both crop farming and fish aquaculture. Other factors preventing algal bio-fertilizer use in agricultural production include residueal nutrition in sewage after algal growth, difficult storage and transportation of solvent bio-fertilizers, unregulated removal efficiency of microalgae biomass for use as a slow-release bio-fertilizer, and reduced consumer acceptance of bio-fertilizers (Win et al. 2018).

10.13 CONCLUSION

Algae-based bio-fertilizers have proven to be useful in the growth of green agriculture. In addition to N_2 fixation, they can improve soil fertility and enable the growth of crops that PGPR benefit. Carrier technologies are widely available. Furthermore, supplying farmers with the technology they require for their individual demands can provide value and assist in the development of small-scale bio-fertilizer production in their area.

Other advantages of algal fertilizers include the expansion of organic agriculture without the need for a large amount of land or even the efficient use of land areas. Finally, algal organic fertilizers can be employed to meet the needs of sustainable agriculture, with three primary aims in mind: environmental wellness, economic welfare, and social justice (Win et al. 2018).

REFERENCES

Abbey L, Abbey J, Leke-Aladekoba A, Iheshiulo EMA, Ijenyo M (2019) Biopesticides and biofertilizers: types, production, benefits, and utilization. In: Benjamin K Simpson, Alberta N Aryee, Fidel Toldrá (Eds.), *Byproducts from Agriculture and Fisheries: Adding Value for Food, Feed, Pharma, and Fuels*. Wiley, pp. 479–500.

Adnan M, Shah Z, Sharif M, Rahman H (2018) Liming induces carbon dioxide (CO_2) emission in PSB inoculated alkaline soil supplemented with different phosphorus sources. *Environ Sci Pol Res* 25:9501–9509.

Agawin NS, Rabouille S, Veldhuis MJ, Servatius L, Hol S, van Overzee HM, Huisman J (2007). Competition and facilitation between unicellular nitrogen-fixing cyanobacteria and non-nitrogen-fixing phytoplankton species. *Limnol & Oceanogr* 52:2233–2248.

Ali J, Muhammad H, Ullah I, Rashid JA, Adnan M, Ali M, Ahmad W, Rehman A, Khan J (2016) Mango seed germination in different media at different depth. *J Nat Sci Res* 6(1):56–59.

Altaf M, Raziq F, Khan I, Hussain H, Shah B, Ullah W, Naeem A, Adnan M, Junaid K, Shah SRA, Attaullah, Iqbal M (2016) Study on the response of different maize cultivars to various inoculum levels of *Bipolarismaydis* (Y. Nisik& C. Miyake) shoemaker under field conditions. *J Ento & Zool Studies* 4(2):533–537.

Arshad M, Adnan M, Ali A, Khan AK, Khan F, Khan A, Kamal MA, Alam M, Ullah H, Saleem A, Hussain A, Shahwar D (2016) Integrated effect of phosphorous and zinc on wheat quality and soil properties. *Adv Environ Biol* 10(2):40–45.

Arun S, Sinharoy A, Pakshirajan K, Lens PN (2020) Algae based microbial fuel cells for wastewater treatment and recovery of value-added products. *Renewable Sustainable Energy Rev* 132:110041.

Besser H, Hamed Y (2021) Environmental impacts of land management on the sustainability of natural resources in Oriental Erg Tunisia, North Africa. *Environ, Dev & Sustainability* 23:11677–11705.

Bhardwaj D, Ansari MW, Sahoo RK, Tuteja N (2014) Biofertilizers function as key player in sustainable agriculture by improving soil fertility, plant tolerance and crop productivity. *Microb Cell Fact* 13:1–10.

Bhat TA, Ahmad L, Ganai MA, Khan O (2015) Nitrogen fixing biofertilizers; mechanism and growth promotion: a review. *J Pure Appl Microbiol* 9:1675–1690.

Bhuyar P, Muniyasamy S, Govindan N (2018) Green revolution to protect environment–an identification of potential micro algae for the biodegradation of plastic waste in Malaysia. *World Congress on Biopolymers and Bioplastics & Recycling. Expert Opin Environ Biol* 7.

Cardoso CKM, Mattedi S, Lobato AK de CL, Moreira ÍTA (2021) Remediation of petroleum contaminated saline water using value-added adsorbents derived from waste coconut fibres. *Chemosphere* 279:130562. doi.org/10.1016/j.chemosphere.2021.130562.

Chakraborty T, Akhtar N (2021). Biofertilizers: prospects and challenges for future. *Biofertilizers: Study and Impact* 575–590.

Chen J, Fan X, Zhang L, Chen X, Sun S, Sun RC (2020) Research progress in lignin-based slow/controlled release fertilizer. *Chem Sus Chem* 13:4356–4366.

Chisti Y (2016) Large-scale production of algal biomass: raceway ponds. In: Faizal Bux, Yusuf Chisti (Eds.), *Algae Biotechnology*. Springer, pp. 21–40.

Cole JC, Smith MW, Penn CJ, Cheary BS, Conaghan KJ (2016) Nitrogen, phosphorus, calcium, and magnesium applied individually or as a slow release or controlled release fertilizer increase growth and yield and affect macronutrient and micronutrient concentration and content of field-grown tomato plants. *Scientia Hortic* 211:420–430.

Coppens J et al. (2016) The use of microalgae as a high-value organic slow-release fertilizer results in tomatoes with increased carotenoid and sugar levels. *J Appl Phycol* 28:2367–2377. doi.org/10.1007/s10811-015-0775-2

Das P, Khan S, Chaudhary AK, AbdulQuadir M, Thaher MI, Al-Jabri H (2019) Potential applications of algae-based bio-fertilizer. In: Bhoopander Giri, Ram Prasad, Qiang-Sheng Wu, Ajit Varma (Eds.), *Biofertilizers for Sustainable Agriculture and Environment*. Springer, pp. 41–65.

Dey A (2021) Liquid biofertilizers and their applications: an overview. In: Bibhuti Bhusan Mishra, Suraja Kumar Nayak, Swati Mohapatra, Deviprasad Samantaray (Eds.), *Environmental and Agricultural Microbiology: Applications for Sustainability*. Wiley, pp. 275–292.

Godlewska K, Michalak I, Pacyga P, Baśladyńska S, Chojnacka K (2019) Potential applications of cyanobacteria: spirulina platensis filtrates and homogenates in agriculture. *World J Microbiol & Biotechnol* 35:1–18.

Gupta PL, Lee S-M, Choi H-J (2015) A mini review: photobioreactors for large scale algal cultivation. *World J Microbiol & Biotechnol* 31:1409–1417.

Han W, Mao Y, Wei Y, Shang P, Zhou X (2020) Bioremediation of aquaculture wastewater with algal-bacterial biofilm combined with the production of selenium rich biofertilizer. *Water* 12:2071.

Holajjer P, Kamra A, Gaur H, Manjunath M (2013) Potential of cyanobacteria for biorational management of plant parasitic nematodes: a review. *Crop Prot* 53:147–151.

Hussain F, Shah SZ, Ahmad H, Abubshait SA, Abubshait HA, Laref A, Manikandan A, Kusuma HS, Iqbal M (2021) Microalgae an ecofriendly and sustainable wastewater treatment option: biomass application in biofuel and bio-fertilizer production. A review. *Renewable & Sustainable Energy Rev* 137:110603.

Hussain I, Alam SS, Khan I, Shah B, Naeem A, Khan N, Ullah W, Iqbal B, Adnan M, Junaid K, Shah SRA, Ahmed N, Iqbal M (2016) Study on the biological control of fusarium wilt of tomato. *J Ento & Zool Studies* 4(2):525–528.

Iqbal MM, Muhammad G, Aslam MS, Hussain MA, Shafiq Z, Razzaq H (2021) Algal biofertilizer. In: Inamuddin, Mohd Imran Ahamed, Rajender Boddula, Mashallah Rezakazemi (Eds.), *Biofertilizers: Study and Impact*. Wiley, pp. 607–635.

Irisarri, P. (2006). Role of cyanobacteria as biofertilizers: potentials and limitations. In: Rai MK (Ed.), *A Handbook of Microbial Biofertilizers*. The Harworth Press Inc, USA, pp. 417–430.

Iyovo GD, Du G, Chen J (2010). Sustainable bioenergy bioprocessing: biomethane production, digestate as biofertilizer and as supplemental feed in algae cultivation to promote algae biofuel commercialization. *J Microb Biochem Technol* 2:100–106.

Kalra A, Khanuja S (2007) Research and development priorities for biopesticide and biofertiliser products for sustainable agriculture in India. In: Teng PS (Ed.), *Business Potential for Agricultural Biotechnology.* Asian Productivity Organisation 96102.

Khan I, Alam S, Hussain H, Shah B, Naeem A, Ullah W, Ali W, Adnan M, Junaid K, Shah SRA, Ahmed N, Iqbal M (2016) Study on the management of potato black scurf disease by using biocontrol agent and phytobiocides. *J Ento & Zool Studies* 4(2):471–475.

Kulikova N, Volkova E, Bondarenko N, Chebykin E, Saibatalova E, Timoshkin O, Suturin A (2018) Element composition and biogeochemical functions of algae Ulothrix zonata (F. Weber et Mohr) Kützing in the coastal zone of the Southern Baikal. *Water Resour* 45:908–919.

Kumar M, Sun Y, Rathour R, Pandey A, Thakur IS, Tsang DC (2020) Algae as potential feedstock for the production of biofuels and value-added products: Opportunities and challenges. *Sci Total Environ* 716:137116.

Lee S-M, Ryu C-M (2021) Algae as new kids in the beneficial plant microbiome. *Front Plant Sci* 12:91.

Lee T-M, Tseng Y-F, Cheng C-L, Chen Y-C, Lin C-S, Su H-Y, Chow T-J, Chen C-Y, Chang J-S (2017) Characterization of a heat-tolerant Chlorella sp. GD mutant with enhanced photosynthetic CO_2 fixation efficiency and its implication as lactic acid fermentation feedstock. *Biotechnol Biofuels* 10:1–12.

Lian J, Wijffels RH, Smidt H, Sipkema D (2018) The effect of the algal microbiome on industrial production of microalgae. *Microb Biotechnol* 11:806–818.

Lin Z, Li J, Luan Y, Dai W (2020) Application of algae for heavy metal adsorption: a 20-year meta-analysis. *Ecotoxicol & Environ Safety* 190:110089.

Mahapatra DM, Chanakya H, Joshi N, Ramachandra T, Murthy G (2018) Algae-based biofertilizers: a biorefinery approach. In: *Microorganisms for Green Revolution*. Springer, pp. 177–196.

Maheswari NU, Elakkiya T (2014) Effect of liquid biofertilizers on growth and yield of Vigna mungo L. *Int J Pharm Sci Rev Res* 29:42–45.

Meeks JC (1998) Symbiosis between nitrogen-fixing cyanobacteria and plants. *BioScience* 48:266–276.

Mohammadi K, Ghalavand A, Aghaalikhani M (2010) Effect of organic matter and biofertilizers on chickpea quality and biological nitrogen fixation. *Int J Agric Biosyst Eng* 4:578–583.

Molinuevo-Salces B, Riaño B, Hernández D, Cruz García-González M (2019) Microalgae and wastewater treatment: advantages and disadvantages. In: Md Asraful Alam, Zhongming Wang (Eds.), *Microalgae Biotechnology for Development of Biofuel and Wastewater Treatment*. Springer, pp. 505–533.

Morales-Borrell D, González-Fernández N, Mora-González N, Pérez-Heredia C, Campal-Espinosa A, Bover-Fuentes E, Salazar-Gómez E, Morales-Espinosa Y (2020) Design of a culture medium for optimal growth of the bacterium Pseudoxanthomonas indica H32 allowing its production as biopesticide and biofertilizer. *AMB Express* 10:1–10.

Park S, Kim J (2018) Red tide algae image classification using deep learning based open source. *Smart Media J* 7:34–39.

Prasanna R, Gupta H, Yadav VK, Gupta K, Buddhadeo R, Gogoi R, Bharti A, Mahawar H, Nain L (2020) Prospecting the promise of cyanobacterial formulations developed using soil-less substrates as carriers. *Environ Technol & Innov* 18:100652.

Prasanna R, Hossain F, Babu S, Bidyarani N, Adak A, Verma S, Shivay YS, Nain L (2015) Prospecting cyanobacterial formulations as plant-growth-promoting agents for maize hybrids. *South African J Plant & Soil* 32:199–207.

Priyadarshani I, Rath B (2012) Commercial and industrial applications of micro algae – A review. *J Algal Biomass Util* 3:89–100.

Ramli RA (2019) Slow release fertilizer hydrogels: a review. *Polymer Chemistry* 10:6073–6090.

Reddy CN, Nguyen HT, Noori MT, Min B (2019) Potential applications of algae in the cathode of microbial fuel cells for enhanced electricity generation with simultaneous nutrient removal and algae biorefinery: current status and future perspectives. *Bioresour Technol* 292:122010.

Saif S, Abid Z, Ashiq MF, Altaf M, Ashraf RS (2021) Biofertilizer Formulations. *Biofertilizers: Study and Impact* 211–256.

Sampathkumar P, Dineshkumar R, Rasheeq AA, Arumugam A, Nambi K (2019) Marine microalgal extracts on cultivable crops as a considerable bio-fertilizer: a review. *Indian J Trad Knowl (IJTK)* 18:849–854.

Saud S, Li X, Jiang Z, Fahad S, Hassan S (2022b) Exploration of the phytohormone regulation of energy storage compound accumulation in microalgae. *Food Energy Secur* 2022:e418

Schaap A, Rohrlack T, Bellouard Y (2012) Optical classification of algae species with a glass lab-on-a-chip. *Lab on a Chip* 12:1527–1532.

Schreiber C, Behrendt D, Huber G, Pfaff C, Widzgowski J, Ackermann B, Müller A, Zachleder V, Moudříková Š, Mojzeš P (2017) Growth of algal biomass in laboratory and in large-scale algal photobioreactors in the temperate climate of western Germany. *Bioresour Technol* 234:140–149.

Shah B, Khan IA, Akbar R, Fayaz W, Zaman M, Din MMU, Ahmad N, Adnan M, Junaid K, Shah SRA, Rahman IU (2016c) Population dynamics of corn flea beetle and fourspotted leaf beetle on different maize cultivars at Peshawar. *J Ento & Zool Studies* 4(1):308–311.

Sharma GK, Khan SA, Shrivastava M, Bhattacharyya R, Sharma A, Gupta DK, Kishore P, Gupta N (2021) Circular economy fertilization: phycoremediated algal biomass as biofertilizers for sustainable crop production. *J Environ Manage* 287:112295.

Shehata FH, Whitton BA (1982) Zinc tolerance in strains of the blue-green alga Anacystis nidulans. *British Phycol J* 17: 5–12.

Singh B, Upadhyay A, Al-Tawaha T, Al-Tawaha A, Sirajuddin S (2020) Biofertilizer as a tool for soil fertility management in changing climate. In: IOP Conference Series: Earth and Environmental Science, Volume 492. IOP Publishing, p. 012158.

Sivamurugan A, Ravikesavan R, Singh A, Jat S (2018) Effect of different levels of P and liquid biofertilizers on growth, yield attributes and yield of maize. *Chem Sci Rev Lett* 7:520–523.

Stella M, Theeba M, Illani Z (2019) Organic fertilizer amended with immobilized bacterial cells for extended shelf-life. *Biocatalysis & Agric Biotechnol* 20:101248.

Tripathi R, Dwivedi S, Shukla M, Mishra S, Srivastava S, Singh R, Rai U, Gupta D (2008) Role of blue green algae biofertilizer in ameliorating the nitrogen demand and fly-ash stress to the growth and yield of rice (Oryza sativa L.) plants. *Chemosphere* 70:1919–1929.

Vishwakarma K, Upadhyay N, Kumar N, Tripathi DK, Chauhan DK, Sharma S, Sahi S (2018) Potential applications and avenues of nanotechnology in sustainable agriculture. In: *Nanomaterials in Plants, Algae, and Microorganisms*. Elsevier, pp. 473–500.

Wijffels RH, Kruse O, Hellingwerf KJ (2013) Potential of industrial biotechnology with cyanobacteria and eukaryotic microalgae. *Curr Opinion Biotechnol* 24:405–413.

Win TT, Barone GD, Secundo F, Fu P (2018) Algal biofertilizers and plant growth stimulants for sustainable agriculture. *Ind Biotechnol* 14:203–211.

Yao Y, Gao B, Chen J, Yang L (2013) Engineered biochar reclaiming phosphate from aqueous solutions: mechanisms and potential application as a slow-release fertilizer. *Environ Sci & Technol* 47:8700–8708.

Zahra Z, Choo DH, Lee H, Parveen A (2020) Cyanobacteria: review of current potentials and applications. *Environ* 7:13.

Zhao H-x, Magoulès F (2012) A review on the prediction of building energy consumption. *Renewable Sustainable Energy Rev* 16:3586–3592.

Zou Y, Zeng Q, Li H, Liu H, Lu Q (2021) Emerging technologies of algae-based wastewater remediation for bio-fertilizer production: a promising pathway to sustainableagriculture. *J Cheml Technol & Biotechnol* 96:551–563.

11 Ectomycorrhizal Fungi
Role as Bio-fertilizers in Forestry

Hafiz Muhammad Rashad Javeed,[1] Mazhar Ali,[1] Muhammad Shahid Ibni Zamir,[2] Rafi Qamar,[3] Muhammad Mubeen,[1] Atique-ur-Rehman,[4] Muhammad Shahzad,[5] Samina Khalid,[1] and Ayman EL Sabagh[6]*

[1]Department of Environmental Sciences, COMSATS University Islamabad, Punjab, Pakistan

[2] Department of Agronomy, University of Agriculture, Faisalabad, Punjab, Pakistan

[3] Department of Agronomy, College of Agriculture, University of Sargodha, Sargodha, Punjab, Pakistan

[4] Department of Agronomy, Bahauddin Zakariya University, Multan, Punjab, Pakistan

[5] Department of Agronomy, University of Poonch Rawalakot, AJK, Pakistan

[6] Department of Agronomy, Faculty of Agriculture, University of Kafrel-sheikh, Kafrel-Sheikh, Egypt

Correspondence: rashadjaveed@cuivehari.edu.pk

CONTENTS

11.1 INTRODUCTION

Fungi have a great role in land colonization, and the colonization of soil by plants without fungi would probably not have been possible (Jansa et al. 2008). Phototrophs face many difficulties while moving from an aquatic environment to an aerial habitat such as due to water shortages and/or a limited supply of soluble minerals, especially phosphorus (Mullineaux and Liu 2020). These

DOI: 10.1201/9781003286233-11

difficulties are overcome by photosynthetic organisms by making a mutualistic relationship with the fungi that are called mycorrhizas (Hoffland et al. 2004). Fungi which have a symbiotic and mildly pathogenic association with the roots of plants are called mycorrhizae (Goltapeh et al. 2008). Mycorrhizae are divided into two classifications: ectomycorrhizae and endomycorrhizae (Bonfante and Anca 2009; Shah et al. 2016). Ectomycorrhizae have an interaction with plant roots that is called symbiosis. Intracellular hyphae make the Hartig and mantle network from the ectomycorrhizae fungi (ECMF) (Mosse et al. 1981). Arbuscular mycorrhizae fungi (AMF) made from the arbuscules, vesicles which variate as compared to ECMF and make association with trees and herbaceous plants. Endomycorrhizae are also subdivided into different classes such as orchid mycorrhizae, monotropoid mycorrhizae, arbutoid mycorrhizae, arbuscular mycorrhizae, and ectendo mycorrhizae (Dickie et al. 2013). These classes are categorized according to the attack of fungal hyphae on plant roots and varies according to the intracellular development of fungal hyphae (Brundrett 2004). Mainly ectomycorrhizal are *Basidiomycetes* and some ectomycorrhizal are *Ascomycetes* (Sharma 2017). In this symbiotic association, the Hartig network is the main one which provides the metabolic exchange between the roots of plants and fungi. The mycorrhizal mantle is linked to the fibres of fungi and spread over the soil. The mycorrhizal mantle performs functions such as captivation, conducting the nutrients from one place to another, helping in mobilization, and providing water to the roots of plants (Horton and Bruns 2001). Ectomycorrhizae include more than 7000 species of fungi, and include significant profitable trees such as oak, pine, and poplar (Rinaldi et al. 2008). Macromycetes are called mushrooms when the reproductive structures are produced on soil, while they are called truffles when they are produced underground (Suz et al. 2012). It is very important to identify a mutualistic partner and its functions to comprehend the biological significance of symbiosis. ECMF were originally classified on the basis of their fruiting bodies but are now divided based on direct identification (Fitter 2005). Mostly fungi are edible, such as saprophytes, and some are ECMF. Some ECMF which have been studied are truffles, chanterelles (*Cantharellus* spp.), and matsutake mushroom. The largely cultivable fungi are tuber melanosporum and black truffle or perigord, however other mushroom species are not cultivated, including white truffles (*T. magnatum*), an expensive Italian fungus (Savoie and Largeteau 2011).

ECMF provide many services in terrestrial ecosystems, especially in forest ecosystems, such as nutrient cycling, etc. The global interest in ECMF is mainly in regard to eatable fruiting bodies and the production of biomass, moreover, restoration of ecosystems and reforestation are also of great interest (Requena 2005). Recently, scientists have been focusing on the application of ECMF in bioremediation, bio-fertilization, and controlling pathogens. In this regard, a novel and innovative step is the valuation of ECMF in forest management. This will ultimately be very helpful in sustainable development (Alizadeh 2011).

11.2 ANALYSING THE DIVERSITY OF ECTOMYCORRHIZAL FUNGI IN FOREST ECOSYSTEMS

To analyse the functional structure of ECMF, describing their biodiversity in forest ecosystems is very important. Firstly, it can be checked by numbering the fruiting fungal entities of recognized ECMF (Buscot et al. 2000). Taxonomic identification can be done by using different mycological identification methods, but most of the fungi forming ECMs do not fulfil this requirement with most forming many unavailable fruitbodies (Agerer 2001). If described morphologically, the root tips of ECMF can be divided into anatomotypes and morphotypes. However, usually, the root tips of ECM are morphologically and anatomically not identified at the fungal species level (Dahlberg 2001). For the past 20 years, PCR-based techniques and methods have been used for these types of analyses. Amplification of the internal transcribed spacer (ITS) area of the ribosomal genes is the most applicable method for species level identification. DNA extraction from only a single mycorrhizal tip allows identification of ECM, then PCR amplification and sequencing are followed by its comparison with the sequences of databases, which has enabled the discovery of many novel ECM fungi such as *Tomentella* sp., *Clavulina* sp., etc. (Courty et al. 2010a,b). The assessment of ECM in many forest ecosystems has been boosted by advancements in sequencing technologies and success in database searches. The analysis of the 16S rNA gene enables scientists to understand and elaborate the bacterial community and ectomycorrhizosphere diversity (Uroz et al. 2007).

11.3 SELECTION OF ECMF FOR SUSTAINABLE DEVELOPMENT

Most preferred ECMF are selected on the basis of the following biotic criteria: isolation, weather or climatic conditions, high or low temperature, insolation, moisture content, properties of soil, such as quality, texture and pore spaces, ability to tolerate the abiotic stress of soil, alleviation of soil contamination, mobility of metal in soil, and/or nutrient cycling (Azul et al. 2014; Khan et al. 2016). ECMF can also be selected on the basis of increased biomass of plants, growth and development of plants, and specification of plant or plant health. There are also different criteria relating to fungi which include fungus availability, effectiveness, ability to grow, and ability to be eaten. Other criteria on which fungus selection is based involve functioning of the environment, human well-being, food availability, and nutraceutical value (Suz et al. 2012).

11.4 ECOLOGICAL FUNCTIONS OF ECMF

ECMF perform many functions such as increasing nutrient and water supply, and increasing the land availability for plants. ECMF increase the availability of those nutrients which are not easily available to plants (Claridge 2002). They also improve the absorption mechanism of P which includes the following: postponement of extrametrical hyphae and Pi (inorganic) transfer, mobility of organic P, release of phosphatases, Pi transporters on the soil edge, and ectomycorrhizal fungi. In forestry, they act as bio-fertilizers and help in the mobilization of insoluble mineral Pi, and release organic acids with low molecular weights. They also improve the N absorption mechanism in the nitrogen cycle (NH_4^+ NO_3) and acclimatization of organic nitrogen (releasing chitinases and protease) (Nara 2008; Ibrahim et al. 2016). Colonies of ECMF around roots provide resistance to soil pathogens. Plants also obtain many non-nutritive benefits due to acclimatization of carbon, variation in the water relation, and the level of phytochromones. ECMF transfer carbon through the fungal mycelium through which different species of plants are connected. Therefore, this improves the stability and variety of ecologies and decreases competition among them (Mello and Balestrini 2018). Placental and marsupial mammals use epigeous and hypogeal sporocarps of ECMF as a food source. The roots of mycorrhizae, their mycelium, and the fruiting bodies provide homes for invertebrates and are a significant food source. The microbial population present in soil is affected by mycorrhizae and the exudates in the mycorrhizosphere and hyphosphere. Soil structure is also improved by the hyphal network formed by ECM (Koide et al. 2014). Mycorrhizal fungi also alter the quality and quantity of soil organic content and subsidize the carbon in soil. They also increase the resistance of plants against biotic and abiotic stresses. There are two types of fungal species, hydrophilic, which are drought sensitive, and hydrophobic, which are drought tolerant. These different species have varying reactions depending upon the hydraulic redistribution pattern (Prieto et al. 2016).

11.5 EVALUATION OF ECM FUNGI

Biodiversity, climatic stability, biogeochemical cycles, and economic growth are regulated and boosted by mycelial chains of ECMF in forest ecosystems. ECMF regulate and enhance the growth of host plants and also create great metabolism networks that play an important role in the functions of an ecosystem. This is ultimately very good for the health of plants (Smith and Read 2008). These fungi are vital drivers for sustainability, innovation, and research in biotechnology, food industry, agroforestry, biomedicine, etc. (Courty et al. 2010a,b). Globally, native forests are under threat from lack of management, pollution, fire, contamination of soil, spread of diseases, and water scarcity. In forest management strategies, the interactions among ECMF and native species are considered to be very important (Dahlberg et al. 2010). However, in a bio-industrial innovation, the contribution of ECMF has been underestimated. The most fundamental services of ECM fungi to mankind are as food, indicators of quality of the environment, enhancement of the value of landscape, ecology tourism, and multifunctionality. In addition, many compounds that are bioactive have been collected from ECMF, and these have many biological activities, properties, and applications. Some of these are organic compounds with low molecular weight that can be utilized in the food production industries to enhance the flavours of mushrooms (Mizuno and Kwai 1992). Some are compounds with antioxidant and anticancer properties (Arshad et al. 2016; Kreisel et al. 1990). The polysaccharides in these compounds could be employed in the diets of diabetic patients and show immunosuppressive action (Huijuan et al. 1994). Some of the lipids and fatty acids from them also have antioxidant, anticancer, immunosuppressive, and anti-inflammatory properties (Reis et al. 2011). Finally, the enzymes have roles in the pollution control, detergent, textile and paper, food, and cosmetic industries (Casieri et al. 2010; Gupta et al. 2003).

11.6 ECMF AND FORESTRY

ECMF have had applications in forestry since the late 1950s, as they are considered as bio-fertilizers. These fungi have the ability to increase the up-take of P, N, water, and some other nutrients by plants, and ultimately boost and nourish plant growth (Feldmann et al. 2009). Studies are being carried out to understand the functions of ECMF and their practical use in forestry. Nursery inoculation and successful inoculation of tree seedlings have been described. Before outplanting, those seedlings which are inoculated in the nursery can form a great system of ECMF (Domínguez-Núñez et al. 2019). The ECMF's interaction with symbiotic fungi expands to larger fungal communities that also have saprotrophic species in rich forest soils. In addition, oxidation of different phenolic substrates may be catalysed by various other soil prokaryotes. All this illustrates the functional complexity of the ectomycorrhizosphere microflora (Frey-Klett et al. 2007).

11.6.1 Applications of ECMF in Forest Nurseries

Producing a good-quality ECMF plant that is inhabited by a specific fungus is a huge challenge in the synthesis of ECMF symbiosis. The essential steps of the production procedure are the accurate inspection of the employed inoculum and limiting adulteration during the development of inoculated plants. These steps can prevent the mix-up of genetic material of unwanted species with the indigenous species and can stop the attack of unwanted species (Murat and Martin 2008). The most important step for successful mycorrhization is proper selection of the appropriate plant-host species. For this purpose, generally, fast-growing fungi are selected as they have a short incubation period (Olivier 2000; Fatima et al. 2016). However, some undesirable kinds of ECMF are found. Therefore, it is preferred to use fresh cultures, rather than repeatedly transferring and storing cultures (Marx 1980). It is suggested that best practice is to pass essential fungus cultures from a host inoculation and mycorrhiza development and then its re-isolation to maintain the fungus-forming capacity (Amaranthus 1998). In addition, these fungi are superior in the uptake of minerals and soil exploration that make large hyphal stands of rhizomorphs in soil cultures, as compared to those that lack rhizomorphic growth. On the other hand, the mycorrhizal state of the seedlings alone does not determine the fruiting of the ECMF species (Hortal et al. 2009). After planting, the spread and persistence of the cultivated fungus is influenced by the already-present native competitors and physiochemical or biotic properties of that soil. The progress of a mycorrhizal inoculation programme is influenced by the kind of ECMF material employed for inoculation (Rossi et al. 2007).

The sources of fungal inoculum include utilizing the spores of fruiting bodies taken from the field. This is a suitable method as the spores do not need the aseptic culture's extension, moreover, the spore inoculum is not heavy (Sharma 2017). The disadvantages of this method are the process's success is determined by the viability of the spores, and a sufficient amount of fruiting entities are needed that might not be available every year, and also, the genetic definition may not be sufficient. Before inoculating in soil, the spores should be mixed with some sort of physical supports such as being soaked in water, suspended in water, sprayed and sprinkled on the ground, embedded in hydrocolloid chips, or coated on the seeds. Their viability can be maintained by freezing and storing them at low temperature in a dark place (Marx and Cordell 1989). The second method is the use of humus or soil obtained from the soil in which mycorrhizae seedlings are going to be embedded. The disadvantage of this method is that there is no control over the species of ECMF present in the soil or of other microbes and harmful pathogens. This method is discarded in the mycorrhization programmes but is still widely used in developing countries. Moreover, a "nurse seedling" of mycorrhizae is planted or may be a chopped ECMF root host that is incorporated into the nursery beds as a sources of fungi for the adjacent young seedlings (Sim and Eom 2006). The third method is the use of hyphae in a substrate, liquid, or solid medium as an inoculum. The success rate of fungal hyphae from mycorrhizae is very low, at about 5–20%. Therefore, it is mainly cultivated from the sterile portions of the fruiting bodies rather than mycorrhizae (Molina 1982). Very often, fungal hyphae are taken from sexual spores or less frequently from sclerotia (Fries and Birraux 1980). This method enables the choice of specific strains of a fungus, which is why it is considered the most suitable method, as firstly it is checked for its suitability to enhance plant growth (Marx 1980). Most of the species of *Suillus*, *Hebeloma*, *Laccaria*, *Amanita*, *Rhizopogon*, and *Pisolithus* grow very well in culture. Liquid substrates are more appropriate as compared to solid ones as they can be mixed very easily, therefore they make a uniform state for crop growth. However, the risk of bacterial attack is very high with this method and it is somewhat costly (Rossi et al. 2007; Hussain et al. 2016). Meanwhile, the major benefit of hard solid media is limited pollution by bacteria because of less moisture content, cheap equipment, i.e. low cost, and the simple and easy design of the bio-reactors (Cannel 1980). The major disadvantage of the utilization of mycelial inoculum is that various varieties of ECMF are not easy to produce in the lab environment, because of a slow growth rate due to the lack of their symbiont, and producing a huge amount of inoculum is also not very easy. Some recent techniques have been invented using mycelium encapsulated in "beads" of calcium alginate, but these must be kept in a refrigerator (Le Tacon et al. 1983). Inoculum beads can remain viable for a few months if kept in refrigerator, however the outcomes vary between fungal species. It has been tested with many plants of economic interests for several species. Therefore, these processes have great ability and potential to be used in reforestation programmes. A bio-reactor was designed by Rossi et al. (2007) which have the capability of make inocula for 300,000 seedlings, which is sufficient for the reforestation of about 200 hectares of land. On the basis of 3 billion cubic metres of global wood demand, approximately 4.3 tons of mycelium would be required to inoculate 12 million seedlings, i.e. 5 g of dry mycelium per plant. The benefit of alginate gel is the possibility of making a multi-microbial inoculant (Rossi et al. 2007).

11.6.2 Application of ECMF in Forest Management

ECMS can be injected to increase the production of edible carpophores but it is also very significant in forest management, especially in reforestation, where it increases the quality and economic productivity of plantations (Tommerup and Bougher 2000; Garbaye 1990). In nurseries, the growth of plants having mycorrhized seeds depends on how quickly they obtian the nutrients and the availability of water in the soil. In mycorrhizal plantations, the diversity of mycorrhizae depends on the introduction of other dominant species from the mixed communities. Instead, unfortunately, the fungi which are chooses from the nursery for the best colonization have been deprived of competitors in the field, particularly in the planting areas where native fungi are present (Dunabeitia et al. 2004). There are many possible ways through which we can explain the failure of inoculation to make positive impacts in the planting areas. Possibly, the fundamental cause of failure is the persistence of inocula in the roots after transfer of plants from the nursery to the field. The soil conditions which plants experience during planting are different from those of plants growing in a nursery or a container and there is also a reduction in the strength of roots and their fungal associates (McAfee and Fortin 1986). The species which show the greatest growth in the inoculation program are *Pisolithus tinctorius* (15 subspecies), in conditions such as a polluted environment or a shortage of an autochthonous mycorrhizal population. In the case of an edible ECMF of interest such as a tuber or black truffle, with an artificially mycorrhized plant, it shows that the formation of plots has always been the chief goal of developing fruiting bodies, leaving in the background the impact of the ecological roles of the symbiosis (Núñez et al. 2006; Garbaye 1990). The prominent and vital service of ECMF forest plants are restoration of the ecosystem by bioremediation, bio-fertilization, and control of soil pathogens. Unplanned industrial activities and unfavourable land management ultimately lead to a degraded ecosystem. The main causes of environmental degradation all over the world include salinity stress, soil erosion, compactness of soil, and shortage of water. A balanced ecosystem contains composed and balanced microbiota in the soil, in such a way that there is harmony among potential pathogenic and mycorrhizal fungi (Geml et al. 2012). ECMF can exist in life-threatening conditions such as extreme temperature, drought, metal and salt concentrations, drought, and other situations that lead to degradation of the ecosystem (Colpaert et al. 2011). ECMF have great significance because they help to balance the ecosystem in many ways, such as by enhancing plants' ability to tolerate abiotic stresses, the ability to ameliorate heavy metals or to reduce organic matter, interactions with pathogenic nematodes, soil microbes, and bacteria, and increased nutritional quality, growth, and development of their associate plants. Furthermore, ECMF's extraradical mycelium offers a path for the translocation of photosynthesized C to microbes and a huge surface area for the microorganisms to contact (Suz et al. 2012; Machuca 2011). It has been discovered that ECMF help in the rhizosphere to remediate from persistent organic pollutants (POPs) (Meharg and Cairney 2000).

11.6.3 Role of Mycorrhizal Networks in Functioning of a Forest Ecosystem

ECMFs' extraradical mycelia join the roots of the same or a distinct ECM tree species, forming common mycorrhizal networks (CMNs) that permit net C and nutrient exchange between host plants (Lindahl and Tunlid 2015). Several review publications have established the role of CMNs in plant community dynamics (Perry et al. 1992). However, problems in researching mycelial organizations in situ in the soil without upsetting or killing them, as well as separating symbiotic from saprotrophic mycelia, have hampered advances in comprehending the arrangement and function of CMNs in the field (Simard and Durall 2004). The application of molecular markers has demonstrated that several ECMF colonize different plant hosts at the same time, implying that the creation of CMNs in forest stands has high potential. That is confirmed by the fact that various ECM fungal taxa may develop extremely enormous genets, up to 10 meters wide, and hence can cover a broad region, spanning many trees of the same or different species (Ishida et al. 2007). Ectomycorrhizal CMNs are created by *Basidiomycetes* and *Ascomycetes* fungal species (Bruns et al. 2002). Although a few ECMF taxa are host-specific, the majority are generalists and, at least in temperate ECM ecosystems, are frequently more numerous than non-generalists (Richard et al. 2005). Furthermore, some *Ascomycetes* that develop ericoid endomycorrhizae with *Ericaceae* plants may generate different kinds of CMN that can be incorporated into ECMF chains: some *Ericaceae* species form equally ericoid endomycorrhizae and ECM networks (Dunham et al. 2003; Selosse et al. 2006). In the same way, ECM forest trees and myco-heterotrophic and mixotrophic orchids might have shared *Basidiomycete* symbionts (Taylor et al. 2002).

11.6.4 Impacts of the Interaction between CMNs and Trees

The importance of resource transmission from one tree to another via CMNs for functional ecology and geochemical cycling has lately piqued researchers' interest. Indeed, CMNs have the potential to create "guilds of mutual help" between synchronized plants, or transfer means along source-sink gradients (Leake et al. 2014; Leake et al. 2004). The predominance of the hyphal route for C transfer from giver to recipient green plants, and in addition to this, the destiny of the C transported and its physiological significance, are all debatable (Fitter et al. 1998). Many orchids, on the other hand, have been shown to transition from saprotrophic and harmful fungal associates like *Rhizoctonia* sp. to ECM fungi that can deliver huge quantities of C over extended periods of time during their growth (Simard et al. 2003). *Cryptothallus mirabilis*, an achlorophyllous liverwort, has been observed forming ECM connections with *Tulasnella*, a genus of *Basidiomycete* that frequently forms orchid mycorrhizas (Bidartondo et al. 2003). Furthermore, current investigations using steady isotope enrichment (13C and 15N) have revealed that certain green forest orchids had 13C enrichments that fell somewhere between entirely autotrophic plants and myco-heterotrophic plants that obtained all their C from their associated fungus (Tedersoo et al. 2007). In a method known as "mixotrophy", these green orchids get their carbon from a combination of myco-heterotrophic and photosynthetic sources. It has been discovered that related fungus provided 38 percent to 80 percent of the carbon to certain pyroloid shrubs (*Ericaceae*) and green forest orchids (Tedersoo et al. 2007). Water transport between conspecific and non-conspecific neighbouring plants is also facilitated by CMNs, as has been demonstrated employing isotopic tracers and fluorescent dyes (Querejeta et al. 2003). Water was also transferred from adult plants to seedlings as a result of the "hydraulic lift". Hydraulic lift, on the other hand, is primarily a physical activity that may be carried out by rotten roots and, probably, fallen rhizomorphs (Egerton-Warburton et al. 2007; Leffler et al. 2005).

11.6.5 Services of CMNs in the Composition and Functioning of Forest Plant Communities

One of the most important factors governing the cohabitation of plant species, and hence the formation of plant networks, is resource competition. Resource sharing has been anticipated as a mechanism for reducing resource rivalry, especially in mycorrhizal plant communities (Hart et al. 2003; Read et al. 2004). This approach is heavily reliant on CMNs as a deciding element in resource distribution. The organization of tree communities is likely to be affected by their membership in a CMN, at least at the local scale (Smith and Read 2008). Indeed, two plants can contribute uneven amounts of C to a shared fungus or obtain unequal amounts of nutrients from a fungus that they mutually support, meaning a net advantage to one species at the expense of the other (Selosse et al. 2006). In other ways, negative feedback might modify dominant populations' competitive ability, promoting plant variety. Furthermore, CMNs can aid seedling recruitment: covered sprouts in the understory can grow effectively alongside competing ECM tree species by joining to established mycorrhizal systems and obtaining carbon from mature overstore trees (Cullings et al. 2000). Furthermore, seedling recruitment is aided by early integration into existing mycorrhizal networks, as seedling survival and establishment rates are determined by the pace at which they become mycorrhizal. For example, it was found in one study that *Pseudotsuga* seedling survival was higher in the ECM *Arctostaphylos* chaparral than in the non-ECM *Adenostoma* chaparral (Horton et al. 1999). Seedlings in touch with mature ECM trees have greater survival and mycorrhizal colonization than seedlings separated from adult plants, according to a study in tropical forests and temperate forests (Dickie et al. 2004; Onguene and Kuyper 2002). Adult trees can therefore serve as "nurse trees" for both conspecific and non-conspecific seedlings, promoting species diversity and cohabitation in forest outlooks (Grime et al. 1987). The selection forces behind this phenomenon, however, remain unknown, as this collaboration may be harmful to mature plants (Newman 1988).

11.7 CONCLUSION

This chapter has described the ECMF's role in ecosystems as a bio-fertilizer and how it could be enhanced, as they significantly contribute to a number of important ecosystem functions, especially nutrient cycling and C fluxes. The main focus of the research should be evaluating a suitable technique for commercial manufacturing and huge-scale inoculation of the ECMF and the use of appropriate arrangements for microbes and plants, carried out in a suitable well-defined soil condition and environment. The role of ECMF in balancing and maintaining the ecosystem is enormous as it considerably enhances plants' tolerance towards biotic and abiotic stresses. The ECMF are fundamental because of their role as a bio-fertilizer in the restoration of the environment and reforestation. In short, ECMF are very important for environmental sustainability and balanced ecology.

REFERENCES

Agerer R (2001) Exploration types of ectomycorrhizae. *Mycorrhiza* 11:107–114.

Alizadeh O (2011) Mycorrhizal symbiosis. *Adv Stud Biol* 6:273–281.

Amaranthus MP (1998) *The Importance and Conservation of Ectomycorrhizal Fungal Diversity in Forest Ecosystems: Lessons from Europe and the Pacific Northwest*. Volume 431. US Department of Agriculture, Forest Service, Pacific Northwest Research Station.

Arshad M, Adnan M, Ahmed S, Khan AK, Ali I, Ali M, Ali A, Khan A, Kamal MA, Gul F, Khan MA (2016) Integrated effect of phosphorus and zinc on wheat crop. *American-Eurasian J Agric & Environ Sci* 16(3):455–459.

Azul AM, Nunes J, Ferreira I, Coelho AS, Veríssimo P, Trovão J, Campos A, Castro P, Freitas H (2014) Valuing native ectomycorrhizal fungi as a Mediterranean forestry component for sustainable and innovative solutions. *Botany* 92:161–171.

Bidartondo MI, Bruns TD, Weiß M, Sérgio C, Read DJ (2003) Specialized cheating of the ectomycorrhizal symbiosis by an epiparasitic liverwort. *Proc Royal Soc B-Biol Sci* 270:835–842.

Bonfante P, Anca I-A (2009) Plants, mycorrhizal fungi, and bacteria: a network of interactions. *Annu Rev Microbiol* 63:363–383.

Brundrett M (2004) Diversity and classification of mycorrhizal associations. *Biol Rev* 79:473–495.

Bruns TD, Bidartondo MI, Taylor DL (2002) Host specificity in ectomycorrhizal communities: what do the exceptions tell us? *Integr & Comp Biol* 42:352–359.

Buscot F, Munch JC, Charcosset J-Y, Gardes M, Nehls U, Hampp R (2000) Recent advances in exploring physiology and biodiversity of ectomycorrhizas highlight the functioning of these symbioses in ecosystems. *FEMS Microbiol Rev* 24:601–614.

Cannel E (1980) Solid-state fermentation systems. *Process Biochem* 15:2–7, 24–28.

Casieri L, Anastasi A, Prigione V, Varese GC (2010) Survey of ectomycorrhizal, litter-degrading, and wood-degrading Basidiomycetes for dye decolorization and ligninolytic enzyme activity. *Antonie van Leeuwenhoek* 98: 483–504.

Claridge AW (2002) Ecological role of hypogeous ectomycorrhizal fungi in Australian forests and woodlands. In: *Diversity and Integration in Mycorrhizas*. Springer, pp. 291–305.

Colpaert JV, Wevers JH, Krznaric E, Adriaensen K (2011) How metal-tolerant ecotypes of ectomycorrhizal fungi protect plants from heavy metal pollution. *Annals Forest Sci* 68.

Courty P-E, Buée M, Diedhiou AG, Frey-Klett P, Le Tacon F, Rineau F, Turpault M-P, Uroz S, Garbaye J (2010a) The role of ectomycorrhizal communities in forest ecosystem processes: new perspectives and emerging concepts. *Soil Biol & Biochem* 42:679–698.

Courty P-E, Franc A, Garbaye J (2010b) Temporal and functional pattern of secreted enzyme activities in an ectomycorrhizal community. *Soil Biol & Biochem* 42:2022–2025.

Cullings KW, Vogler DR, Parker VT, Finley SK (2000) Ectomycorrhizal specificity patterns in a mixed Pinus contorta and Picea engelmannii forest in Yellowstone National Park. *Appl & Environ Microbiol* 66:4988–4991.

Dahlberg A (2001) Community ecology of ectomycorrhizal fungi: an advancing interdisciplinary field. *New Phytol* 150:555–562.

Dahlberg A, Genney DR, Heilmann-Clausen J (2010) Developing a comprehensive strategy for fungal conservation in Europe: current status and future needs. *Fungal Ecol* 3:50–64.

Dickie IA, Guza RC, Krazewski SE, Reich PB (2004) Shared ectomycorrhizal fungi between a herbaceous perennial (Helianthemum bicknellii) and oak (Quercus) seedlings. *New Phytol* 164:375–382.

Dickie IA, Martínez-García LB, Koele N, Grelet G-A, Tylianakis JM, Peltzer DA, Richardson SJ (2013) Mycorrhizas and mycorrhizal fungal communities throughout ecosystem development. *Plant & Soil* 367:11–39.

Domínguez-Núñez JA, Berrocal-Lobo M, Albanesi AS (2019) Ectomycorrhizal fungi: role as biofertilizers in forestry. In: *Biofertilizers for Sustainable Agriculture and Environment*. Springer, pp. 67–82.

Dunabeitia M, Rodríguez N, Salcedo I, Sarrionandia E (2004) Field mycorrhization and its influence on the establishment and development of the seedlings in a broadleaf plantation in the Basque Country. *Forest Ecol & Manage* 195:129–139.

Dunham SM, Kretzer A, Pfrender ME (2003) Characterization of Pacific golden chanterelle (Cantharellus formosus) genet size using co-dominant microsatellite markers. *Mol Ecol* 12:1607–1618.

Egerton-Warburton LM, Querejeta JI, Allen MF (2007) Common mycorrhizal networks provide a potential pathway for the transfer of hydraulically lifted water between plants. *J Exp Botany* 58:1473–1483.

Fatima, Shah M, Usman A, Sohail K, Afzaal M, Shah B, Adnan M, Ahmed N, Junaid K, Shah SRA, Rahman LU (2016) Rearing and identification of Callosobruchusmaculatus (Bruchidae: Coleoptera) in Chickpea. *J Ento & Zoo Studies* 4(2):264–266.

Feldmann F, Hutter I, Schneider C (2009) Best production practice of arbuscular mycorrhizal inoculum. In: Md Asraful Alam, Zhongming Wang (Eds.), *Symbiotic Fungi*. Springer, pp. 319–336.

Fitter A, Graves J, Watkins N, Robinson D, Scrimgeour C (1998) Carbon transfer between plants and its control in networks of arbuscular mycorrhizas. *Funct Ecol* 12:406–412.

Fitter AH (2005) Darkness visible: reflections on underground ecology. *J Ecol* 93:231–243.

Frey-Klett P, Garbaye J, Tarkka M (2007) The mycorrhiza helper bacteria revisited. *New Phytol* 176:22–36.

Fries N, Birraux D (1980) Spore germination in Hebeloma stimulated by living plant roots. *Experientia* 36:1056–1057.

Garbaye J (1990) Use of mycorhizas in forestry. *Les Mycorhizes des Arbres et Plantes Cultivées* 197–248.

Geml J, Timling I, Robinson CH, Lennon N, Nusbaum HC, Brochmann C, Noordeloos ME, Taylor DL (2012) An arctic community of symbiotic fungi assembled by long-distance dispersers: phylogenetic diversity of ectomycorrhizal basidiomycetes in Svalbard based on soil and sporocarp DNA. *J Biogeography* 39:74–88.

Goltapeh EM, Danesh YR, Prasad R, Varma A (2008) Mycorrhizal fungi: what we know and what should we know? In: Sally E Smith and David Read (Eds.), *Mycorrhiza*. Springer, pp. 3–27.

Grime J, Mackey J, Hillier S, Read D (1987). Floristic diversity in a model system using experimental microcosms. *Nature* 328 420–422.

Gupta R, Gigras P, Mohapatra H, Goswami VK, Chauhan B (2003) Microbial α-amylases: a biotechnological perspective. *Process Biochem* 38:1599–1616.

Hart MM, Reader RJ, Klironomos JN (2003) Plant coexistence mediated by arbuscular mycorrhizal fungi. *Trends Ecol & Evol* 18:418–423.

Hoffland E, Kuyper TW, Wallander H, Plassard C, Gorbushina AA, Haselwandter K, Holmström S, Landeweert R, Lundström US, Rosling A (2004) The role of fungi in weathering. *Front Ecol & Environ* 2:258–264.

Hortal S, Pera J, Parladé J (2009) Field persistence of the edible ectomycorrhizal fungus Lactarius deliciosus: effects of inoculation strain, initial colonization level, and site characteristics. *Mycorrhiza* 19:167–177.

Horton TR, Bruns TD, Parker VT (1999) Ectomycorrhizal fungi associated with Arctostaphylos contribute to Pseudotsuga menziesii establishment. *Can J Botany* 77:93–102.

Horton TR, Bruns TD (2001) The molecular revolution in ectomycorrhizal ecology: peeking into the black-box. *Mol Ecol* 10:1855–1871.

Huijuan H, Peizhen L, Tao L, Bingqian H, Yaowei G (1994) Effects of polysaccharide of Tuber sinica on tumor and immune system of mice. *J China Pharm Univ* 5:289–292.

Hussain H, Raziq F, Khan I, Shah B, Altaf M, Attaullah, Ullah W, Naeem A, Adnan M, Junaid K, Shah SRA, Iqbal M (2016) Effect of Bipolaris maydis (Y. Nisik & C. Miyake) shoemaker at various growth stages of different maize cultivars. *J Ento & Zool Studies* 4(2):439–444.

Ibrahim M, Jamal Y, Basir A, Adnan M, I-u-Rahman, Khan IA, Attaullah (2016) Response of Sesame (Sesamumindicuml.) to various levels of nitrogen and phosphorus in agro-climatic condition of Peshawar. *Pure Appl Biol* 5(1):121–126.

Ishida TA, Nara K, Hogetsu T (2007) Host effects on ectomycorrhizal fungal communities: insight from eight host species in mixed conifer–broadleaf forests. *New Phytol* 174:430–440.

Jansa J, Smith FA, Smith SE (2008) Are there benefits of simultaneous root colonization by different arbuscular mycorrhizal fungi? *New Phytol* 177:779–789.

Khan IA, Shah B, Khan A, Zaman M, Din MMU, Shah SRA, Junaid K, Adnan M, Ahmad N, Akbar R, Fayaz W, Rahman IU (2016) Screening of different maize Cultivars against maize shootfly and red pumpkin beetle at Peshawar. *J Ento & Zool Studies* 4(1):324–327.

Koide RT, Fernandez C, Malcolm G (2014) Determining place and process: functional traits of ectomycorrhizal fungi that affect both community structure and ecosystem function. *New Phytol* 201:433–439.

Kreisel H, Lindequist U, Horak M (1990) Distribution, ecology, and immunosuppressive properties of Tricholoma populinum (Basidiomycetes). *Zentralblatt für Mikrobiologie* 145:393–396.

Le Tacon F, Jung G, Michelot P, Mugnier M (1983) Efficiency in a forest nursery of an inoculum of an ectomycorrhizal fungus produced in a fermentor and entrapped in polymetric gels. *Canad J Bot* 40(2):165–176.

Leake J, Johnson D, Donnelly D, Muckle G, Boddy L, Read D (2004) Networks of power and influence: the role of mycorrhizal mycelium in controlling plant communities and agroecosystem functioning. *Canad J Bot* 82:1016–1045.

Leake J, Johnson D, Donnelly D, Muckle G, Boddy L, Read D (2014) Erratum: Networks of power and influence: the role of mycorrhizal mycelium in controlling plant communities and agroecosystem functioning. *Botany* 92:83.

Leffler AJ, Peek MS, Ryel RJ, Ivans CY, Caldwell MM (2005) Hydraulic redistribution through the root systems of senesced plants. *Ecol* 86:633–642.

Lindahl BD, Tunlid A (2015) Ectomycorrhizal fungi – potential organic matter decomposers, yet not saprotrophs. *New Phytol* 205:1443–1447.

Machuca Á (2011) Metal-chelating agents from ectomycorrhizal fungi and their biotechnological potential. In: Mahendra Rai, Ajit Varma (Eds.), *Diversity and Biotechnology of Ectomycorrhizae*. Springer, pp. 347–369.

Marx D (1980) Ectomycorrhizal fungus inoculations: a tool for improving forestation practices. *Tropical Mycorrhiza Res* 13–71.

Marx D, Cordell C (1989) The use of specific ectomycorrhizas to improve artificial forestation practices. *Proc of the Brit Mycol Soc* 1–25.

McAfee B, Fortin J (1986) Competitive interactions of ectomycorrhizal mycobionts under field conditions. *Can J Botany* 64:848–852.

Meharg AA, Cairney JW (2000) Ectomycorrhizas – extending the capabilities of rhizosphere remediation? *Soil Biol & Biochem* 32:1475–1484.

Mello A, Balestrini R (2018) Recent insights on biological and ecological aspects of ectomycorrhizal fungi and their interactions. *Front Microbiol* 9:216.

Mizuno T, Kwai M (1992) *Chemistry and Biochemistry of Mushroom Fungi*. Gakai-shupan Center, Tokyo.

Molina R (1982) Isolation, maintenance, and pure culture manipulation of ectomycorrhizal fungi. *Methods and Principles of Mycorrhizal Research* 115–129.

Mosse B, Stribley D, LeTacon F (1981) Ecology of mycorrhizae and mycorrhizal fungi. In: Larry L Barton, Diana E Northup (Eds.), *Advances in Microbial Ecology*. Springer, pp. 137–210.

Mullineaux CW, Liu L-N (2020) Membrane dynamics in phototrophic bacteria. *Annu Rev Microbiol* 74:633–654.

Murat C, Martin F (2008) Sex and truffles: first evidence of Perigord black truffle outcrosses. *New Phytol* 180(2): 260–263.

Nara K (2008) Community developmental patterns and ecological functions of ectomycorrhizal fungi: implications from primary succession. In: Sally E Smith and David Read (Eds.), *Mycorrhiza*. Springer, pp. 581–599.

Newman E (1988) Mycorrhizal links between plants: their functioning and ecological significance. *Adv Ecol Res* 18:243–270.

Núñez JAD, Serrano JS, Barreal JAR, de Omeñaca González JAS (2006) The influence of mycorrhization with Tuber melanosporum in the afforestation of a Mediterranean site with Quercus ilex and Quercus faginea. *Forest Ecol & Manage* 231:226–233.

Olivier J (2000) Progress in the cultivation of truffles. *Mushroom Science XV: Sci & Cultiv Edible Fungi* 2:937–942.

Onguene N, Kuyper T (2002) Importance of the ectomycorrhizal network for seedling survival and ectomycorrhiza formation in rain forests of south Cameroon. *Mycorrhiza* 12:13–17.

Perry DA, Bell T, Amaranthus M (1992) Mycorrhizal fungi in mixed-species forests and other tales of positive feedback, redundancy and stability. Special publications of the British Ecological Society.

Prieto I, Roldán A, Huygens D, del Mar Alguacil M, Navarro-Cano JA, Querejeta JI (2016) Species-specific roles of ectomycorrhizal fungi in facilitating interplant transfer of hydraulically redistributed water between Pinus halepensis saplings and seedlings. *Plant & Soil* 406:15–27.

Querejeta J, Egerton-Warburton LM, Allen MF (2003) Direct nocturnal water transfer from oaks to their mycorrhizal symbionts during severe soil drying. *Oecologia* 134:55–64.

Read DJ, Leake JR, Perez-Moreno J (2004) Mycorrhizal fungi as drivers of ecosystem processes in heathland and boreal forest biomes. *Can J Botany* 82:1243–1263.

Reis FS, Pereira E, Barros L, Sousa MJ, Martins A, Ferreira IC (2011) Biomolecule profiles in inedible wild mushrooms with antioxidant value. *Molecules* 16:4328–4338.

Requena N (2005) Measuring quality of service: phosphate "a la carte" by arbuscular mycorrhizal fungi. *New Phytol* 268–271.

Richard F, Millot S, Gardes M, Selosse MA (2005) Diversity and specificity of ectomycorrhizal fungi retrieved from an old-growth Mediterranean forest dominated by Quercus ilex. *New Phytol* 166:1011–1023.

Rinaldi A, Comandini O, Kuyper TW (2008) Ectomycorrhizal fungal diversity: seperating the wheat from the chaff. *Fungal Diversity* 33:1–45.

Rossi MJ, Furigo Jr A, Oliveira VL (2007) Inoculant production of ectomycorrhizal fungi by solid and submerged fermentations. *Food Technol & Biotechnol* 45:277–286.

Savoie J-M, Largeteau ML (2011) Production of edible mushrooms in forests: trends in development of a mycosilviculture. *Appl Microbiol & Biotechnol* 89:971–979.

Selosse M-A, Richard F, He X, Simard SW (2006) Mycorrhizal networks: des liaisons dangereuses? *Trends Ecol & Evol* 21:621–628.

Shah B, Khan IA, Khan A, Din MMU, Junaid K, Adnan M, Shah SRA, Zaman M, Ahmad N, Akbar R, Fayaz W, Rahman IU (2016) Correlation between proximate chemical composition and insect pests of maize cultivars in Peshawar. *J Ento & Zool Studies* 4(1):312–316.

Sharma R (2017) Ectomycorrhizal mushrooms: their diversity, ecology and practical applications. In: Ajit Varma, Ram Prasad, Narendra Tuteja (Eds.), *Mycorrhiza – Function, Diversity, State of the Art*. Springer, pp. 99–131.

Sim M-Y, Eom A-H (2006) Effects of ectomycorrhizal fungi on growth of seedlings of Pinus densiflora. *Mycobiol* 34:191–195.

Simard SW, Jones MD, Durall DM (2003) Carbon and nutrient fluxes within and between mycorrhizal plants. In: Marcel GA Heijden, Ian R Sanders (Eds.), *Mycorrhizal Ecol*. Springer, pp. 33–74.

Simard SW, Durall DM (2004) Mycorrhizal networks: a review of their extent, function, and importance. *Can J Botany* 82:1140–1165.

Smith S, Read D (2008) Mycorrhizal symbiosis, 3rd edn. Academic Press, Elsevier, London.

Suz LM, Azul AM, Bodas RP, Martín MP (2012) Ectomycorrhizal fungi in biotechnology: present and future perspectives. In: Shahedur Rahman (Ed.), *Environment & Biotechnology*. Lambert Academic Publishing AG & Co, p. 17.

Taylor D, Bruns T, Leake J, Read D (2002) Mycorrhizal specificity and function in myco-heterotrophic plants. In: Marcel GA Heijden, Ian R Sanders (Eds.), *Mycorrhizal Ecology*. Springer, pp. 375–413.

Tedersoo L, Pellet P, Koljalg U, Selosse M-A (2007) Parallel evolutionary paths to mycoheterotrophy in understorey Ericaceae and Orchidaceae: ecological evidence for mixotrophy in Pyroleae. *Oecologia* 151:206–217.

Tommerup I, Bougher N (2000) The role of ectomycorrhizal fungi in nutrient cycling in temperate Australian woodlands. In: Richard J Hobbs and Colin J Yates (Eds.), *Temperate Eucalypt Woodlands in Australia: Biology, Conservation, Management and Restoration*. Surrey Beatty & Sons, Chipping Norton, NSW, pp. 190–224.

Uroz S, Calvaruso C, Turpault M-P, Pierrat J-C, Mustin C, Frey-Klett P (2007) Effect of the mycorrhizosphere on the genotypic and metabolic diversity of the bacterial communities involved in mineral weathering in a forest soil. *Appl & Environm Microbiol* 73:33019–33027.

12 Plant Growth-promoting Rhizobacteria/*Pseudomonas* as a Biofertilizer

Rafiq Ahmad,[1] Sohail Khan,[2] Muhammad Hayat,[3] Syed Muhammad Afzal,[4] and Javed Muhammad[5]**

[1] Department of Microbiology, The University of Haripur, KP, Pakistan
[2] College of Life Science and Technology, State Key Laboratory for Conservation and Utilization of Subtropical Agro-bioresources, Guangxi University, Nanning, Guangxi, China
[3] State Key Laboratory of Microbial Technology, Institute of Microbial Technology, Shandong University Qingdao, China
[4] Center of Biotechnology and Microbiology, University of Peshawar, KP, Pakistan
[5] Department of Microbiology, The University of Haripur, KP, Pakistan
* Correspondence: javed.muhammad@uoh.edu.pk

CONTENTS

12.1 INTRODUCTION

The rhizosphere is the layer of soil surrounding the plant roots which plays a critical role in plant growth and development. This tiny zone surrounding the plant roots is a key hotspot for microorganisms and is considered one of the most complicated ecosystems (Mendes et al. 2013; Raaijmakers et al. 2009). Hiltner (1904) used the term “rhizosphere” to characterize the limited soil zone where the root boosts microbial populations (Drogue et al. 2013; Meena et al. 2020; Shukla et al. 2011; Zehra et al. 2021; Al-Zahrani et al. 2022; Rajesh et al. 2022; Anam et al. 2021; Deepranjan et al. 2021; Haider et al. 2021; Amjad et al. 2021; Sajjad et al. 2021a; Fakhre et al.

DOI: 10.1201/9781003286233-12

2021; Khatun et al. 2021; Ibrar et al. 2021; Bukhari et al. 2021; Haoliang et al. 2022; Sana et al. 2022; Abid et al. 2021; Zaman et al. 2021; Sajjad et al. 2021b; Rehana et al. 2021; Yang et al. 2022; Ahmad et al. 2022; Shah et al. 2022). Plant roots produce diverse metabolites with a plentiful supply of amino acids, vitamins, sugars, and organic matter. The rhizosphere microbiome is the total microbial population present in the rhizosphere and significantly differs from the nearby soil microbial population (Chaparro et al. 2013; Kumar & Dubey 2020). Microbial taxa inhabit the rhizosphere, including prokaryotes (bacteria, archaea) and diverse eukaryotes, including (fungi, algae, protozoa, nematode, arthropods, and viruses. However, bacteria and fungi are the most prevalent (Buee et al. 2009; Kalam et al. 2017).

According to Kloepper (1980), plant growth-promoting rhizobacteria (PGPR) are soil bacteria that thrive in the rhizosphere, colonize plant roots aggressively, and aid plant development (Dutta and Podile 2010; Adnan et al. 2016). PGPR bacteria play a significant role in plant growth and help prevent the spread of phytopathogenic microorganisms (Kloepper et al. 1980; Son et al. 2014). Microbes may indirectly impact plant phenotypic plasticity and plant health by modulating plant growth and defensive responses due to their lengthy co-evolution with plants (Gao et al. 2015). They penetrate and grow in the rhizosphere environment inside the plant microbiome, gaining traction in agricultural operations as traditional chemical fertilizers (Compant et al. 2019; Muhammad et al. 2022; Wiqar et al. 2022; Farhat et al. 2022; Niaz et al. 2022; Ihsan et al. 2022; Chao et al. 2022, Qin et al. 2022; Xue et al. 2022; Ali et al. 2022; Mehmood et al. 2022; El Sabagh et al. 2022; Ibad et al. 2022). PGPR plays a vital role in sustainable agriculture, enhancing crop production, soil fertility, biodiversity, and interaction with other beneficial microbes and reducing pathogen development and infection (Tapia-Vázquez et al. 2020). As a result of these processes, PGPR could be used as a biofertilizer (Vessey et al. 2003). A biofertilizer is a substance that includes live microorganisms that colonize either the internal plant tissues or the rhizosphere and enhance plant growth when applied to soil or plant surfaces (Verma et al. 2019). Their rise to prominence as a viable alternative stems from the abuse of agrochemicals such as fertilizers and pesticides, which pollute the soil, fruits, and vegetables (Adesemoye et al. 2009). Among the PGPR, *Pseudomonas*, *Klebsiella*, *Alcaligenes*, *Arthrobacter*, *Azospirillum*, *Enterobacter*, *Burkholderia*, *Serratia*, and *Bacillus* are the most significant (Bhattacharyya et al. 2012; Kumari et al. 2018).

Pseudomonas is a genus that contains the most varied group of bacteria on the planet, with over 100 distinct species (Heiman et al. 2020). It belongs to the gamma subclass of proteobacteria and is found in diverse habitats, including marine ecosystems, freshwater, and terrestrial environments. They have a close relationship with higher life forms and are among the most studied bacterial species (Selvakumar et al. 2015; Saleem et al. 2015). The properties of plant growth-promoting (PGP) *Pseudomonas* strains that may improve agricultural production in both standard and stressed settings are discussed in this chapter.

12.2 RHIZOBACTERIA THAT PROMOTE PLANT GROWTH

To achieve sustainable agriculture, plants should be disease-resistant, drought-tolerant, and metal stress-tolerant and have a higher nutritional value (Deepranjan et al. 2021; Haider et al. 2021; Huang Li et al. 2021; Ikram et al. 2021; Jabborova et al. 2021; Khadim et al. 2021a,b; Manzer et al. 2021; Muzammal et al. 2021; Abdul et al. 2021a,b; Ashfaq et al. 2021; Amjad et al. 2021; Atif et al. 2021; Athar et al. 2021; Adnan et al. 2018a,b; Adnan et al. 2019; Akram et al. 2018a,b; Aziz et al. 2017a,b; Chang et al. 2021; Chen et al. 2021; Emre et al. 2021; Habib et al. 2017; Hafiz et al. 2016; Hafiz et al. 2019; Ghulam et al. 2021; Guofu et al. 2021; Hafeez et al. 2021; Khan et al. 2021; Kamaran et al. 2017; Muhmmad et al. 2019; Safi et al. 2021; Sajjad et al. 2019; Saud et al. 2013; Saud et al. 2014; Saud et al. 2017; Saud et al. 2016; Shah et al. 2013; Saud et al. 2020; Saud et al. 2022a,b). The use of soil microorganisms (bacteria, fungi, algae, etc.) that boost nutrient absorption capacity and water use efficiency (Armada et al. 2014) is one option for achieving the aforementioned desirable crop attributes. The microbial populations greatly influence a plant's health. Diverse microbiomes populate different environments in plants, including the phyllosphere, endosphere (tissues), and rhizosphere (Berendsen et al. 2012). The rhizosphere is described as soil adjacent to plant roots and contains a lot of plant rhizobacteria (Bakker et al. 2013). The rhizosphere contains up to 30,000 diverse prokaryotic cells, and it includes 10^{11} microbial cells per gram, which positively impact plant growth (Egamberdieva et al. 2008; Mendes et al. 2013). The procedure of nutrient efficiency enhancement in soil to increase plant growth and production is known as rhizosphere management (Shah et al. 2015; Zia et al. 2021; Qamar et al. 2017; Hamza et al. 2021; Irfan et al. 2021;Wajid et al. 2017; Yang et al. 2017; Zahida et al. 2017; Depeng et al. 2018; Hussain et al. 2020; Hafiz et al. 2020 a,b; Shafi et al. 2020; Wahid et al. 2020; Subhan et al. 2020; Zafar-ul-Hye et al. 2020a,b; Zafar et al. 2021; Adnan et al. 2020; Ilyas et al. 2020; Saleem et al 2020a,b,c; Rehman 2020; Farhat et al. 2020; Wu et al. 2020; Mubeen et al. 2020; Farhana 2020; Jan et al. 2019; Wu et al. 2019; Ahmad et al. 2019; Baseer et al. 2019; Hafiz et al. 2018; Tariq et al. 2018). Plant roots produce and accumulate a variety of chemicals and exudates and provide mechanical support and allow water and nutrient intake (Walker et al. 2003). These various chemicals and exudates change the physical and chemical properties of the soil and regulate the microbial community structure at the root surface (Dakora and Phillips 2002). Bacteria are the most common microbes in the rhizosphere, coexisting alongside fungi, protozoa, and algae.

Consequently, they probably substantially impact the physiology and competitiveness of root colonization in plants (Glick 2012). Their diversity remains powerful with repeated alterations in community structure and species abundance. PGPR include *Bacillus*, *Serratia burkholderia*, *Alcaligenes*, *Enterobacter*, *Azotobacter*, *Azospirillum*, *Arthrobacter*, *Klebsiella*, and *Pseudomonas*, which either live freely or in an association such as symbiotic, parasitic, or saprophytic (Bhattacharyya et al. 2012). However, PGPR in agriculture account for just a tiny percentage of all agricultural practices globally. This is due to the inconsistencies in the inoculated PGPR's characteristics, which can impact crop output. The longevity of PGPR in soil depends on the interaction of crops with indigenous microflora and ecological conditions.

12.2.1 Properties of Perfect PGPR

The following characteristics should be present in an optimal PGPR strain:

(1) It should be rhizosphere-friendly and rhizosphere-competent.
(2) Upon injection, it should colonize the plant roots in sufficient numbers.
(3) It should be able to encourage the development of plants.
(4) It should have a wide range of actions.
(5) It needs to get along with the other bacteria in the rhizosphere.
(6) It must be resistant to physicochemical elements such as heat, desiccation, radiation, and oxidants.
(7) It should outperform current rhizobacterial communities in terms of competitive abilities.

For decades, effective biocontrol agents and biofertilizers that have been capable of promoting plant growth and health, respectively, have been produced. Biofertilizers offer a lot of promise for increasing crop yields in an ecologically sustainable way. These biofertilizers are used in composting sites or soil, which enhances the desired microbial population and plays a key role in nutrient recycling and its availability for plants (Mahmood et al. 2021; Farah et al. 2020; Sadam et al. 2020; Unsar et al. 2020; Fazli et al. 2020; Enamul et al. 2020; Gopakumar et al. 2020; Zia-ur-Rehman 2020; Ayman et al. 2020; Mohammad I. Al-Wabel et al. 2020a,b; Senol 2020; Amjad et al. 2020; Ibrar et al. 2020; Sajid et al. 2020; Muhammad et al. 2021; Sidra et al. 2021; Zahir et al. 2021; Sahrish et al. 2022) (Figure 12.1).

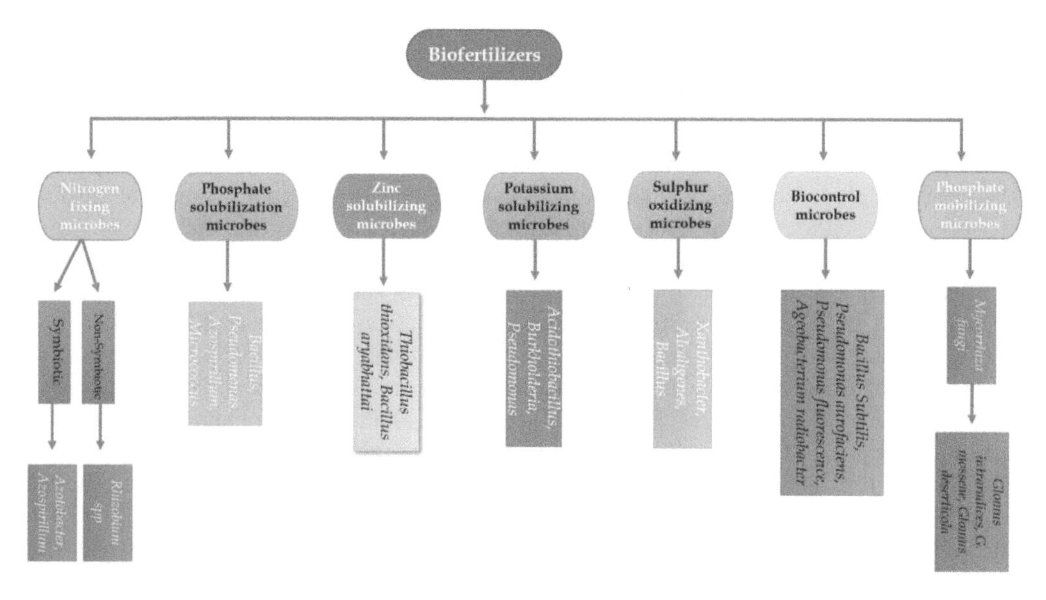

FIGURE 12.1 Graphic representation of microorganisms used as biofertilizer. Reprinted with permission from Fasusi et al. (2021).

12.3 WHAT ARE BIOFERTILIZERS AND HOW DO THEY WORK?

The usage of chemical fertilizers in agriculture has negative consequences for ecosystems (Iqra et al. 2020; Akbar et al. 2020; Mahar et al. 2020; Noor et al. 2020; Bayram et al. 2020; Amanullah, Fahad 2018a,b; Amanullah, Fahad 2017; Amanullah et al. 2020; Amanullah et al. 2021; Rashid et al. 2020; Arif et al. 2020; Amir et al. 2020; Saman et al. 2020; Muhammad Tahir et al. 2020; Jakirand Allah 2020). They leache into water bodies and cause eutrophication in plants due to poor absorption by plants. They may also have various negative consequences on soils, such as loss of water storage capacity and uneven soil fertility. They are expensive, and they also increase greenhouse gas emissions due to the usage of fossil fuels in their manufacturing. There is a need to discover environmentally friendly, low-cost, effective fertilizers. The biofertilizers and biocontrol agents have shown a lot of promise in increasing crop yields in ecologically benign ways. Biofertilizer research began in 1895 with Nobbe and Hiltner's "Nitragin," the first commercially produced and patented *Rhizobium* culture. Biofertilizers are microbial formulations that include beneficial microbial strains that have been immobilized or trapped on inert carrier materials and could be used to improve plant development and soil fertility.

Biofertilizers are microorganisms found in the rhizosphere that help in plant development by increasing nutrient accessibility to plants. Formulating microbes and their products to assist in fixing atmospheric nitrogen, phosphate solubilization, or PGP compounds leads to nutrient solubility that can boost agricultural plant production.

12.3.1 Advantages of Biofertilizers

1) Biological fertilizers help soils establish biological activity by mobilizing nutrients.
2) The provision of appropriate nutrients helps to maintain plant health.
3) Food is supplied, and beneficial soil worms and microbes are encouraged to develop.
4) Root development is stimulated because of the excellent structure offered to the soil.
5) They encourage the formation of mycorrhizal connections, which increases the amount of phosphorus (P) available in the soil.

TABLE 12.1
Role of *Pseudomonas* Species Involved in Plant and Crop Improvement, Growth, and Production

	Isolate	Host	Function	References
1	*P. fluorescens* (R68)	Wheat	Enhance leaf and root number, root length, and fresh weight	Jimtha et al. 2017
2	*P. trivialis* Fs9	Sunflower	Promote shoot length and weight	Majeed et al. 2018
3	*Pseudomonas* spp. AF-54	Sunflower	Enhance plant height	Majeed et al. 2018
4	*P. fluorescens*	Sweet potato	Increase weight, length of roots and density of Santana roots	Santana-Fernández et al. 2021
5	*P. flourescens*	Rice	Increase seed germination	Elekhtyar et al. 20165
6	*P. flourescens*	Melon	Effect soil nutrient distribution and increase yield quality	Martinez et al. 2019
7	*P. aeruginosa* RRALC3	*Pongamia pinnata*	Increase biomass,dry matter uptake, sugar content, amino acid content, carbon content	Radhapriya et al. 2015
8	*Psudomonas* spp.		Improve plant biomass, relative water content, leaf water potential, and root adherence	Sandhya et al. 2010
9	*Pseudomonas putida*	Wheat	Improve wheat yield 96%	Selvakumar et al. 2014

6) They assist in the elimination of plantar illnesses and the provision of a steady supply of micronutrients to the soil.
7) They assist in the maintenance of nitrogen (N) and phosphorus (P) concentrations that are stable.

12.3.2 *Pseudomonas* as a Biofertilizer

Pseudomonas is a Gram-negative, non-spore-forming, aerobic bacteria with straight or slightly curved rods. This bacterial genus is found in the Proteobacteria phylum, Gamma proteobacteria class, and Pseudomonadaceae family. *Pseudomonas* is also a noteworthy bacterial genus since several species are clinically relevant opportunistic human pathogens, plant pathogens, and biocontrol agents. *Pseudomonas aeruginosa* is a human pathogen, *Pseudomonas syringae* is a plant pathogen, and *Pseudomonas putida* and *Pseudomonas fluorescens* are non-pathogenic biocontrol agents.

Pseudomonas species develop in a variety of ways in agro-fields, demonstrating their capacity to stimulate, sustain, and remediate. They produce diverse antifungal metabolites, mediate quorum sensing, having a synergistic association with the plant root. The majority of *Pseudomonas* species live as endophytes and colonize the rhizosphere, promoting plant health by controlling plant-pathogenic bacteria and increasing disease tolerance. Different *Pseudomonas* species involved in plant and crop improvement, growth, and production are included in Table 12.1.

12.4 MECHANISMS OF ACTION OF THE *PSEUDOMONAS* BIOFERTILIZER

Pseudomonas are ecologically essential bacteria that plays a vital role in nitrogen and carbon cycles, because of their capacity to secrete specific chemicals that play a critical role in phosphate solubilization, siderophore synthesis, and nitrogen fixation for promoting plant growth. *Pseudomonas* species may operate as both a PGR and in phytopathogen control. *Pseudomonas* species generate antagonistic substances for plant disease control, such as cell wall breakdown enzymes and antibiotics, and maintain a mutualistic relationship with the linked plant. All of the growth-promoting characteristics listed above may be unusual among strains belonging to the same species. Most *Pseudomonas* spp., on the other hand, have been shown to have these basic growth-promoting properties.

12.4.1 Solubilization of Phosphate

Phosphorus is a critical macronutrient that plants need for optimal development. It is an essential nutrient since it is involved in metabolic activities, signal transduction, macromolecule production, and photosynthesis. It was detected in soil at concentrations of 400–1200 mg/kg. It has lower solubility, rendering it inaccessible to plants despite its high concentration level.

Due to the fast fixation, phosphorus becomes limited due to the formation of oxides like aluminium hydroxide $Al(OH)_3$, calcium hydroxide $Ca(OH)_2$, and ferrous hydroxide $Fe(OH)_2$. Agricultural areas are often fertilized with phosphate fertilizers to address soil phosphorus shortage. Therefore, the phosphate fertilizers have weak absorption by plants, and the remaining P fertilizer is rapidly converted into insoluble complexes in the soil. However, to address this issue, phosphate-solubilizing bacteria (PSB) transform insoluble phosphorus into a soluble form. *Pseudomonas*, *Bacillus*, *Rhizobium*, and *Azotobacter* are among the most significant PSBs in the industrial world. The phosphate-solubilizing bacteria solubilize and mineralize phosphate, as well as produce organic acids like citric acid and gluconic acid. Similarly, these bacteria also produce phosphatase enzymes that convert inorganic phosphorus to mono- or dibasic ions. In *Triticum aestivum* treated with *Pseudomonas* spp., Ma et al. (2012) found an increase in biomass output and phosphorus consumption. PSBs that generate cyanide as a secondary metabolite include *Pseudomonas aeruginosa* and *Pseudomonas fluorescens*. This metabolite is important because it aids PSBs by controlling phytopathogens and diseases. According to Srivastava (2020), *Arabidopsis thaliana* infected with the phosphorus-solubilizing bacterium *P. putida* MTCC 5279 flourished under salt stress and P-deficient conditions, with considerably greater acidic and alkaline phosphatases activity and biomass.

12.4.2 Siderophore Biosynthesis

Iron (Fe) is another important nutrient for plants. In an aerobic environment, it is found in Fe^{2+} and Fe^{3+}, oxyhydroxides, and insoluble hydroxides, making it unavailable for plant absorption. Siderophores are low-molecular-weight water-soluble organic compounds that have an affinity for iron carriers or iron-binding molecules. They are classified into extracellular or intracellular siderophores. Rhizobacteria utilize the siderophores produced by other rhizobacteria of different genera (heterologous siderophores), though some can use siderophores produced by the same genus (known as homologous siderophores). Siderophores are mostly produced by Gram-negative bacteria, like *Pseudomonas* and *Enterobacter* genera, which have a significant role in controlling rhizospheric phytopathogens. Plants use a variety of methods such as releasing or chelating iron from siderophores, direct absorption of siderophore–iron complexes, and a ligand exchange process. *Pseudomonas* sp., such as PGPR, meets their iron needs by consuming siderophores produced by a variety of different rhizosphere microbes. *P. putida* uses heterologous siderophores made by other microorganisms to increase the amount of iron accessible in its natural environment. A luminous siderophore (pyoverdine) plays a key role in the stimulation of plant growth (Rehman et al. 2016). Pattnaik et al. (2019) found that a mutant strain of *Pseudomonas aeruginosa* with no ability for siderophore production while its biocontrol activity against *Pythium sp.* in tomato plants showed infection rendering with inactive *Pseudomonas aeruginosa* siderophore complex. Sharma et al. (2003) investigated the effect of the *Pseudomonas strain* GRP3 producing siderophore in *Vigna radiata* iron feeding. When compared to control plants, chlorotic symptoms decreased after 45 days, and iron, chlorophyll a, and chlorophyll b levels increased in strain GRP3-infected plants.

12.4.3 Production of Phytohormones

Plant hormones are plant growth regulators (PGRs) that works as chemical messengers through which plants respond to their surroundings (Fahad and Bano 2012; Fahad et al. 2017; Fahad et al. 2013; Fahad et al. 2014a,b; Fahad et al. 2016a,b,c,d; Fahad et al. 2015a,b; Fahad et al. 2018a,b; Fahad et al. 2019a,b; Fahad et al. 2020; Fahad et al. 2021a,b,c,d,e,f; Fahad et al. 2022a,b; Hesham and Fahad 2020). These chemical compounds are effective at extremely low concentrations and are primarily generated in certain areas of the plant before being transferred to another location. Cross-talk signalling is the term for this kind of communication. Phytohormones start cell development, reproduction, and specialization in the root system. Phytohormones help with root development by invading roots directly or indirectly through 1-aminocyclopropane-1-carboxylic acid (ACC) catalysis. The ACC deaminase hydrolyses the plant ACC and prevents the ethylene at a phytohormone level, acting antagonistically with the plant, while plant ACC hydrolysed by ACC deaminase prevents the phytohormone ethylene level, working antagonistically with the plant. Bacteria, both free-living and symbiotic, aid plant development by generating compounds that are functionally equivalent to phytohormones produced by the plant. Auxins, cytokinins, gibberellins, and ethylene are some of the chemicals involved in the regulation of biological processes important for plant growth and development (Figure 12.2).

12.4.3.1 Phytohormones

12.4.3.1.1 Auxin

Auxin is a chemical that regulates most plant functions, either directly or indirectly. It plays a key role in cell division and differentiation of plants, germination, geotropism, phototropism, metabolite biosynthesis, and stress tolerance. Indole-3-acetic acid (IAA) is the most active and well-known auxin produced by plants (IAA). The most well-known strains that produce auxin are *Pseudomonas putida*, *Pseudomonas fluorescens*, and *Pseudomonas syringae*. These strains use L-Trp (precursor molecule) for PGPR activities. When comparing plants injected with an IAA-deficient *P. putida* mutant to plants treated with *P. putida* strain GR12-2, Patten and Glick (2002) discovered improved root growth in *Brassica napus*.

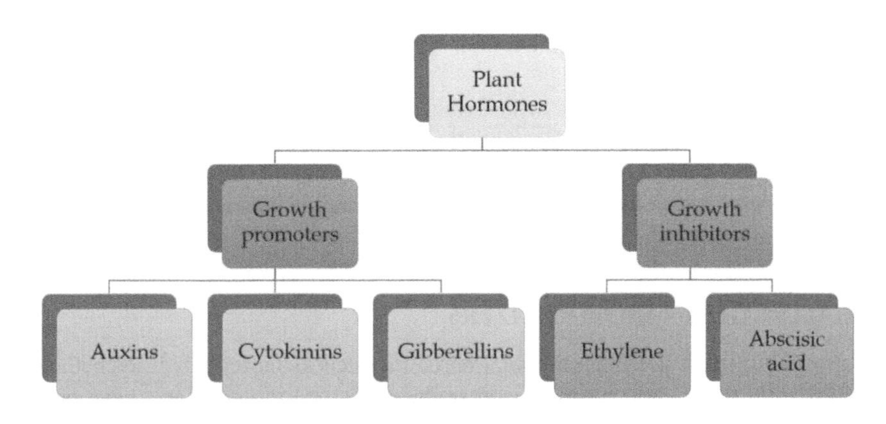

FIGURE 12.2 Plant growth regulators.

12.4.3.1.2 Cytokinins

Cytokinins are a class of plant hormones that stimulate cell division and vascular differentiation in plants, as well as root hair proliferation, and inhibit lateral root growth and elongation of the root. Several species of *Pseudomonas* produce cytokinins that help in the growth of roots and shoots. Salt stress also increases the proline content in *Glycine max* tissues (Naz et al. 2009; Ali et al. 2016).

12.4.3.1.3 Gibberellin

Gibberellin is a plant hormone that helps in seed germination, floral stimulation, development of flowers and fruits, and increases leaf growth (GA). Several PGPR bacteria, including *Pseudomonas* sp., *Bacillus* spp., and *Enterococcus faecium*, stimulate the synthesis of gibberellins. Similarly, *P. fluorescens* and *P. koreensis* have been reported to produce gibberellin-like compounds in MU2 cell-free medium.

12.4.3.1.4 Ethylene

Ethylene is a plant growth hormone that regulates fruit ripening, leaf senescence, root gravitropism, and biotic and abiotic stress responses. High concentrations, on the other hand, may limit root extension, nodule formation, and nitrogen fixation by beneficial microbes in leguminous plants, inducing hypertrophy and hastening maturity and leaf senescence. Bacterial genera seldom generate ethylene, but they do have an enzyme called ACC (1-aminocyclopropane-1-carboxylate) deaminase that may reduce ethylene's harmful effects on plants. In the presence of the *P. stutzeri* A1501 strain, the ACC deaminase enzyme was revealed to be critical in boosting rice growth under salt and heavy metal stress. Ahmad et al. (2013) showed that ACC-deaminase-producing *Rhizobium* and *Pseudomonas* strains might enhance mung bean growth, physiology, and quality in salt-affected conditions.

12.5 USE OF PGPRS AS BIOFERTILIZERS: GUIDELINES AND PRECAUTIONS

(1) It is critical that the given biofertilizer is of acceptable quality, includes 10^7 live cells per gram as an inoculum, and be acquired solely from a reputable source when employing PGPRs as biofertilizers.
(2) Due to its specificity, the biofertilizer should only be used for the crop(s) listed on the commercially available product package.
(3) The name of the crop for which the culture bag is to be used should be written on the bag.
(4) Excess culture should be injected, and any remnants/remaining culture should be promptly placed in field grooves so that inoculum bacteria begin interacting with other microbiota in the rhizosphere and colonize the rhizosphere.
(5) Because biofertilizers are microbial products, they should be kept in cool, shaded areas, ideally at room temperature (25–28°C), before being applied to fields to ensure longer shelf life.
(6) Direct contact of biofertilizers with agrochemicals (herbicides, weedicides, insecticides) should be avoided during storage or application.
(7) In general, 200 grams of biofertilizer may efficiently treat 10 kilograms of seeds.
(8) Soil additions such as lime or rock phosphate are frequently recommended in adverse soil conditions, particularly if the soil is extremely acidic.

12.6 CHALLENGES IN BIOFERTILIZERS

There is a growing interest in using microbial-based products as biofertilizers. Nonetheless, transferring from the lab to the field poses several problems. These bioinoculants have been used on agricultural plants such as legumes and grains in the past. Furthermore, once injected into the soil, plant growth-promoting microorganisms (PGPM) are subjected to competitive situations, which

may significantly limit their favourable effects. Therefore, the positive benefits of applying a certain biofertilizer may vary substantially depending on the agri-environmental circumstances, leading to debate about the effectiveness of microbial-based solutions. *Pseudomonas aeruginosa* is a powerful strain for decomposing hydrocarbons and is often utilized in environmental research. It is also an opportunistic pathogen that causes bloodstream, skin, and soft tissue infections, as well as otitis externa and pneumonia.
Some of the major challenges can be summarized as (Timmusk et al. 2017):

- Lack of understanding of the environmentally friendly value of microbial biofertilizers among farmer communities is amongst the significant obstacles.
- Farmer communities are unaware of the value of microbial biofertilizers for the environment.
- Inadequate agricultural extension worker promotion and encouragement of farmers to utilize biofertilizer products.
- Lack of acceptable carriers for biofertilizer formulation.
- Lack of storage facilities to avoid biofertilizer product contamination and harsh weather conditions.

12.7 CURRENT CHALLENGES DUE TO *PSEUDOMONAS* BIOFERTILIZERS

- It remains problematic to apply our knowledge about biofertilizer effects on microbiome function and plant nutrition in an agroecosystem.
- Work on biofertilizers, which have limited usage because of uneven reactions across varied soisl, crops, and environmental circumstances, as well as practical issues such as mass production, shelf life, proper advice, and convenience of use for farmers (Debnath et al. 2019)
- This is especially problematic in the lab-to-field transfer, since it is usual for a microbial strain that performs well in vitro to function badly in a greenhouse or in field trails (Parnell et al. 2016; Hart et al. 2018; Keswani et al. 2019)
- It is difficult to anticipate the result of an immunization since we normally only evaluate and control for a small number of factors, ignoring their complex relationships (Moutia et al. 2010; Busby et al. 2017).
- Negative interactions with the microbiota in the area (e.g., competition, predation, antagonism).
- The shelf life and survival of inoculated strains in a dramatically diverse environment are restricted by certain biofertilizers and their long-term effects.
- Because of the complexity and genetic variety of the soil and plant microbiome, a single formulation is unlikely to be successful in all areas. However, designing customized biofertilizers for each place is impracticable and unrealistic (Bell et al. 2019).

REFERENCES

Abdul S, Muhammad AA, Shabir H, Hesham A El E, Sajjad H, Niaz A, Abdul G, RZ Sayyed, Fahad S, Subhan D, Rahul D (2021a) Zinc nutrition and arbuscular mycorrhizal symbiosis effects on maize (Zea mays L.) growth and productivity. *J Saudi Soc Agri Sci* 28(11):6339–6351. https://doi.org/10.1016/j.sjbs.2021.06.096

Abdul S, Muhammad AA, Subhan D, Niaz A, Fahad S, Rahul D, Mohammad JA, Omaima N, Muhammad Habib ur R, Bernard RG (2021b) Effect of arbuscular mycorrhizal fungi on the physiological functioning of maize under zinc-deficient soils. *Sci Rep* 11:18468.

Abid M, Khalid N, Qasim A, Saud A, Manzer HS, Chao W, Depeng W, Shah S, Jan B, Subhan D, Rahul D, Hafiz MH, Wajid N, Muhammad M, Farooq S, Fahad S (2021) Exploring the potential of moringa leaf extract as bio stimulant for improving yield and quality of black cumin oil. *Sci Rep* 11:24217. https://doi.org/10.1038/s41598-021-03617-w

Adesemoye AO, Torbert HA, Kloepper JW (2009) Plant growth-promoting rhizobacteria allow reduced application rates of chemical fertilizers. *Microb Ecol* 58(4):921–929. doi.org/10.1007/s00248-009-9531-y

Adnan M, Shah Z, Ullah H, Khan B, Arshad M, Mian IA, Khan GA, Alam M, Basir A, Rahman IU, Ali M, Khan WU (2016) Yield response of wheat to nitrogen and potassium fertilization. *Pure and Appli Bio* 5(4):868–875.

Adnan M, Zahir S, Fahad S, Arif M, Mukhtar A, Imtiaz AK, Ishaq AM, Abdul B, Hidayat U, Muhammad A, Inayat-Ur R, Saud S, Muhammad ZI, Yousaf J, Amanullah, Hafiz MH, Wajid N (2018a) Phosphate-solubilizing bacteria nullify the antagonistic effect of soil calcification on bioavailability of phosphorus in alkaline soils. *Sci Rep* 8:4339. https://doi.org/10.1038/s41598-018-22653-7

Adnan M, Shah Z, Sharif M, Rahman H (2018b) Liming induces carbon dioxide (CO_2) emission in PSB inoculated alkaline soil supplemented with different phosphorus sources. *Environ Sci Poll Res* 25(10):9501–9509.

Adnan M, Fahad S, Khan IA, Saeed M, Ihsan MZ, Saud S, Riaz M, Wang D, Wu C (2019) Integration of poultry manure and phosphate solubilizing bacteria improved availability of Ca bound P in calcareous soils. *3 Biotech* 9(10):368.

Adnan M, Fahad S, Muhammad Z, Shahen S, Ishaq AM, Subhan D, Zafar-ul-Hye M, Martin LB, Raja MMN, Beena S, Saud S, Imran A, Zhen Y, Martin B, Jiri H, Rahul D (2020) Coupling phosphate-solubilizing bacteria with phosphorus supplements improves maize phosphorus acquisition and growth under lime induced salinity stress. *Plants* 9(7):900. doi: 10.3390/plants9070900

Ahmad M, Zahir ZA, Khalid M, Nazli F, Arshad M (2013) Efficacy of Rhizobium and Pseudomonas strains to improve physiology, ionic balance and quality of mung bean under salt-affected conditions on farmer's fields. *Plant Physiol Biochem* 63:170–176. doi.org/10.1016/j.plaphy.2012.11.024.

Ahmad N, Hussain S, Ali MA, Minhas A, Waheed W, Danish S, Fahad S, Ghafoor U, Baig, KS, Sultan H, Muhammad IH, Mohammad JA, Theodore DM (2022) Correlation of soil characteristics and citrus leaf nutrients contents in current scenario of Layyah District. *Hortic* 8:61. https://doi.org/10.3390/horticulturae8010061

Ahmad S, Kamran M, Ding R, Meng X, Wang H, Ahmad I, Fahad S, Han Q (2019) Exogenous melatonin confers drought stress by promoting plant growth, photosynthetic capacity and antioxidant defense system of maize seedlings. *Peer J* 7:e7793. http://doi.org/10.7717/peerj.7793

Akbar H, Timothy JK, Jagadish T, Golam M, Apurbo KC, Muhammad F, Rajan B, Fahad S, Hasanuzzaman M (2020) Agricultural land degradation: processes and problems undermining future food security. In: Fahad S, Hasanuzzaman M, Alam M, Ullah H,Saeed M, Khan AK, Adnan M (Eds.) *Environment, Climate, Plant and Vegetation Growth*. Springer Publ Ltd, Springer Nature Switzerland, pp. 17–62. https://doi.org/10.1007/978-3-030-49732-3

Akram R, Turan V, Hammad HM, Ahmad S, Hussain S, Hasnain A, Maqbool MM, Rehmani MIA, Rasool A, Masood N, Mahmood F, Mubeen M, Sultana SR, Fahad S, Amanet K, Saleem M, Abbas Y, Akhtar HM, Waseem F, Murtaza R, Amin A, Zahoor SA, ul Din MS, Nasim W (2018a) Fate of organic and inorganic pollutants in paddy soils. In: Hashmi MZ, Varma A (Eds.) *Environmental Pollution of Paddy Soils, Soil Biology*. Springer International Publishing, Switzerland, pp. 197–214.

Akram R, Turan V, Wahid A, Ijaz M, Shahid MA, Kaleem S, Hafeez A, Maqbool MM, Chaudhary HJ, Munis, MFH, Mubeen M, Sadiq N, Murtaza R, Kazmi DH, Ali S, Khan N, Sultana SR, Fahad S, Amin A, Nasim W (2018b) Paddy land pollutants and their role in climate change. In: Hashmi MZ, Varma A (Eds.) *Environmental Pollution of Paddy Soils, Soil Biology*. Springer International Publishing, Switzerland, pp. 113–124.

Ali A, Imtiyaz M, Adnan M, Arshad M, Rahman IU, Jamal Y, Muhammad H, Saleem N, Rahman Z (2016) Effect of zinc activities on shoot, root biomass and phosphorus uptake in wheat gynotypes. *American-Eurasian J Agric & Environ Sci* 16(1):204–208.

Ali S, Hameed G, Muhammad A, Depeng W, Fahad S (2022) Comparative genetic evaluation of maize inbred lines at seedling and maturity stages under drought stress. *J Plant Growth Regul* 42, 989–1005. https://doi.org/10.1007/s00344-022-10608-2

Al-Zahrani HS, Alharby HF and Fahad S (2022) Antioxidative defense system, hormones, and metabolite accumulation in different plant parts of two contrasting rice cultivars as influenced by plant growth regulators under heat stress. *Front Plant Sci* 13:911846. doi: 10.3389/fpls.2022.911846

Amanullah, Fahad S (Eds.) (2017) Rice – Technology and Production. IntechOpen Croatia 2017. http://dx.doi.org/10.5772/64480

Amanullah, Fahad S (Eds.) (2018a) Corn – Production and Human Health in Changing Climate. *IntechOpen United Kingdom* 2018. http://dx.doi.org/10.5772/intechopen.74074

Amanullah, Fahad S (Eds.) (2018b) Nitrogen in Agriculture – Updates. IntechOpen Croatia 2018. http://dx.doi.org/10.5772/65846

Amanullah, Muhammad I, Haider N, Shah K, Manzoor A, Asim M, Saif U, Izhar A, Fahad S, Adnan M et al. (2021) Integrated foliar nutrients application improve wheat (Triticum aestivum L.) productivity under calcareous soils in drylands. *Commun Soil Sci Plant Anal* 52(21):2748–2766. https://Doi.Org/10.1080/00103624.2021.1956521

Amanullah, Shah K, Imran, Hamdan AK, Muhammad A, Abdel RA, Muhammad A, Fahad S, Azizullah S, Brajendra P (2020) Effects of climate change on irrigation water quality. In: Fahad S, Hasanuzzaman M, Alam M, Ullah H, Saeed M, Khan AK, Adnan M (Eds.) *Environment, Climate, Plant and Vegetation Growth*. Springer Publ Ltd, Springer Nature Switzerland, pp. 123–132. https://doi.org/10.1007/978-3-030-49732-3

Amir M, Muhammad A, Allah B, Sevgi Ç, Haroon ZK, Muhammad A, Emre A (2020) Bio fortification under climate change: the fight between quality and quantity. In: Fahad S, Hasanuzzaman M, Alam M, Ullah H, Saeed M, Khan AK, Adnan M (Eds.) *Environment, Climate, Plant and Vegetation Growth*. Springer Publ Ltd, Springer Nature Switzerland, pp. 173–228. https://doi.org/10.1007/978-3-030-49732-3

Amjad I, Muhammad H, Farooq S, Anwar H (2020) Role of plant bioactives in sustainable agriculture. In: Fahad S, Hasanuzzaman M, Alam M, Ullah H, Saeed M, Khan AK, Adnan M (Eds.) *Environment, Climate, Plant and Vegetation Growth*. Springer Publ Ltd, Springer Nature Switzerland, pp. 591–606. https://doi.org/10.1007/978-3-030-49732-3

Amjad SF, Mansoora N, Din IU, Khalid IR, Jatoi GH, Murtaza G, Yaseen S, Naz M, Danish S, Fahad S et al. (2021) Application of zinc fertilizer and mycorrhizal inoculation on physio-biochemical parameters of wheat grown under water-stressed environment. *Sustainability* 13:11007. https://doi.org/10.3390/su131911007

Anam I, Huma G, Ali H, Muhammad K, Muhammad R, Aasma P, Muhammad SC, Noman W, Sana F, Sobia A, Fahad S (2021) Ameliorative mechanisms of turmeric-extracted curcumin on arsenic (As)-induced biochemical alterations, oxidative damage, and impaired organ functions in rats. *Environ Sci Pollut Res* 28(46):66313–66326. https://doi.org/10.1007/s11356-021-15695-4

Arif M, Talha J, Muhammad R, Fahad S, Muhammad A, Amanullah, Kawsar A, Ishaq AM, Bushra K, Fahd R (2020) Biochar: a remedy for climate change. In: Fahad S, Hasanuzzaman M, Alam M, Ullah H, Saeed M, Khan AK, Adnan M (Eds.) *Environment, Climate, Plant and Vegetation Growth*. Springer Publ Ltd, Springer Nature Switzerland, pp. 151–172. https://doi.org/10.1007/978-3-030-49732-3

Armada E, Portela G, Roldán A, Azcón R (2014) Combined use of beneficial soil microorganism and agrowaste residue to cope with plant water limitation under semiarid conditions. *Geoderma* 1(232):640–648. doi.org/10.1016/j.geoderma.2014.06.025

Ashfaq AR, Uzma Y, Niaz A, Muhammad AA, Fahad S, Haider S, Tayebeh Z, Subhan D, Süleyman T, Hesham AElE, Pramila T, Jamal MA, Sulaiman AA, Rahul D (2021) Toxicity of cadmium and nickel in the context of applied activated carbon biochar for improvement in soil fertility. *Saudi Society Agric Sci* 29(2):743–750. https://doi.org/10.1016/j.sjbs.2021.09.035

Athar M, Masood IA, Sana S, Ahmed M, Xiukang W, Sajid F, Sher AK, Habib A, Faran M, Zafar H, Farhana G, Fahad S (2021) Bio-diesel production of sunflower through sulphur management in a semi-arid subtropical environment. *Environ Sci Pollution Res* 29(9):13268–13278. https://doi.org/10.1007/s11356-021-16688-z

Atif B, Hesham A, Fahad S (2021) Biochar coupling with phosphorus fertilization modifies antioxidant activity, osmolyte accumulation and reactive oxygen species synthesis in the leaves and xylem sap of rice cultivars under high-temperature stress. *Physiol Mol Biol Plants* 27(9):2083–2100. https://doi.org/10.1007/s12298-021-01062-7

Ayman El Sabagh, Akbar Hossain, Celaleddin Barutçular, Muhammad Aamir Iqbal, M Sohidul Islam, Shah Fahad, Oksana Sytar, Fatih Çig, Ram Swaroop Meena, and Murat Erman (2020) Consequences of salinity stress on the quality of crops and its mitigation strategies for sustainable crop production: an outlook of arid and semi-arid regions. In: Fahad S, Hasanuzzaman M, Alam M, Ullah H, Saeed M, Khan AK, Adnan M (Eds.) *Environment, Climate, Plant and Vegetation Growth*. Springer Publ Ltd, Springer Nature Switzerland, pp. 503–534. https://doi.org/10.1007/978-3-030-49732-3

Aziz K, Daniel KYT, Fazal M, Muhammad ZA, Farooq S, FanW, Fahad S, Ruiyang Z (2017a) Nitrogen nutrition in cotton and control strategies for greenhouse gas emissions: a review. *Environ Sci Pollut Res* 24:23471–23487. https://doi.org/10.1007/s11356-017-0131-y

Aziz K, Daniel KYT, Muhammad ZA, Honghai L, Shahbaz AT, Mir A, Fahad S (2017b) Nitrogen fertility and abiotic stresses management in cotton crop: a review. *Environ Sci Pollut Res* 24:14551–14566. https://doi.org/10.1007/s11356-017-8920-x

Bakker PA, Berendsen RL, Doornbos RF, Wintermans PC, Pieterse CM (2013) The rhizosphere revisited: root microbiomics. *Frontiers in Plant Science* 4:165. doi.org/10.3389/fpls.2013.00165

Baseer M, Adnan M, Fazal M, Fahad S, Muhammad S, Fazli W, Muhammad A, Jr Amanullah, Depeng W, Saud S, Muhammad N, Muhammad Z, Fazli S, Beena S, Mian AR, Ishaq AM (2019) Substituting urea by organic wastes for improving maize yield in alkaline soil. *J Plant Nutrition* 42(19):2423–2434. doi.org/10.1080/01904167.2019.1659344

Bayram AY, Seher Ö, Nazlican A (2020) Climate change forecasting and modeling for the year of 2050. In: Fahad S, Hasanuzzaman M, Alam M, Ullah H, Saeed M, Khan AK, Adnan M (Eds.) *Environment, Climate, Plant and Vegetation Growth*. Springer Publ Ltd, Springer Nature Switzerland, pp. 109–122. https://doi.org/10.1007/978-3-030-49732-3

Bell TH, Kaminsky LM, Gugino BK, Carlson JE, Malik RJ, Hockett KL, Trexler RV (2019) Factoring ecological, societal, and economic considerations into inoculant development. *Trends Biotechnol* 37(6):572–573. doi.org/10.1016/j.tibtech.2019.02.009

Berendsen RL, Pieterse CM, Bakker PA (2012) The rhizosphere microbiome and plant health. *Trends Plant Sci* 17(8):478–486 doi.org/10.1016/j.tplants.2012.04.001

Bhattacharyya PN, Jha DK (2012) Plant growth-promoting rhizobacteria (PGPR): emergence in agriculture. *World J Microbiol Biotechnol* 28(4):1327–1350 doi.org/10.1007/s11274-011-0979-9

Buee M, De Boer W, Martin F, Van Overbeek L, Jurkevitch E (2009). The rhizosphere zoo: an overview of plant-associated communities of microorganisms, including phages, bacteria, archaea, and fungi, and of some of their structuring factors. *Plant Soil* 321:189–212. doi.org/10.1007/s11104-009-9991-3

Bukhari MA, Adnan NS, Fahad S, Javaid I, Fahim N, Abdul M, Mohammad SB (2021) Screening of wheat (*Triticum aestivum* L.) genotypes for drought tolerance using polyethylene glycol. *Arabian Journal of Geosci* 14:2808

Busby PE, Soman C, Wagner MR, Friesen ML, Kremer J, Bennett A et al. (2017) Research priorities for harnessing plant microbiomes in sustainable agriculture. *PLoS Biol* 15:e2001793. doi: 10.1371/journal.pbio.2001793.

Chang W, Qiujuan J, Evgenios A, Haitao L, Gezi L, Jingjing Z, Fahad S, Ying J (2021) Hormetic effects of zinc on growth and antioxidant defense system of wheat plants. *Sci Total Environ* 807, 150992. https://doi.org/10.1016/j.scitotenv.2021.150992

Chao W, Youjin S, Beibei Q, Fahad S (2022) Effects of asymmetric heat on grain quality during the panicle initiation stage in contrasting rice genotypes. *J Plant Growth Regul* 42:630–636. https://doi.org/10.1007/s00344-022-10598-1

Chaparro JM, Badri DV, Bakker MG, Sugiyama A, Manter DK, Vivanco JM (2013) Root exudation of phytochemicals in Arabidopsis follows specific patterns that are developmentally programmed and correlate with soil microbial functions. *PloS One* 8(2):e55731. doi.org/10.1371/journal.pone.0055731

Chen Y, Guo Z, Dong L, Fu Z, Zheng Q, Zhang G, Qin L, Sun X, Shi Z, Fahad S, Xie F, Saud S (2021) Turf performance and physiological responses of native Poa species to summer stress in Northeast China. *Peer J* 9:e12252. http://doi.org/10.7717/peerj.12252

Compant S, Samad A, Faist H, Sessitsch A (2019) A review on the plant microbiome: ecology, functions, and emerging trends in microbial application. *J Adv Res* 19:29–37. doi.org/10.1016/j.jare.2019.03.004

Dakora FD, Phillips DA (2002) Root exudates as mediators of mineral acquisition in low-nutrient environments. *Food Security in Nutrient-Stressed Environments: Exploiting Plants' Genetic Capabilities* 245(1): 201–213. doi.org/10.1007/978-94-017-1570-6_23

Debnath S, Rawat D, Kumar Mukherjee A, Adhikary S, Kundu R (2019) Applications and constraints of plant beneficial microorganisms in agriculture, biostimulants. In: *Plant Science* Mirmajlessi SM, Radhakrishnan R (Eds.) (IntechOpen), pp. 1–25. doi: 10.5772/intechopen.89190

Deepranjan S, Ardith S, Siva OD, Sonam S, Shikha, Manoj P, Amitava R, Sayyed RZ, Abdul G, Mohammad JA, Subhan D, Fahad S, Rahul D (2021) Optimizing nutrient use efficiency, productivity, energetics, and economics of red cabbage following mineral fertilization and biopriming with compatible rhizosphere microbes. *Sci Rep* 11:15680 https://doi.org/10.1038/s41598-021-95092-6

Depeng W, Fahad S, Saud S, Muhammad K, Aziz K, Mohammad NK, Hafiz MH, Wajid N (2018) Morphological acclimation to agronomic manipulation in leaf dispersion and orientation to promote "Ideotype" breeding: Evidence from 3D visual modeling of "super" rice (*Oryza sativa* L.). *Plant Physiol Biochem* 135:499–510. https://doi.org/10.1016/j.plaphy.2018.11.010 doi.org/10.1128/MRA.01305-19

Drogue B, Combes-Meynet E, Moënne-Loccoz Y, Wisniewski-Dyé F, Prigent-Combaret C (2013) Control of the cooperation between plant growth-promoting rhizobacteria and crops by rhizosphere signals. *Mol Microb Ecol Rhizo* 1:279–293. doi.org/10.1002/9781118297674.ch27

Dutta S, Podile AR (2010) Plant growth promoting rhizobacteria (PGPR): the bugs to debug the root zone. *Crit Rev Microbiol* 36(3):232–244. doi.org/10.3109/10408411003766806

Egamberdieva D, Kamilova F, Validov S, Gafurova L, Kucharova Z, Lugtenberg B (2008) High incidence of plant growth-stimulating bacteria associated with the rhizosphere of wheat grown on salinated soil in Uzbekistan. *Environ Microbiol* 10(1):1–9. doi.org/10.1111/j.1462-2920.2007.01424.x

EL Sabagh A, Islam MS, Hossain A, Iqbal MA, Mubeen M, Waleed M, Reginato M, Battaglia M, Ahmed S, Rehman A, Arif M, Athar H-U-R, Ratnasekera D, Danish S, Raza MA, Rajendran K, Mushtaq M, Skalicky M, Brestic M, Soufan W, Fahad S, Pandey S, Kamran M, Datta R, Abdelhamid MT (2022) Phytohormones as growth regulators during abiotic stress tolerance in plants. Front *Agron* 4:765068. doi: 10.3389/fagro.2022.765068

Elekhtyar NMIII (2016) Influence of different plant growth promoting rhizobacteria (PGPR) strains on rice promising line. *The Sixth Field Crops Conference*, FCRI, ARC, Giza, Egypt, 22–23.

Emre B, Ömer SU, Martín LB, Andre D, Fahad S, Rahul D, Muhammad Z-ul-H, Ghulam SH, Subhan D (2021) Studying soil erosion by evaluating changes in physico-chemical properties of soils under different land-use types. *J Saudi Society Agricultural Sci* 20:190–197.

Enamul MdH, Shoeb AZM, Mallik AH, Fahad S, Kamruzzaman MM, Akib J, Nayyer S, Mehedi KMA, Swati AS, Yeamin MdA, Most SS (2020) Measuring vulnerability to environmental hazards: qualitative to quantitative. In: Fahad S, Hasanuzzaman M, Alam M, Ullah H, Saeed M, Khan AK, Adnan M (Eds.) *Environment, Climate, Plant and Vegetation Growth.* Springer Publ Ltd, Springer Nature Switzerland, pp. 421–452. https://doi.org/10.1007/978-3-030-49732-3

Fahad S, Bano A (2012) Effect of salicylic acid on physiological and biochemical characterization of maize grown in saline area. *Pak J Bot* 44:1433–1438.

Fahad S, Chen Y, Saud S,Wang K, Xiong D, Chen C,Wu C, Shah F, Nie L, Huang J (2013) Ultraviolet radiation effect on photosynthetic pigments, biochemical attributes, antioxidant enzyme activity and hormonal contents of wheat. *J Food, Agri Environ* 11(3&4):1635–1641.

Fahad S, Hussain S, Bano A, Saud S, Hassan S, Shan D, Khan FA, Khan F, Chen Y, Wu C, Tabassum MA, Chun MX, Afzal M, Jan A, Jan MT, Huang J (2014a) Potential role of phytohormones and plant growth-promoting rhizobacteria in abiotic stresses: consequences for changing environment. *Environ Sci Pollut Res* 22(7):4907–4921. https://doi.org/10.1007/s11356-014-3754-2

Fahad S, Hussain S, Matloob A, Khan FA, Khaliq A, Saud S, Hassan S, Shan D, Khan F, Ullah N, Faiq M, Khan MR, Tareen AK, Khan A, Ullah A, Ullah N, Huang J (2014b) Phytohormones and plant responses to salinity stress: A review. *Plant Growth Regul* 75(2):391–404. https://doi.org/10.1007/s10725-014-0013-y

Fahad S, Hussain S, Saud S, Tanveer M, Bajwa AA, Hassan S, Shah AN, Ullah A,Wu C, Khan FA, Shah F, Ullah S, Chen Y, Huang J (2015a) A biochar application protects rice pollen from high-temperature stress. *Plant Physiol Biochem* 96:281–287.

Fahad S, Nie L, Chen Y, Wu C, Xiong D, Saud S, Hongyan L, Cui K, Huang J (2015b) Crop plant hormones and environmental stress. *Sustain Agric Rev* 15:371–400.

Fahad S, Hussain S, Saud S, Hassan S, Chauhan BS, Khan F et al. (2016a) Responses of rapid viscoanalyzer profile and other rice grain qualities to exogenously applied plant growth regulators under high day and high night temperatures. *PLoS One* 11(7):e0159590. https://doi.org/10.1371/journal.pone.0159590

Fahad S, Hussain S, Saud S, Khan F, Hassan S, Jr A, Nasim W, Arif M, Wang F, Huang J (2016b) Exogenously applied plant growth regulators affect heat-stressed rice pollens. *J Agron Crop Sci* 202:139–150

Fahad S, Hussain S, Saud S, Hassan S, Ihsan Z, Shah AN,Wu C, Yousaf M, Nasim W, Alharby H, Alghabari F, Huang J (2016c) Exogenously applied plant growth regulators enhance the morphophysiological growth and yield of rice under high temperature. *Front Plant Sci* 7:1250. https://doi.org/10.3389/fpls.2016. 01250

Fahad S, Hussain S, Saud S, Hassan S, Tanveer M, Ihsan MZ, Shah AN, Ullah A, Nasrullah KF, Ullah S, AlharbyH NW, Wu C, Huang J (2016d) A combined application of biochar and phosphorus alleviates heat-induced adversities on physiological, agronomical and quality attributes of rice. *Plant Physiol Biochem* 103:191–198.

Fahad S, Bajwa AA, Nazir U, Anjum SA, Farooq A, Zohaib A, Sadia S, NasimW, Adkins S, Saud S, Ihsan MZ, Alharby H,Wu C,Wang D, Huang J (2017) Crop production under drought and heat stress: Plant responses and management options. *Front Plant Sci* 8:1147. https://doi.org/10.3389/fpls.2017.01147

Fahad S, Muhammad ZI, Abdul K, Ihsanullah D, Saud S, Saleh A, Wajid N, Muhammad A, Imtiaz AK, Chao W, Depeng W, Jianliang H (2018a) Consequences of high temperature under changing climate optima for rice pollen characteristics-concepts and perspectives, *Archives Agron Soil Sci* 64(11):1473–1488. doi: 10.1080/03650340.2018.1443213

Fahad S, Abdul B, Adnan M (Eds.) (2018b) Global Wheat Production. *IntechOpen United Kingdom* 2018. http://dx.doi.org/10.5772/intechopen.72559

Fahad S, Rehman A, Shahzad B, Tanveer M, Saud S, Kamran M, Ihtisham M, Khan SU, Turan V, Rahman MHU (2019a) Rice responses and tolerance to metal/metalloid toxicity. In: Hasanuzzaman M, Fujita M, Nahar K, Biswas JK (Eds.) *Advances in Rice Research for Abiotic Stress Tolerance*. Woodhead Publishing Lth, UK, pp. 299–312.

Fahad S, Adnan M, Hassan S, Saud S, Hussain S, Wu C, Wang D, Hakeem KR, Alharby HF, Turan V, Khan MA, Huang J (2019b) Rice responses and tolerance to high temperature. In: Hasanuzzaman M, Fujita M, Nahar K, Biswas JK (Eds.) *Advances in Rice Research for Abiotic Stress Tolerance*. Woodhead Publishing Ltd, UK, pp. 201–224.

Fahad S, Hasanuzzaman M, Alam M, Ullah H, Saeed M, Ali Khan I, Adnan M (Eds.) (2020) *Environment, Climate, Plant and Vegetation Growth*. Springer Nature Switzerland. doi: https://doi.org/10.1007/978-3-030-49732-3

Fahad S, Sönmez O, Saud S, Wang D, Wu C, Adnan M, Turan V (Eds.) (2021a) Plant growth regulators for climate-smart agriculture. First edition. *Footprints of Climate Variability on Plant Diversity*. CRC Press, Boca Raton.

Fahad S, Sonmez O, Saud S, Wang D, Wu C, Adnan M, Turan V (Eds.) (2021b) Climate change and plants: biodiversity, growth and interactions. First edition. *Footprints of Climate Variability on Plant Diversity*. CRC Press, Boca Raton.

Fahad S, Sonmez O, Saud S, Wang D, Wu C, Adnan M, Turan V (Eds.) (2021c) Developing climate resilient crops: improving global food security and safety. First edition. *Footprints of Climate Variability on Plant Diversity*. CRC Press, Boca Raton.

Fahad S, Sönmez O, Turan V, Adnan M, Saud S, Wu C, Wang D (Eds.) (2021d) Sustainable soil and land management and climate change. First edition. *Footprints of Climate Variability on Plant Diversity*. CRC Press, Boca Raton.

Fahad S, Sönmez O, Saud S, Wang D, Wu C, Adnan M, Arif M, Amanullah (Eds.) (2021e) Engineering tolerance in crop plants against abiotic stress. First edition. *Footprints of Climate Variability on Plant Diversity*. CRC Press, Boca Raton.

Fahad S, Saud S, Yajun C, Chao W, Depeng W (Eds.) (2021f) *Abiotic Stress in Plants*. IntechOpen United Kingdom 2021. http://dx.doi.org/10.5772/intechopen.91549

Fahad S, Adnan M, Saud S (Eds.) (2022a) Improvement of plant production in the era of climate change. First edition. *Footprints of Climate Variability on Plant Diversity*. CRC Press, Boca Raton.

Fahad S, Adnan M, Saud S, Nie L (Eds.) (2022b) Climate change and ecosystems: challenges to sustainable development. First edition. *Footprints of Climate Variability on Plant Diversity*. CRC Press, Boca Raton.

Fakhre A, Ayub K, Fahad S, Sarfraz N, Niaz A, Muhammad AA, Muhammad A, Khadim D, Saud S, Shah H, Muhammad ASR, Khalid N, Muhammad A, Rahul D, Subhan D (2021) Phosphate solubilizing bacteria optimize wheat yield in mineral phosphorus applied alkaline soil. *J Saudi Soc Agric Sci* 21(5):339–348. https://doi.org/10.1016/j.jssas.2021.10.007

Farah R, Muhammad R, Muhammad SA, Tahira Y, Muhammad AA, Maryam A, Shafaqat A, Rashid M, Muhammad R, Qaiser H, Afia Z, Muhammad AA, Muhammad A, Fahad S (2020) Alternative and non-conventional soil and crop panagement strategies for increasing water use efficiency. In: Fahad S, Hasanuzzaman M, Alam M, Ullah H, Saeed M, Khan AK, Adnan M (Eds.) *Environment, Climate, Plant and Vegetation Growth*. Springer Publ Ltd, Springer Nature Switzerland, pp. 323–338. https://doi.org/10.1007/978-3-030-49732-3

Farhana G, Ishfaq A, Muhammad A, Dawood J, Fahad S, Xiuling L, Depeng W, Muhammad F, Muhammad F, Syed AS (2020) Use of crop growth model to simulate the impact of climate change on yield of various wheat cultivars under different agro-environmental conditions in Khyber Pakhtunkhwa, Pakistan. *Arabian J Geosci* 13:112. https://doi.org/10.1007/s12517-020-5118-1

Farhat A, Hafiz MH, Wajid I, Aitazaz AF, Hafiz FB, Zahida Z, Fahad S, Wajid F, Artemi C (2020) A review of soil carbon dynamics resulting from agricultural practices. *J Environ Manage* 268: 110319.

Farhat UK, Adnan AK, Kai L, Xuexuan X, Muhammad A, Fahad S, Rafiq A, Mushtaq AK, Taufiq N, Faisal Z (2022) Influences of long-term crop cultivation and fertilizer management on soil aggregates stability and fertility in the Loess Plateau, *Northern China J Soil Sci Plant Nutri* 22:1446–1457. https://doi.org/10.1007/s42729-021-00744-1

Fasusi OA, Cruz C, Babalola OO (2021) Agricultural sustainability: microbial biofertilizers in rhizosphere management. *Agric* 11(2):163. doi.org/10.3390/agriculture11020163

Fazli W, Muhmmad S, Amjad A, Fahad S, Muhammad A, Muhammad N, Ishaq AM, Imtiaz AK, Mukhtar A, Muhammad S, Muhammad I, Rafi U, Haroon I, Muhammad A (2020) Plant–microbes interactions and functions in changing climate. In: Fahad S, Hasanuzzaman M, Alam M, Ullah H, Saeed M, Khan AK, Adnan M (Eds.), *Environment, Climate, Plant and Vegetation Growth*. Springer Publ Ltd, Springer Nature Switzerland, pp. 397–420. https://doi.org/10.1007/978-3-030-49732-3

Gao QM, Zhu S, Kachroo P, Kachroo A (2015) Signal regulators of systemic acquired resistance. *Front Plant Sci* 6:228. doi.org/10.3389/fpls.2015.00228

Ghulam M, Muhammad AA, Donald LS, Sajid M, Muhammad FQ, Niaz A, Ateeq ur R, Shakeel A, Sajjad H, Muhammad A, Summia M, Aqib HAK, Fahad S, Rahul D, Mazhar I, Timothy DS (2021) Formalin fumigation and steaming of various composts differentially influence the nutrient release, growth and yield of muskmelon (*Cucumis melo* L.). *Sci Rep* 11:21057.

Glick BR (2012) Plant growth-promoting bacteria: mechanisms and applications. *Scientifica* 2012:963401. doi:10.6064/2012/963401

Gopakumar L, Bernard NO, Donato V (2020) Soil microarthropods and nutrient cycling. In: Fahad S, Hasanuzzaman M, Alam M, Ullah H, Saeed M, Khan AK, Adnan M (Eds.) *Environment, Climate, Plant and Vegetation Growth*. Springer Publ Ltd, Springer Nature Switzerland, pp. 453–472. https://doi.org/10.1007/978-3-030-49732-3

Guofu L, Zhenjian B, Fahad S, Guowen C, Zhixin X, Hao G, Dandan L, Yulong L, Bing L, Guoxu J, Saud S (2021) Compositional and structural changes in soil microbial communities in response to straw mulching and plant revegetation in an abandoned artificial pasture in Northeast China. *Glob Ecol Conserv* 31:e01871.

Habib ur R, Ashfaq A, Aftab W, Manzoor H, Fahd R, Wajid I, Md Aminul I, Vakhtang S, Muhammad A, Asmat U, Abdul W, Syeda RS, Shah S, Shahbaz K, Fahad S, Manzoor H, Saddam H, Wajid N (2017) Application of CSM-CROPGRO-cotton model for cultivars and optimum planting dates: evaluation in changing semi-arid climate. *Field Crops Res* 238:139–152. http://dx.doi.org/10.1016/j.fcr.2017.07.007

Hafeez M, Farman U, Muhammad MK, Xiaowei L, Zhijun Z, Sakhawat S, Muhammad I, Mohammed AA, Mandela F-G, Nicolas D, Muzammal R, Fahad S, Yaobin L (2021) Metabolic-based insecticide resistance mechanism and ecofriendly approaches for controlling of beet armyworm Spodoptera exigua:a review. *Environ Sci Pollution Res* 29:1746–1762. https://doi.org/10.1007/s11356-021-16974-w

Hafiz MH, Wajid F, Farhat A, Fahad S, Shafqat S, Wajid N, Hafiz FB (2016) Maize plant nitrogen uptake dynamics at limited irrigation water and nitrogen. *Environ Sci Pollut Res* 24(3):2549–2557. https://doi.org/10.1007/s11356-016-8031-0

Hafiz MH, Farhat A, Shafqat S, Fahad S, Artemi C, Wajid F, Chaves CB, Wajid N, Muhammad M, Hafiz FB (2018) Offsetting land degradation through nitrogen and water management during maize cultivation under arid conditions. *Land Degrad Dev* 29(5): 1–10. doi: 10.1002/ldr.2933

Hafiz MH, Muhammad A, Farhat A, Hafiz FB, Saeed AQ, Muhammad M, Fahad S, Muhammad A (2019) Environmental factors affecting the frequency of road traffic accidents: a case study of sub-urban area of Pakistan. *Environ Sci Pollut Res* 26:11674–11685. https://doi.org/10.1007/s11356-019-04752-8

Hafiz MH, Farhat A, Ashfaq A, Hafiz FB, Wajid F, Carol Jo W, Fahad S, Gerrit H (2020a) Predicting kernel growth of maize under controlled water and nitrogen applications. *Int J Plant Prod* 14:609–620 https://doi.org/10.1007/s42106-020-00110-8

Hafiz MH, Abdul K, Farhat A, Wajid F, Fahad S, Muhammad A, Ghulam MS, Wajid N, Muhammad M, Hafiz FB (2020b) Comparative effects of organic and inorganic fertilizers on soil organic carbon and wheat productivity under arid region. *Commun Soil Sci Plant Anal* 51:1406–1422. doi: 10.1080/00103624.2020.1763385

Haider SA, Lalarukh I, Amjad SF, Mansoora N, Naz M, Naeem M, Bukhari SA, Shahbaz M, Ali SA, Marfo TD, Subhan D, Rahul D, Fahad S (2021) Drought stress alleviation by potassium-nitrate-containing chitosan/montmorillonite microparticles confers changes in Spinacia oleracea L. *Sustain* 13: 9903. https://doi.org/10.3390/su13179903

Hamza SM, Xiukang W, Sajjad A, Sadia Z, Muhammad N, Adnan M, Fahad S et al. (2021) Interactive effects of gibberellic acid and NPK on morpho-physio-biochemical traits and organic acid exudation pattern in coriander (Coriandrum sativum L.) grown in soil artificially spiked with boron. *Plant Physiol Biochem* 167(2021):884–900.

Haoliang Y, Matthew TH, Ke L, Bin W, Puyu F, Fahad S, Holger M, Rui Y, De LL, Sotirios A, Isaiah H, Xiaohai T, Jianguo M, Yunbo Z, Meixue Z (2022) Crop traits enabling yield gains under more frequent extreme climatic events. *Sci Total Environ* 808 (2022):152170.

Hart MM, Antunes PM, Chaudhary VB, Abbott LK (2018) Fungal inoculants in the field: is the reward greater than the risk? *Funct Ecol* 32: 126–135. doi: 10.1111/1365-2435.12976.

Heiman CM, Wiese J, Kupferschmied P, Maurhofer M, Keel C, Vacheron J (2020) Draft genome sequence of Pseudomonas sp. strain LD120, isolated from the marine alga *Saccharina latissima. Microbiol Resour Announc* 9(8):e01305–19. doi.org/10.1128/MRA.01305-19

Hesham FA, Fahad S (2020) Melatonin application enhances biochar efficiency for drought tolerance in maize varieties: modifications in physio-biochemical machinery. *Agron J* 112(4):1–22.

Hiltner L (1904) Uber nevere Erfahrungen und Probleme auf dem Gebiet der Boden Bakteriologie und unter besonderer Beurchsichtigung der Grundungung und Broche. *Arbeit Deut Landw Ges Berlin* 98:59–78.

Huang Li-Y, Li Xiao-X, Zhang Yun-B, Fahad S, Wang F (2021) dep1 improves rice grain yield and nitrogen use efficiency simultaneously by enhancing nitrogen and dry matter translocation. *J Integrative Agri* 21(11):3185–3198. doi: 10.1016/S2095-3119(21)63795-4

Hussain MA, Fahad S, Rahat S, Muhammad FJ, Muhammad M, Qasid A, Ali A, Husain A, Nooral A, Babatope SA, Changbao S, Liya G, Ibrar A, Zhanmei J, Juncai H (2020) Multifunctional role of brassinosteroid and its analogues in plants. *Plant Growth Regul* 92:141–156. https://doi.org/10.1007/s10725-020-00647-8

Ibad U, Dost M, Maria M, Shadman K, Muhammad A, Fahad S, Muhammad I, Ishaq AM, Aizaz A, Muhammad HS, Muhammad S, Farhana G, Muhammad I, Muhammad ASR, Hafiz MH, Wajid N, Shah S, Jabar ZKK, Masood A, Naushad A, Rasheed Akbar M, Shah MK, Jan B (2022) Comparative effects of biochar and NPK on wheat crops under different management systems. *Crop Pasture Sci* 74:31–40. https://doi.org/10.1071/CP21146

Ibrar H, Muqarrab A, Adel MG, Khurram S, Omer F, Shahid I, Fahim N, Shakeel A, Viliam B, Marian B, Sami Al Obaid, Fahad S, Subhan D, Suleyman T, Hanife AKÇA, Rahul D (2021) Improvement in growth and yield attributes of cluster bean through optimization of sowing time and plant spacing under climate change Scenario. *Saudi J Bio Sci* 29(2):781–792. https://doi.org/10.1016/j.sjbs.2021.11.018

Ibrar K, Aneela R, Khola Z, Urooba N, Sana B, Rabia S, Ishtiaq H, Mujaddad Ur Rehman, Salvatore M (2020) Microbes and environment: global warming reverting the frozen zombies. In: Fahad S, Hasanuzzaman M, Alam M, Ullah H, Saeed M, Khan AK, Adnan M (Eds.) *Environment, Climate, Plant and Vegetation Growth.* Springer Publ Ltd, Springer Nature Switzerland, pp. 607–634. https://doi.org/10.1007/978-3-030-49732-3

Ihsan MZ, Abdul K, Manzer HS, Liaqat A, Ritesh K, Hayssam MA, Amar M, Fahad S (2022) The response of Triticum aestivum treated with plant growth regulators to acute day/night temperature rise. *J Plant Growth Regul* 41:2020–2033. https://doi.org/10.1007/s00344-022-10574-9

Ikram U, Khadim D, Muhammad T, Muhammad S, Fahad S (2021) Gibberellic acid and urease inhibitor optimize nitrogen uptake and yield of maize at varying nitrogen levels under changing climate. *Environ Sci Pollution Res* 29(5):6568–6577. https://doi.org/10.1007/s11356-021-16049-w

Ilyas M, Mohammad N, Nadeem K, Ali H, Aamir HK, Kashif H, Fahad S, Aziz K, Abid U (2020) Drought tolerance strategies in plants: a mechanistic approach. *J Plant Growth Regul* 40:926–944. https://doi.org/10.1007/s00344-020-10174-5

Iqra M, Amna B, Shakeel I, Fatima K, Sehrish L, Hamza A, Fahad S (2020) Carbon cycle in response to global warming. In: Fahad S, Hasanuzzaman M, Alam M, Ullah H, Saeed M, Khan AK, Adnan M (Eds.)

Environment, Climate, Plant and Vegetation Growth. Springer Publ Ltd, Springer Nature Switzerland, pp. 1–16. https://doi.org/10.1007/978-3-030-49732-3

Irfan M, Muhammad M, Muhammad JK, Khadim MD, Dost M, Ishaq AM, Waqas A, Fahad S, Saud S et al. (2021) Heavy metals immobilization and improvement in maize (Zea mays L.) growth amended with biochar and compost. *Sci Rep* 11:18416.

Jabborova D, Sulaymanov K, Sayyed RZ, Alotaibi SH, Enakiev Y, Azimov A, Jabbarov Z, Ansari MJ, Fahad S, Danish S et al. (2021) Mineral fertilizers improve the quality of turmeric and soil. *Sustain* 13:9437. https://doi.org/10.3390/su13169437

Jan M, Muhammad Anwar-ul-Haq, Adnan NS, Muhammad Y, Javaid I, Xiuling L, Depeng W, Fahad S (2019) Modulation in growth, gas exchange, and antioxidant activities of salt-stressed rice (Oryza sativa L.) genotypes by zinc fertilization. *Arabian J Geosci* 12:775. https://doi.org/10.1007/s12517-019-4939-2

Jimtha JC, Jishma P, Karthika NR, Nidheesh KS, Ray JG, Mathew J, Radhakrishnan EK (2017). *Pseudomonas fluorescens* R68 assisted enhancement in growth and fertilizer utilization of Amaranthus tricolor (L.). *3 Biotech* 7(4):1–6. doi.org/10.1007/s13205-017-0887-2

Kalam S, Das SN, Basu A, Podile AR (2017) Population densities of indigenous Acidobacteria change in the presence of plant growth promoting rhizobacteria (PGPR) in rhizosphere. *J Basic Microbiol* 57(5):376–385. doi.org/10.1002/jobm.201600588

Kamaran M, Wenwen C, Irshad A, Xiangping M, Xudong Z, Wennan S, Junzhi C, Shakeel A, Fahad S, Qingfang H, Tiening L (2017) Effect of paclobutrazol, a potential growth regulator on stalk mechanical strength, lignin accumulation and its relation with lodging resistance of maize. *Plant Growth Regul* 84:317–332. https://doi.org/10.1007/ s10725-017-0342-8

Keswani C, Prakash O, Bharti N, Vílchez JI, Sansinenea E, Lally RD, ... Singh HB (2019) Re-addressing the biosafety issues of plant growth promoting rhizobacteria. *Sci Total Environ* 690:841–852. doi.org/10.1016/j.scitotenv.2019.07.046

Khadim D, Fahad S, Jahangir MMR, Iqbal M, Syed SA, Shah AK, Ishaq AM, Rahul D et al. (2021a) Biochar and urease inhibitor mitigate NH_3 and N_2O emissions and improve wheat yield in a urea fertilized alkaline soil. *Sci Rep* 11:17413.

Khadim D, Saif-ur-R, Fahad S, Syed SA, Shah AK et al. (2021b) Influence of variable biochar concentration on yield-scaled nitrous oxide emissions, wheat yield and nitrogen use efficiency. *Sci Rep* 11:16774.

Khan MMH, Niaz A, Umber G, Muqarrab A, Muhammad AA, Muhammad I, Shabir H, Shah F, Vibhor A, Shams HA-H, Reham A, Syed MBA, Nadiyah MA, Ali TKZ, Subhan D, Rahul D (2021) Synchronization of boron application methods and rates is environmentally friendly approach to improve quality attributes of *Mangifera indica* L. on sustainable basis. *Saudi J Bio Sci* 29(3):1869–1880. https://doi.org/10.1016/j.sjbs.2021.10.036

Khatun M, Sarkar S, Era FM, Islam AKMM, Anwar MP, Fahad S, Datta R, Islam AKMA (2021) Drought stress in grain legumes: effects, tolerance mechanisms and management. *Agron* 11:2374. https://doi.org/10.3390/agronomy11122374

Kloepper JW, Leong J, Teintze M, Schroth MN (1980) Enhanced plant growth by siderophores produced by plant growth-promoting rhizobacteria. *Nature* 286(5776):885–886. doi.org/10.1038/286885a0

Kumar A, Dubey A (2020) Rhizosphere microbiome: Engineering bacterial competitiveness for enhancing crop production. *J Adv Res* 24:337–352. doi.org/10.1016/j.jare.2020.04.014

Kumari P, Meena M, Gupta P, Dubey MK, Nath G, Upadhyay RS (2018) Plant growth promoting rhizobacteria and their biopriming for growth promotion in mung bean (Vigna radiata (L.) R. Wilczek). *Biocatal Agric Biotechnol* 16:163–171. doi.org/10.1016/j.bcab.2018.07.030

Ma Z, Peng C, Zhu Q, Chen H, Yu G, Li W, ... Zhang W (2012) Regional drought-induced reduction in the biomass carbon sink of Canada's boreal forests. *PNAS* 109(7):2423–2427. doi.org/10.1073/pnas.1111576109

Mahar A, Amjad A, Altaf HL, Fazli W, Ronghua L, Muhammad A, Fahad S, Muhammad A, Rafiullah, Imtiaz AK, Zengqiang Z (2020) Promising technologies for Cd-contaminated soils: drawbacks and possibilities. In: Fahad S, Hasanuzzaman M, Alam M, Ullah H,Saeed M, Khan AK, Adnan M (Eds.) *Environment, Climate, Plant and Vegetation Growth*. Springer Publ Ltd, Springer Nature Switzerland, pp. 63–92. https://doi.org/10.1007/978-3-030-49732-3

Mahmood Ul H, Tassaduq R, Chandni I, Adnan A, Muhammad A, Muhammad MA, Muhammad H-ur-R, Mehmood AN, Alam S, Fahad S (2021) Linking plants functioning to adaptive responses under heat stress conditions: a mechanistic review. *J Plant Growth Regul* 41:2596–2613. https://doi.org/10.1007/s00344-021-10493-1

Majeed A, Abbasi MK, Hameed S, Yasmin S, Hanif MK, Naqqash T, Imran A (2018) *Pseudomonas* sp. AF-54 containing multiple plant beneficial traits acts as growth enhancer of Helianthus annuus L. under reduced fertilizer input. *Microbiol Res* 216:56–69. doi.org/10.1016/j.micres.2018.08.006

Manzer HS, Saud A, Soumya M, Abdullah A, Al-A, Qasi DA, Bander MA, Al-M, Hayssam MA, Hazem MK, Fahad S, Vishnu DR, Om PN (2021) Molybdenum and hydrogen sulfide synergistically mitigate arsenic toxicity by modulating defense system, nitrogen and cysteine assimilation in faba bean (*Vicia faba* L.) seedlings. *Environ Pollut* 290:117953.

Martínez JI, Gómez-Garrido M, Gómez-López MD, Faz Á, Martínez-Martínez S, Acosta JA (2019) Pseudomonas fluorescens affects nutrient dynamics in plant-soil system for melon production. *Chil J Agric Res* 79(2):223–233. doi:10.4067/S0718-58392019000200223

Md Jakir H, Allah B (2020) Development and applications of transplastomic plants: a way towards eco-friendly agriculture. In: Fahad S, Hasanuzzaman M, Alam M, Ullah H, Saeed M, Khan AK, Adnan M (Eds.) *Environment, Climate, Plant and Vegetation Growth*. Springer Publ Ltd, Springer Nature Switzerland, pp. 285–322. https://doi.org/10.1007/978-3-030-49732-3

Meena RS, Kumar S, Datta R, Lal R, Vijayakumar V, Brtnicky M, Sharma MP, Yadav GS, Jhariya MK, Jangir CK, Pathan SI, Marfo TD (2020) Impact of agrochemicals on soil microbiota and management: a review. *Land* 9(2):34 doi.org/10.3390/land9020034

Mehmood K, Bao Y, Saifullah, Bibi S, Dahlawi S, Yaseen M, Abrar MM, Srivastava P, Fahad S, Faraj TK (2022) Contributions of open biomass burning and crop straw burning to air quality: current research paradigm and future outlooks. *Front Environ Sci* 10:852492. doi: 10.3389/fenvs.2022.852492

Mendes R, Garbeva P, Raaijmakers JM (2013) The rhizosphere microbiome: significance of plant beneficial, plant pathogenic, and human pathogenic microorganisms. *FEMS Microbial Rev* 37(5):634–663. doi.org/10.1111/1574-6976.12028

Mohammad I, Al-Wabel, Munir Ahmad, Adel RA Usman, Mutair Akanji, and Muhammad Imran Rafique (2020a) Advances in pyrolytic technologies with improved carbon capture and storage to combat climate change. In: Fahad S, Hasanuzzaman M, Alam M, Ullah H, Saeed M, Khan AK, Adnan M (Eds.) *Environment, Climate, Plant and Vegetation Growth*. Springer Publ Ltd, Springer Nature Switzerland, pp. 535–576. https://doi.org/10.1007/978-3-030-49732-3

Mohammad I, Al-Wabel, Abdelazeem S, Munir A, Khalid E, Adel RAU (2020b) Extent of climate change in Saudi Arabia and its impacts on agriculture: a case study from Qassim Region. In: Fahad S, Hasanuzzaman M, Alam M, Ullah H, Saeed M, Khan AK, Adnan M (Eds.) *Environment, Climate, Plant and Vegetation Growth*. Springer Publ Ltd, Springer Nature Switzerland, pp. 635–658. https://doi.org/10.1007/978-3-030-49732-3

Moutia JFY, Saumtally S, Spaepen S, Vanderleyden J (2010) Plant growth promotion by Azospirillum sp. in sugarcane is influenced by genotype and drought stress. *Plant Soil* 337(1):233–242. doi: 10.1007/s11104-010-0519-7

Mubeen M, Ashfaq A, Hafiz MH, Muhammad A, Hafiz UF, Mazhar S, Muhammad Sami ul Din, Asad A, Amjed A, Fahad S, Wajid N (2020) Evaluating the climate change impact on water use efficiency of cotton-wheat in semi-arid conditions using DSSAT model. *J Water Climate Change* 11(4):1661–1675. doi/10.2166/wcc.2019.179/622035/jwc2019179.pdf

Muhammad I, Khadim D, Fahad S, Imran M, Saud A, Manzer HS, Shah S, Jabar ZKK, Shamsher A, Shah H, Taufiq N, Hafiz MH, Jan B, Wajid N (2022) Exploring the potential effect of *Achnatherum splendens* L.–derived biochar treated with phosphoric acid on bioavailability of cadmium and wheat growth in contaminated soil. *Environ Sci Pollut Res* 29(25):37676–37684. https://doi.org/10.1007/s11356-021-17950-0.

Muhammad N, Muqarrab A, Khurram S, Fiaz A, Fahim N, Muhammad A, Shazia A, Omaima N, Sulaiman AA, Fahad S, Subhan D, Rahul D (2021) Kaolin and jasmonic acid improved cotton productivity under water stress conditions. *J Saudi Society Agricultural Sci* 28 (2021):6606–6614. https://doi.org/10.1016/j.sjbs.2021.07.043

Muhammad Tahir ul Qamar, Amna F, Amna B, Barira Z, Xitong Z, Ling-Ling C (2020) Effectiveness of conventional crop improvement strategies vs. omics. In: Fahad S, Hasanuzzaman M, Alam M, Ullah H, Saeed M, Khan AK, Adnan M (Eds.) *Environment, Climate, Plant and Vegetation Growth*. Springer Publ Ltd, Springer Nature Switzerland, pp. 253–284. https://doi.org/10.1007/978-3-030-49732-3

Muhammad Z, Abdul MK, Abdul MS, Kenneth BM, Muhammad S, Shahen S, Ibadullah J, Fahad S (2019) Performance of Aeluropus lagopoides (mangrove grass) ecotypes, a potential turfgrass, under high saline conditions. *Environ Sci Pollut Res* 26(13):13410–13412. https://doi.org/10.1007/s11356-019-04838-3

Muzammal R, Fahad S, Guanghui D, Xia C, Yang Y, Kailei T, Lijun L, Fei-Hu L, Gang D (2021) Evaluation of hemp (Cannabis sativa L.) as an industrial crop: a review. *Environ Sci Pollution Res* 28(38):52832–52843. https://doi.org/10.1007/s11356-021-16264-5

Naz I, Bano A, Ul-Hassan T (2009) Isolation of phytohormones producing plant growth promoting rhizobacteria from weeds growing in Khewra salt range, Pakistan and their implication in providing salt tolerance to Glycine max L. *African J Biotechnol* 8(21):5762–5766. doi: 10.5897/AJB09.1176

Niaz A, Abdullah E, Subhan D, Muhammad A, Fahad S, Khadim D, Suleyman T, Hanife A, Anis AS, Mohammad JA, Emre B, Ömer SU, Rahul D, Bernard RG (2022) Mitigation of lead (Pb) toxicity in rice cultivated with either ground water or wastewater by application of acidified carbon. *J Environ Manage* 307:114521.

Noor M, Naveed ur R, Ajmal J, Fahad S, Muhammad A, Fazli W, Saud S, Hassan S (2020) Climate change and coastal plant lives. In: Fahad S, Hasanuzzaman M, Alam M, Ullah H, Saeed M, Khan AK, Adnan M (Eds.) *Environment, Climate, Plant and Vegetation Growth.* Springer Publ Ltd, Springer Nature Switzerland, pp. 93–108. https://doi.org/10.1007/978-3-030-49732-3

Parnell JJ, Berka R, Young HA, Sturino JM, Kang Y, Barnhart DM, DiLeo MV (2016) From the lab to the farm: an industrial perspective of plant beneficial microorganisms. *Front Plant Sci* 7:1110. doi: 10.3389/fpls.2016.01110

Patten CL, Glick BR (2002). Role of *Pseudomonas putida* indoleacetic acid in development of the host plant root system. *Appl Environ Microbiol* 68(8):3795–3801. doi.org/10.1128/AEM.68.8.3795-3801.2002

Pattnaik S, Mohapatra B, Kumar U, Pattnaik M, Samantaray D (2019) Microbe-mediated plant growth promotion: a mechanistic overview on cultivable plant growth-promoting members. *Biofertilizer Sustainable Agricul Environ* 55:435–463. doi: 10.1007/978-3-030-18933-4_19

Qamar-uz Z, Zubair A, Muhammad Y, Muhammad ZI, Abdul K, Fahad S, Safder B, Ramzani PMA, Muhammad N (2017) Zinc biofortification in rice: leveraging agriculture to moderate hidden hunger in developing countries. *Arch Agron Soil Sci* 64:147–161. https://doi.org/10.1080/03650340.2017.1338343

Qin ZH, Nasib ur Rahman, Ahmad A, Yu-pei Wang, Sakhawat S, Ehmet N, Wen-juan Shao, Muhammad I, Kun S, Rui L, Fazal S, Fahad S (2022) Range expansion decreases the reproductive fitness of *Gentiana officinalis* (*Gentianaceae*). *Sci Rep* 12:2461. https://doi.org/10.1038/s41598-022-06406-1

Raaijmakers JM, Paulitz TC, Steinberg C, Alabouvette C, Moënne-Loccoz Y (2009) The rhizosphere: a playground and battlefield for soilborne pathogens and beneficial microorganisms. *Plant and Soil* 321(1):341–361. doi.org/ 10.1007/s11104-008-9568-6

Radhapriya P, Ramachandran A, Anandham R, Mahalingam S (2015) *Pseudomonas aeruginosa* RRALC3 enhances the biomass, nutrient and carbon contents of *Pongamia pinnata* seedlings in degraded forest soil. *PLoS One* 10(10):e0139881. doi.org/10.1371/journal.pone.0139881

Rajesh KS, Fahad S, Pawan K, Prince C, Talha J, Dinesh J, Prabha S, Debanjana S, Prathibha MD, Bandana B, Akash H, Gupta NK, Rekha S, Devanshu D, Dalpat LS, Ke L, Matthew TH, Saud S, Adnan NS, Taufiq N (2022) Beneficial elements: new players in improving nutrient use efficiency and abiotic stress tolerance. *Plant Growth Regul* 100:237–265. https://doi.org/10.1007/s10725-022-00843-8

Rashid M, Qaiser H, Khalid SK, Mohammad I, Al-Wabel, Zhang A, Muhammad A, Shahzada SI, Rukhsanda A, Ghulam AS, Shahzada MM, Sarosh A, Muhammad FQ (2020) Prospects of biochar in alkaline soils to mitigate climate change. In: Fahad S, Hasanuzzaman M, Alam M, Ullah H,Saeed M, Khan AK, Adnan M (Eds.) *Environment, Climate, Plant and Vegetation Growth.* Springer Publ Ltd, Springer Nature Switzerland, pp. 133–150. https://doi.org/10.1007/978-3-030-49732-3

Rehana S, Asma Z, Shakil A, Anis AS, Rana KI, Shabir H, Subhan D, Umber G, Fahad S, Jiri K, Sami Al Obaid, Mohammad JA, Rahul D (2021) Proteomic changes in various plant tissues associated with chromium stress in sunflower. *Saudi J Bio Sci* 29(4):2604–2612. https://doi.org/10.1016/j.sjbs.2021.12.042

Rehman M, Fahad S, Saleem MH, Hafeez M, Muhammad Habib ur Rahman, Liu F, Deng G (2020) Red light optimized physiological traits and enhanced the growth of ramie (Boehmeria nivea L.). *Photosynthetica* 58(4):922–931.

Sadam M, Muhammad Tahir ul Qamar, Ghulam M, Muhammad SK, Faiz AJ (2020) Role of biotechnology in climate resilient agriculture. In: Fahad S, Hasanuzzaman M, Alam M, Ullah H, Saeed M, Khan AK, Adnan M (Eds.) *Environment, Climate, Plant and Vegetation Growth.* Springer Publ Ltd, Springer Nature Switzerland, pp. 339–366. https://doi.org/10.1007/978-3-030-49732-3

Safi UK, Ullah F, Mehmood S, Fahad S, Ahmad Rahi A, Althobaiti F et al. (2021) Antimicrobial, antioxidant and cytotoxic properties of Chenopodium glaucum L. *PLoS One* 16(10):e0255502. https://doi.org/10.1371/journal.pone.0255502

Sahrish N, Shakeel A, Ghulam A, Zartash F, Sajjad H, Mukhtar A, Muhammad AK, Ahmad K, Fahad S, Wajid N, Sezai E, Carol Jo W, Gerrit H (2022) Modeling the impact of climate warming on potato phenology. *European J AgroN* 132:126404.

Sajid H, Jie H, Jing H, Shakeel A, Satyabrata N, Sumera A, Awais S, Chunquan Z, Lianfeng Z, Xiaochuang C, Qianyu J, Junhua Z (2020) Rice production under climate change: Adaptations and mitigating ttrategies. In: Fahad S, Hasanuzzaman M, Alam M, Ullah H, Saeed M, Khan AK, Adnan M (Eds.) *Environment, Climate, Plant and Vegetation Growth.* Springer Publ Ltd, Springer Nature Switzerland, pp. 659–686. https://doi.org/10.1007/978-3-030-49732-3

Sajjad H, Muhammad M, Ashfaq A, Waseem A, Hafiz MH, Mazhar A, Nasir M, Asad A, Hafiz UF, Syeda RS, Fahad S, Depeng W, Wajid N (2019) Using GIS tools to detect the land use/land cover changes during forty years in Lodhran district of Pakistan. *Environ Sci Pollut Res* 27:39676–39692. https://doi.org/10.1007/s11356-019-06072-3

Sajjad H, Muhammad M, Ashfaq A, Fahad S, Wajid N, Hafiz MH, Ghulam MS, Behzad M, Muhammad T, Saima P (2021a) Using space–time scan statistic for studying the effects of COVID-19 in Punjab, Pakistan: a guideline for policy measures in regional Agriculture. *Environ Sci Pollut Res* 30:42495–42508. https://doi.org/10.1007/s11356-021-17433-2

Sajjad H, Muhammad M, Ashfaq A, Nasir M, Hafiz MH, Muhammad A, Muhammad I, Muhammad U, Hafiz UF, Fahad S, Wajid N, Hafiz MRJ, Mazhar A, Saeed AQ, Amjad F, Muhammad SK, Mirza W (2021b) Satellite-based evaluation of temporal change in cultivated land in Southern Punjab (Multan region) through dynamics of vegetation and land surface temperature. *Open Geo Sci* 13:1561–1577.

Saleem MH, Fahad S, Adnan M, Mohsin A, Muhammad SR, Muhammad K, Qurban A, Inas AH, Parashuram B, Mubassir A, Reem MH (2020a) Foliar application of gibberellic acid endorsed phytoextraction of copper and alleviates oxidative stress in jute (Corchorus capsularis L.) plant grown in highly copper-contaminated soil of China. *Environ Sci Pollution Res* 27:37121–37133. https://doi.org/10.1007/s11356-020-09764-3

Saleem MH, Rehman M, Fahad S, Tung SA, Iqbal N, Hassan A, Ayub A, Wahid MA, Shaukat S, Liu L, Deng G (2020b) Leaf gas exchange, oxidative stress, and physiological attributes of rapeseed (Brassica napus L.) grown under different light-emitting diodes. *Photosynthetica* 58(3): 836–845.

Saleem MH, Fahad S, Shahid UK, Mairaj D, Abid U, Ayman ELS, Akbar H, Analía L, Lijun L (2020c) Copper-induced oxidative stress, initiation of antioxidants and phytoremediation potential of flax (Linum usitatissimum L.) seedlings grown under the mixing of two different soils of China. *Environ Sci Poll Res* 27:5211–5221. https://doi.org/10.1007/s11356-019-07264-7

Saleem N, Adnan M, Khan NA, Zaheer S, Jalal F, Amin M, Khan WM, Arif M, Rahman I, Ibrahim M, Jamal Y, Shah SRA, Junaid K, Ali M (2015) Dual purpose canola: grazing and grains options. *Pak J Weed Sci Res* 21(2):295–304.

Saman S, Amna B, Bani A, Muhammad Tahir ul Qamar, Rana MA, Muhammad SK (2020) QTL mapping for abiotic stresses in cereals. In: Fahad S, Hasanuzzaman M, Alam M, Ullah H, Saeed M, Khan AK, Adnan M (Eds.) *Environment, Climate, Plant and Vegetation Growth.* Springer Publ Ltd, Springer Nature Switzerland, pp. 229–252. https://doi.org/10.1007/978-3-030-49732-3

Sana U, Shahid A, Yasir A, Farman UD, Syed IA, Mirza MFAB, Fahad S, Al-Misned F, Usman A, Xinle G, Ghulam N, Kunyuan W (2022) Bifenthrin induced toxicity in Ctenopharyngodon idella at an acute concentration: a multi-biomarkers based study. *J King Saud Uni – Sci* 34 (2022):101752.

Sandhya VSKZ, Ali SZ, Grover M, Reddy G, Venkateswarlu B (2010) Effect of plant growth promoting Pseudomonas spp. on compatible solutes, antioxidant status and plant growth of maize under drought stress. *Plant Growth Regul* 62(1):21–30. doi.org/10.1007/s10725-010-9479-4

Santana-Fernández A, Beovides-García Y, Simó-González JE, Pérez-Peñaranda MC, López-Torres J, Rayas-Cabrera A, ... Basail-Pérez M (2021) Effect of a Pseudomonas fluorescens-based biofertilizer on sweet potato yield components. *Asian J Appl Sci* 9(2):105–113.

Saud S, Chen Y, Long B, Fahad S, Sadiq A (2013) The different impact on the growth of cool season turf grass under the various conditions on salinity and drought stress. *Int J Agric Sci Res* 3:77–84.

Saud S, Li X, Chen Y, Zhang L, Fahad S, Hussain S, Sadiq A, Chen Y (2014) Silicon application increases drought tolerance of Kentucky bluegrass by improving plant water relations and morph physiological functions. *Sci World J* 2014:1–10. https://doi.org/10.1155/2014/ 368694

Saud S, Chen Y, Fahad S, Hussain S, Na L, Xin L, Alhussien SA (2016) Silicate application increases the photosynthesis and its associated metabolic activities in Kentucky bluegrass under drought stress and post-drought recovery. *Environ Sci Pollut Res* 23(17):17647–17655. https://doi.org/10.1007/s11356-016-6957-x

Saud S, Fahad S, Yajun C, Ihsan MZ, Hammad HM, Nasim W, Amanullah Jr, Arif M and Alharby H (2017) Effects of nitrogen supply on water stress and recovery mechanisms in Kentucky bluegrass plants. *Front Plant Sci* 8:983. doi: 10.3389/fpls.2017.00983

Saud S, Fahad S, Cui G, Chen Y, Anwar S (2020) Determining nitrogen isotopes discrimination under drought stress on enzymatic activities, nitrogen isotope abundance and water contents of Kentucky bluegrass. *Sci Rep* 10:6415. https://doi.org/10.1038/s41598-020-63548-w

Saud S, Fahad S, Hassan S (2022a) Developments in the investigation of nitrogen and oxygen stable isotopes in atmospheric nitrate. *Sustain Chem Climate Action* 1:100003.

Saud S, Li X, Jiang Z, Fahad S, Hassan S (2022b) Exploration of the phytohormone regulation of energy storage compound accumulation in microalgae. *Food Energy Secur* 2022:e418.

Selvakumar G, Kim K, Hu S, Sa T (2014) Effect of salinity on plants and the role of arbuscular mycorrhizal fungi and plant growth-promoting rhizobacteria in alleviation of salt stress. In: Parvaiz Ahmad, Mohd Rafiq Wani (Eds.), *Physiological Mechanisms and Adaptation Strategies in Plants under Changing Environment*. Springer, pp. 115–144. doi.org/10.1007/978-1-4614-8591-9_6

Selvakumar G, Panneerselvam P, Bindu GH, Ganeshamurthy AN (2015) Pseudomonads: plant growth promotion and beyond. In: Naveen Kumar Arora (Ed.), *Plant Microbes Symbiosis: Applied Facets*, pp. 193–208. doi.org/10.1007/978-81-322-2068-8_10

Senol C (2020) The effects of climate change on human behaviors. In: Fahad S, Hasanuzzaman M, Alam M, Ullah H, Saeed M, Khan AK, Adnan M (Eds.) *Environment, Climate, Plant and Vegetation Growth*. Springer Publ Ltd, Springer Nature Switzerland, pp. 577–590. https://doi.org/10.1007/978-3-030-49732-3

Shafi MI, Adnan M, Fahad S, Fazli W, Ahsan K, Zhen Y, Subhan D, Zafar-ul-Hye M, Martin B, Rahul D (2020) Application of single superphosphate with humic acid improves the growth, yield and phosphorus uptake of wheat (Triticum aestivum L.) in calcareous soil. *Agron* (10):1224. doi:10.3390/agronomy10091224

Shah F, Lixiao N, Kehui C, Tariq S, Wei W, Chang C, Liyang Z, Farhan A, Fahad S, Huang J (2013) Rice grain yield and component responses to near 2°C of warming. *Field Crop Res* 157:98–110.

Shah S, Shah H, Liangbing X, Xiaoyang S, Shahla A, Fahad S (2022) The physiological function and molecular mechanism of hydrogen sulfide resisting abiotic stress in plants. *Brazilian J Botany* 45:563–572 https://doi.org/10.1007/s40415-022-00785-5

Shah SRA, Khan SA, Saljoqi AUR, Junaid K, Khan I, Zaman M, Said F, Adnan M, Saleem N (2015) Assessment of the varietal preference and resistance against Lipaphis erysimi (K.) in segregating mustard genotypes under agro-ecological conditions of Peshawar, Pakistan. *J Ento & Zool Studies* 3(5):100–103.

Sharma A, Johri BN, Sharma AK, Glick BR (2003) Plant growth-promoting bacterium *Pseudomonas* sp. strain GRP3 influences iron acquisition in mung bean (Vigna radiata L. Wilzeck). *Soil Biol Biochem* 35(7):887–894. doi.org/10.1016/S0038-0717(03)00119-6

Shukla KP, Sharma S, Singh NK, Singh V, Tiwari K, Singh S (2011) Nature and role of root exudates: efficacy in bioremediation. *Afr J Biotechnol* 10(48): 9717–9724. doi: 10.5897/AJB10.2552

Sidra K, Javed I, Subhan D, Allah B, Syed IUSB, Fatma B, Khaled DA, Fahad S, Omaima N, Ali TKZ, Rahul D (2021) Physio-chemical characterization of indigenous agricultural waste materials for the development of potting media. *J Saudi Society Agricultural Sci.* 28(12):7491–7498 https://doi.org/10.1016/j.sjbs.2021.08.058

Son JS, Sumayo M, Hwang YJ, Kim BS, Ghim SY (2014) Screening of plant growth-promoting rhizobacteria as elicitor of systemic resistance against gray leaf spot disease in pepper. *Appl Soil Ecol* 73:1–8. doi.org/10.1016/j.apsoil.2013.07.016

Srivastava S, Srivastava S (2020) Prescience of endogenous regulation in Arabidopsis thaliana by *Pseudomonas putida* MTCC 5279 under phosphate starved salinity stress condition. *Sci Rep* 10(1):1–15. doi.org/10.1038/s41598-020-62725-1

Subhan D, Zafar-ul-Hye M, Fahad S, Saud S, Martin B, Tereza H, Rahul D (2020) Drought stress alleviation by ACC deaminase producing *Achromobacter xylosoxidans* and *Enterobacter cloacae*, with and without timber waste biochar in maize. *Sustain* 12(15):6286. doi:10.3390/su12156286

Tapia-Vázquez I, Sánchez-Cruz R, Arroyo-Domínguez M, Lira-Ruan V, Sánchez-Reyes A, del Rayo Sánchez-Carbente M, ... Folch-Mallol JL (2020) Isolation and characterization of psychrophilic and psychrotolerant plant-growth promoting microorganisms from a high-altitude volcano crater in Mexico. *Microbiol Res* 232:126394. doi.org/10.1016/j.micres.2019.126394

Tariq M, Ahmad S, Fahad S, Abbas G, Hussain S, Fatima Z, Nasim W, Mubeen M, ur Rehman MH, Khan MA, Adnan M (2018) The impact of climate warming and crop management on phenology of sunflower-based cropping systems in Punjab, Pakistan. *Agri and Forest Met* 15(256):270–282.

Timmusk S, Behers L, Muthoni J, Muraya A, Aronsson AC (2017) Perspectives and challenges of microbial application for crop improvement. *Front Plant Sci* 8:49. doi: 10.3389/fpls.2017.00049

Unsar Naeem-U, Muhammad R, Syed HMB, Asad S, Mirza AQ, Naeem I, Muhammad H ur R, Fahad S, Shafqat S (2020) Insect pests of cotton crop and management under climate change scenarios. In: Fahad S, Hasanuzzaman M, Alam M, Ullah H, Saeed M, Khan AK, Adnan M (Eds.) *Environment, Climate, Plant and Vegetation Growth*. Springer Publ Ltd, Springer Nature Switzerland, pp. 367–396. https://doi.org/10.1007/978-3-030-49732-3

Verma M, Mishra J, Arora NK (2019) Plant growth-promoting rhizobacteria: diversity and applications. In: Naga Raju Maddela, Luz Cecilia García (Eds.), *Environ Biotech for Sustainable Future*, Springer, pp. 129–173. doi.org/10.1007/978-981-10-7284-0_6

Vessey JK (2003) Plant growth promoting rhizobacteria as biofertilizers. *Plant Soil* 255(2):571–586. doi.org/10.1023/A:1026037216893

Wahid F, Fahad S, Subhan D, Adnan M, Zhen Y, Saud S, Manzer HS, Martin B, Tereza H, Rahul D (2020) Sustainable management with mycorrhizae and phosphate solubilizing bacteria for enhanced phosphorus uptake in calcareous soils. *Agri* 10(8):334. doi:10.3390/agriculture10080334

Wajid N, Ashfaq A, Asad A, Muhammad T, Muhammad A, Muhammad S, Khawar J, Ghulam MS, Syeda RS, Hafiz MH, Muhammad IAR, Muhammad ZH, Muhammad Habib ur R, Veysel T, Fahad S, Suad S, Aziz K, Shahzad A (2017) Radiation efficiency and nitrogen fertilizer impacts on sunflower crop in contrasting environments of Punjab. *Pakistan Environ Sci Pollut Res* 25:1822–1836. https://doi.org/10.1007/s11356-017-0592-z

Walker TS, Bais HP, Grotewold E, Vivanco JM (2003) Root exudation and rhizosphere biology. *Plant Physiol* 132(1):44–51. doi.org/10.1104/pp.102.019661

Wiqar A, Arbaz K, Muhammad Z, Ijaz A, Muhammad A, Fahad S (2022) Relative efficiency of biochar particles of different sizes for immobilising heavy metals and improving soil properties. *Crop Pasture Sci.* 42(2):112–120 https://doi.org/10.1071/CP20453.

Wu C, Tang S, Li G, Wang S, Fahad S, Ding Y (2019) Roles of phytohormone changes in the grain yield of rice plants exposed to heat: a review. *Peer J* 7:e7792. doi 10.7717/peerj.7792

Wu C, Kehui C, She T, Ganghua L, Shaohua W, Fahad S, Lixiao N, Jianliang H, Shaobing P, Yanfeng D (2020) Intensified pollination and fertilization ameliorate heat injury in rice (Oryza sativa L.) during the flowering stage. *Field Crops Res* 252: 107795.

Xue B, Huang L, Li X, Lu J, Gao R, Kamran M, Fahad S (2022) Effect of clay mineralogy and soil organic carbon in aggregates under straw incorporation. *Agron* 12:534. https://doi.org/10.3390/ agronomy12020534

Yang R, Dai P, Wang B, Jin T, Liu K, Fahad S, Harrison MT, Man J, Shang J, Meinke H, Deli L, Xiaoyan W, Yunbo Z, Meixue Z, Yingbing T, Haoliang Y (2022) Over-optimistic projected future wheat yield potential in the North China Plain: the role of future climate extremes. *Agron* 12:145. https://doi.org/10.3390/agronomy12010145

Yang Z, Zhang Z, Zhang T, Fahad S, Cui K, Nie L, Peng S, Huang J (2017) The effect of season-long temperature increases on rice cultivars grown in the central and southern regions of China. *Front Plant Sci* 8:1908. https://doi.org/10.3389/fpls.2017.01908

Zafar-ul-Hye M, Muhammad N, Subhan D, Fahad S, Rahul D, Mazhar A, Ashfaq AR, Martin B, Jiˇrí H, Zahid HT, Muhammad N (2020a) Alleviation of cadmium adverse effects by improving nutrients uptake in bitter gourd through cadmium tolerant rhizobacteria. *Environ* 7(8), 54. doi:10.3390/environments7080054

Zafar-ul-Hye M, Muhammad Tahzeeb-ul-Hassan, Muhammad A, Fahad S, Martin B, Tereza D, Rahul D, Subhan D (2020b) Potential role of compost mixed biochar with rhizobacteria in mitigating lead toxicity in spinach. *Scientific Rep* 10:12159. https://doi.org/10.1038/s41598-020-69183-9.

Zafar-ul-Hye M, Akbar MN, Iftikhar Y, Abbas M, Zahid A, Fahad S, Datta R, Ali M, Elgorban AM, Ansari MJ et al. (2021) Rhizobacteria inoculation and caffeic acid alleviated drought stress in lentil plants. *Sustain* 13:9603. https://doi.org/10.3390/su13179603

Zahida Z, Hafiz FB, Zulfiqar AS, Ghulam MS, Fahad S, Muhammad RA, Hafiz MH, Wajid N, Muhammad S (2017) Effect of water management and silicon on germination, growth, phosphorus and arsenic uptake in rice. *Ecotoxicol Environ Saf* 144:11–18.

Zahir SM, Zheng-HG, Ala Ud D, Amjad A, Ata Ur R, Kashif J, Shah F, Saud S, Adnan M, Fazli W, Saud A, Manzer HS, Shamsher A, Wajid N, Hafiz MH, Fahad S (2021) Synthesis of silver nanoparticles using Plantago lanceolata extract and assessing their antibacterial and antioxidant activities. *Sci Rep* 11:20754.

Zaman I, Ali M, Shahzad K, Tahir, MS, Matloob A, Ahmad W, Alamri S, Khurshid MR, Qureshi MM, Wasaya A, Khurram SB, Manzer HS, Fahad S, Rahul D (2021) Effect of plant spacings on growth, physiology, yield and fiber quality attributes of cotton genotypes under nitrogen fertilization. *Agron* 11:2589. https://doi.org/10.3390/agronomy11122589

Zehra A, Raytekar NA, Meena M, Swapnil P (2021) Efficiency of microbial bio-agents as elicitors in plant defense mechanism under biotic stress. *Rev Current Res Microb Sc* 2:100054. doi.org/10.1016/j.crmicr.2021.100054

Zia R, Nawaz MS, Siddique MJ, Hakim S, Imran A (2021) Plant survival under drought stress: implications, adaptive responses, and integrated rhizosphere management strategy for stress mitigation. *Microbiol Res* 242:126626. doi.org/10.1016/j.micres.2020.126626

Zia-ur-Rehman M (2020) Environment, climate change and biodiversity. In: Fahad S, Hasanuzzaman M, Alam M, Ullah H, Saeed M, Khan AK, Adnan M (Eds.) *Environment, Climate, Plant and Vegetation Growth*. Springer Publ Ltd, Springer Nature Switzerland, pp. 473–502. https://doi.org/10.1007/978-3-030-49732-3

13 Nitrogen-fixing Biofertilizers

Muhammad Romman[*1], *Rainaz Parvez*[2], *Muhammad Adnan*[*3], *Farhana Gul*[3], *Muhammad Haroon*[3], *Rafi Ullah*[3], *Shah Saud*[4], *Nazeer Ahmed*[3], *Ishfaq Hameed*[1], *Taufiq Nawaz*[5], *Muhammad Hamzah Saleem*[6], *Sahar Mumtaz*[7], *Amanullah*[8], *Muhammad Arif*[8], *Maid Zaman*[9], *Abdel Rahman Altawaha*[10], *and Shah Hassan*[11]

[1] Department of Botany, University of Chitral, KP, Pakistan
[2] Department of Botany, Government Girls Degree College Dargai, Malakand, KP, Pakistan
[3] Department of Agriculture, University of Swabi, Swabi, Pakistan
[4] College of Life Science, Linyi University, Linyi, Shandong, China
[5] Department of Food Science and Technology, The University of Agriculture, Peshawar, Pakistan
[6] MOA Key Laboratory of Crop Ecophysiology and Farming System in the Middle Reaches of the Yangtze River, College of Plant Science and Technology, Huazhong, China
[7] Division of Science and Technology, Department of Botany University of Education, Lahore, Pakistan
[8] Department of Agronomy, the University of Agriculture Peshawar, Khyber Pakhtunkhwa, Pakistan
[9] Department of Entomology, The University of Haripur, Khyber Pakhtunkhwa, Pakistan
[10] Department of Biological Sciences Al Hussein Bin Talal University, Ma'an, Jordan
[11] Department of Agricultural Extension Education and Communication, The University of Agriculture, Peshawar, Pakistan
* Correspondence: dr.romman.uom@gmail.com (M.R): madnanses@gmail.com (M.A)

CONTENTS

DOI: 10.1201/9781003286233-13

13.1 INTRODUCTION

Nitrogen (N) is a notable major supplement required by plants for their development. To accomplish a high return, cultivating practices require synthetic fertilizers that are expensive and may also cause environmental damage (Adnan et al. 2018a,b; Adnan et al. 2019; Amanullah and Fahd 2017; Akram et al. 2018a,b; Aziz et al. 2017a,b; Chang et al. 2021; Chen et al. 2021; Emre et al. 2021; Habib et al. 2017; Hafiz et al. 2016; Hafiz et al. 2019; Ghulam et al. 2021; Guofu et al. 2021; Hafeez et al. 2021; Khan et al. 2021; Kamaran et al. 2017; Muhmmad et al. 2019; Safi et al. 2021; Sajjad et al. 2019; Saud et al. 2013; Saud et al. 2014; Saud et al. 2017; Saud et al. 2016). Due to ecological risks and customer well-being concerns, the utilization of compound composts in agribusiness is under-used. Consequently, some fertilizers, known as biofertilizers or bioinoculants, that contain microorganisms with significant plant growth improvement impacts are used. Some of these microbial strains are good for phosphorus solubilizing, nitrogen fixing from the air and, producing cellulytic proteins. Biofertilizers are applied in a number of ways to soil, to enhance plant nourishment. One method is their direct application in soil; another way is seed treatment or application with compost. However these biofertilizers are used, they increase the amounts of favorable microorganisms in the soil to further enhance plant requirements for different free-living nitrogen-fixing organisms (Kloepper and Beauchamp 1992; Narula et al. 2000; Phillips et al. 2011; Al-Zahrani et al. 2022; Rajesh et al. 2022; Anam et al. 2021; Deepranjan et al. 2021; Haider et al. 2021; Amjad et al. 2021; Sajjad et al. 2021; Fakhre et al. 2021; Khatun et al. 2021; Ibrar et al. 2021; Bukhari et al. 2021; Haoliang et al. 2022; Sana et al. 2022; Abid et al. 2021; Zaman et al. 2021). These microorganisms greatly influence plant growth when used as seed inoculants (Davison 1988; Reed et al. 2011; Richardson et al. 2009). Some could impact plant growth through a mixture of advanced synthetics, fixing nitrogen, and solubilizing rock phosphates, when used as biofertilizers (Amer and Utkhede 2000;Adnan et al. 2016; Kumawat et al. 2017; Rana et al. 2020; Sajjad et al. 2021; Rehana et al. 2021; Yang et al. 2022; Ahmad et al. 2022; Shah et al. 2022; Muhammad et al. 2022; Wiqar et al. 2022; Farhat et al. 2022; Niaz et al. 2022; Ihsan et al. 2022; Chao et al. 2022; Qin et al. 2022; Xue et al. 2022; Ali et al. 2022; Mehmood et al. 2022; El Sabagh et al. 2022; Ibad et al. 2022).

Biological nitrogen-fixing micro-organisms enable regular uptake of nitrogen (N) in normal and horticultural biological systems. *Azotobacter* species have important features of both significant free-living N_2-fixing microscopic organisms and potential bacterial biofertilizers with demonstrated viability for plant nourishment and natural soil richness. Likewise, this micro-organism has shown advanced attributes, e.g., supplement use effectiveness, security against phytopathogens, phytohormone biosynthesis, and so on (National Research Council 1994; Deepranjan et al. 2021; Haider et al. 2021; Huang Li et al. 2021; Ikram et al. 2021; Jabborova et al. 2021; Khadim et al. 2021a,b; Manzer et al. 2021; Muzammal et al. 2021; Abdul et al. 2021a,b; Ashfaq et al. 2021; Amjad et al. 2021; Atif et al. 2021; Athar et al. 2021; Shah et al. 2013; Saud et al. 2020; Saud et al. 2022a,b; Qamar et al. 2017; Hamza et al. 2021; Irfan et al. 2021;Wajid et al. 2017; Yang et al. 2017; Zahida et al. 2017; Depeng et al. 2018; Hussain et al. 2020; Hafiz et al. 2020 a,b; Shafi et al. 2020; Wahid et al. 2020; Subhan et al. 2020; Zafar-ul-Hye et al. 2020a,b; Zafar et al. 2021; Adnan et al. 2020; Ilyas et al. 2020; Saleem et al. 2020a,b,c; Rehman 2020; Farhat et al. 2020; Wu et al. 2020; Mubeen et al. 2020; Farhana 2020; Jan et al. 2020; Wu et al. 2019; Ahmad et al. 2019; Baseer et al. 2019; Hafiz et al. 2018; Tariq et al. 2018).

Azotobacter, as a biofertilizer, has interesting properties, for example, blister development protection. Such advantageous qualities could be investigated significantly for the highest level of plan to explore the importance of *Azotobacter* (Figure 13.1). Moreover, *Azotobacter* species are also used address specific horticultural difficulties (e.g., supplement inadequacies, biotic and abiotic limitations) due to their natural capacities, collaborative and multi-trophic communication, biogeography, and overflow appropriation (Aasfar et al. 2021; Kizilkaya 2009; Ladha et al. 2016) (Tables 13.1 and 13.2).

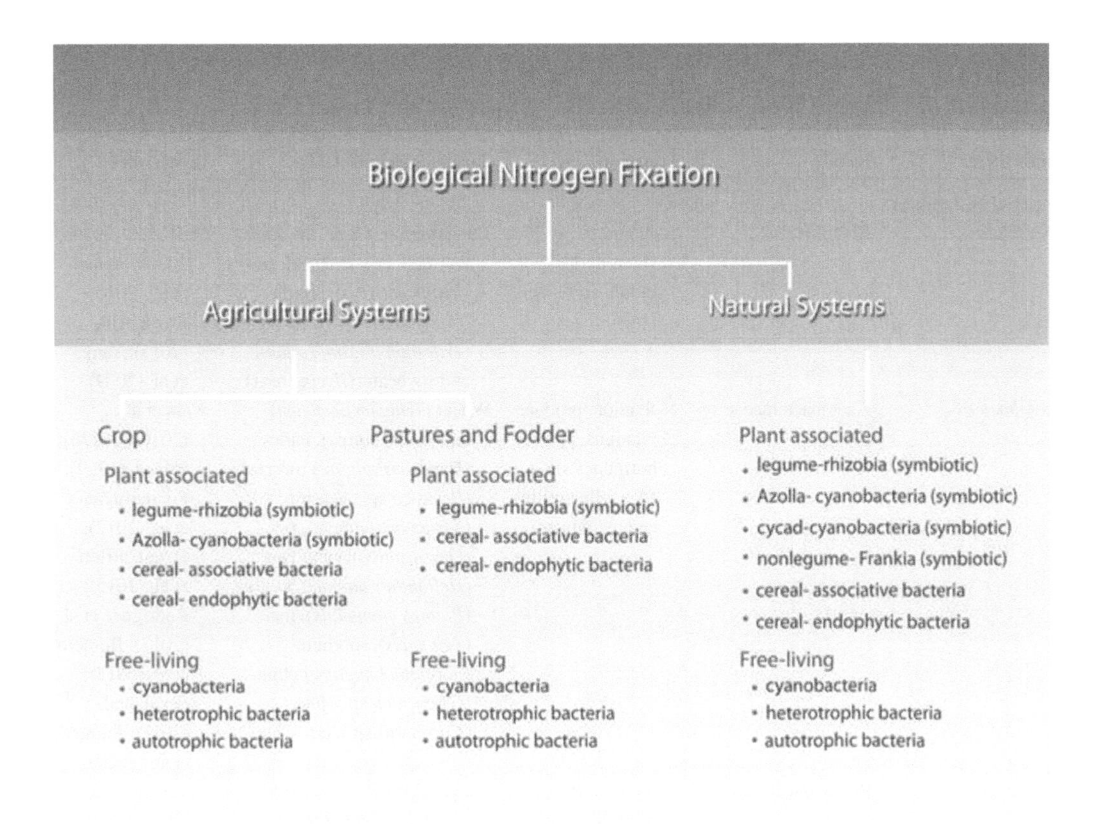

FIGURE 13.1 Classification of biological nitrogen fixation.

TABLE 13.1
The Important Groups of Biofertilizers

Group of Biofertilizers	Nature of Organisms	Examples	References
Nitrogen-fixing biofertilizers	Symbiotic	*Rhizobia*, *Frankia*, *Anabaena azollae*	Gaur (2010), Sharma et al. (2012), Singh et al. (2015a), and Mathivanan et al. (2015)
	Free-living	*Azotobacter*, *Beijerinkia*, *Anabaena*, *Clostridium*, *Klebsiella*, *Nostoc*, *Desulfovibrio*, *Bacillus polymyxa*	Sharma et al. (2012) and Singh et al. (2020b)
	Associative symbiotic	*Azospirillum* sp., *Gluconoacetobacter diazotrophicus*, *Enterobacter*	Sharma et al. (2012) and Singh et al. (2015a); Okon et al. (1995)

13.2 BIOFERTILIZERS: MECHANISMS AND APPLICATION

Currently, climate change is a major threat to the agriculture sector (Fahad and Bano 2012; Fahad et al. 2017; Fahad et al. 2013; Fahad et al. 2014a,b; Fahad et al. 2016a,b,c,d; Fahad et al. 2015a,b; Fahad et al. 2018a,b; Fahad et al. 2019a,b; Fahad et al. 2020; Fahad et al. 2021a,b,c,d,e,f; Fahad et al. 2022a,b; Hesham and Fahad 2020; Iqra et al. 2020; Akbar et al. 2020; Mahar et al. 2020; Noor et al.

TABLE 13.2
Partial List of Different Biofertilizers with Their Function and Target Crops

Biofertilizer	Family	Function	Target/Research Crops	References
Bacterial biofertilizer				
Rhizobium	Rhizobiaceae	N-fixation, siderophore, indole acetic acid (IAA)	Bengal gram (*Cicer arietinum*) lentil (*Lens esculenta*), pea (*Pisum sativum*), alfalfa (*Medicago sativa*), berseem (*Trifolium alexandrinum*), and soybean (*Glycine max*)	Patil et al. (2016a, 2020b); Omer et al. (2016); Ruzzi et al. (2015); Yang et al. (2009)
Azotobacter	Azotobacteraceae	N-fixation, produce vitamins, plant hormones viz., IAA, gibberellins, and cytokinins	Wheat (*Triticum aestivum*), oat (*Avena sativa*), barley (*Hordeum vulgare*) mustard (*Brassica* sp.), seasum (*Sesamum indicum*), rice (*Oryza sativa*), sunflower (*Helianthus annuus*), castor (*Ricinus communis*), maize (*Zea mays*), sorghum (*Sorghum bicolor*), cotton (*Gossypium* sp.), jute (*Corchorus* sp.), etc.	Wani et al. (2016a, 2013b); Núñez, (1999); Ponmurugan et al. (2012); Poza-Carrión et al. (2015); Rodriguez et al. (2017); Romero et al. (2017); Segal et al. (2017); Sumbul et al. (2020)
Azospirillum	Rhodospirillaceae	N-fixation, P-solubilization, IAA and siderophore production, increases the number of lateral roots and enhances root hairs formation	Maize (*Zea mays*), rice (*Oryza sativa*), wheat (*Triticum aestivum*)	Rodrigues et al. (2004), Sahoo et al. (2014)
Acetobacter	Acetobacteraceae	Nitrogen fixation, produce auxins, synthesis of IAA	Sugarcane (*Saccharum officinarum*), sugar beet (*Beta vulgaris*) and pearl millet (*Pennisetum glaucum* L.)	Glick (2012); Soumare et al. (2020); Yadav et al. (1991); Gupta et al. (2012)
Paenibacillus polymyxa	*Paenibacilliaceae*	Biological N-fixation, produce plant growth regulators, control plant ethylene levels, enhances root permeability and biocontrol plant pathogens	Wheat (*Triticum aestivum*), cucumber (*Cucumis sativus*),watermelon (*Citrullus lanatus*), soybean (*Glycine max*), tomato (*Solanum lycopersicum*),apple (*Malus domestica*)	Chen et al. (2006); Wu et al. (2006a, 2006b); Yanni et al. (1999)

TABLE 13.2 (Continued)
Partial List of Different Biofertilizers with Their Function and Target Crops

Biofertilizer	Family	Function	Target/Research Crops	References
Pseudomonas sp.	Pseudomonadaceae	Phosphate solubilization, biocontrol, natural suppressor of specific soil borne fungal pathogens, produce phytohormones (auxins and cytokinins) and secondary metabolites	Wheat (*Triticum aestivum*), cucumber (*Cucumis sativus*), mustard (*Brassica* sp.), rice (*Oryza sativa*), tomato (*Solanum lycopersicum*), turmeric (*Curcuma longa*)	Shaharoona et al. (2006), Zahir et al. (2010), Sharma et al. (2012); Trivedi et al. (2008)
Bacillus sp. (*B. subtilis*, *B. cereus*, *B. thuringiensis*, *B. pumilus*, *B. megaterium*)	Paenibacillaceae/ Bacillaceae	Solubilize the P, synthesis of plant growth hormones (IAA, GAs, cytokinins and spermidines)	Wheat (*Triticum aestivum*), cucumber (*Cucumis sativus*), mustard (*Brassica* sp.), rice (*Oryza sativa*), tomato (*Solanum lycopersicum*)	Kang et al. (2010), Swain and Ray (2009); Sabier et al. (2015)
Bacillus mucilaginosus, *Paenibacillus glucanolyticus*, *Bacillus edaphicus*	Paenibacillaceae/ Bacillaceae	Potash solubilization, nutrient uptake, and biomass	Maize (Zea mays), cucumber (*Cucumis sativus* L.), Pepper (*Capsicum annum* L.), Cotton (*Gossypium hirsutum*), rape (*Brassica napus* L), sesame (*Sesamum indicum*), wheat (*Triticum aestivum*), black pepper (*Piper nigrum* L.)	Sangeeth et al. (2012); Makawi (1973); Qaisrani et al. (2019); Eshaghi et al. (2019)
Thiobacillus ferrooxidans, *T. thiooxidans*, *T. thioparus*	Thiobacillaceae	Sulphur-oxidizing	Various crops	Sharma et al. (2012); Mal (2014); Karuppiah et al. (2012)
Actinobacteria, Firmicutes, Proteobacteria	–	Manganese-oxidizing bacteria	Various crops	Sharma et al. (2012); Malik et al.;(2002)
Thiobacillus thioxidans, *T. ferrooxidans*, *Microbacterium saperdae*, *Pseudomonas monteilli*, *Enterobacter cancerogens*	–	Zinc solubilization	Rice (*Oryza sativa*), maize (*Zea mays*), cotton (*Gossypium hirsutum*), sorghum (*Sorghum bicolor*)	Wang et al. (2015), Saravanan et al. (2011); Zayadan et al. (2014)

2020; Bayram et al. 2020; Amanullah, Fahad 2018a,b; Amanullah, Fahad 2017; Amanullah et al. 2020; Amanullah et al. 2021; Rashid et al. 2020; Arif et al. 2020; Amir et al. 2020; Saman et al. 2020; Muhammad Tahir et al. 2020; Jakir and Allah 2020; Mahmood et al. 2021; Farah et al. 2020; Sadam et al. 2020; Unsar et al. 2020; Fazli et al. 2020; Enamul et al. 2020; Gopakumar et al. 2020; Zia-ur-Rehman 2020; Ayman et al. 2020; Mohammad I. Al-Wabel et al. 2020a,b; Senol 2020; Amjad et al.

2020; Ibrar et al. 2020; Sajid et al. 2020; Muhammad et al. 2021; Sidra et al. 2021; Zahir et al. 2021; Sahrish et al. 2022).

This issue can be solved to some extent by introducing several biofertilizers in the market. Biofertilizers are the essential contributions of supplements for practical and natural cultivation. Microbial strains which enable greater plant development ought to be segregated from various agroecological zones and marketed. Fluid biofertilizers are superior to transporter-based biofertilizers. Consequently, they energize the development of fluid biofertilizers. Just 0.1% of biofertilizer strains formed have been marketed to date, which shows that markedly less work has been done in the field of detailing of microbial strains for biofertilizers. Be that as it may, farmers have not benefited on the ground as the planned work has not been progressed. The interest in biofertilizers is exceptionally high yet their manufacture is restricted. In this way, it ought to work in the field of microbed related to foster biofertilizers as per biological zones to provide benefits to people (Dasgupta et al. 2021; Shah et al. 2015; Nosrati et al. 2014; Rojas-Tapias et al. 2012).

13.3 QUALITY CONTROL AND GUIDELINES FOR BIOFERTILIZERS

Biofertilizers have turned into excellent contenders avoiding damage to the ecosystem with compound-free techniques for horticulture. The developing interest and government support in the natural agribusiness has provided an impetus in the biofertilizer business; notwithstanding, the developing interest and time-restricted production of organism-based biofertilizer has raised concerns over their quality control. There are concerns related to the production, storage, and use of biofertilizers. In the future, bio-inoculants should not be restricted only to the use of microorganisms but their utilization must include the utilization of organisms, green growth, and infections. Standardized processes have been developed by numerous research centres for biomass age and bioactive atom creation from the various selected organic strains. Biomass creation and dynamic particle detachments are the conclusive variables for quality control. The types of strains and bio-partitions in processes are significant observations in quality testing. The method involves collecting, detailing, and the time span of usability of the final product, which are the properties recorded in the value control documentation. The positive change in the acknowledgment and utilization of biofertilizers in traditional farming requires improvements of value ideas, manuals, and guidelines. The ongoing actions are an endeavor to comprehend the flows status of the quality control and guidelines of biofertilizer applications, and proposes the future endeavors expected in the improvement of biofertilizer manufacture and applications (Bharti and Suryavanshi 2021; Pandey et al. 1991; Ahmad et al. 2015; Saeed et al. 2015).

13.4 MAJOR MECHANISMS IN BIOFERTILIZERS

Biofertilizers are viewed as a natural option in contrast to synthetic compounds and substance-based composts, the abuse of which has caused real harm to the global climate and environment. The creation of biofertilizers doesn't rely just upon information on plant physiology and soil microorganisms, but also a few innovative difficulties such as the definitions, the different species of microflora, and their communication framework. Along these lines, the improvements in a steady bioformulation are conceivable by joining the information on plant physiology with microbial and specialized perspectives. Accordingly, biofertilizers have an exceptionally huge capacity in managing horticulture as they are naturally harmless to the ecosystem (Ali et al. 2022; Mrkovacki et al. 2002; Tilman et al. 2011).

13.5 MICROBIAL BIOFERTILIZERS: TYPES, APPLICATIONS, AND CURRENT DIFFICULTIES FOR ECONOMICAL RURAL PRODUCTION

Biofertilizers can be used to support farming and to help meet fulfill the needs of the expanding global population for horticultural items while preserving and supporting the planet for future generations. The significance of biofertilizers in increasing the efficiency and nature of agricultural products has proactively been demonstrated through different research works carried out around the world. Despite demonstrating their true capacity, biofertilizers remain underutilized to an enormous degree. Along these lines, there is an urgent need to advance the use of biofertilizers among farmers to ensure higher farming productivity which can be accomplished through the following:

1. Advantages of biofertilizers in ensuring soil health, supporting the efficiency of regular assets, and achieving high efficiency and a greater cost–benefit ratio.
2. Primary emphasis should be on the quality control during the production of biofertilizers to safeguard their effectiveness for a long time.
3. Research into biofertilizers with multi-strain and multi-microorganism consortia should be carried out for improvements in crop productivity. In contrast to single-strain biofertilizers, multi-strain and multi-microorganism consortia can achieve higher results under more hostile conditions.
4. Biofertilizers should be made effectively accessible for all researchers in order to improve their quality and for better utilization (Gautam et al., 2021; Marciano et al., 2012; Ortiz-Marquez et al., 2012).

13.6 PRODUCTION INNOVATION, PROPERTIES, AND QUALITY ADMINISTRATION OF BIOFERTILIZERS

Biofertilizers are naturally effective microorganisms that improve plant growth by obtaining supplements that are otherwise difficult for plants to obtain. They upgrade soil efficiency by fixing barometrical nitrogen, solubilizing soil phosphorus, and increasing plant development. Biofertilizer innovations have incorporated plant supplements for practical agribusiness through biological nitrogen fixing. The plan for biofertilizer production is a pivotal multistep process that incorporates blending of an appropriate medium with an inoculant, giving ideal circumstances for the capacity, bundling, and dispatch, and guaranteeing endurance and foundation after placement in soil. One of the main points of contention is the quality control of biofertilizers, which is regulated under the Fertilizer Control Order (FCO) (Singh et al. 2021; Mirzakhani et al. 2014; Basir et al. 2015; Rovira et al. 1991). Quality confirmation of bioproducts can go far in completing the picture for biofertilizers. The creation of microbial plans, transportation, and commercialization are currently at an early phase and thus production by typical means is underscored (Maan and Garcha 2021; Rashid et al. 2004; Velmourougane et al. 2019).

13.7 MULTILEGUME BIOFERTILIZERS

Biofertilizers are indispensable in supplementing cycling in the biosphere. The biggest role is that for natural nitrogen absorption as carried out by *Rhizobium* in relation to vegetables. The majority of rhizobia are well defined for have associated plant species, however most do not act indiscriminately. Such rhizobia with expansive hosts could be utilized for the production of multilegume biofertilizers. Immunizing vegetables with rhizobia could accomplish significant growth in vegetable nodulation, biomass yield, nitrogen absorption, and post-crop soil nitrate levels. This development could help to meet the dietary needs of the world's constantly growing population. Multilegume biofertilizers could be amongst the most significant advances towards economical cultivation in light of the fact that selecting the best inoculant combination for a specific vegetable host is difficult. The prospects of this framework are engaging in light of the fact that the whole world is trying to embrace more natural methods of cultivation. Introduction of these biofertilizer will not only improve the environment, but also will help reducing compost data sources (Ikbal et al. 2020; Lazali and Bargaz 2017; Mishra et al. 2016).

13.8 BIOFERTILIZERS IN AGRIBUSINESS

Biofertilizers are a promising area in farming biological systems as a valuable, inexhaustible, and ecofriendly opportunity for plant supplements. As they have the capacity to change healthful significant components from non-useable forms to exceptionally assimilable structures without pernicious consequences for the environment, they are a significant part of an integrated plant nutrient system (IPNS) (Alley and Vanlauwe 2009; Malusá et al. 2012). The use of organic manures is believed to be a critical component in maintaining soil health and yield efficiency at an adequate level, which is fundamental to accomplishing on-going cultivation. Biofertilizers may also assist with relieving problems emerging from the growing global population for food and from adverse effects of the far-reaching chemicalization of agroecosystems. The changing methods to update horticultural practices makes biofertilizers a fundamental piece of cutting-edge crop production which stresses of the use of organic inoculants in the futures. Various rhizosphere microorganisms are known to have diverse plant development advancement factors, however not many have been used as biofertilizers so far. Consequently, new procedures, taking into consideration their lengthy application, are expected to achieve the goals of more natural cultivation (Mącik et al. 2020; Abdul Latef et al. 2020; Ledbetter et al. 2017).

13.9 CONCLUSION

Nitrogen is a fundamental supplement for plant development and improvement, however it is inaccessible in its most pervasive form as air nitrogen. Plants rather rely on joined, or fixed, types of nitrogen, such as ammonia and nitrate. Management of nitrogen is a challenging task and several methods individually and in combination are in use to manage its efficiency. The use of properly managed nitrogen manures has been prompted around the world. Organic nitrogen absorption offers a specific method for supplying plants with nitrogen. It is a basic part of numerous aquatic, as well as land-based environments across our biosphere.

ACKNOWLEDGMENT

We would like to thank to the staff of the Department of Botany, University of Chitral, Khybar Pakhtunkhwa, Pakistan.

REFERENCES

Aasfar A, Bargaz A, Yaakoubi K, Hilali A, Bennis I, Zeroual Y, Meftah Kadmiri I (2021) Nitrogen fixing azotobacter species as potential soil biological enhancers for crop nutrition and yield stability. *Front Microbiol* 12. https://doi.org/10.3389/fmicb.2021.628379

Abdel Latef AA, Abu Alhmad MF, Kordrostami M, Abo–Baker AA, Zakir A (2020) Inoculation with Azospirillum lipoferum or azotobacter chroococcum reinforces maize growth by improving physiological activities under saline conditions. *J Plant Growth Regul* 39(3):1293–1306. doi:10.1007/s00344-020-10065-9

Abdul S, Muhammad AA, Shabir H, Hesham A El E, Sajjad H, Niaz A, Abdul G, RZ Sayyed, Fahad S, Subhan D, Rahul D (2021a) Zinc nutrition and arbuscular mycorrhizal symbiosis effects on maize (Zea mays L.) growth and productivity. *J Saudi Soc Agri Sci* 28(11), 6339–6351 https://doi.org/10.1016/j.sjbs.2021.06.096

Abdul S, Muhammad AA, Subhan D, Niaz A, Fahad S, Rahul D, Mohammad JA, Omaima N, Muhammad Habib ur R, Bernard RG (2021b) Effect of arbuscular mycorrhizal fungi on the physiological functioning of maize under zinc-deficient soils. *Sci Rep.* 11:18468

Abid M, Khalid N, Qasim A, Saud A, Manzer HS, Chao W, Depeng W, Shah S, Jan B, Subhan D, Rahul D, Hafiz MH, Wajid N, Muhammad M, Farooq S, Fahad S (2021) Exploring the potential of moringa leaf extract as bio stimulant for improving yield and quality of black cumin oil. *Sci Rep.* 11:24217 https://doi.org/10.1038/s41598-021-03617-w

Adnan M, Fahad S, Muhammad Z, Shahen S, Ishaq AM, Subhan D, Zafar-ul-Hye M, Martin LB, Raja MMN, Beena S, Saud S, Imran A, Zhen Y, Martin B, Jiri H, Rahul D (2020) Coupling phosphate-solubilizing bacteria with phosphorus supplements improve maize phosphorus acquisition and growth under lime induced salinity stress. *Plants* 9(7):900. doi: 10.3390/plants9070900

Adnan M, Fahad S, Khan IA, Saeed M, Ihsan MZ, Saud S, Riaz M, Wang D, Wu C. (2019). Integration of poultry manure and phosphate solubilizing bacteria improved availability of Ca bound P in calcareous soils. *3 Biotech* 9(10):368

Adnan M, Shah Z, Sharif M, Rahman H (2018b) Liming induces carbon dioxide (CO_2) emission in PSB inoculated alkaline soil supplemented with different phosphorus sources. *Environ Sci Poll Res* 25(10):9501–9509.

Adnan M, Zahir S, Fahad S, Arif M, Mukhtar A, Imtiaz AK, Ishaq AM, Abdul B, Hidayat U, Muhammad A, Inayat-Ur R, Saud S, Muhammad ZI, Yousaf J, Amanullah, Hafiz MH, Wajid N (2018a) Phosphate-solubilizing bacteria nullify the antagonistic effect of soil calcification on bioavailability of phosphorus in alkaline soils. *Sci Rep* 8:4339. https://doi.org/10.1038/s41598-018-22653-7

Adnan, M., Z. Shah., N. Saleem, A. Basir, I. U. Rahman., H. Ullah., M. Ibrahim., J. A. Shah, Muhammad, A. Khan and S. R. Ali Shah (2016). Isolation and evaluation of summer legumes rhizobia as PGPR. *Pure App Bio* 5 (1):127–133.

Ahmad K, Adnan M, Khan MA, Hussain Z, Junaid K, Saleem N, Ali M, Basir A (2015) Bioactive neem leaf powder enhances the shelf life of stored mungbean grains and extends protection from pulse beetle. *Pak J Weed Sci Res* 21(1):71–81.

Ahmad N, Hussain S, Ali MA, Minhas A, Waheed W, Danish S, Fahad S, Ghafoor U, Baig, KS, Sultan H, Muhammad IH, Mohammad JA, Theodore DM (2022) Correlation of soil characteristics and citrus leaf nutrients contents in current scenario of Layyah District. *Horticultureae* 8:61. https://doi.org/10.3390/horticulturae8010061

Ahmad S, Kamran M, Ding R, Meng X, Wang H, Ahmad I, Fahad S, Han Q (2019) Exogenous melatonin confers drought stress by promoting plant growth, photosynthetic capacity and antioxidant defense system of maize seedlings. *Peer J* 7:e7793. http://doi.org/10.7717/peerj.7793

Akbar H, Timothy JK, Jagadish T, M. Golam M, Apurbo KC, Muhammad F, Rajan B, Fahad S, Hasanuzzaman M (2020) Agricultural land degradation: Processes and problems undermining future food security. In: Fahad S, Hasanuzzaman M, Alam M, Ullah H, Saeed M, Khan AK, Adnan M (Eds.) *Environment, Climate, Plant and Vegetation Growth.* Springer Switzerland, pp. 17–62. https://doi.org/10.1007/978-3-030-49732-3

Akram R, Turan V, Hammad HM, Ahmad S, Hussain S, Hasnain A, Maqbool MM, Rehmani MIA, Rasool A, Masood N, Mahmood F, Mubeen M, Sultana SR, Fahad S, Amanet K, Saleem M, Abbas Y, Akhtar HM, Waseem F, Murtaza R, Amin A, Zahoor SA, ul Din MS, Nasim W (2018a) Fate of organic and inorganic pollutants in paddy soils. In: Hashmi MZ, Varma A (Eds.) *Environmental Pollution of Paddy Soils, Soil Biology*. Springer International Publishing, Switzerland, pp. 197–214

Akram R, Turan V, Wahid A, Ijaz M, Shahid MA, Kaleem S, Hafeez A, Maqbool MM, Chaudhary HJ, Munis, MFH, Mubeen M, Sadiq N, Murtaza R, Kazmi DH, Ali S, Khan N, Sultana SR, Fahad S, Amin A, Nasim W (2018b) Paddy land pollutants and their role in climate change. In: Hashmi MZ, Varma A (Eds.) *Environmental Pollution of Paddy Soils, Soil Biology*. Springer International Publishing, Switzerland, pp. 113–124.

Ali S, Hameed G, Muhammad A, Depeng W, Fahad S (2022) Comparative genetic evaluation of maize inbred lines at seedling and maturity stages under drought stress. *J Plant Growth Regul* 42:989–1005. https://doi.org/10.1007/s00344-022-10608-2

Alley, M. M., & Vanlauwe, B. (2009). *The Role of Fertilizers in Integrated Plant management.* International Fertilizer Industry Association (IFA)

Al-Zahrani HS, Alharby HF, Fahad S (2022) Antioxidative defense system, hormones, and metabolite accumulation in different plant parts of two contrasting rice cultivars as influenced by plant growth regulators under heat stress. *Front Plant Sci* 13:911846. doi: 10.3389/fpls.2022.911846

Amanullah, Muhammad I, Haider N, Shah K, Manzoor A, Asim M, Saif U, Izhar A, Fahad S, Adnan M et al. (2021) Integrated foliar nutrients application improve wheat (Triticum aestivum L.) productivity under calcareous soils in drylands. *Commun Soil Sci Plant Anal* 52(21):2748–2766. https://Doi.Org/10.1080/00103624.2021.1956521

Amanullah, Fahad S (Eds.) (2017) Rice – Technology and Production. *IntechOpen Croatia* 2017 http://dx.doi.org/10.5772/64480

Amanullah, Fahad S (Eds.) (2018a) Corn – Production and Human Health in Changing Climate. *IntechOpen United Kingdom* 2018. http://dx.doi.org/10.5772/intechopen.74074

Amanullah, Fahad S (Eds.) (2018b) Nitrogen in Agriculture – Updates. *IntechOpen Croatia* 2018. http://dx.doi.org/10.5772/65846

Amanullah, Shah K, Imran, Hamdan AK, Muhammad A, Abdel RA, Muhammad A, Fahad S, Azizullah S, Brajendra P (2020) Effects of climate change on irrigation water quality. In: Fahad S, Hasanuzzaman M, Alam M, Ullah H, Saeed M, Khan AK, Adnan M (Eds.) *Environment, Climate, Plant and Vegetation Growth*. Springer Nature Switzerland, pp. 123–132. https://doi.org/10.1007/978-3-030-49732-3

Amer GA, Utkhede RS (2000) Development of formulations of biological agents for management of root rot of lettuce and cucumber. *Can J Microbiol* 46(9):809–816. doi:10.1139/w00-063

Amir M, Muhammad A, Allah B, Sevgi Ç, Haroon ZK, Muhammad A, Emre A (2020) Biofortification under climate change: The fight between quality and quantity. In: Fahad S, Hasanuzzaman M, Alam M, Ullah H, Saeed M, Khan AK, Adnan M (Eds.) *Environment, Climate, Plant and Vegetation Growth*. Springer Nature Switzerland, pp. 173–228. https://doi.org/10.1007/978-3-030-49732-3

Amjad I, Muhammad H, Farooq S, Anwar H (2020) Role of plant bioactives in sustainable agriculture. In: Fahad S, Hasanuzzaman M, Alam M, Ullah H, Saeed M, Khan AK, Adnan M (Eds.), *Environment, Climate, Plant and Vegetation Growth*. Springer Nature Switzerland, pp. 591–606. https://doi.org/10.1007/978-3-030-49732-3

Amjad SF, Mansoora N, Din IU, Khalid IR, Jatoi GH, Murtaza G, Yaseen S, Naz M, Danish S, Fahad S et al. (2021) Application of zinc fertilizer and mycorrhizal inoculation on physio-biochemical parameters of wheat grown under water-stressed environment. *Sustainability* 13:11007. https://doi.org/10.3390/su131911007

Anam I, Huma G, Ali H, Muhammad K, Muhammad R, Aasma P, Muhammad SC, Noman W, Sana F, Sobia A, Fahad S (2021) Ameliorative mechanisms of turmeric-extracted curcumin on arsenic (As)-induced biochemical alterations, oxidative damage, and impaired organ functions in rats. *Environ Sci Pollut Res* 28(46):66313–66326. https://doi.org/10.1007/s11356-021-15695-4

Arif M, Talha J, Muhammad R, Fahad S, Muhammad A, Amanullah, Kawsar A, Ishaq AM, Bushra K, Fahd R (2020) Biochar; a remedy for climate change. In: Fahad S, Hasanuzzaman M, Alam M, Ullah H, Saeed M, Khan AK, Adnan M (Eds.) *Environment, Climate, Plant and Vegetation Growth*. Springer Nature Switzerland, pp. 151–172. https://doi.org/10.1007/978-3-030-49732-3

Ashfaq AR, Uzma Y, Niaz A, Muhammad AA, Fahad S, Haider S, Tayebeh Z, Subhan D, Süleyman T, Hesham AElE, Pramila T, Jamal MA, Sulaiman AA, Rahul D (2021) Toxicity of cadmium and nickel in the context of applied activated carbon biochar for improvement in soil fertility. *Saudi Soc Agri Sci*. 29(2):743–750 https://doi.org/10.1016/j.sjbs.2021.09.035

Athar M, Masood IA, Sana S, Ahmed M, Xiukang W, Sajid F, Sher AK, Habib A, Faran M, Zafar H, Farhana G, Fahad S (2021) Bio-diesel production of sunflower through sulphur management in a semi-arid subtropical environment. *Environ Sci Pollut Res*. 29(9):13268–13278. https://doi.org/10.1007/s11356-021-16688-z

Atif B, Hesham A, Fahad S (2021) Biochar coupling with phosphorus fertilization modifies antioxidant activity, osmolyte accumulation and reactive oxygen species synthesis in the leaves and xylem sap of rice cultivars under high-temperature stress. *Physiol Mol Biol Plant*. 27(9):2083–2100. https://doi.org/10.1007/s12298-021-01062-7

Ayman EL Sabagh, Akbar Hossain, Celaleddin Barutçular, Muhammad Aamir Iqbal, M Sohidul Islam, Shah Fahad, Oksana Sytar, Fatih Çig, Ram Swaroop Meena, Murat Erman (2020) Consequences of salinity stress on the quality of crops and its mitigation strategies for sustainable crop production: An outlook of arid and semi-arid regions. In: Fahad S, Hasanuzzaman M, Alam M, Ullah H, Saeed M, Khan AK, Adnan M (Eds.) *Environment, Climate, Plant and Vegetation Growth*. Springer Nature Switzerland, pp. 503–534. https://doi.org/10.1007/978-3-030-49732-3

Aziz K, Daniel KYT, Fazal M, Muhammad ZA, Farooq S, Fan W, Fahad S, Ruiyang Z (2017a) Nitrogen nutrition in cotton and control strategies for greenhouse gas emissions: a review. *Environ Sci Pollut Res* 24:23471–23487. https://doi.org/10.1007/s11356-017-0131-y

Aziz K, Daniel KYT, Muhammad ZA, Honghai L, Shahbaz AT, Mir A, Fahad S (2017b) Nitrogen fertility and abiotic stresses management in cotton crop: a review. *Environ Sci Pollut Res* 24:14551–14566. https://doi.org/10.1007/s11356-017-8920-x

Baseer M, Adnan M, Fazal M, Fahad S, Muhammad S, Fazli W, Muhammad A, Jr. Amanullah, Depeng W, Saud S, Muhammad N, Muhammad Z, Fazli S, Beena S, Mian AR, Ishaq AM (2019) Substituting urea by organic wastes for improving maize yield in alkaline soil. *J Plant Nutr* 42(19):2423–2434. doi.org/10.1080/01904167.2019.1659344

Basir A, Ali R, Alam M, Shah AS, Khilwat A, Adnan M, Ibrahim M, Rehman I, Adnan T (2015) Potential of wheat (Tritium aestivum L.) advanced lines for yield and yield attributes under different planting dates in Peshawar Valley. *Am-Eurasian J Agri Environ Sci* 15(12):2484–2488.

Bayram AY, Seher Ö, Nazlican A (2020) Climate change forecasting and modeling for the year of 2050. In: Fahad S, Hasanuzzaman M, Alam M, Ullah H, Saeed M, Khan AK, Adnan M (Eds.) *Environment, Climate, Plant and Vegetation Growth.* Springer Nature Switzerland, pp. 109–122. https://doi.org/10.1007/978-3-030-49732-3

Bharti N, Suryavanshi M (2021) Quality control and regulations of biofertilizers: Current scenario and future prospects. *Biofertilizers* 133–141. doi:10.1016/b978-0-12-821667-5.00018-x

Bukhari MA, Adnan NS, Fahad S, Javaid I, Fahim N, Abdul M, Mohammad SB (2021) Screening of wheat (*Triticum aestivum* L.) genotypes for drought tolerance using polyethylene glycol. *Arabian J Geosci* 14:2808

Chang W, Qiujuan J, Evgenios A, Haitao L, Gezi L, Jingjing Z, Fahad S, Ying J (2021) Hormetic effects of zinc on growth and antioxidant defense system of wheat plants. *Sci Total Environ.* 807, 150992 https://doi.org/10.1016/j.scitotenv.2021.150992

Chao W, Youjin S, Beibei Q, Fahad S (2022) Effects of asymmetric heat on grain quality during the panicle initiation stage in contrasting rice genotypes. *J Plant Growth Regul* 42:630–636. https://doi.org/10.1007/s00344-022-10598-1

Chen Y, Guo Z, Dong L, Fu Z, Zheng Q, Zhang G, Qin L, Sun X, Shi Z, Fahad S, Xie F, Saud S (2021) Turf performance and physiological responses of native Poa species to summer stress in Northeast China. *Peer J* 9:e12252 http://doi.org/10.7717/peerj.12252

Chen Y, Rekha P, Arun A, Shen F, Lai W, Young C (2006) Phosphate solubilizing bacteria from subtropical soil and their tricalcium phosphate solubilizing abilities. *Appl Soil Ecol* 34(1):33–41. doi:10.1016/j.apsoil.2005.12.002

Dasgupta D, Kumar K, Miglani R, Mishra R, Panda AK, Bisht SS (2021) Microbial biofertilizers: Recent trends and future outlook. *Recent Advancement in Microbial Biotechnology* 1–26. doi:10.1016/b978-0-12-822098-6.00001-x

Davison, J. (1988). Plant beneficial bacteria. *Nat Biotechnol* 6(3):282–286. doi:10.1038/nbt0388-282

Deepranjan S, Ardith S, O. Siva D, Sonam S, Shikha, Manoj P, Amitava R, Sayyed RZ, Abdul G, Mohammad JA, Subhan D, Fahad S, Rahul D (2021) Optimizing nutrient use efficiency, productivity, energetics, and economics of red cabbage following mineral fertilization and biopriming with compatible rhizosphere microbes. *Sci Rep* 11:15680 https://doi.org/10.1038/s41598-021-95092-6

Depeng W, Fahad S, Saud S, Muhammad K, Aziz K, Mohammad NK, Hafiz MH, Wajid N (2018) Morphological acclimation to agronomic manipulation in leaf dispersion and orientation to promote "Ideotype" breeding: Evidence from 3D visual modeling of "super" rice (*Oryza sativa* L.). *Plant Physiol Biochem* 135:499–510 https://doi.org/10.1016/j.plaphy.2018.11.010

EL Sabagh A, Islam MS, Hossain A, Iqbal MA, Mubeen M, Waleed M, Reginato M, Battaglia M, Ahmed S, Rehman A, Arif M, Athar H-U-R, Ratnasekera D, Danish S, Raza MA, Rajendran K, Mushtaq M, Skalicky M, Brestic M, Soufan W, Fahad S, Pandey S, Kamran M, Datta R. Abdelhamid MT (2022) Phytohormones as growth regulators during abiotic stress tolerance in plants. *Front Agron* 4:765068. doi: 10.3389/fagro.2022.765068

Emre B, Ömer SU, Martín LB, Andre D, Fahad S, Rahul D, Muhammad Z-ul-H, Ghulam SH, Subhan D (2021) Studying soil erosion by evaluating changes in physico-chemical properties of soils under different land-use types. *J Saudi Soc Agri Sci* 20:190–197

Enamul H, Shoeb AZM, Mallik AH, Fahad S, Kamruzzaman MM, Akib J, Nayyer S, Mehedi KMA, Swati AS, Yeamin MdA, Most SS (2020) Measuring vulnerability to environmental hazards: Qualitative to quantitative. In: Fahad S, Hasanuzzaman M, Alam M, Ullah H, Saeed M, Khan AK, Adnan M (Eds.) *Environment, Climate, Plant and Vegetation Growth.* Springer Nature Switzerland, pp. 421–452. https://doi.org/10.1007/978-3-030-49732-3

Eshaghi E, Nosrati R, Owlia P, Malboobi MA, Ghaseminejad P, Ganjali MR (2019) Zinc solubilization characteristics of efficient siderophore-producing soil bacteria. *Iranian J Microbiol.* doi:10.18502/ijm.v11i5.1961

Fahad S, Abdul B, Adnan M (Eds.) (2018b) Global Wheat Production. *IntechOpen United Kingdom* 2018. http://dx.doi.org/10.5772/intechopen.72559

Fahad S, Bajwa AA, Nazir U, Anjum SA, Farooq A, Zohaib A, Sadia S, NasimW, Adkins S, Saud S, Ihsan MZ, Alharby H,Wu C,Wang D, Huang J (2017) Crop production under drought and heat stress: Plant responses and management options. *Front Plant Sci* 8:1147. https://doi.org/10.3389/fpls.2017.01147

Fahad S, Bano A (2012) Effect of salicylic acid on physiological and biochemical characterization of maize grown in saline area. *Pak J Bot* 44:1433–1438

Fahad S, Chen Y, Saud S, Wang K, Xiong D, Chen C,Wu C, Shah F, Nie L, Huang J (2013) Ultraviolet radiation effect on photosynthetic pigments, biochemical attributes, antioxidant enzyme activity and hormonal contents of wheat. *J Food Agri Environ* 11(3&4):1635–1641

Fahad S, Hussain S, Bano A, Saud S, Hassan S, Shan D, Khan FA, Khan F, Chen Y, Wu C, Tabassum MA, Chun MX, Afzal M, Jan A, Jan MT, Huang J (2014a) Potential role of phytohormones and plant growth-promoting rhizobacteria in abiotic stresses: consequences for changing environment. *Environ Sci Pollut Res* 22(7):4907–4921. https://doi.org/10.1007/s11356-014-3754-2

Fahad S, Hussain S, Matloob A, Khan FA, Khaliq A, Saud S, Hassan S, Shan D, Khan F, Ullah N, Faiq M, Khan MR, Tareen AK, Khan A, Ullah A, Ullah N, Huang J (2014b) Phytohormones and plant responses to salinity stress: A review. *Plant Growth Regul* 75(2):391–404. https://doi.org/10.1007/s10725-014-0013-y

Fahad S, Hussain S, Saud S, Hassan S, Chauhan BS, Khan F et al (2016a) Responses of rapid viscoanalyzer profile and other rice grain qualities to exogenously applied plant growth regulators under high day and high night temperatures. *PLoS One* 11(7):e0159590. https://doi.org/10.1371/journal.pone.0159590

Fahad S, Hussain S, Saud S, Hassan S, Ihsan Z, Shah AN, Wu C, Yousaf M, Nasim W, Alharby H, Alghabari F, Huang J (2016c) Exogenously applied plant growth regulators enhance the morphophysiological growth and yield of rice under high temperature. *Front Plant Sci* 7:1250. https://doi.org/10.3389/fpls.2016. 01250

Fahad S, Hussain S, Saud S, Hassan S, Tanveer M, Ihsan MZ, Shah AN, Ullah A, Nasrullah KF, Ullah S, Alharby HNW, Wu C, Huang J (2016d) A combined application of biochar and phosphorus alleviates heat-induced adversities on physiological, agronomical and quality attributes of rice. *Plant Physiol Biochem* 103:191–198

Fahad S, Hussain S, Saud S, Khan F, Hassan S, Jr A, Nasim W, Arif M, Wang F, Huang J (2016b) Exogenously applied plant growth regulators affect heat-stressed rice pollens. *J Agron Crop Sci* 202:139–150

Fahad S, Hussain S, Saud S, Tanveer M, Bajwa AA, Hassan S, Shah AN, Ullah A, Wu C, Khan FA, Shah F, Ullah S, Chen Y, Huang J (2015a) A biochar application protects rice pollen from high-temperature stress. *Plant Physiol Biochem* 96:281–287

Fahad S, Muhammad ZI, Abdul K, Ihsanullah D, Saud S, Saleh A, Wajid N, Muhammad A, Imtiaz AK, Chao W, Depeng W, Jianliang H (2018a): Consequences of high temperature under changing climate optima for rice pollen characteristics-concepts and perspectives, *Arch Agron Soil Sci* 64(11):1473–1488. DOI: 10.1080/03650340.2018.1443213

Fahad S, Nie L, Chen Y, Wu C, Xiong D, Saud S, Hongyan L, Cui K, Huang J (2015b) Crop plant hormones and environmental stress. *Sustain Agric Rev* 15:371–400

Fahad S, Saud S, Yajun C, Chao W, Depeng W (Eds.) (2021f) Abiotic Stress in Plants. *IntechOpen United Kingdom* 2021. http://dx.doi.org/10.5772/intechopen.91549

Fahad S, Hasanuzzaman M, Alam M, Ullah H, Saeed M, Ali Khan I, Adnan M. (Eds.) (2020) *Environment, Climate, Plant and Vegetation Growth*. Springer Nature Switzerland. DOI: https://doi.org/10.1007/978-3-030-49732-3

Fahad S, Adnan M, Hassan S, Saud S, Hussain S, Wu C, Wang D, Hakeem KR, Alharby HF, Turan V, Khan MA, Huang J (2019b) Rice responses and tolerance to high temperature. In: Hasanuzzaman M, Fujita M, Nahar K, Biswas JK (Eds.) *Advances in Rice Research for Abiotic Stress Tolerance*. Woodhead Publishing, UK, pp. 201–224

Fahad S, Adnan M, Saud S, Nie L (Eds.) (2022b) *Climate Change and Ecosystems: Challenges to Sustainable Development*. First edition. Footprints of Climate Variability on Plant Diversity. CRC Press, Boca Raton

Fahad S, Adnan M, Saud S (Eds.) (2022a) *Improvement of Plant Production in the Era of Climate Change*. First edition. Footprints of Climate Variability on Plant Diversity. CRC Press, Boca Raton

Fahad S, Rehman A, Shahzad B, Tanveer M, Saud S, Kamran M, Ihtisham M, Khan SU, Turan V, Rahman MHU (2019a) Rice responses and tolerance to metal/metalloid toxicity. In: Hasanuzzaman M, Fujita

M, Nahar K, Biswas JK (Eds.) *Advances in Rice Research for Abiotic Stress Tolerance*. Woodhead Publishing, UK, pp. 299–312.

Fahad S, Sönmez O, Saud S, Wang D, Wu C, Adnan M, Arif M, Amanullah (Eds.) (2021e) *Engineering Tolerance in Crop Plants against Abiotic Stress*. First edition. Footprints of Climate Variability on Plant Diversity. CRC Press, Boca Raton

Fahad S, Sonmez O, Saud S, Wang D, Wu C, Adnan M, Turan V (Eds.) (2021b) *Climate Change and Plants: Biodiversity, Growth and Interactions*. First edition. Footprints of Climate Variability on Plant Diversity. CRC Press, Boca Raton

Fahad S, Sonmez O, Saud S, Wang D, Wu C, Adnan M, Turan V (Eds.) (2021c) *Developing Climate Resilient Crops: Improving Global Food Security and Safety*. First edition. Footprints of Climate Variability on Plant Diversity. CRC Press, Boca Raton

Fahad S, Sönmez O, Saud S, Wang D, Wu C, Adnan M, Turan V (Eds.) (2021a) *Plant Growth Regulators for Climate-Smart Agriculture*. First edition. Footprints of Climate Variability on Plant Diversity. CRC Press, Boca Raton

Fahad S, Sönmez O, Turan V, Adnan M, Saud S, Wu C, Wang D (Eds.) (2021d) *Sustainable Soil and Land Management and Climate Change*. First edition. Footprints of Climate Variability on Plant Diversity. CRC Press, Boca Raton

Fakhre A, Ayub K, Fahad S, Sarfraz N, Niaz A, Muhammad AA, Muhammad A, Khadim D, Saud S, Shah H, Muhammad ASR, Khalid N, Muhammad A, Rahul D, Subhan D (2021) Phosphate solubilizing bacteria optimize wheat yield in mineral phosphorus applied alkaline soil. *J Saudi Soc Agric Sci* 21(5):339–348. https://doi.org/10.1016/j.jssas.2021.10.007

Farah R, Muhammad R, Muhammad SA, Tahira Y, Muhammad AA, Maryam A, Shafaqat A, Rashid M, Muhammad R, Qaiser H, Afia Z, Muhammad AA, Muhammad A, Fahad S (2020) Alternative and non-conventional soil and crop management strategies for increasing water use efficiency. In: Fahad S, Hasanuzzaman M, Alam M, Ullah H, Saeed M, Khan AK, Adnan M (Eds.), *Environment, Climate, Plant and Vegetation Growth*. Springer Nature Switzerland, pp. 323–338. https://doi.org/10.1007/978-3-030-49732-3

Farhana G, Ishfaq A, Muhammad A, Dawood J, Fahad S, Xiuling L, Depeng W, Muhammad F, Muhammad F, Syed AS (2020) Use of crop growth model to simulate the impact of climate change on yield of various wheat cultivars under different agro-environmental conditions in Khyber Pakhtunkhwa, Pakistan. *Arabian J Geosci* 13:112. https://doi.org/10.1007/s12517-020-5118-1

Farhat A, Hafiz MH, Wajid I, Aitazaz AF, Hafiz FB, Zahida Z, Fahad S, Wajid F, Artemi C (2020) A review of soil carbon dynamics resulting from agricultural practices. *J Environ Manage* 268 (2020):110319

Farhat UK, Adnan AK, Kai L, Xuexuan X, Muhammad A, Fahad S, Rafiq A, Mushtaq AK, ·Taufiq N, Faisal Z (2022) Influences of long-term crop cultivation and fertilizer management on soil aggregates stability and fertility in the Loess Plateau, Northern China. *J Soil Sci Plant Nutri* 22:1446–1457. https://doi.org/10.1007/s42729-021-00744-1

Fazli W, Muhmmad S, Amjad A, Fahad S, Muhammad A, Muhammad N, Ishaq AM, Imtiaz AK, Mukhtar A, Muhammad S, Muhammad I, Rafi U, Haroon I, Muhammad A (2020) Plant–microbes interactions and functions in changing climate. In: Fahad S, Hasanuzzaman M, Alam M, Ullah H, Saeed M, Khan AK, Adnan M (Eds.) *Environment, Climate, Plant and Vegetation Growth*. Springer Nature Switzerland, pp. 397–420. https://doi.org/10.1007/978-3-030-49732-3

Gaur V (2010) Biofertilizer–necessity for sustainability. *J Adv Dev* 1:7–8

Gautam K, Sirohi C, Singh NR, Thakur Y, Jatav SS, Rana K, … Parihar M (2021) Microbial biofertilizer: Types, applications, and current challenges for sustainable agricultural production. *Biofertilizers* 3–19. doi:10.1016/b978-0-12-821667-5.00014-2

Ghulam M, Muhammad AA, Donald LS, Sajid M, Muhammad FQ, Niaz A, Ateeq ur R, Shakeel A, Sajjad H, Muhammad A, Summia M, Aqib HAK, Fahad S, Rahul D, Mazhar I, Timothy DS (2021) Formalin fumigation and steaming of various composts differentially influence the nutrient release, growth and yield of muskmelon (*Cucumis melo* L.). *Sci Rep* 11:21057.

Glick, B. R. (2012). Plant growth-promoting bacteria: Mechanisms and applications. *Scientifica* 2012:1–15. doi:10.6064/2012/963401

Gopakumar L, Bernard NO, Donato V (2020) Soil microarthropods and nutrient cycling. In: Fahad S, Hasanuzzaman M, Alam M, Ullah H, Saeed M, Khan AK, Adnan M (Eds.), *Environment, Climate,*

Plant and Vegetation Growth. Springer Nature Switzerland, pp. 453–472. https://doi.org/10.1007/978-3-030-49732-3

Guofu L, Zhenjian B, Fahad S, Guowen C, Zhixin X, Hao G, Dandan L, Yulong L, Bing L, Guoxu J, Saud S (2021) Compositional and structural changes in soil microbial communities in response to straw mulching and plant revegetation in an abandoned artificial pasture in Northeast China. *Global Ecol Conserv* 31(2021):e01871

Gupta G, Panwar J, Akhtar MS, Jha PN (2012) Endophytic nitrogen-fixing bacteria as Biofertilizer. *Sustainable Agric Rev*. 183–221. doi:10.1007/978-94-007-5449-2_8

Habib ur R, Ashfaq A, Aftab W, Manzoor H, Fahd R, Wajid I, Md. Aminul I, Vakhtang S, Muhammad A, Asmat U, Abdul W, Syeda RS, Shah S, Shahbaz K, Fahad S, Manzoor H, Saddam H, Wajid N (2017) Application of CSM-CROPGRO-Cotton model for cultivars and optimum planting dates: Evaluation in changing semi-arid climate. *Field Crops Res* 238:139–152. http://dx.doi.org/10.1016/j.fcr.2017.07.007

Hafeez M, Farman U, Muhammad MK, Xiaowei L, Zhijun Z, Sakhawat S, Muhammad I, Mohammed AA, G. Mandela F-G, Nicolas D, Muzammal R, Fahad S, Yaobin L (2021) Metabolic-based insecticide resistance mechanism and ecofriendly approaches for controlling of beet armyworm Spodoptera exigua: a review. *Environ Sci Pollution Res* 29:1746–1762. https://doi.org/10.1007/s11356-021-16974-w

Hafiz MH, Abdul K, Farhat A, Wajid F, Fahad S, Muhammad A, Ghulam MS, Wajid N, Muhammad M, Hafiz FB (2020b) Comparative effects of organic and inorganic fertilizers on soil organic carbon and wheat productivity under arid region. *Commun Soil Sci Plant Anal* 51:1406–1422. DOI: 10.1080/00103624.2020.1763385

Hafiz MH, Farhat A, Ashfaq A, Hafiz FB, Wajid F, Carol Jo W, Fahad S, Gerrit H (2020a) Predicting kernel growth of maize under controlled water and nitrogen applications. *Int J Plant Prod* 14:609–620. https://doi.org/10.1007/s42106-020-00110-8

Hafiz MH, Farhat A, Shafqat S, Fahad S, Artemi C, Wajid F, Chaves CB, Wajid N, Muhammad M, Hafiz FB (2018) Offsetting land degradation through nitrogen and water management during maize cultivation under arid conditions. *Land Degrad Dev* 29(5), 1–10. DOI: 10.1002/ldr.2933

Hafiz MH, Muhammad A, Farhat A, Hafiz FB, Saeed AQ, Muhammad M, Fahad S, Muhammad A (2019) Environmental factors affecting the frequency of road traffic accidents: a case study of sub-urban area of Pakistan. *Environ Sci Pollut Res* 26:11674–11685. https://doi.org/10.1007/s11356-019-04752-8

Hafiz MH, Wajid F, Farhat A, Fahad S, Shafqat S, Wajid N, Hafiz FB (2016) Maize plant nitrogen uptake dynamics at limited irrigation water and nitrogen. *Environ Sci Pollut Res* 24(3):2549–2557. https://doi.org/10.1007/s11356-016-8031-0

Haider SA, Lalarukh I, Amjad SF, Mansoora N, Naz M, Naeem M, Bukhari SA, Shahbaz M, Ali SA, Marfo TD, Subhan D, Rahul D, Fahad S (2021) Drought stress alleviation by potassium-nitrate-containing chitosan/montmorillonite microparticles confers changes in Spinacia oleracea L. *Sustainabiliy* 13:9903. https://doi.org/10.3390/su13179903

Hamza SM, Xiukang W, Sajjad A, Sadia Z, Muhammad N, Adnan M, Fahad S et al. (2021) Interactive effects of gibberellic acid and NPK on morpho-physio-biochemical traits and organic acid exudation pattern in coriander (Coriandrum sativum L.) grown in soil artificially spiked with boron. *Plant Physiol Biochem* 167 (2021):884–900

Haoliang Y, Matthew TH, Ke L, Bin W, Puyu F, Fahad S, Holger M, Rui Y, De LL, Sotirios A, Isaiah H, Xiaohai T, Jianguo M, Yunbo Z, Meixue Z (2022) Crop traits enabling yield gains under more frequent extreme climatic events. *Sci Total Environ* 808 (2022):152170

Hesham FA and Fahad S (2020) Melatonin application enhances biochar efficiency for drought tolerance in maize varieties: Modifications in physio-biochemical machinery. *Agron J* 112(4):1–22

Huang Li-Y, Li Xiao-X, Zhang Yun-B, Fahad S, Wang F (2021) dep1 improves rice grain yield and nitrogen use efficiency simultaneously by enhancing nitrogen and dry matter translocation. *J Integrative Agri* 21(11):3185–3198. doi: 10.1016/S2095-3119(21)63795-4

Hussain MA, Fahad S, Rahat S, Muhammad FJ, Muhammad M, Qasid A, Ali A, Husain A, Nooral A, Babatope SA, Changbao S, Liya G, Ibrar A, Zhanmei J, Juncai H (2020) Multifunctional role of brassinosteroid and its analogues in plants. *Plant Growth Regul* 92:141–156. https://doi.org/10.1007/s10725-020-00647-8

Ibad U, Dost M, Maria M, Shadman K, Muhammad A, Fahad S, Muhammad I, Ishaq AM, Aizaz A, Muhammad HS, Muhammad S, Farhana G, Muhammad I, Muhammad ASR, Hafiz MH, Wajid N, Shah S, Jabar ZKK, Masood A, Naushad A, Rasheed Akbar M, Shah MK Jan B (2022) Comparative effects of biochar and NPK on wheat crops under different management systems. *Crop Pasture Sci* 74:31–40. https://doi.org/10.1071/CP21146

Ibrar H, Muqarrab A, Adel MG, Khurram S, Omer F, Shahid I, Fahim N, Shakeel A, Viliam B, Marian B, Sami Al Obaid, Fahad S, Subhan D, Suleyman T, Hanife AKÇA, Rahul D (2021) Improvement in growth and yield attributes of cluster bean through optimization of sowing time and plant spacing under climate change scenario. *Saudi J Bio Sci* 29(2):781–792. https://doi.org/10.1016/j.sjbs.2021.11.018

Ibrar K, Aneela R, Khola Z, Urooba N, Sana B, Rabia S, Ishtiaq H, Mujaddad Ur Rehman, Salvatore M (2020) Microbes and environment: Global warming reverting the frozen zombies. In: Fahad S, Hasanuzzaman M, Alam M, Ullah H, Saeed M, Khan AK, Adnan M (Eds.) *Environment, Climate, Plant and Vegetation Growth*. Springer Nature Switzerland, pp. 607–634. https://doi.org/10.1007/978-3-030-49732-3

Ihsan MZ, Abdul K, Manzer HS, Liaqat A, Ritesh K, Hayssam MA, Amar M, Fahad S, (2022) The response of Triticum aestivum treated with plant growth regulators to acute day/night temperature rise. *J Plant Growth Regul* 41:2020–2033. https://doi.org/10.1007/s00344-022-10574-9

Ikbal, Passricha N, Saifi SK, Sikka VK, Tuteja N (2020) Multilegume biofertilizer: A dream. In: Sharma V, Salwan R, Al-Ani LKT, (Eds.) *Molecular Aspects of Plant Beneficial Microbes in Agriculture*. Acadmic Press, London, pp. 35–45. doi:10.1016/b978-0-12-818469-1.00003-1

Ikram U, Khadim D, Muhammad T, Muhammad S, Fahad S (2021) Gibberellic acid and urease inhibitor optimize nitrogen uptake and yield of maize at varying nitrogen levels under changing climate. *Environ Sci Pollution Res* 29(5):6568–6577. https://doi.org/10.1007/s11356-021-16049-w

Ilyas M, Mohammad N, Nadeem K, Ali H, Aamir HK, Kashif H, Fahad S, Aziz K, Abid U (2020) Drought tolerance strategies in plants: A mechanistic approach. *J Plant Growth Regul* 40:926–944. https://doi.org/10.1007/s00344-020-10174-5

Iqra M, Amna B, Shakeel I, Fatima K, Sehrish L, Hamza A, Fahad S (2020) Carbon cycle in response to global warming. In: Fahad S, Hasanuzzaman M, Alam M, Ullah H, Saeed M, Khan AK, Adnan M (Eds.) *Environment, Climate, Plant and Vegetation Growth*. Springer Nature Switzerland, pp. 1–16. https://doi.org/10.1007/978-3-030-49732-3

Irfan M, Muhammad M, Muhammad JK, Khadim MD, Dost M, Ishaq AM, Waqas A, Fahad S, Saud S et al. (2021) Heavy metals immobilization and improvement in maize (*Zea mays* L.) growth amended with biochar and compost. *Sci Rep* 11:18416

Jabborova D, Sulaymanov K, Sayyed RZ, Alotaibi SH, Enakiev Y, Azimov A, Jabbarov Z, Ansari MJ, Fahad S, Danish S et al. (2021) Mineral fertilizers improve the quality of turmeric and soil. *Sustainability* 13:9437. https://doi.org/10.3390/su13169437

Jakir H, Allah B (2020) Development and applications of transplastomic plants; A way towards eco-friendly agriculture. In: Fahad S, Hasanuzzaman M, Alam M, Ullah H, Saeed M, Khan AK, Adnan M (Eds.) *Environment, Climate, Plant and Vegetation Growth*. Springer Nature Switzerland, pp. 285–322. https://doi.org/10.1007/978-3-030-49732-3

Jan M, Muhammad Anwar-ul-Haq, Adnan NS, Muhammad Y, Javaid I, Xiuling L, Depeng W, Fahad S (2019) Modulation in growth, gas exchange, and antioxidant activities of salt-stressed rice (Oryza sativa L.) genotypes by zinc fertilization. *Arabian J Geosci* 12:775. https://doi.org/10.1007/s12517-019-4939-2

Kamaran M, Wenwen C, Irshad A, Xiangping M, Xudong Z, Wennan S, Junzhi C, Shakeel A, Fahad S, Qingfang H, Tiening L (2017) Effect of paclobutrazol, a potential growth regulator on stalk mechanical strength, lignin accumulation and its relation with lodging resistance of maize. *Plant Growth Regul* 84:317–332. https://doi.org/10.1007/ s10725-017-0342-8

Kang BG, Kim WT, Yun HS, Chang SC (2010) Use of plant growth-promoting rhizobacteria to control stress responses of plant roots. *Plant Biotechnol Rep* 4(3):179–183. doi:10.1007/s11816-010-0136-1

Karuppiah P, Rajaram S (2012) Exploring the potential of chromium reducing Bacillus sp. and their plant growth promoting activities. *J Microbiol Res* 1(1):17–23. doi:10.5923/j.microbiology.20110101.04

Khadim D, Fahad S, Jahangir MMR, Iqbal M, Syed SA, Shah AK, Ishaq AM, Rahul D et al. (2021a) Biochar and urease inhibitor mitigate NH_3 and N_2O emissions and improve wheat yield in a urea fertilized alkaline soil. *Sci Rep* 11:17413

Khadim D, Saif-ur-R, Fahad S, Syed SA, Shah AK et al. (2021b) Influence of variable biochar concentration on yield-scaled nitrous oxide emissions, wheat yield and nitrogen use efficiency. *Sci Rep* 11:16774

Khan MMH, Niaz A, Umber G, Muqarrab A, Muhammad AA, Muhammad I, Shabir H, Shah F, Vibhor A, Shams HA-H, Reham A, Syed MBA, Nadiyah MA, Ali TKZ, Subhan D, Rahul D (2021) Synchronization of boron application methods and rates is environmentally friendly approach to improve quality attributes of *Mangifera indica* L. on sustainable basis. *Saudi J Bio Sci* 29(3):1869–1880. https://doi.org/10.1016/j.sjbs.2021.10.036

Khatun M, Sarkar S, Era FM, Islam AKMM, Anwar MP, Fahad S, Datta R, Islam AKMA (2021) Drought stress in grain legumes: Effects, tolerance mechanisms and management. *Agronomy* 11: 2374. https://doi.org/10.3390/agronomy11122374

Kizilkaya R (2009) Nitrogen fixation capacity of *Azotobacter* spp. strains isolated from soils in different ecosystems and relationship between them and the microbiological properties of soils. *J Environ Biol* 30:73–82.

Kloepper JW, Beauchamp CJ (1992) A review of issues related to measuring colonization of plant roots by bacteria. *Can J Microbiol* 38(12):1219–1232. doi:10.1139/m92-202

Kumawat N, Kumar R, Kumar S, Meena VS (2017) Nutrient solubilizing microbes (NSMs): Its role in sustainable crop production. *Agricultural Imp Microbes Sustainable Agri* 25–61. doi:10.1007/978-981-10-5343-6_2

Ladha JK, Tirol-Padre A, Reddy CK, Cassman KG, Verma S, Powlson DS, … Pathak H (2016) Global nitrogen budgets in cereals: A 50-year assessment for maize, rice and wheat production systems. *Sci Rep* 6(1). doi:10.1038/srep19355

Lazali M, Bargaz A (2017). Examples of belowground mechanisms enabling legumes to mitigate phosphorus deficiency. In: Sulieman L.-S.P. (Ed.), *Legume Nitrogen Fixation in Soils with Low Phosphorus Availability*. Berlin: Springer International Publishing, pp. 135–152. doi:10.1007/978-3-319-55729-8_7

Ledbetter RN, Garcia Costas AM, Lubner CE, Mulder DW, Tokmina-Lukaszewska M, Artz JH, … Seefeldt LC (2017) The electron bifurcating FixABCX protein complex from *Azotobacter vinelandii*: Generation of low-potential reducing equivalents for nitrogenase catalysis. *Biochemistry* 56(32):4177–4190. doi:10.1021/acs.biochem.7b00389

Maan PK, Garcha S (2021) Production technology, properties, and quality management. *Biofertilizers* 31–43. doi:10.1016/b978-0-12-821667-5.00013-0

Mącik M, Gryta A, Frąc M (2020) Biofertilizers in agriculture: An overview on concepts, strategies and effects on soil microorganisms. *Adv Agron* 31–87. doi:10.1016/bs.agron.2020.02.001

Mahar A, Amjad A, Altaf HL, Fazli W, Ronghua L, Muhammad A, Fahad S, Muhammad A, Rafiullah, Imtiaz AK, Zengqiang Z (2020) Promising technologies for Cd-contaminated soils: Drawbacks and possibilities. In: Fahad S, Hasanuzzaman M, Alam M, Ullah H, Saeed M, Khan AK, Adnan M (Eds.) *Environment, Climate, Plant and Vegetation Growth*. Springer Nature Switzerland, pp. 63–92. https://doi.org/10.1007/978-3-030-49732-3

Mahmood Ul H, Tassaduq R, Chandni I, Adnan A, Muhammad A, Muhammad MA, Muhammad H-ur-R, Mehmood AN, Alam S, Fahad S (2021) Linking plants functioning to adaptive responses under heat stress conditions: a mechanistic review. *J Plant Growth Regul* 41:2596–2613. https://doi.org/10.1007/s00344-021-10493-1

Makavvi A (1973) The density of azotobacter in root-free soils and in the rhizosphere of several plants in arid and semi-arid Libyan regions. *Zentralbl Bakteriol, Parasitenkd, Infektionskrankh Hyg Zweite Naturwiss Abt: Allg, Landwirtsch Tech Mikrobiol* 128(1–2):135–139. doi:10.1016/s0044-4057(73)80051-6

Mal B, Mahapatra P, Mohanty S (2014) Effect of Diazotrophs and chemical fertilizers on production and economics of okra *Abelmoschus esculentus* L.) cultivars. *Am J Plant Sci* 05(01):168–174. doi:10.4236/ajps.2014.51022

Malik AI, Colmer TD, Lambers H, Setter TL, Schortemeyer M (2002) Short-term waterlogging has long-term effects on the growth and physiology of wheat. *New Phytol* 153(2):225–236. doi:10.1046/j.0028-646x.2001.00318.x

Malusá E, Sas-Paszt L, Ciesielska J (2012) Technologies for beneficial microorganisms inocula used as biofertilizers. *Sci World J* 2012:1–12. doi:10.1100/2012/491206

Manzer HS, Saud A, Soumya M, Abdullah A. Al-A, Qasi DA, Bander MA . Al-M, Hayssam MA, Hazem MK, Fahad S, Vishnu DR, Om PN (2021) Molybdenum and hydrogen sulfide synergistically mitigate arsenic toxicity by modulating defense system, nitrogen and cysteine assimilation in faba bean (*Vicia faba* L.) seedlings. *Environ Pollut* 290:117953

Marciano Marra L, Fonsêca Sousa Soares CR, De Oliveira SM, Avelar Ferreira PA, Lima Soares B, De Fráguas Carvalho R, … De Souza Moreira FM (2012) Biological nitrogen fixation and phosphate solubilization by bacteria isolated from tropical soils. *Plant Soil* 357(1–2):289–307. doi:10.1007/s11104-012-1157-z

Mehmood K, Bao Y, Saifullah, Bibi S, Dahlawi S, Yaseen M, Abrar MM, Srivastava P, Fahad S, Faraj TK (2022) Contributions of open biomass burning and crop straw burning to air quality: Current research paradigm and future outlooks. *Front Environ Sci* 10:852492. doi: 10.3389/fenvs.2022.852492

Mirzakhani M, Ardakani MR, Rejali F, Rad AH, Miransari M (2014) Safflower (*Carthamus tinctorius* L.) oil content and yield components as affected by Co-inoculation with azotobacter chroococcum and Glomus intraradices at various N and P levels in a dry climate. In: Miransari M. (Ed.) *Use of Microbes for the Alleviation of Soil Stresses. New York: Springer, pp* 153–164. doi:10.1007/978-1-4939-0721-2_9

Mishra J, Arora NK (2016) Bioformulations for plant growth promotion and combating phytopathogens: A sustainable approach. *Bioformulations* 3–33. doi:10.1007/978-81-322-2779-3_1

Mohammad I Al-Wabel, Abdelazeem S, Munir A, Khalid E, Adel RAU (2020b) Extent of climate change in Saudi Arabia and its impacts on agriculture: A case study from Qassim Region. In: Fahad S, Hasanuzzaman M, Alam M, Ullah H, Saeed M, Khan AK, Adnan M (Eds.) *Environment, Climate, Plant and Vegetation Growth.* Springer Nature Switzerland, pp. 635–658. https://doi.org/10.1007/978-3-030-49732-3

Mohammad I Al-Wabel, Munir Ahmad, Adel R. A. Usman, Mutair Akanji, and Muhammad Imran Rafique (2020a) Advances in pyrolytic technologies with improved carbon capture and storage to combat climate change. In: Fahad S, Hasanuzzaman M, Alam M, Ullah H, Saeed M, Khan AK, Adnan M (Eds.) *Environment, Climate, Plant and Vegetation Growth.* Springer Nature Switzerland, pp. 535–576. https://doi.org/10.1007/978-3-030-49732-3

Mubeen M, Ashfaq A, Hafiz MH, Muhammad A, Hafiz UF, Mazhar S, Muhammad Sami ul Din, Asad A, Amjed A, Fahad S, Wajid N (2020) Evaluating the climate change impact on water use efficiency of cotton-wheat in semi-arid conditions using DSSAT model. *J Water Climate Change* 11(4):1661–1675 doi/10.2166/wcc.2019.179/622035/jwc2019179.pdf

Muhammad I, Khadim D, Fahad S, Imran M, Saud A, Manzer HS, Shah S, Jabar ZKK, Shamsher A, Shah H, Taufiq N, Hafiz MH, Jan B, Wajid N (2022) Exploring the potential effect of *Achnatherum splendens* L.-derived biochar treated with phosphoric acid on bioavailability of cadmium and wheat growth in contaminated soil. *Environ Sci Pollut Res* 29(25):37676–37684. https://doi.org/10.1007/s11356-021-17950-0.

Muhammad N, Muqarrab A, Khurram S, Fiaz A, Fahim N, Muhammad A, Shazia A, Omaima N, Sulaiman AA, Fahad S, Subhan D, Rahul D (2021) Kaolin and jasmonic acid improved cotton productivity under water stress conditions. *J Saudi Soc Agri Sci* 28 (2021) 6606–6614. https://doi.org/10.1016/j.sjbs.2021.07.043

Muhammad Tahir ul Qamar, Amna F, Amna B, Barira Z, Xitong Z, Ling-Ling C (2020) Effectiveness of conventional crop improvement strategies vs. omics. In: Fahad S, Hasanuzzaman M, Alam M, Ullah H, Saeed M, Khan AK, Adnan M (Eds.) *Environment, Climate, Plant and Vegetation Growth.* Springer Nature Switzerland, pp. 253–284. https://doi.org/10.1007/978-3-030-49732-3

Muhammad Z, Abdul MK, Abdul MS, Kenneth BM, Muhammad S, Shahen S, Ibadullah J, Fahad S (2019) Performance of Aeluropus lagopoides (mangrove grass) ecotypes, a potential turfgrass, under high saline conditions. *Environ Sci Pollut Res* 26(13):13410–13412. https://doi.org/10.1007/s11356-019-04838-3

Muzammal R, Fahad S, Guanghui D, Xia C, Yang Y, Kailei T, Lijun L, Fei-Hu L, Gang D (2021) Evaluation of hemp (Cannabis sativa L.) as an industrial crop: a review. *Environ Sci Pollution Res* 28(38):52832–52843. https://doi.org/10.1007/s11356-021-16264-5

Narula N, Kumar V, Behl RK, Deubel A, Gransee A, Merbach W (2000) Effect of P-solubilizingazotobacter chroococcum on N, P, K uptake in P-responsive wheat genotypes grown under greenhouse conditions. *J Plant Nutr Soil Sci* 163(4):393–398. doi:10.1002/1522-2624(200008)163:4<393::aid-jpln393>3.0.co;2-w

National Research Council. Biological Nitrogen Fixation: Research Challenges.Washington, DC: National Academy Press, 1994.

Niaz A, Abdullah E, Subhan D, Muhammad A, Fahad S, Khadim D, Suleyman T, Hanife A, Anis AS, Mohammad JA, Emre B, ¨Omer SU, Rahul D, Bernard RG (2022) Mitigation of lead (Pb) toxicity in rice cultivated with either ground water or wastewater by application of acidified carbon. *J Environ Manage* 307:114521

Noor M, Naveed ur R, Ajmal J, Fahad S, Muhammad A, Fazli W, Saud S, Hassan S (2020) Climate change and coastal plant lives. In: Fahad S, Hasanuzzaman M, Alam M, Ullah H,Saeed M, Khan AK, Adnan M (Eds.) *Environment, Climate, Plant and Vegetation Growth.* Springer Nature Switzerland, pp. 93–108. https://doi.org/10.1007/978-3-030-49732-3

Nosrati R, Owlia P, Saderi H, Rasooli I, Ali Malboobi M (2014) Phosphate solubilization characteristics of efficient nitrogen fixing soil *Azotobacter* strains. *Iran J Microbiol* 6: 285–295.

Nùñez C, Moreno S, Soberón-Chávez G, Espín G (1999) The *Azotobacter vinelandii* response regulator AlgR is essential for cyst formation. *J Bacteriol* 181(1):141–148. doi:10.1128/jb.181.1.141-148.1999

Okon Y, Itzigsohn R (1995) The development of Azospirillum as a commercial inoculant for improving crop yields. *Biotechnol Adv* 13(3):415–424. doi:10.1016/0734-9750(95)02004-m

Omer A, Emara H, Zaghloul R, Abdel M, Dawwam G (2016) Potential of *Azotobacter salinestris* as plant growth promoting rhizobacteria under saline stress conditions. *Res J Pharm Biol Chem Sci* 7: 2572–2583

Ortiz-Marquez JCF, Do Nascimento M, de los Angeles Dublan M, Curatti L (2012) Association with an ammonium-excreting bacterium allows diazotrophic culture of oil-rich eukaryotic microalgae. *Appl Environ Microbiol* 78: 2345–2352.

Pandey A, Shende S (1991) Effect of Azotobacter chroococcum inoculation on yield and post harvest seed quality of wheat (Triticum aestivum). *Zentralblatt für Mikrobiologie* 146(7–8):489–494. doi:10.1016/s0232-4393(11)80236-2

Patil HJ, Solanki MK (2016) Microbial inoculant: Modern era of fertilizers and pesticides. In D.P. Singh et al. (eds.), *Microbial Inoculants in Sustainable Agricultural Productivity. India: Springer.* 319–343. doi:10.1007/978-81-322-2647-5_19

Phillips KA, Skirpan AL, Liu X, Christensen A, Slewinski TL, Hudson C, … McSteen P (2011) *vanishing tassel2* encodes a grass-specific tryptophan aminotransferase required for vegetative and reproductive development in maize. *Plant Cell* 23(2):550–566. doi:10.1105/tpc.110.075267

Ponmurugan K, Sankaranarayanan A, Al-Dharbi NA (2012) Biological activities of plant growth promoting Azotobacter sp. isolated from vegetable crops rhizosphere soils. *J Pure Appl Microbiol* 6: 1689–1698.

Poza-Carrión C, Echavarri-Erasun C, Rubio LM (2015) Regulation of nif gene expression in Azotobacter vinelandii. *Biol Nitrogen Fixation* 99–108. doi:10.1002/9781119053095.ch9

Qaisrani MM, Zaheer A, Mirza MS, Naqqash T, Qaisrani TB, Hanif MK, … Rasool M (2019) A comparative study of bacterial diversity based on culturable and culture-independent techniques in the rhizosphere of maize (Zea mays L.). *Saudi Jo Bio Sci* 26(7):1344–1351. doi:10.1016/j.sjbs.2019.03.010

Qamar-uz Z, Zubair A, Muhammad Y, Muhammad ZI, Abdul K, Fahad S, Safder B, Ramzani PMA, Muhammad N (2017) Zinc biofortification in rice: leveraging agriculture to moderate hidden hunger in developing countries. *Arch Agron Soil Sci* 64:147–161. https://doi.org/10.1080/03650340.2017.1338343

Qin ZH, Nasib ur Rahman, Ahmad A, Yu pei Wang, Sakhawat S, Ehmet N, Wen juan Shao, Muhammad I, Kun S, Rui L, Fazal S, Fahad S (2022) Range expansion decreases the reproductive fitness of *Gentiana officinalis* (*Gentianaceae*). *Sci Rep* 12:2461. https://doi.org/10.1038/s41598-022-06406-1

Rajesh KS, Fahad S, Pawan K, Prince C, Talha J, Dinesh J, Prabha S, Debanjana S, Prathibha MD, Bandana B, Akash H, Gupta NK, Rekha S, Devanshu D, Dalpat LS, Ke L, Matthew TH, Saud S, Adnan NS, Taufiq N (2022) Beneficial elements: New players in improving nutrient use efficiency and abiotic stress tolerance. *Plant Growth Reg* 100:237–265. https://doi.org/10.1007/s10725-022-00843-8

Rana KL, Kour D, Kaur T, Devi R, Yadav AN, Yadav N, … Saxena AK (2020) Endophytic microbes: Biodiversity, plant growth-promoting mechanisms and potential applications for agricultural sustainability. *Antonie van Leeuwenhoek* 113(8):1075–1107. doi:10.1007/s10482-020-01429-y

Rashid M, Qaiser H, Khalid SK, Mohammad I. Al-Wabel, Zhang A, Muhammad A, Shahzada SI, Rukhsanda A, Ghulam AS, Shahzada MM, Sarosh A, Muhammad FQ (2020) Prospects of biochar in alkaline soils to mitigate climate change. In: Fahad S, Hasanuzzaman M, Alam M, Ullah H, Saeed M, Khan AK, Adnan M (Eds.), *Environment, Climate, Plant and Vegetation Growth.* Springer Nature Switzerland, pp. 133–150. https://doi.org/10.1007/978-3-030-49732-3

Rashid M, Samina K, Najma A, Sadia A, Farooq L (2004) Organic acids production and phosphate solubilization by phosphate solubilizing microorganisms (PSM) under in vitro conditions. *Pak J Biol Sci* 7(2):187–196. doi:10.3923/pjbs.2004.187.196

Reed SC, Cleveland CC, Townsend AR (2011) Functional ecology of free-living nitrogen fixation: A contemporary perspective. *Ann Rev Ecol Evol Syst* 42(1):489–512. doi:10.1146/annurev-ecolsys-102710-145034

Rehana S, Asma Z, Shakil A, Anis AS, Rana KI, Shabir H, Subhan D, Umber G, Fahad S, Jiri K, Sami Al Obaid, Mohammad JA, Rahul D (2021) Proteomic changes in various plant tissues associated with chromium stress in sunflower. *Saudi J Bio Sci* 29(4):2604–2612. https://doi.org/10.1016/j.sjbs.2021.12.042

Rehman M, Fahad S, Saleem MH, Hafeez M, Muhammad Habib ur Rahman, Liu F, Deng G (2020) Red light optimized physiological traits and enhanced the growth of ramie (Boehmeria nivea L.). *Photosynthetica* 58 (4):922–931

Richardson AE, Barea J, McNeill AM, Prigent-Combaret C (2009) Acquisition of phosphorus and nitrogen in the rhizosphere and plant growth promotion by microorganisms. *Plant Soil* 321(1–2):305–339. doi:10.1007/s11104-009-9895-2

Rodriguez H, Gonzalez T, Goire I, Bashan Y (2004) Gluconic acid production and phosphate solubilization by the plant growth-promoting bacterium *Azospirillum* spp. *Naturwissenschaften* 91:552–555. doi: 10.1007/s00114-004-0566-0.

Rodriguez-Salazar J, Moreno S, Espín G (2017) LEA proteins are involved in cyst desiccation resistance and other abiotic stresses in Azotobacter vinelandii. *Cell Stress Chaperones* 22(3):397–408. doi:10.1007/s12192-017-0781-1

Rojas-Tapias D, Moreno-Galván A, Pardo-Díaz S, Obando M, Rivera D, Bonilla R (2012) Effect of inoculation with plant growth-promoting bacteria (PGPB) on amelioration of saline stress in maize (Zea mays). *Appl Soil Ecol* 61: 264–272. doi:10.1016/j.apsoil.2012.01.006

Romero-Perdomo F, Abril J, Camelo M, Moreno-Galván A, Pastrana I, Rojas-Tapias D, Bonilla R (2017) Azotobacter chroococcum as a potentially useful bacterial biofertilizer for cotton (gossypium hirsutum): Effect in reducing N fertilization. *Rev Argen Microbiol* 49(4):377–383. doi:10.1016/j.ram.2017.04.006

Rovira AD (1991). Rhizosphere research – 85 years of progress and frustration. In: DL Keister, PB Cregan (Eds.), *The Rhizosphere and Plant Growth: Beltsville Agricultural Research Center (BARC), Beltsville, Maryland.* Dordrecht: Springer, pp. 3–13. doi: 10.1007/978-94-011-3336-4_1

Ruzzi M, Aroca R (2015) Plant growth-promoting rhizobacteria act as biostimulants in horticulture. *Sci Horti* 196: 124–134. doi:10.1016/j.scienta.2015.08.042

Sabier Sae K, Abdulla Ah S, Ahmaed Has I, Hamed Ahme P (2015) Effect of bio-fertilizer and chemical fertilizer on growth and yield in cucumber (Cucumis sativus) in green house condition. *Pak J Biol Sci* 18(3):129–134. doi:10.3923/pjbs.2015.129.134

Sadam M, Muhammad Tahir ul Qamar, Ghulam M, Muhammad SK, Faiz AJ (2020) Role of biotechnology in climate resilient agriculture. In: Fahad S, Hasanuzzaman M, Alam M, Ullah H, Saeed M, Khan AK, Adnan M (Eds.) *Environment, Climate, Plant and Vegetation Growth.* Springer Nature Switzerland, pp. 339–366. https://doi.org/10.1007/978-3-030-49732-3

Saeed K, Ahmed SA, Hassan IA, Ahmed PH (2015) Effect of bio-fertilizer and chemical fertilizer on growth and yield in cucumber (*Cucumis sativus* L.) in green house condition. *Am Eurasian J Agric Environ Sci* 15: 353–358.

Safi UK, Ullah F, Mehmood S, Fahad S, Ahmad Rahi A, Althobaiti F, et al. (2021) Antimicrobial, antioxidant and cytotoxic properties of Chenopodium glaucum L. *PLoS One* 16(10):e0255502. https://doi.org/10.1371/journal. pone.0255502

Sahrish N, Shakeel A, Ghulam A, Zartash F, Sajjad H, Mukhtar A, Muhammad AK, Ahmad K, Fahad S, Wajid N, Sezai E, Carol Jo W, Gerrit H (2022) Modeling the impact of climate warming on potato phenology. *European J Agro N* 132:126404

Sajid H, Jie H, Jing H, Shakeel A, Satyabrata N, Sumera A, Awais S, Chunquan Z, Lianfeng Z, Xiaochuang C, Qianyu J, Junhua Z (2020) Rice production under climate change: Adaptations and mitigating strategies. In: Fahad S, Hasanuzzaman M, Alam M, Ullah H, Saeed M, Khan AK, Adnan M (Eds.), *Environment, Climate, Plant and Vegetation Growth.* Springer Nature Switzerland, pp. 659–686. https://doi.org/10.1007/978-3-030-49732-3

Sajjad H, Muhammad M, Ashfaq A, Nasir M, Hafiz MH, Muhammad A, Muhammad I, Muhammad U, Hafiz UF, Fahad S, Wajid N, Hafiz MRJ, Mazhar A, Saeed AQ, Amjad F, Muhammad SK, Mirza W (2021) Satellite-based evaluation of temporal change in cultivated land in Southern Punjab (Multan region) through dynamics of vegetation and land surface temperature. *Open Geo Sci* 13:1561–1577

Sajjad H, Muhammad M, Ashfaq A, Waseem A, Hafiz MH, Mazhar A, Nasir M, Asad A, Hafiz UF, Syeda RS, Fahad S, Depeng W, Wajid N (2019) Using GIS tools to detect the land use/land cover changes during forty years in Lodhran district of Pakistan. *Environ Sci Pollut Res* 27:39676–39692. https://doi.org/10.1007/s11356-019-06072-3

Saleem MH, Fahad S, Adnan M, Mohsin A, Muhammad SR, Muhammad K, Qurban A, Inas AH, Parashuram B, Mubassir A, Reem MH (2020a) Foliar application of gibberellic acid endorsed phytoextraction of copper and alleviates oxidative stress in jute (Corchorus capsularis L.) plant grown in highly copper-contaminated soil of China. *Environ Sci Pollution Res* 27:37121–37133. https://doi.org/10.1007/s11356-020-09764-3

Saleem MH, Fahad S, Shahid UK, Mairaj D, Abid U, Ayman ELS, Akbar H, Analía L, Lijun L (2020c) Copper-induced oxidative stress, initiation of antioxidants and phytoremediation potential of flax (Linum usitatissimum L.) seedlings grown under the mixing of two different soils of China. *Environ Sci Poll Res* 27:5211–5221. https://doi.org/10.1007/s11356-019-07264-7

Saleem MH, Rehman M, Fahad S, Tung SA, Iqbal N, Hassan A, Ayub A, Wahid MA, Shaukat S, Liu L, Deng G (2020b) Leaf gas exchange, oxidative stress, and physiological attributes of rapeseed (*Brassica napus* L.) grown under different light-emitting diodes. *Photosynthetica* 58 (3): 836–845.

Saman S, Amna B, Bani A, Muhammad Tahir ul Qamar, Rana MA, Muhammad SK (2020) QTL mapping for abiotic stresses in cereals. In: Fahad S, Hasanuzzaman M, Alam M, Ullah H, Saeed M, Khan AK, Adnan M (Eds.) *Environment, Climate, Plant and Vegetation Growth.* Springer Nature Switzerland, pp. 229–252. https://doi.org/10.1007/978-3-030-49732-3

Sana U, Shahid A, Yasir A, Farman UD, Syed IA, Mirza MFAB, Fahad S, F. Al-Misned, Usman A, Xinle G, Ghulam N, Kunyuan W (2022) Bifenthrin induced toxicity in Ctenopharyngodon idella at an acute concentration: A multi-biomarkers based study. *J King Saud Uni Sci* 34 (2022):101752

Sangeeth KP, Bhai RS, Srinivasan V (2012) *Paenibacillus glucanolyticus*, a promising potassium solubilizing bacterium isolated from black pepper (*Piper nigrum* L.) rhizosphere. *J Spices Aromat Crops* 21: 118–124.

Saravanan VS, Kumar MR, Sa TM (2011) Microbial zinc solubilization and their role on plants. In: Maheshwari D. (Ed), *Bacteria in Agrobiology: Plant Nutrient Management.* Berlin: Springer, pp. 47–63. doi:10.1007/978-3-642-21061-7_3

Saud S, Fahad S, Hassan S (2022a) Developments in the investigation of nitrogen and oxygen stable isotopes in atmospheric nitrate. *Sustainable Chem Climate Action* 1:100003

Saud S, Li X, Jiang Z, Fahad S, Hassan S (2022b) Exploration of the phytohormone regulation of energy storage compound accumulation in microalgae. *Food Energy Secur* 2022;00:e418

Saud S, Chen Y, Fahad S, Hussain S, Na L, Xin L, Alhussien SA (2016) Silicate application increases the photosynthesis and its associated metabolic activities in Kentucky bluegrass under drought stress and post-drought recovery. *Environ Sci Pollut Res* 23(17):17647–17655. https://doi.org/10.1007/s11356-016-6957-x

Saud S, Chen Y, Long B, Fahad S, Sadiq A (2013) The different impact on the growth of cool season turf grass under the various conditions on salinity and drought stress. *Int J Agric Sci Res* 3:77–84

Saud S, Fahad S, Cui G, Chen Y, Anwar S (2020) Determining nitrogen isotopes discrimination under drought stress on enzymatic activities, nitrogen isotope abundance and water contents of Kentucky bluegrass. *Sci Rep* 10:6415. https://doi.org/10.1038/s41598-020-63548-w

Saud S, Fahad S, Yajun C, Ihsan MZ, Hammad HM, Nasim W, Amanullah Jr, Arif M, Alharby H (2017) Effects of nitrogen supply on water stress and recovery mechanisms in Kentucky bluegrass plants. Front *Plant Sci* 8:983. doi: 10.3389/fpls.2017.00983

Saud S, Li X, Chen Y, Zhang L, Fahad S, Hussain S, Sadiq A, Chen Y (2014) Silicon application increases drought tolerance of Kentucky bluegrass by improving plant water relations and morph physiological functions. *Sci World J* 2014:1–10. https://doi.org/10.1155/2014/ 368694

Segal HM, Spatzal T, Hill MG, Udit AK, Rees DC (2017) Electrochemical and structural characterization of *Azotobacter vinelandii* flavodoxin II. *Protein Sci* 26(10):1984–1993. doi:10.1002/pro.3236

Senol C (2020) The effects of climate change on human behaviors. In: Fahad S, Hasanuzzaman M, Alam M, Ullah H, Saeed M, Khan AK, Adnan M (Eds.) *Environment, Climate, Plant and Vegetation Growth.* Springer Nature Switzerland, pp. 577–590. https://doi.org/10.1007/978-3-030-49732-3

Shafi MI, Adnan M, Fahad S, Fazli W, Ahsan K, Zhen Y, Subhan D, Zafar-ul-Hye M, Martin B, Rahul D (2020) Application of single superphosphate with humic acid improves the growth, yield and phosphorus uptake of wheat (Triticum aestivum L.) in calcareous soil. *Agronomy* (10):1224. doi:10.3390/agronomy10091224

Shah F, Lixiao N, Kehui C, Tariq S, Wei W, Chang C, Liyang Z, Farhan A, Fahad S, Huang J (2013) Rice grain yield and component responses to near 2°C of warming. *Field Crop Res* 157:98–110

Shah S, Shah H, Liangbing X, Xiaoyang S, Shahla A, Fahad S (2022) The physiological function and molecular mechanism of hydrogen sulfide resisting abiotic stress in plants. *Braz J Bot* 45:563–572. https://doi.org/10.1007/s40415-022-00785-5

Shah SAR, Khan SA, Junaid K, Sattar S, Zaman M, Saleem N, Adnan M (2015) Screening of mustard genotypes for antixenosis and multiplication against mustard aphid, *Lipaphiserysimi*(Kalt) (Aphididae: Homoptera). *J Ento Zool Stud* 3(6):84–87.

Shaharoona B, Arshad M, Zahir Z (2006) Effect of plant growth promoting rhizobacteria containing ACC-deaminase on maize (Zea mays L.) growth under axenic conditions and on nodulation in mung bean (Vigna radiata L.). *Lett Appl Microbiol* 42(2):155–159. doi:10.1111/j.1472-765x.2005.01827.x

Sharma S, Gupta R, Dugar G, Srivastava AK (2012) Impact of application of biofertilizers on soil structure and resident microbial community structure and function. In: Maheshwari D. (Ed), *Bacteria in Agrobiology: Plant Probiotics*. Berlin: Springer, pp. 65–77. doi:10.1007/978-3-642-27515-9_4

Sidra K, Javed I, Subhan D, Allah B, Syed IUSB, Fatma B, Khaled DA, Fahad S, Omaima N, Ali TKZ, Rahul D (2021) Physio-chemical characterization of indigenous agricultural waste materials for the development of potting media. *J Saudi Soci Agri Sci* 28(12):7491–7498. https://doi.org/10.1016/j.sjbs.2021.08.058

Singh B, Ryan J, Singh B, Ryan J (2015a). *Managing Fertilizers to Enhance Soil Health Managing Fertilizers to Enhance Soil Health*. Paris: IFA, pp. 1–24.

Singh G, Biswas DR, Marwaha TS (2010b) Mobilization of potassium from waste mica by plant growth promoting rhizobacteria and its assimilation by maize (*Zea mays*) and wheat (*Triticum aestivum.*): a hydroponics study under phytotron growth chamber. *J Plant Nutr* 33: 1236–1251. doi: 10.1080/01904161003765760

Singh D, Thapa S, Geat N, Mehriya ML, Rajawat MV (2021) Biofertilizers: Mechanisms and application. *Biofertilizers* 151–166. doi:10.1016/b978-0-12-821667-5.00024-5

Subhan D, Zafar-ul-Hye M, Fahad S, Saud S, Martin B, Tereza H, Rahul D (2020) Drought stress alleviation by ACC deaminase producing *Achromobacter xylosoxidans* and *Enterobacter cloacae*, with and without timber waste biochar in maize. *Sustainability* 12(15):6286. doi:10.3390/su12156286

Sumbul A, Ansari RA, Rizvi R, Mahmood I (2020) Azotobacter: A potential bio-fertilizer for soil and plant health management. *Saudi J Biol Sci* 27(12):3634–3640. doi:10.1016/j.sjbs.2020.08.004

Swain M, Ray R (2009) Biocontrol and other beneficial activities of bacillus subtilis isolated from cowdung microflora. *Microbiol Res* 164(2):121–130. doi:10.1016/j.micres.2006.10.009

Tariq M, Ahmad S, Fahad S, Abbas G, Hussain S, Fatima Z, Nasim W, Mubeen M, ur Rehman MH, Khan MA, Adnan M (2018) The impact of climate warming and crop management on phenology of sunflower-based cropping systems in Punjab, Pakistan. *Agri Forest Met* 15;256:270–282.

Tilman D, Balzer C, Hill J, Befort BL (2011) Global food demand and the sustainable intensification of agriculture. *Proceedings of the National Academy of Sciences* 108(50):20260–20264. doi:10.1073/pnas.1116437108

Unsar Naeem-U, Muhammad R, Syed HMB, Asad S, Mirza AQ, Naeem I, Muhammad H ur R, Fahad S, Shafqat S (2020) Insect pests of cotton crop and management under climate change scenarios. In: Fahad S, Hasanuzzaman M, Alam M, Ullah H, Saeed M, Khan AK, Adnan M (Eds.) *Environment, Climate, Plant and Vegetation Growth*. Springer Nature Switzerland, pp. 367–396. https://doi.org/10.1007/978-3-030-49732-3

Velmourougane K, Prasanna R, Chawla G, Nain L, Kumar A, Saxena AK (2019) Trichoderma–*Azotobacter* biofilm inoculation improves soil nutrient availability and plant growth in wheat and cotton. *J Basic Microbiol* 59: 632–644. doi: 10.1002/jobm.201900009

Wahid F, Fahad S, Subhan D, Adnan M, Zhen Y, Saud S, Manzer HS, Martin B, Tereza H, Rahul D (2020) Sustainable management with mycorrhizae and phosphate solubilizing bacteria for enhanced phosphorus uptake in calcareous soils. *Agriculture* 10(8):334. doi:10.3390/agriculture10080334

Wajid N, Ashfaq A, Asad A, Muhammad T, Muhammad A,Muhammad S, Khawar J, Ghulam MS, Syeda RS, Hafiz MH, Muhammad IAR, Muhammad ZH, Muhammad Habib ur R, Veysel T, Fahad S, Suad S, Aziz K, Shahzad A (2017) Radiation efficiency and nitrogen fertilizer impacts on sunflower crop in contrasting environments of Punjab. *Pak Environ Sci Pollut Res* 25:1822–1836. https://doi.org/10.1007/s11356-017-0592-z

Wani SA, Chand S, Wani MA, Ramzan M, Hakeem KR (2016) Azotobacter chroococcum – A potential biofertilizer in agriculture: An overview. In Hakeem K.R. et al. (Eds.), *Soil Science: Agricultural and Environmental Prospectives*. Switzerland: Springer, pp. 333–348. doi:10.1007/978-3-319-34451-5_15

Wiqar A, Arbaz K, Muhammad Z, Ijaz A, Muhammad A, Fahad S (2022) Relative efficiency of biochar particles of different sizes for immobilising heavy metals and improving soil properties. *Crop Pasture Sci* 42(2):112–120. https://doi.org/10.1071/CP20453.

Wu C, Kehui C, She T, Ganghua L, Shaohua W, Fahad S, Lixiao N, Jianliang H, Shaobing P, Yanfeng D (2020) Intensified pollination and fertilization ameliorate heat injury in rice (Oryza sativa L.) during the flowering stage. *Field Crops Res* 252:107795

Wu C, Tang S, Li G, Wang S, Fahad S, Ding Y (2019) Roles of phytohormone changes in the grain yield of rice plants exposed to heat: a review. *Peer J* 7:e7792. DOI 10.7717/peerj.7792

Wu S, Luo Y, Cheung K, Wong M (2006) Influence of bacteria on PB and Zn speciation, mobility and bioavailability in soil: A laboratory study. *Environ Pollut* 144(3):765–773. doi:10.1016/j.envpol.2006.02.022

Xue B, Huang L, Li X, Lu J, Gao R, Kamran M, Fahad S (2022) Effect of clay mineralogy and soil organic carbon in aggregates under straw incorporation. *Agronomy* 12:534. https://doi.org/10.3390/ agronomy12020534

Yang R, Dai P, Wang B, Jin T, Liu K, Fahad S, Harrison MT, Man J, Shang J, Meinke H, Deli L, Xiaoyan W, Yunbo Z, Meixue Z, Yingbing T, Haoliang Y (2022) Over-optimistic projected future wheat yield potential in the North China Plain: The role of future climate extremes. *Agronomy* 12:145. https://doi.org/ 10.3390/ agronomy12010145

Yang Z, Zhang Z, Zhang T, Fahad S, Cui K, Nie L, Peng S, Huang J (2017) The effect of season-long temperature increases on rice cultivars grown in the central and southern regions of China. *Front Plant Sci* 8:1908. https://doi.org/10.3389/fpls.2017.01908

Yang J, Kloepper JW, Ryu C (2009). Rhizosphere bacteria help plants tolerate abiotic stress. *Trends Plant Sci* 14(1):1–4. doi:10.1016/j.tplants.2008.10.004

Zafar-ul-Hye M, Muhammad T ahzeeb-ul-Hassan, Muhammad A, Fahad S,, Martin B, Tereza D, Rahul D, Subhan D (2020b) Potential role of compost mixed biochar with rhizobacteria in mitigating lead toxicity in spinach. *Sci Rep* 10:12159. https://doi.org/10.1038/s41598-020-69183-9.

Zafar-ul-Hye M, Muhammad N, Subhan D, Fahad S, Rahul D, Mazhar A, Ashfaq AR, Martin B, Ji˘rí H, Zahid HT, Muhammad N (2020a) Alleviation of cadmium adverse effects by improving nutrients uptake in bitter gourd through cadmium tolerant rhizobacteria. *Environments* 7(8), 54. doi:10.3390/ environments7080054

Zafar-ul-Hye M, Akbar MN, Iftikhar Y, Abbas M, Zahid A, Fahad S, Datta R, Ali M, Elgorban AM, Ansari MJ et al. (2021) Rhizobacteria inoculation and caffeic acid alleviated drought stress in lentil plants. *Sustainability* 13:9603. https://doi.org/10.3390/su13179603

Zahida Z, Hafiz FB, Zulfiqar AS, Ghulam MS, Fahad S, Muhammad RA, Hafiz MH, Wajid N, Muhammad S (2017) Effect of water management and silicon on germination, growth, phosphorus and arsenic uptake in rice. *Ecotoxicol Environ Saf* 144:11–18

Zahir SM, Zheng-HG, Ala Ud D, Amjad A, Ata Ur R, Kashif J, Shah F, Saud S, Adnan M, Fazli W, Saud A, Manzer HS, Shamsher A, Wajid N, Hafiz MH, Fahad S (2021) Synthesis of silver nanoparticles using Plantago lanceolata extract and assessing their antibacterial and antioxidant activities. *Sci Rep* 11:20754

Zaman I, Ali M, Shahzad K, Tahir, MS, Matloob A, Ahmad W, Alamri S, Khurshid MR, Qureshi MM, Wasaya A, Khurram SB, Manzer HS, Fahad S, Rahul D (2021) Effect of plant spacings on growth, physiology, yield and fiber quality attributes of cotton genotypes under nitrogen fertilization. *Agronomy* 11: 2589. https://doi.org/ 10.3390/agronomy11122589

Zayadan BK, Matorin DN, Baimakhanova GB, Bolathan K, Oraz GD, Sadanov AK (2014) Promising microbial consortia for producing biofertilizers for rice fields. *Microbiology* 83: 391–397. doi: 10.1134/ S0026261714040171

Zia-ur-Rehman M (2020) Environment, climate change and biodiversity. In: Fahad S, Hasanuzzaman M, Alam M, Ullah H, Saeed M, Khan AK, Adnan M (Eds.), *Environment, Climate, Plant and Vegetation Growth.* Springer Nature Switzerland, pp. 473–502. https://doi.org/10.1007/978-3-030-49732-3

14 Status of Research and Applications of Bio-fertilizers
Global Scenario

Rafi Ullah[1*], *Muhammad Junaid*[2], *Mehwish Kanwal*[3], *Muhammad Adnan* *[1], *Taufiq Nawaz*[4], *Nazeer Ahmed*[1], *Fazli Subhan*[1], *Muhammad Romman*[5], *Shah Saud*[6], *Shakeel Ahmad*[7], *Anas Iqbal*[8], *Fazli Wahid*[1], *Muhammad Haroon*[1], *Muhammad Zamin*[1], *Nazish Huma Khan*[9], *Jamal Nasar*[7], *Shah Hassan*[10]

[1] Department of Agriculture, University of Swabi, Anbar, Swabi, Khyber Pakhtunkhwa, Pakistan
[2] Graduate School of Life and Environmental Sciences, University of Tsukuba, Tsukuba, Ibaraki, Japan
[3] Ministry of National Food Security & Research, Pakistan Tobacco Board, Peshawar, Khyber Pakhtunkhwa, Pakistan
[4] Department of Food Science and Technology, The University of Agriculture, Peshawar, Pakistan
[5] Department of Botany, University of Chitral, Khyber Pakhtunkhwa, Pakistan
[6] College of Life Science, Linyi University, Linyi, Shandong, China
[7] Guangxi Colleges and Universities, Key Laboratory of Crop Cultivation and Tillage, National Demonstration Center for Experimental Plant Science Education, Agricultural College of Guangxi University, Nanning, China
[8] College of Agriculture, Guangxi University, Nanning, China
[9] Department of Environmental Sciences, University of Swabi, Pakistan
[10] Department of Agricultural Extension Education and Communication, The University of Agriculture, Peshawar, Pakistan
* **Correspondence:** rafiullah@uoswabi.edu.pk (R.U); madnan@uoswabi.edu.pk (M.A)

CONTENTS

DOI: 10.1201/9781003286233-14

14.1 INTRODUCTION

With greater health awareness and pollution constraints, organic farming is garnering the interest of academics and the general public. The negative impacts of pest-resisting agrochemicals that can build up in soil, water, and air, as well as the negative consequences of secondary pest outbreaks, make bio-pesticide manufacture imperative. Bio-fertilizers are naturally dynamic additives that can boost plant development and yield by increasing soil nutrient availability in the rhizosphere. The most often utilized bio-fertilizers include bacteria, blue-green algae, and arbuscular mycorrhiza fungi. In the case of microbial bio-pesticides, the most often utilized bio-fertilizers include bacteria, algae, and fungi. *Bacillus thuringiensis* (Bt) is the most often employed microbial bio-pesticide strain, accounting for about 90% of all bio-pesticides. Fertilizers and insecticides made from chemicals are currently being phased out in favour of bio-fertilizers and bio-pesticides, which also are popular because they are simple to use, environmentally friendly, economically effective, and nontoxic (Al-Zaidi et al. 2011; Forlani et al. 2014; Adnan et al. 2016). The public's growing knowledge of health and ecology, as well as the dangers of uncritical use of commercial fertilizers and fungicides, is driving demand for bio-pesticides. Bio-alternatives, on the other hand, account for only 1–2% of the whole crop protection industry, and they, too, rely on Bt. Bio-products that are widely used, can decrease the negative effects of phyto-pathogens, and cover a wider range of target crops are being developed by researchers all over the world (Mazid et al. 2011a; Arriola et al. 2015).

The world's population is growing, which means that agriculture must improve in order to boost food productivity and sustainability (Al-Zahrani et al. 2022; Rajesh et al. 2022; Anam et al. 2021; Deepranjan et al. 2021; Haider et al. 2021; Amjad et al. 2021; Sajjad et al. 2021a; Fakhre et al. 2021; Khatun et al. 2021; Ibrar et al. 2021; Bukhari et al. 2021; Haoliang et al. 2022; Sana et al.2022; Abid et al. 2021; Zaman et al. 2021; Sajjad et al. 2021b; Rehana et al. 2021; Yang et al. 2022; Ahmad et al. 2022; Shah et al. 2022; Muhammad et al. 2022; Wiqar et al. 2022; Farhat et al. 2022; Niaz et al. 2022; Ihsan et al. 2022; Chao et al. 2022, Qin et al. 2022; Xue et al. 2022; Ali et al. 2022; Mehmood et al. 2022; El Sabagh et al. 2022; Ibad et al. 2022).

The most extensively utilized fertilizers are chemical fertilizers. However, to achieve this increase, their continued and excessive usage has resulted in environmental contamination, producing significant ecological harm as well as insect resistance and health concerns, resulting in a decrease in crop output (Youssef & Eissa 2014; Deepranjan et al. 2021; Haider et al. 2021; Huang Li et al. 2021; Ikram et al. 2021; Jabborova et al. 2021; Khadim et al. 2021a,b; Manzer et al. 2021; Muzammal et al. 2021; Abdul et al. 2021a,b; Ashfaq et al. 2021; Amjad et al. 2021; Atif et al. 2021; Athar et al. 2021; Adnan et al. 2018a,b; Jan et al. 2019; Adnan et al. 2019; Akram et al. 2018a,b; Aziz et al. 2017a,b; Chang et al. 2021; Chen et al. 2021; Emre et al. 2021; Habib et al. 2017; Hafiz et al. 2016; Hafiz et al. 2019; Ghulam et al. 2021; Guofu et al. 2021; Hafeez et al. 2021; Khan et al. 2021; Kamaran et al. 2017; Muhmmad et al. 2019; Safi et al. 2021). Bio-fertilizers can be used instead of commercial fertilizers to increase crop production while being environmentally friendly. A "bio-fertilizer" is a commercial product that contains microorganisms within the plant when applied to seed It can increase the delivery or obtainability of essential nutrients to the host plant on plant surfaces or in the soil to boost growth (Bhattacharjee & Dey 2014). Bio-fertilizers use natural mechanisms to fix N_2, solubilize P, and stimulate plant development by combining growth-promoting substances and compounds (Malusa' et al., 2016). Depending on their nature and purpose, they can be categorized in a variety of ways. As an unavoidable aspect of agriculture sustainability, bio-fertilizer technology needs to meet the minimum necessities for its primary uses. End-users' social and infrastructure situations must be accommodated by bio-fertilizer technology. They must also be economically practical and profitable in terms of return on investment for all farmers, regardless of their financial situation and position; be ecologically pleasant, unchanging, and well-organized; and be adaptive to existing circumstances in the area and acceptable by many social sectors while meeting personal demands. Furthermore, bio-fertilizers must be practical to apply within a specific governmental system, conform to varied societal cultural patterns, be simple to use and repurpose without requiring large extra inputs, and be productive in the long term (Mushtaq et al. 2014; Suhag 2016). This chapter discusses the present state of bio-fertilizer research and implementation, as well as the global scenario.

14.2 USING BIO-BASED PRODUCTS

Despite the improved food grain production brought about by the ecological imbalance caused by the green revolution and the usage of chemical fertilizers, natural resource depletion poses a severe threat to humanity (Sajjad et al. 2019; Saud et al. 2013; Saud et al. 2014; Saud et al. 2017; Saud et al. 2016; Shah et al. 2013; Saud et al. 2020; Saud et al. 2022a,b; Qamar et al. 2017; Hamza et al. 2021; Irfan et al. 2021; Wajid et al. 2017; Yang et al. 2017; Zahida et al. 2017; Depeng et al. 2018; Hussain et al. 2020; Hafiz et al. 2020 a,b; Shafi et al. 2020; Wahid et al. 2020; Subhan et al. 2020; Zafar-ul-Hye et al. 2020a,b; Zafar et al. 2021). Chemical fertilizers are becoming increasingly expensive as they are mostly made of petroleum products and feedstock that deteriorate. The energy required for preparation of 1 kg of commercial fertilizers needs more energy and resources, such as 11.2 kWh for nitrogen (N), 1.1 kWh for phosphorus (P), and 1 kWh for potash, respectively (Adnan et al. 2020; Ilyas et al. 2020; Saleem et al. 2020a,b,c; Rehman 2020; Farhat et al. 2020; Wu et al. 2020; Mubeen et al. 2020; Farhana 2020; Jan et al. 2019; Wu et al. 2019; Ahmad et al. 2019; Baseer et al. 2019; Hafiz et al. 2018; Tariq et al. 2018). Apart from climate change it also affects the agriculture sector significantly (Fahad and Bano 2012; Fahad et al. 2017; Fahad et al. 2013; Fahad et al. 2014a,b; Fahad et al. 2016a,b,c,d; Fahad et al. 2015a,b; Fahad et al. 2018a,b; Fahad et al. 2019a,b; Fahad et al. 2020; Fahad et al. 2021a,b,c,d,e,f; Fahad et al. 2022a,b; Hesham and Fahad 2020; Iqra et al. 2020; Akbar et al. 2020; Mahar et al. 2020; Noor et al. 2020; Bayram et al. 2020; Amanullah, Fahad 2018a,b; Amanullah, Fahad 2017; Amanullah et al. 2020; Amanullah et al. 2021; Rashid et al. 2020; Arif et al. 2020; Amir et al. 2020; Saman et al. 2020; Muhammad Tahir et al. 2020; Md Jakirand Allah 2020; Mahmood et al. 2021; Farah et al. 2020; Sadam et al. 2020). Therefore bio-based fertilizers are becoming increasingly needed by growers to minimize the constraints. These bio-based fertilizers can also be used in conjunction with agrochemicals to help to create nutrients in the soil medium and combat pests that cause crop damage and poor yields. As a result, research and development into bio-fertilizers and bio-pesticides can only be justified if farmers are willing to use them and eventually produce them. Biological N fixation is used by N-fixing bacteria in mutual relationships with plants to convert elemental N to NO_2^- that are conveniently accessible to them (Malusá and Vassilev 2014).

14.3 N-FIXING INOCULANTS AS BIO-FERTILIZERS

Using bio-fertilizers to fix nitrogen (N) using plant growth-promoting Rhizobacteria (PGPR) has an advantage over bacteria that live in the soil in terms of minimizing plant pathogen impacts and improving crop rotation by increasing N, P, and iron availability (Fe). Nitrogen is required for photosynthesis, as an element of protein and chlorophyll, which affects crop output, as well as vegetative and reproductive growth (Marek-Kozaczuk et al., 2014; Sana at al. 2015; Unsar et al. 2020; Fazli et al. 2020; Enamul et al. 2020; Gopakumar et al. 2020; Zia-ur-Rehman 2020; Ayman et al. 2020; Mohammad I. Al-Wabel et al. 2020a,b; Senol 2020; Amjad et al. 2020; Ibrar et al. 2020; Sajid et al. 2020; Muhammad et al. 2021; Sidra et al. 2021; Zahir et al. 2021; Sahrish et al. 2022). *Azotobacter*, a nitrogen-fixing bacteria, is the best as a bio-fertilizer for cereal crops, assisting in the production of PGPR, enzymes, and phytohormones such as auxins, gibberellins, and cytokinins. Inoculating maize seeds with *Azospirillum brasilense* increased root and shoot length as well as the plant's dry weight (Forlain et al. 1998; Braccini et al. 2012; Puente et al. 2009). Farmers generally have only marginal interest in the use of bio-fertilizers because of the high price compared with low-value commodity crops.

14.3.1 Bio-fertilizers in Liquid Form

These are unique liquid formulations containing the appropriate microorganisms and nutrients, to increase the shelf life of resting spores/cysts, as well as cell protectants that withstand harsh environments (Sumita et al. 2015). They can also overcome the limits of traditional (solid) bio-fertilizers, such as their resilience to increased temperature and UV radiation (Hegde 2008).

They have many enzyme activities that farmers prefer, and they are easy to apply as a powder in foliar applications, hand sprayers, and as a base manure combined with poultry and farmyard manures. The efficiency of bacteria other than *Rhizobium* is unknown, hence microbial inoculants are of great interest.

14.3.2 Bioengineered Microbes

These are bacteria that have been found to have specific functional genes using rDNA technology. The introduction of nif genes in *Rhizobium meliloti* protects crops from disease and degrades bio-based products, further improving nitrogen fixation.

14.3.3 Potassium-Solubilizing Bacteria (PSB)

Potassium is one of the most important macronutrients, in addition to nitrogen and phosphorus, for a plant's biological processes to flourish. As a result, rhizotrophic bacteria's potassium-solubilizing activity is crucial. In diverse crops, the genera *Aspergillus*, *Bacillus*, and *Clostridium* are effective in solubilizing K mobility in the soil (Mohammadi and Sohrabi 2012).

14.3.4 Phosphate-Solubilizing Bacteria (Ph.SB)

The major bacteria that solubilize phosphate (Ph.SB) present in soil are *Bacillus*, *Micrococcus*, *Aspergillus*, *Fusarium*, *Pseudomonas*, and others. Organic acids such as citric, oxalic, succinic, tartaric, and malic acids are secreted in the soil rhizosphere to convert non-available inorganic phosphate to soluble form, hence boosting phosphorus availability. They can increase agricultural yields by 200–500 kg per hectare (Ambrosini et al. 2012).

14.3.5 Mycorrhiza

Mycorrhiza can create symbiotic relationships with host plants for carbohydrates in exchange for zinc (Zn) and phosphorus (P) and is known as a disease carrier of crops. Mycorrhiza use their mycelia to create a connection with fertile soil to absorb water and macronutrients like N, P, K, and Ca, as well as producing growth stimulants like cytokinins and secreting antimicrobial compounds to protect plants from invading pathogens.

14.3.6 Blue-Green Algae (BGA)

These are free-living photosynthetic N-fixing organisms that promote soil porosity by contributing growth-promoting materials such as vitamin B12. They have heterocysts, which behave as micronodules and aid in the fixation of nitrogen. They live in symbiotic relationships with fungi, ferns, flowering plants, and liverworts, but especially with *Azolla* for nitrogen fixation.

14.3.7 *Azolla*

This is a free-floating symbiotic fern from the Azollaceae family that can be utilized as a dual crop or green manuring. When *Azolla* inocula are soaked in super-phosphate solution and inoculated, paddy yield is increased, and it has been found to fix nitrogen at 45–50 kg/ha. The most popular species utilized as a bio-fertilizer in paddy crops is *Azolla pinnata*, which decomposes swiftly in soil and provides the most N availability to rice and other crops (Ghosh 2004).

14.3.8 *Acetobacter*

This is ideal for sugarcane cultivation as it can withstand high sugar levels. It may fix up to 15 kg of nitrogen per hectare per year.

14.3.9 *Azospirillum*

This is a nitrogen-fixing bacteria known as *B. polymyxa*. It is a member of the Spirillaceae family, and is good for non-leguminous plants. This genus has a variety of species. *Azospirillum amazonense*, *Azospirillum halopraeferens*, and *Azospirillum brasilense* are all *Azospirillum* species having a 20–40 kg/ha N-fixing capacity. These make a large difference in terms of leaf area and grain yield of millet, sugarcane, sorghum, and maize (C4 plants), as well as other crops. They use organic acid salts such as malic and aspartic acid to fix nitrogen (Brusamarello-Santos et al. 2017).

14.3.10 *Herbspirillum*

This is an N-fixing symbiont that lives in sugarcane roots and improves hormone production as well as N, K, and P uptake (Khan et al. 2011a; Rasheed et al. 2015). There are bio-protective benefits of inoculating drought-stricken corn with two types of bacteria of PGPR, *Azospirillum brasilense* strain SP-7 and *Erbaspirillum* sp. These bacteria were discovered to be useful in mitigating drought stress that has an adverse effect on plants (Curá et al. 2017).

14.3.11 *Azotobacter*

The Azotobacteriaceae family includes this non-symbiotic, aerobic, and free-living bacterium, although the root nodules are not evident. It sets N at 40–200 kg/ha and can meet the crop's N requirements to the tune of 80–90%. Some of the organisms found in the rhizosphere of crops

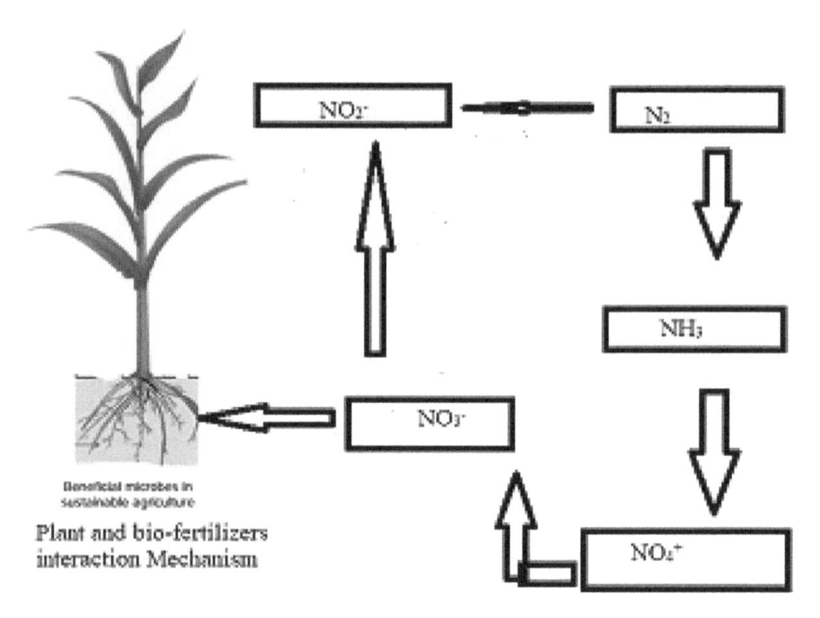

FIGURE 14.1 Mechanism of plant and bio-fertilizer interaction.

such as rice, maize, sugarcane, bajra, and vegetables include *Azotobacter vinelandii*, *Acinetobacter beijerinckii*, *Agrionoptera insignis*, and *Azomonas macrocytogenes*, and plantations, where yields can rise by up to 50%. These species aid in the inhibition of some root diseases through seed germination and plant growth (Mazid et al. 2011b).

14.3.12 *Rhizobium*

This N-fixing symbiotic bacterium is a member of the Rhizobiaceae family and has the ability to fix nitrogen at a rate of 50–100 kg/ha in pulse crops, making it a must-have for all legumes except mung beans. *Parasponia* is the only lineage outside the legume family to be able to form a nodule symbiosis with *Rhizobium*. *Sesbania rostrate* has stem nodules as well as root nodules (Saikia and Jain, 2007).

14.4 PLANT AND MICROORGANISM INTERACTION MECHANISM

Plants must defend themselves against infections and devise techniques to deal with them. In this regard they are aided by beneficial microorganisms. Understanding the connection between plant and microorganism is becoming increasingly important as the world's population grows (Figure 14.1). Changes in the ecology and physical environment have a significant impact on human health. Plant–microbe interactions can be both beneficial and harmful. Plants have a sophisticated immune system to protect themselves from viruses and pests, as well as to avoid crop losses in agriculture. They also increase the number of helpful microbes in their root system. Immune responses also form a rationale for building long-term disease resistance measures in next-generation crops.

14.5 APPLICATIONS AND TYPES OF BIO-FERTILIZERS

Under optimal conditions, bio-fertilizers can be used to inoculate the seeds and roots of a variety of agricultural crops and they can also be sprayed directly onto the soil. Nitrogenous and phosphatic fertilizers are combined in water during seed preparation. After that, the seeds are immersed in the mixture (Abbasniayzare et al. 2012; Rai 2006). The seeds are dried after this paste has been applied to them. They must be sowed as soon as possible after drying off to avoid being destroyed by hazardous germs. On the soil where the crop is to be grown, a bed of water is placed during which the roots of seedlings are dipped in the chemical solution for treatment. The seedlings are submerged in bio-fertilizer-treated water for 8–10 hours to allow the roots to absorb the inoculum. After that, the seedlings are transferred. Tomatoes, rice, onions, and flowers have all been reported to benefit from this procedure. The most common therapy is seed treatment, which accounts for 66% of the global market. All bio-fertilizers, as well as compost fertilizers, are combined together in the soil treatment. They are held for one night only. The mixture is then placed on the soil where seeds must be sowed the next day. This method is most commonly employed on fruit trees, sugarcane, and other crops that require localized application (Bhattacharjee & Dey 2014; Qadeem et al. 2015). There are currently a number of bio-fertilizers available, each with its own set of functions and crop types. To augment chemical fertilizers, bio-fertilizers are a low-cost, effective, and sustainable source of plant nutrients. Bacteria, fungi, and blue-green algae are examples of microorganisms that can be employed as bio-fertilizers (Singh et al. 2011; Bhattacharjee & Dey 2014).

The most popular type of bio-fertilizer is bacterial bio-fertilizer (Suhag 2016). These are bacteria that aid in the fixation of several nutrients required for plant growth in soil. They fix nitrogen, solubilize phosphorus, and create other growth-promoting compounds to increase plant growth. *Azotobacter*, *Azospirillum*, and *Rhizobium* are examples of popular bacterial bio-fertilizers (Bhardwaj et al. 2014). Nonlegume crops use *Azotobacter* and *Azospirillum*, while legume crops need *Rhizobium*. *Acetobacter* is more sugarcane-specific under various agro-climatic circumstances, and field experiments on *Azotobacter* have revealed that it is acceptable when treated with agricultural crops under normal field circumstances (Gupta et al. 2015). *Azotobacter* inoculation reduces the need for nitrogenous fertilizers by 12% to 22% (El-Fattah et al. 2013). The inoculation of *Azospirillum* improves plant vegetative growth while reducing nitrogen fertilizer use by 25–30%. *Azospirillum lipoferum*, *A. brasilense*, *A. amazonense*, and A. *iraquense* are the only species known so far (Saikia et al. 2013). The production of sugarcane only be increased by the use of *Acetobacter* (Raja 2013). Auxins and antibiotic-like compounds have also been detected following its application (Berg et al. 2013).

Bio-fertilizers such as phosphor-bacteria are a form of bio-fertilizer. Phosphorus is an important nutrient for plants, as it promotes strong development and helps plants resist disease. Phosphorus aids in the production of roots and the growth of plants. When phosphatic fertilizers are applied to soil, the plants absorb only 15–20% and the remaining 80–85 % make complexes with other nutrients and remains in the soil for long a time. *Bacillus megaterium* (PSB) in the bio-promoters develops and releases organic acids, which dissolve the unavailable phosphate into soluble form and make it available to plants. As a result, the soil's remaining phosphate fertilizers can be fully utilized, and external application can be optimized. All crops, including paddy, millets, oilseeds, pulses, and vegetables, can benefit from PSB (Park et al. 2010).

Fungi are non-green microbes that help in the aggregation of soil structure and availability of phosphatic fertilizers to plants. Fungal bio-fertilizers must create a symbiotic interaction with plant roots in order to supply the promised nutrients. Mycorrhiza is a type of association in which the fungal bio-fertilizer successfully allows the nutrients to be absorbed and released, particularly phosphatic nutrients (Smith et al. 2011). Mycorrhiza partnerships have about eight different forms, but arbuscular mycorrhiza is one of the most important in agriculture. Arbuscular mycorrhiza between the roots of vascular plants and fungi is the most prevalent type of synergetic association (Akyol et al. 2019). The accumulation of pathogens, nematodes, and heavy metals in the root zone of plants is prevented. The soil condition is improved, making it well aerated and allowing for easy nutrient transport. Plants develop quickly due to appropriate phosphorus availability and the generation of phytohormones like cytokinin, because of which plant physiology has a progressive effect (Smith et al. 2011).

The following are some crucial guidelines for getting the best reaction from bio-fertilizer application (Simarmata et al., 2016):

- Bio-fertilizers must be devoid of contaminating microorganisms and possess a strong population of suitable efficient strains.
- Choose the proper mix of bio-fertilizers and utilize them before the expiration date.
- Use the recommended application methods and apply at the proper time as directed on the label.
- A good adhesive should be used for seed treatment to obtain the best outcomes.
- Use corrective treatments for difficult soils, such as seed pelleting with lime or gypsum, or soil pH adjustment with lime.
- Ensure that phosphorus and other nutrients are available.

14.6 MARKET FOR BIO-FERTILIZERS

There are certain drivers and constraints in the global bio-fertilizer industry. In these certain drivers three are some main drivers; the first one is the growing popularity of organic agriculture. The rise in demand for organic products is attributable to an increase in consumer awareness as well as rising earnings, allowing individuals to consume more food. The increased demand for bio-fertilizers has been fueled by an increase in the eating of organic vegetables and fruits, as well as a rise in consumer health consciousness, particularly in advanced countries such as those in Europe and North America. Furthermore, the bio-fertilizer market has benefited from the growing land area of organic farming, easily availability of the product, and economic and other improvements (Malusa' et al. 2016). Furthermore, growing worries about synthetic agrochemicals' adverse environmental consequences, such as food chain contamination and pollution of soil, have prompted the use of bio-products, such as bio-fertilizers. Chemical fertilizer use is limited and restricted by the common agricultural policy, which promotes the use of bio-fertilizers (Sheraz et al. 2010). As a result, the second driving force is the requirement to improve soil organic matter. The final market driver is a favourable regulatory framework that permits the bio-fertilizer market to be regulated (Malusa' & Vassilev 2014).

There are also some limitations to this market. The first is a lack of product efficacy in adverse situations. The second factor is the current strong demand for synthetic fertilizers due to their lower costs and lack of innovation compared to bio-fertilizers, which has an impact on the product's price. Chemical fertilizers also produce results quickly after application. As a result, farmers prefer chemical fertilizers over bio-fertilizers in order to obtain quick results. The fourth constraint is microbes' short shelf life, which necessitates extra caution during storing (Bodake et al. 2009).

The bio-fertilizers market is divided into microorganisms such as *Rhizobium*, *Azotobacter*, *Azospirillum*, blue-green algae, phosphate-solubilizing bacteria, mycorrhiza, and other microorganisms, as well as technology types such as carrier-enriched bio-fertilizers, liquid bio-fertilizers, and others, application types such as seed treatment and soil treatment, and crop types such as cereals and legume (Mordor Intelligence LLP 2019). The present worldwide bio-fertilizer market is depicted in Figure 14.2 based on the microorganisms utilized as bio-fertilizers. As can be observed, the most widely used agricultural bio-fertilizer is *Rhizobium*-based agricultural bio-fertilizer, which was valued at $256.8 million in 2018. Carrier-enriched bio-fertilizers are the most widely used technology, with a market value of $797.5 million in 2018, followed by liquid bio-fertilizers. The main issue with carrier-enriched bio-fertilizers is that they are not UV-resistant and cannot withstand temperatures beyond 30–32°C; as a result, the microbial population density of these bio-fertilizers falls every day as they are applied, and their marketability is likely to reduce in the near future (Mordor Intelligence LLP 2019).

When looking at the global market for bio-fertilizers by topography, especially by continents, it can be observed that North America is in first place, closely followed by Europe in second place, and

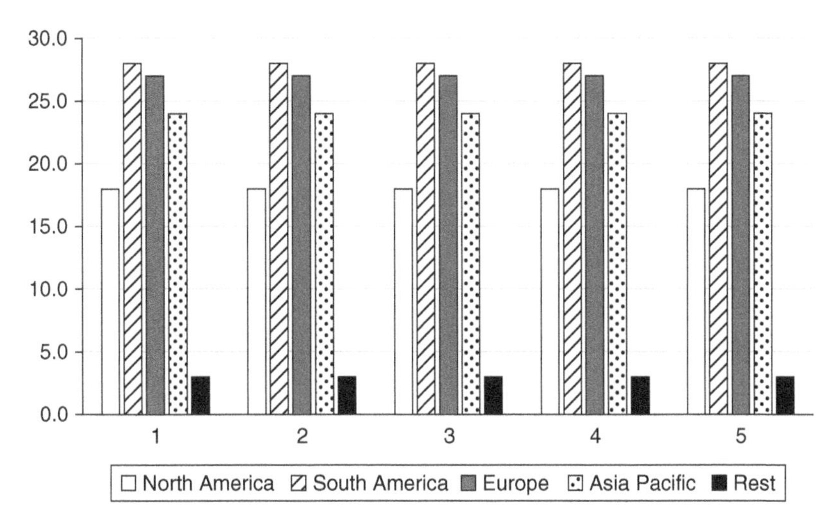

FIGURE 14.2 Global bio-fertilizer market.

Asia-Pacific in third place. South America has moved up to fourth place, and the rest of the world has now been grouped together. For the predicted period of 2019–2024, it is expected that the global economy will grow steadily. Due to the rising cost of commercial fertilizers and their negative effects on agricultural fields and the environment, as well as biotechnology advancements in the field of fermentation technology, North America is the largest market, followed by Canada. However, in the North American bio-fertilizer taxonomy, Canada and Mexico remain new markets. As a result, bio-fertilizers' numerous benefits contribute to their widespread application, as well as greater adoption and use in sustainable agriculture. The country's good agricultural outlook has boosted demand for bio-based products in the region. Because of the increased usage of bio-based crop nutrition products and rising food quality requirements, the North American region is anticipated to maintain its market in the future (Mordor Intelligence LLP 2019) (Figure 14.2).

Bayer Crop Science AG, one of the world's most inventive agricultural corporations, purchased the gigantic Monsanto Company in 2018, making it the largest German acquisition to date. Customers can choose from a wide choice of products, and the company offers them complete services for modern and viable agriculture. Among the company's most notable products are Optimize (*Rhizobium* inoculant), Nitragin Gold (*Rhizobium*), Cell-Tech (*Rhizobium* inoculant), JumpStart (*Penicillium bilaii*), and Biodoz (*Rhizobium* inoculant). Similarly, the aforementioned businesses strive to provide simple but innovative items that benefit Mother Nature, plants, humans, and animals. Because these companies are concerned about environmental issues and the need for green technology, they focus on natural products of microbial or plant origin for the majority of their applications (Mordor Intelligence LLP 2019).

14.7 LIMITS OF BIO-FERTILIZER PROMOTION (TABLE 14.1)

14.7.1 Limited Resources

- Inadequate resource constraints/unavailability of appropriate strains
- Manufacturers' obliviousness to product quality
- Lack of quality resources and quality control procedures
- Infrastructural obstacles, such as the lack of room for a laboratory, production, and storage
- •A shortage of inoculum cold storage equipment

14.7.2 Market Restrictions

- Farmers' lack of awareness/quality assurance, as well as marketing constraints
- Bank-related financial limitations
- Smaller production units suffer as a result of lower sales returns compared to loan payments

14.7.3 Production Limitations

- A lack of quality assurance research and development, resulting in limited output
- Inability to meet seasonal demand/requirements due to a skilled workforce shortage

14.7.4 Constraints in the Field

- Unfavourable environmental conditions such as soil and climate factors, microbiological factors, and population density
- There are no quick visible consequences, unlike with inorganic fertilizers
- There is a lack of knowledge in agricultural activities, as well as a short period of sowing/planting in a given area

14.7.5 Technical Limitations

- Mutations, mishandled methods, and unauthenticated inoculum/microbes
- A lack of understanding of the advantages of technology
- Issues with technology uptake due to differing inoculation methods

TABLE 14.1
Bio-Fertilizer Advantages and Disadvantages

Benefits	Limitations
Bio-fertilizers can help soils establish biological activities by mobilizing nutrients	Nutrient contents in compost products are quite varied
The addition of appropriate nutrients to the soil improves plant health	Extensive and long-term use could lead to an addition of salts, minerals, and toxic elements, which could harm plant growth, soil organism development, quality of water, and human health
Nutrient availability, and beneficial soil worms and microbes are encouraged to develop	In comparison to chemical fertilizers, large amounts are required for application to the desire land to overcome nutritional content deficits
Root growth is improved as it increases the porosity and fertility of the soil	Plant growth and development may be hampered by a lack of essential macronutrients
The organic matter level of the soil increases to more than the normal level	Low translocation of micro- and macronutrients could result in nutritional deficits
Promotes the formation of mycorrhizal connections, which enhances phosphorus (P) availability in the soil	The expense of implementation is higher than those of some chemical fertilizers
Assists in the elimination of plantar illnesses and the provision of a steady supply of essential nutrients to the soil	For the manufacturing of bio-fertilizers, there is a lack of quality assurance and restricted resource generation

14.8 CONCLUSION

Bio-fertilizers are naturally dynamic foodstuffs that can boost plant development and yield by increasing soil nutrient availability in the rhizosphere. With the intensive use of chemical fertilizers by farmers, nutrient accumulation in soil increases, and ultimately destroys the soil fertility and quality. As a result, developing efficient and long-lasting bio-fertilizers for crop plants, where inorganic fertilizer use can be decreased greatly to avoid further pollution issues, is a major research priority. Bio-fertilizers are live, high-cell-density microbial preparations, and the necessary microorganisms must be carefully monitored throughout the manufacturing process. This makes sense, given that the quality of inoculants in a bio-fertilizer is one of the most crucial aspects determining its success or failure, as well as farmer acceptance or rejection. The presence of the proper type of microbe in an active form and in the desired quantities is a measure of its quality. Given these challenges, it is clear that thorough and extensive education of professionals, dealers, and farmers regarding the importance of bio-fertilizer technology and the economic feasibility of application is required. As a result, in order to better understand and apply bio-fertilizer technology, substantial information, practical training, adoption, and perception are required.

ACKNOWLEDGMENT

We would like to thank to staff of the Department of Agriculture, University of Swabi, KP, Pakistan.

REFERENCES

Abbasniayzare SK, Sedaghathoor S, Dahkaei MNP (2012) Effect of biofertilizer application on growth parameters of spathiphyllum illusion. *Am Eurasian J Agric Environ Sci* 12(5):669–673

Abdul S, Muhammad AA, Shabir H, Hesham A El E, Sajjad H, Niaz A, Abdul G, RZ Sayyed, Fahad S, Subhan D, Rahul D (2021a) Zinc nutrition and arbuscular mycorrhizal symbiosis effects on maize (Zea mays L.) growth and productivity. *J Saudi Soc Agri Sci* 28(11), 6339–6351. https://doi.org/10.1016/j.sjbs.2021.06.096

Abdul S, Muhammad AA, Subhan D, Niaz A, Fahad S, Rahul D, Mohammad JA, Omaima N, Muhammad Habib ur R, Bernard RG (2021b) Effect of arbuscular mycorrhizal fungi on the physiological functioning of maize under zinc-deficient soils. *Sci Rep* 11:18468

Abid M, Khalid N, Qasim A, Saud A, Manzer HS, Chao W, Depeng W, Shah S, Jan B, Subhan D, Rahul D, Hafiz MH, Wajid N, Muhammad M, Farooq S, Fahad S (2021) Exploring the potential of moringa leaf extract as bio stimulant for improving yield and quality of black cumin oil. *Sci Rep* 11:24217. https://doi.org/10.1038/s41598-021-03617-w

Adnan M, Fahad S, Muhammad Z, Shahen S, Ishaq AM, Subhan D, Zafar-ul-Hye M, Martin LB, Raja MMN, Beena S, Saud S, Imran A, Zhen Y, Martin B, Jiri H, Rahul D (2020) Coupling phosphate-solubilizing bacteria with phosphorus supplements improve maize phosphorus acquisition and growth under lime induced salinity stress. *Plants* 9(7):900. doi: 10.3390/plants9070900

Adnan M, Fahad S, Khan IA, Saeed M, Ihsan MZ, Saud S, Riaz M, Wang D, Wu C. (2019). Integration of poultry manure and phosphate solubilizing bacteria improved availability of Ca bound P in calcareous soils. *3 Biotech* 9(10):368

Adnan M, Shah Z, Sharif M, Rahman H. (2018b). Liming induces carbon dioxide (CO_2) emission in PSB inoculated alkaline soil supplemented with different phosphorus sources. *Environ Sci Poll Res* 25(10):9501–9509.

Adnan M, Zahir S, Fahad S, Arif M, Mukhtar A, Imtiaz AK, Ishaq AM, Abdul B, Hidayat U, Muhammad A, Inayat-Ur R, Saud S, Muhammad ZI, Yousaf J, Amanullah, Hafiz MH, Wajid N (2018a) Phosphate-solubilizing bacteria nullify the antagonistic effect of soil calcification on bioavailability of phosphorus in alkaline soils. *Sci Rep* 8:4339. https://doi.org/10.1038/s41598-018-22653-7

Adnan M, Shah Z, Arif M, Khan MJ, Mian IA, Sharif M, Alam M, Basir A, Ullah H, Rahman I, Saleem N (2016). Impact of rhizobial inoculum and inorganic fertilizers on nutrients (NPK) availability and uptake in wheat crop. *Can J Soil Sci* 96:169–176.

Ahmad N, Hussain S, Ali MA, Minhas A, Waheed W, Danish S, Fahad S, Ghafoor U, Baig, KS, Sultan H, Muhammad IH, Mohammad JA, Theodore DM (2022) Correlation of soil characteristics and citrus leaf nutrients contents in current scenario of Layyah District. *Horticulturae* 8: 61. https://doi.org/10.3390/horticulturae8010061

Ahmad S, Kamran M, Ding R, Meng X, Wang H, Ahmad I, Fahad S, Han Q (2019) Exogenous melatonin confers drought stress by promoting plant growth, photosynthetic capacity and antioxidant defense system of maize seedlings. *Peer J* 7:e7793. http://doi.org/10.7717/peerj.7793

Akbar H, Timothy JK, Jagadish T, M. Golam M, Apurbo KC, Muhammad F, Rajan B, Fahad S, Hasanuzzaman M (2020) Agricultural land degradation: Processes and problems undermining future food security. In: Fahad S, Hasanuzzaman M, Alam M, Ullah H, Saeed M, Khan AK, Adnan M (Eds.) *Environment, Climate, Plant and Vegetation Growth.* Springer Nature Switzerland, pp. 17–62. https://doi.org/10.1007/978-3-030-49732-3

Akram R, Turan V, Hammad HM, Ahmad S, Hussain S, Hasnain A, Maqbool MM, Rehmani MIA, Rasool A, Masood N, Mahmood F, Mubeen M, Sultana SR, Fahad S, Amanet K, Saleem M, Abbas Y, Akhtar HM, Waseem F, Murtaza R, Amin A, Zahoor SA, ul Din MS, Nasim W (2018a) Fate of organic and inorganic pollutants in paddy soils. In: Hashmi MZ, Varma A (Eds.) *Environmental Pollution of Paddy Soils, Soil Biology.* Springer International Publishing, Switzerland , pp. 197–214

Akram R, Turan V, Wahid A, Ijaz M, Shahid MA, Kaleem S, Hafeez A, Maqbool MM, Chaudhary HJ, Munis, MFH, Mubeen M, Sadiq N, Murtaza R, Kazmi DH, Ali S, Khan N, Sultana SR, Fahad S, Amin A, Nasim W (2018b) Paddy land pollutants and their role in climate change. In: Hashmi MZ, Varma A (Eds.) *Environmental Pollution of Paddy Soils, Soil Biology.* Springer International Publishing, Switzerland, pp. 113–124.

Akyol TY, Niwa R, Hirakawa H, Maruyama H, Sato T, Suzuki T, Fukunaga A, Sato T, Yoshida S, Tawaraya K, Saito M, Ezawa T, Sato S (2019) Impact of introduction of arbuscular mycorrhizal fungi on the root microbial community in agricultural fields. *Microbes Environ* 34(1):23–32

Ali S, Hameed G, Muhammad A, Depeng W, Fahad S (2022) Comparative genetic evaluation of maize inbred lines at seedling and maturity stages under drought stress. *J Plant Growth Regul* 42, 989–1005. https://doi.org/10.1007/s00344-022-10608-2

Al-Zahrani HS, Alharby HF, Fahad S (2022) Antioxidative defense system, hormones, and metabolite accumulation in different plant parts of two contrasting rice cultivars as influenced by plant growth regulators under heat stress. *Front Plant Sci* 13:911846. doi: 10.3389/fpls.2022.911846

Al-Zaidi AA, Elhag EA, Al-Otaibi SH, Baig MB (2011) Negative effects of pesticides on the environment and the farmer's awareness in Saudi Arabia: a case study. *J Anim Plant Sci* 21:605–611.

Amanullah, Muhammad I, Haider N, Shah K, Manzoor A, Asim M, Saif U, Izhar A, Fahad S,, Adnan M et al. (2021) Integrated foliar nutrients application improve wheat (Triticum aestivum L.) productivity under calcareous soils in drylands. *Commun Soil Sci Plant Anal* 52(21):2748–2766. https://Doi.Org/10.1080/00103624.2021.1956521

Amanullah, Fahad S (Eds.) (2017) Rice – Technology and Production. IntechOpen Croatia 2017 http://dx.doi.org/10.5772/64480

Amanullah, Fahad S (Eds.) (2018a) Corn – Production and Human Health in Changing Climate. *IntechOpen United Kingdom* 2018. http://dx.doi.org/10.5772/intechopen.74074

Amanullah, Fahad S (Eds.) (2018b) Nitrogen in Agriculture – Updates. *IntechOpen Croatia* 2018. http://dx.doi.org/10.5772/65846

Amanullah, Shah K, Imran, Hamdan AK, Muhammad A, Abdel RA, Muhammad A, Fahad S, Azizullah S, Brajendra P (2020) Effects of climate change on irrigation water quality. In: Fahad S, Hasanuzzaman M, Alam M, Ullah H, Saeed M, Khan AK, Adnan M (Eds.), *Environment, Climate, Plant and Vegetation Growth.* Springer Nature Switzerland, pp. 123–132. https://doi.org/10.1007/978-3-030-49732-3

Ambrosini A, Beneduzi A, Stefanski T, Pinheiro FG, Vargas LK, Passaglia LM (2012) Screening of plant growth promoting rhizobacteria isolated from sunflower (Helianthus annuus L.). *Plant Soil* 356: 245–264

Amir M, Muhammad A, Allah B, Sevgi Ç, Haroon ZK, Muhammad A, Emre A (2020) Biofortification under climate change: The fight between quality and quantity. In: Fahad S, Hasanuzzaman M, Alam M, Ullah H, Saeed M, Khan AK, Adnan M (Eds.) *Environment, Climate, Plant and Vegetation Growth.* Springer Nature Switzerland, pp. 173–228. https://doi.org/10.1007/978-3-030-49732-3

Amjad I, Muhammad H, Farooq S, Anwar H (2020) Role of plant bioactives in sustainable agriculture. In: Fahad S, Hasanuzzaman M, Alam M, Ullah H, Saeed M, Khan AK, Adnan M (Eds.) *Environment,*

Climate, Plant and Vegetation Growth. Springer Nature Switzerland, pp. 591–606. https://doi.org/10.1007/978-3-030-49732-3

Amjad SF, Mansoora N, Din IU, Khalid IR, Jatoi GH, Murtaza G, Yaseen S, Naz M, Danish S, Fahad S et al. (2021) Application of zinc fertilizer and mycorrhizal inoculation on physio-biochemical parameters of wheat grown under water-stressed environment. *Sustainability* 13: 11007. https://doi.org/10.3390/su131911007

Anam I, Huma G, Ali H, Muhammad K, Muhammad R, Aasma P, Muhammad SC, Noman W, Sana F, Sobia A, Fahad S (2021) Ameliorative mechanisms of turmeric-extracted curcumin on arsenic (As)-induced biochemical alterations, oxidative damage, and impaired organ functions in rats. *Environ Sci Pollut Res* 28(46): 66313–66326 https://doi.org/10.1007/s11356-021-15695-4

Arif M, Talha J, Muhammad R, Fahad S, Muhammad A, Amanullah, Kawsar A, Ishaq AM, Bushra K, Fahd R (2020) Biochar; a remedy for climate change. In: Fahad S, Hasanuzzaman M, Alam M, Ullah H, Saeed M, Khan AK, Adnan M (Eds.) *Environment, Climate, Plant and Vegetation Growth.* Springer Nature Switzerland, pp. 151–172. https://doi.org/10.1007/978-3-030-49732-3

Arriola KG, Queiroz OC, Romero JJ, Casper D, Muniz E, Hamie J, Adesogan AT (2015) Effect of microbial inoculants on the quality and aerobic stability of bermudagrass round-bale haylage. *J Dairy Sci* 98: 478–485

Ashfaq AR, Uzma Y, Niaz A, Muhammad AA, Fahad S, Haider S, Tayebeh Z, Subhan D, Süleyman T, Hesham AElE, Pramila T, Jamal MA, Sulaiman AA, Rahul D (2021) Toxicity of cadmium and nickel in the context of applied activated carbon biochar for improvement in soil fertility. *Saudi Society Agricultural Sci* 29(2):743–750. https://doi.org/10.1016/j.sjbs.2021.09.035

Athar M, Masood IA, Sana S, Ahmed M, Xiukang W, Sajid F, Sher AK, Habib A, Faran M, Zafar H, Farhana G, Fahad S (2021) Bio-diesel production of sunflower through sulphur management in a semi-arid subtropical environment. *Environ Sci Pollution Res* 29(9):13268–13278. https://doi.org/10.1007/s11356-021-16688-z

Atif B, Hesham A, Fahad S (2021) Biochar coupling with phosphorus fertilization modifies antioxidant activity, osmolyte accumulation and reactive oxygen species synthesis in the leaves and xylem sap of rice cultivars under high-temperature stress. *Physiol Mol Biol Plants* 27(9):2083–2100. https://doi.org/10.1007/s12298-021-01062-7

Ayman EL Sabagh, Akbar Hossain, Celaleddin Barutçular, Muhammad Aamir Iqbal, M Sohidul Islam, Shah Fahad, Oksana Sytar, Fatih Çig, Ram Swaroop Meena, and Murat Erman (2020) Consequences of salinity stress on the quality of crops and its mitigation strategies for sustainable crop production: An outlook of arid and semi-arid regions. In: Fahad S, Hasanuzzaman M, Alam M, Ullah H, Saeed M, Khan AK, Adnan M (Eds.) *Environment, Climate, Plant and Vegetation Growth.* Springer Nature Switzerland, pp. 503–534. https://doi.org/10.1007/978-3-030-49732-3

Aziz K, Daniel KYT, Fazal M, Muhammad ZA, Farooq S, Fan W, Fahad S, Ruiyang Z (2017a) Nitrogen nutrition in cotton and control strategies for greenhouse gas emissions: a review. *Environ Sci Pollut Res* 24:23471–23487. https://doi.org/10.1007/s11356-017-0131-y

Aziz K, Daniel KYT, Muhammad ZA, Honghai L, Shahbaz AT, Mir A, Fahad S (2017b) Nitrogen fertility and abiotic stresses management in cotton crop: a review. *Environ Sci Pollut Res* 24:14551–14566. https://doi.org/10.1007/s11356-017-8920-x

Baseer M, Adnan M, Fazal M, Fahad S, Muhammad S, Fazli W, Muhammad A, Jr. Amanullah, Depeng W, Saud S, Muhammad N, Muhammad Z, Fazli S, Beena S, Mian AR, Ishaq AM (2019) Substituting urea by organic wastes for improving maize yield in alkaline soil. *J Plant Nutr* 42(19):2423–2434. doi.org/10.1080/01904167.2019.1659344

Bayram AY, Seher Ö, Nazlican A (2020) Climate change forecasting and modeling for the year of 2050. In: Fahad S, Hasanuzzaman M, Alam M, Ullah H, Saeed M, Khan AK, Adnan M (Eds.) *Environment, Climate, Plant and Vegetation Growth.* Springer Nature Switzerland, pp. 109–122. https://doi.org/10.1007/978-3-030-49732-3

Berg G, Zachow C, Műller H, Phillips J, Tilcher R (2013) Next-generation bio-products sowing the seeds of success for sustainable agriculture. *Agronomy* 3: 648–656

Bhardwaj D, Ansari MW, Sahoo RK, Tuteja N (2014) Bio-fertilizers function as key player in sustainable agriculture by improving soil fertility, plant tolerance and crop productivity. *Microb Cell Factories* 13(66):1–10

Bhattacharjee R, Dey U (2014) Biofertilizer, a way towards organic agriculture: a review. Afr *J Microbiol Res* 8(24):2232–2342

Bodake HD, Gaikwad SP, Shirke VS (2009) Study of constraints faced by the farmers in adoption of biofertilizers. *Int J Agric Sci* 5(1):292–294

Braccini AL, Demoraes Dan LG, Piccinin GG, Albrecht LP, Barbosa MC, Ortiz AHT (2012) Seed inoculation with *Azospirillum brasilense* associated with the use of bioregulators in maize. *Rev Caatinga* 25: 58–64

Brusamarello-Santos LC, Gilard F, Brulé L, Quilleré I, Gourion B, Ratet P, Maltempi de Souza E, Lea PJ, Hirel B (2017) Metabolic profiling of two maize (Zea mays L.) inbred lines inoculated with the nitrogen fixing plant-interacting bacteria *Herbaspirillum seropedicae* and *Azospirillum brasilense*. *PLoS One* 12: e0174576

Bukhari MA, Adnan NS, Fahad S, Javaid I, Fahim N, Abdul M, Mohammad SB (2021) Screening of wheat (*Triticum aestivum* L.) genotypes for drought tolerance using polyethylene glycol. *Arabian J Geosci* 14:2808

Chang W, Qiujuan J, Evgenios A, Haitao L, Gezi L, Jingjing Z, Fahad S, Ying J (2021) Hormetic effects of zinc on growth and antioxidant defense system of wheat plants. *Sci Total Environ* 807, 150992. https://doi.org/10.1016/j.scitotenv.2021.150992

Chao W, Youjin S, Beibei Q, Fahad S (2022) Effects of asymmetric heat on grain quality during the panicle initiation stage in contrasting rice genotypes. *J Plant Growth Regul* 42:630–636. https://doi.org/10.1007/s00344-022-10598-1

Chen Y, Guo Z, Dong L, Fu Z, Zheng Q, Zhang G, Qin L, Sun X, Shi Z, Fahad S, Xie F, Saud S. (2021) Turf performance and physiological responses of native Poa species to summer stress in Northeast China. *Peer J* 9:e12252. http://doi.org/10.7717/peerj.12252

Curá JA, Franz DR, Filosofía JE, Balestrasse KB, Burgueño LE (2017) Inoculation with Azospirillum sp. and Herbaspirillum sp. bacteria increases the tolerance of maize to drought stress. *Microorganisms* 5 (3). pii: E41

Deepranjan S, Ardith S, O. Siva D, Sonam S, Shikha, Manoj P, Amitava R, Sayyed RZ, Abdul G, Mohammad JA, Subhan D, Fahad S, Rahul D (2021) Optimizing nutrient use efficiency, productivity, energetics, and economics of red cabbage following mineral fertilization and biopriming with compatible rhizosphere microbes. *Sci Rep* 11:15680. https://doi.org/10.1038/s41598-021-95092-6

Depeng W, Fahad S, Saud S, Muhammad K, Aziz K, Mohammad NK, Hafiz MH, Wajid N (2018) Morphological acclimation to agronomic manipulation in leaf dispersion and orientation to promote "Ideotype" breeding: Evidence from 3D visual modeling of "super" rice (*Oryza sativa* L.). *Plant Physiol Biochem* 135:499–510. https://doi.org/10.1016/j.plaphy.2018.11.010

EL Sabagh A, Islam MS, Hossain A, Iqbal MA, Mubeen M, Waleed M, Reginato M, Battaglia M, Ahmed S, Rehman A, Arif M, Athar H-U-R, Ratnasekera D, Danish S, Raza MA, Rajendran K, Mushtaq M, Skalicky M, Brestic M, Soufan W, Fahad S, Pandey S, Kamran M, Datta R. Abdelhamid MT (2022) Phytohormones as growth regulators during abiotic stress tolerance in plants. Front *Agronomy* 4:765068. doi: 10.3389/fagro.2022.765068

El-Fattah DAA, Ewedab WE, Zayed MS, Hassaneina MK (2013) Effect of carrier materials, sterilization method, and storage temperature on survival and biological activities of Azotobacter chroococcum inoculants. *Ann Agric Sci* 58: 111–118

Emre B, Ömer SU, Martín LB, Andre D, Fahad S, Rahul D, Muhammad Z-ul-H, Ghulam SH, Subhan D (2021) Studying soil erosion by evaluating changes in physico-chemical properties of soils under different land-use types. *J Saudi Society Agricultural Sci* 20:190–197

Enamul H, Shoeb AZM, Mallik AH, Fahad S, Kamruzzaman MM, Akib J, Nayyer S, Mehedi KSA, Swati AS, Yeamin MdA, Most SS (2020) Measuring vulnerability to environmental hazards: Qualitative to quantitative. In: Fahad S, Hasanuzzaman M, Alam M, Ullah H, Saeed M, Khan AK, Adnan M (Eds.) *Environment, Climate, Plant and Vegetation Growth*. Springer Nature Switzerland, pp. . 421–452. https://doi.org/10.1007/978-3-030-49732-3

Fahad S, Abdul B, Adnan M (Eds.) (2018b) Global wheat production. *IntechOpen United Kingdom* 2018. http://dx.doi.org/10.5772/intechopen.72559

Fahad S, Bajwa AA, Nazir U, Anjum SA, Farooq A, Zohaib A, Sadia S, NasimW, Adkins S, Saud S, Ihsan MZ, Alharby H,Wu C,Wang D, Huang J (2017) Crop production under drought and heat stress: Plant responses and management options. *Front Plant Sci.* 8:1147. https://doi.org/10.3389/fpls.2017.01147

Fahad S, Bano A (2012) Effect of salicylic acid on physiological and biochemical characterization of maize grown in saline area. *Pak J Bot.* 44:1433–1438

Fahad S, Chen Y, Saud S, Wang K, Xiong D, Chen C, Wu C, Shah F, Nie L, Huang J (2013) Ultraviolet radiation effect on photosynthetic pigments, biochemical attributes, antioxidant enzyme activity and hormonal contents of wheat. *J Food, Agri Environ.* 11(3&4):1635–1641

Fahad S, Hussain S, Bano A, Saud S, Hassan S, Shan D, Khan FA, Khan F, Chen Y, Wu C, Tabassum MA, Chun MX, Afzal M, Jan A, Jan MT, Huang J (2014a) Potential role of phytohormones and plant growth-promoting rhizobacteria in abiotic stresses: consequences for changing environment. *Environ Sci Pollut Res.* 22(7):4907–4921. https://doi.org/10.1007/s11356-014-3754-2

Fahad S, Hussain S, Matloob A, Khan FA, Khaliq A, Saud S, Hassan S, Shan D, Khan F, Ullah N, Faiq M, Khan MR, Tareen AK, Khan A, Ullah A, Ullah N, Huang J (2014b) Phytohormones and plant responses to salinity stress: A review. *Plant Growth Regul.* 75(2):391–404. https://doi.org/10.1007/s10725-014-0013-y

Fahad S, Hussain S, Saud S, Hassan S, Chauhan BS, Khan F et al (2016a) Responses of rapid viscoanalyzer profile and other rice grain qualities to exogenously applied plant growth regulators under high day and high night temperatures. *PLoS One* 11(7):e0159590. https://doi.org/10.1371/journal.pone.0159590

Fahad S, Hussain S, Saud S, Hassan S, Ihsan Z, Shah AN, Wu C, Yousaf M, Nasim W, Alharby H, Alghabari F, Huang J (2016c) Exogenously applied plant growth regulators enhance the morphophysiological growth and yield of rice under high temperature. *Front Plant Sci.* 7:1250. https://doi.org/10.3389/fpls.2016. 01250

Fahad S, Hussain S, Saud S, Hassan S, Tanveer M, Ihsan MZ, Shah AN, Ullah A, Nasrullah KF, Ullah S, AlharbyH NW, Wu C, Huang J (2016d) A combined application of biochar and phosphorus alleviates heat-induced adversities on physiological, agronomical and quality attributes of rice. *Plant Physiol Biochem.* 103:191–198

Fahad S, Hussain S, Saud S, Khan F, Hassan S, Jr A, Nasim W, Arif M, Wang F, Huang J (2016b) Exogenously applied plant growth regulators affect heat-stressed rice pollens. *J Agron Crop Sci.* 202:139–150

Fahad S, Hussain S, Saud S, Tanveer M, Bajwa AA, Hassan S, Shah AN, Ullah A, Wu C, Khan FA, Shah F, Ullah S, Chen Y, Huang J (2015a) A biochar application protects rice pollen from high-temperature stress. *Plant Physiol Biochem* 96:281–287

Fahad S, Muhammad ZI, Abdul K, Ihsanullah D, Saud S, Saleh A, Wajid N, Muhammad A, Imtiaz AK, Chao W, Depeng W, Jianliang H (2018a): Consequences of high temperature under changing climate optima for rice pollen characteristics-concepts and perspectives, *Arch Agron Soil Sci* 64(11):1473–1488. DOI: 10.1080/03650340.2018.1443213

Fahad S, Nie L, Chen Y, Wu C, Xiong D, Saud S, Hongyan L, Cui K, Huang J (2015b) Crop plant hormones and environmental stress. *Sustain Agric Rev* 15:371–400

Fahad S, Saud S, Yajun C, Chao W, Depeng W (Eds.) (2021f) Abiotic Stress in Plants. *IntechOpen United Kingdom* 2021. http://dx.doi.org/10.5772/intechopen.91549

Fahad S, Hasanuzzaman M, Alam M, Ullah H, Saeed M, Ali Khan I, Adnan M (Eds.) (2020) *Environment, Climate, Plant and Vegetation Growth.* Springer Nature Switzerland AG 2020. DOI: https://doi.org/10.1007/978-3-030-49732-3

Fahad S, Adnan M, Hassan S, Saud S, Hussain S, Wu C, Wang D, Hakeem KR, Alharby HF, Turan V, Khan MA, Huang J (2019b) Rice responses and tolerance to high temperature. In: Hasanuzzaman M, Fujita M, Nahar K, Biswas JK (Eds.) *Advances in Rice Research for Abiotic Stress Tolerance.* Woodhead Publishing, UK, pp. 201–224

Fahad S, Adnan M, Saud S, Nie L (Eds.) (2022b) Climate change and ecosystems: challenges to sustainable development. First edition. *Footprints of Climate Variability on Plant Diversity.* CRC Press, Boca Raton

Fahad S, Adnan M, Saud S (Eds.) (2022a) Improvement of plant production in the era of climate change. First edition. *Footprints of Climate Variability on Plant Diversity.* CRC Press, Boca Raton

Fahad S, Rehman A, Shahzad B, Tanveer M, Saud S, Kamran M, Ihtisham M, Khan SU, Turan V, Rahman MHU (2019a) Rice responses and tolerance to metal/metalloid toxicity. In: Hasanuzzaman M, Fujita M, Nahar K, Biswas JK (Eds.) *Advances in Rice Research for Abiotic Stress Tolerance.* Woodhead Publishing, UK, pp. 299–312

Fahad S, Sönmez O, Saud S, Wang D, Wu C, Adnan M, Arif M, Amanullah (Eds.) (2021e) Engineering tolerance in crop plants against abiotic stress. First edition. *Footprints of Climate Variability on Plant Diversity.* CRC Press, Boca Raton

Fahad S, Sonmez O, Saud S, Wang D, Wu C, Adnan M, Turan V (Eds.) (2021b) Climate change and plants: biodiversity, growth and interactions. First edition. *Footprints of Climate Variability on Plant Diversity.* CRC Press, Boca Raton

Fahad S, Sonmez O, Saud S, Wang D, Wu C, Adnan M, Turan V (Eds.) (2021c) Developing climate resilient crops: improving global food security and safety. First edition. *Footprints of Climate Variability on Plant Diversity.* CRC Press, Boca Raton

Fahad S, Sönmez O, Saud S, Wang D, Wu C, Adnan M, Turan V (Eds.) (2021a) Plant growth regulators for climate-smart agriculture. First edition. *Footprints of Climate Variability on Plant Diversity.* CRC Press, Boca Raton

Fahad S, Sönmez O, Turan V, Adnan M, Saud S, Wu C, Wang D (Eds.) (2021d) Sustainable soil and land management and climate change. First edition. *Footprints of Climate Variability on Plant Diversity.* CRC Press, Boca Raton

Fakhre A, Ayub K, Fahad S, Sarfraz N, Niaz A, Muhammad AA, Muhammad A, Khadim D, Saud S, Shah H, Muhammad ASR, Khalid N, Muhammad A, Rahul D, Subhan D (2021) Phosphate solubilizing bacteria optimize wheat yield in mineral phosphorus applied alkaline soil. *J Saudi Soc Agric Sci* 21(5):339–348. https://doi.org/10.1016/j.jssas.2021.10.007

Farah R, Muhammad R, Muhammad SA, Tahira Y, Muhammad AA, Maryam A, Shafaqat A, Rashid M, Muhammad R, Qaiser H, Afia Z, Muhammad AA, Muhammad A, Fahad S (2020) Alternative and non-conventional soil and crop management strategies for increasing water use efficiency. In: Fahad S, Hasanuzzaman M, Alam M, Ullah H, Saeed M, Khan AK, Adnan M (Eds.) *Environment, Climate, Plant and Vegetation Growth.* Springer Nature Switzerland, pp. 323–338. https://doi.org/10.1007/978-3-030-49732-3

Farhana G, Ishfaq A, Muhammad A, Dawood J, Fahad S, Xiuling L, Depeng W, Muhammad F, Muhammad F, Syed AS (2020) Use of crop growth model to simulate the impact of climate change on yield of various wheat cultivars under different agro-environmental conditions in Khyber Pakhtunkhwa, Pakistan. *Arabian J Geosci* 13:112. https://doi.org/10.1007/s12517-020-5118-1

Farhat A, Hafiz MH, Wajid I, Aitazaz AF, Hafiz FB, Zahida Z, Fahad S, Wajid F, Artemi C (2020) A review of soil carbon dynamics resulting from agricultural practices. *J Environ Manage* 268 (2020):110319

Farhat UK, Adnan AK, Kai L, Xuexuan X, Muhammad A, Fahad S, Rafiq A, Mushtaq AK, Taufiq N, Faisal Z (2022) Influences of long-term crop cultivation and fertilizer management on soil aggregates stability and fertility in the Loess Plateau, Northern China. *J Soil Sci Plant Nutri* 22:1446–1457. https://doi.org/10.1007/s42729-021-00744-1

Fazli W, Muhmmad S, Amjad A, Fahad S, Muhammad A, Muhammad N, Ishaq AM, Imtiaz AK, Mukhtar A, Muhammad S, Muhammad I, Rafi U, Haroon I, Muhammad A (2020) Plant–microbes interactions and functions in changing climate. In: Fahad S, Hasanuzzaman M, Alam M, Ullah H, Saeed M, Khan AK, Adnan M (Eds.) *Environment, Climate, Plant and Vegetation Growth.* Springer Nature Switzerland, pp. 397–420. https://doi.org/10.1007/978-3-030-49732-3

Forlain G, Pastorelli R, Branzoni M, Favilli F (1998) Root colonization efficiency, plant growth promoting activity and potentially related properties in plant associated bacteria. *J Genet Breed* 49: 343–351

Forlani L, Juárez MP, Lavarías S, Pedrini N (2014) Toxicological and biochemical response of the entomopathogenic fungus Beauveria bassiana after exposure to deltamethrin. *Pest Manag Sci* 70: 751–756

Ghosh N (2004) Promoting biofertilisers in Indian agriculture. *Econ Polit Wkly* 5: 5617–5625

Ghulam M, Muhammad AA, Donald LS, Sajid M, Muhammad FQ, Niaz A, Ateeq ur R, Shakeel A, Sajjad H, Muhammad A, Summia M, Aqib HAK, Fahad S, Rahul D, Mazhar I, Timothy DS (2021) Formalin fumigation and steaming of various composts differentially influence the nutrient release, growth and yield of muskmelon (*Cucumis melo* L.). *Sci Rep* 11:21057.

Gopakumar L, Bernard NO, Donato V (2020) Soil microarthropods and nutrient cycling. In: Fahad S, Hasanuzzaman M, Alam M, Ullah H, Saeed M, Khan AK, Adnan M (Eds.) *Environment, Climate, Plant and Vegetation Growth.* Springer Nature Switzerland, pp. 453–472. https://doi.org/10.1007/978-3-030-49732-3

Guofu L, Zhenjian B, Fahad S, Guowen C, Zhixin X, Hao G, Dandan L, Yulong L, Bing L, Guoxu J, Saud S (2021) Compositional and structural changes in soil microbial communities in response to straw mulching and plant revegetation in an abandoned artificial pasture in Northeast China. *Global Ecol Conserv* 31 (2021):e01871

Gupta G, Parihar SS, Ahirwar NK, Snehi SK, Singh V (2015) Plant growth promoting Rhizobacteria (PGPR): current and future prospects for development of sustainable agriculture. *J Microb Biochem Technol* 7(2):96–102

Habib ur R, Ashfaq A, Aftab W, Manzoor H, Fahd R, Wajid I, Md. Aminul I, Vakhtang S, Muhammad A, Asmat U, Abdul W, Syeda RS, Shah S, Shahbaz K, Fahad S, Manzoor H, Saddam H, Wajid N (2017) Application of CSM-CROPGRO-cotton model for cultivars and optimum planting dates: Evaluation in changing semi-arid climate. *Field Crops Res* 238:139–152. http://dx.doi.org/10.1016/j.fcr.2017.07.007

Hafeez M, Farman U, Muhammad MK, Xiaowei L, Zhijun Z, Sakhawat S, Muhammad I, Mohammed AA, G. Mandela F-G, · Nicolas D, Muzammal R, Fahad S, Yaobin L (2021) Metabolic-based insecticide resistance mechanism and ecofriendly approaches for controlling of beet armyworm Spodoptera exigua: a review. *Environ Sci Pollut Res* 29:1746–1762. https://doi.org/10.1007/s11356-021-16974-w

Hafiz MH, Abdul K, Farhat A, Wajid F, Fahad S, Muhammad A, Ghulam MS, Wajid N, Muhammad M, Hafiz FB (2020b) Comparative effects of organic and inorganic fertilizers on soil organic carbon and wheat productivity under arid region. *Commun Soil Sci Plant Anal* 51:1406–1422. DOI: 10.1080/00103624.2020.1763385

Hafiz MH, Farhat A, Ashfaq A, Hafiz FB, Wajid F, Carol Jo W, Fahad S, Gerrit H (2020a) Predicting kernel growth of maize under controlled water and nitrogen applications. *Int J Plant Prod* 14:609–620. https://doi.org/10.1007/s42106-020-00110-8

Hafiz MH, Farhat A, Shafqat S, Fahad S, Artemi C, Wajid F, Chaves CB, Wajid N, Muhammad M, Hafiz FB (2018) Offsetting land degradation through nitrogen and water management during maize cultivation under arid conditions. *Land Degrad Dev* 29(5), 1–10. DOI: 10.1002/ldr.2933

Hafiz MH, Muhammad A, Farhat A, Hafiz FB, Saeed AQ, Muhammad M, Fahad S, Muhammad A (2019) Environmental factors affecting the frequency of road traffic accidents: a case study of sub-urban area of Pakistan. *Environ Sci Pollut Res* 26:11674–11685. https://doi.org/10.1007/s11356-019-04752-8

Hafiz MH, Wajid F, Farhat A, Fahad S, Shafqat S, Wajid N, Hafiz FB (2016) Maize plant nitrogen uptake dynamics at limited irrigation water and nitrogen. *Environ Sci Pollut Res* 24(3):2549–2557. https://doi.org/10.1007/s11356-016-8031-0

Haider SA, Lalarukh I, Amjad SF, Mansoora N, Naz M, Naeem M, Bukhari SA, Shahbaz M, Ali SA, Marfo TD, Subhan D, Rahul D, Fahad S (2021) Drought stress alleviation by potassium-nitrate-containing chitosan/montmorillonite microparticles confers changes in *Spinacia oleracea* L. *Sustainability* 13:9903. https://doi.org/10.3390/su13179903

Hamza SM, Xiukang W, Sajjad A, Sadia Z, Muhammad N, Adnan M, Fahad S et al. (2021) Interactive effects of gibberellic acid and NPK on morpho-physio-biochemical traits and organic acid exudation pattern in coriander (Coriandrum sativum L.) grown in soil artificially spiked with boron. *Plant Physiol Biochem.* 167 (2021):884–900

Haoliang Y, Matthew TH, Ke L, Bin W, Puyu F, Fahad S, Holger M, Rui Y, De LL, Sotirios A, Isaiah H, Xiaohai T, Jianguo M, Yunbo Z, Meixue Z (2022) Crop traits enabling yield gains under more frequent extreme climatic events. *Sci Total Environ* 808 (2022):152170

Hegde SV (2008) Liquid bio-fertilizers in Indian agriculture. *Biofert Newsl* 17–22

Hesham FA, Fahad S (2020) Melatonin application enhances biochar efficiency for drought tolerance in maize varieties: Modifications in physio-biochemical machinery. *Agron J.* 112(4):1–22

Huang Li-Y, Li Xiao-X, Zhang Yun-B, Fahad S, Wang F (2021) dep1 improves rice grain yield and nitrogen use efficiency simultaneously by enhancing nitrogen and dry matter translocation. *J Integrative Agri* 21(11):3185–3198. doi: 10.1016/S2095-3119(21)63795-4

Hussain MA, Fahad S, Rahat S, Muhammad FJ, Muhammad M, Qasid A, Ali A, Husain A, Nooral A, Babatope SA, Changbao S, Liya G, Ibrar A, Zhanmei J, Juncai H (2020) Multifunctional role of brassinosteroid and its analogues in plants. *Plant Growth Regul* 92:141–156. https://doi.org/10.1007/s10725-020-00647-8

Ibad U, Dost M, Maria M, Shadman K, Muhammad A, Fahad S, Muhammad I, Ishaq AM, Aizaz A, Muhammad HS, Muhammad S, Farhana G, Muhammad I, Muhammad ASR, Hafiz MH, Wajid N, Shah S, Jabar ZKK, Masood A, Naushad A, Rasheed Akbar M, Shah MK Jan B (2022) Comparative effects of biochar and NPK on wheat crops under different management systems. *Crop Pasture Sci* 74:31–40. https://doi.org/10.1071/CP21146

Ibrar H, Muqarrab A, Adel MG, Khurram S, Omer F, Shahid I, Fahim N, Shakeel A, Viliam B, Marian B, Sami Al Obaid, Fahad S, Subhan D, Suleyman T, Hanife AKÇA, Rahul D (2021) Improvement in growth and yield attributes of cluster bean through optimization of sowing time and plant spacing under climate change scenario. *Saudi J Bio Sci* 29(2):781–792. https://doi.org/10.1016/j.sjbs.2021.11.018

Ibrar K, Aneela R, Khola Z, Urooba N, Sana B, Rabia S, Ishtiaq H, Mujaddad Ur Rehman, Salvatore M (2020) Microbes and environment: Global warming reverting the frozen zombies. In: Fahad S, Hasanuzzaman M, Alam M, Ullah H, Saeed M, Khan AK, Adnan M (Eds.) *Environment, Climate, Plant and Vegetation Growth*. Springer Nature Switzerland, pp. 607–634. https://doi.org/10.1007/978-3-030-49732-3

Ihsan MZ, Abdul K, Manzer HS, Liaqat A, Ritesh K, Hayssam MA, Amar M, Fahad S (2022) The response of Triticum aestivum treated with plant growth regulators to acute day/night temperature rise. *J Plant Growth Regul* 41:2020–2033. https://doi.org/10.1007/s00344-022-10574-9

Ikram U, Khadim D, Muhammad T, Muhammad S, Fahad S (2021) Gibberellic acid and urease inhibitor optimize nitrogen uptake and yield of maize at varying nitrogen levels under changing climate. *Environ Sci Pollution Res* 29(5):6568–6577. https://doi.org/10.1007/s11356-021-16049-w

Ilyas M, Mohammad N, Nadeem K, Ali H, Aamir HK, Kashif H, Fahad S, Aziz K, Abid U (2020) Drought tolerance strategies in plants: A mechanistic approach. *J Plant Growth Regul* 40:926–944. https://doi.org/10.1007/s00344-020-10174-5

Iqra M, Amna B, Shakeel I, Fatima K,Sehrish L, Hamza A, Fahad S (2020) Carbon cycle in response to global warming. In: Fahad S, Hasanuzzaman M, Alam M, Ullah H,Saeed M, Khan AK, Adnan M (Eds.) *Environment, Climate, Plant and Vegetation Growth*. Springer Nature Switzerland, pp. 1–16. https://doi.org/10.1007/978-3-030-49732-3

Irfan M, Muhammad M, Muhammad JK, Khadim MD, Dost M, Ishaq AM, Waqas A, Fahad S, Saud S et al.(2021) Heavy metals immobilization and improvement in maize (Zea mays L.) growth amended with biochar and compost. *Sci Rep* 11:18416

Jabborova D, Sulaymanov K, Sayyed RZ, Alotaibi SH, Enakiev Y, Azimov A, Jabbarov Z, Ansari MJ, Fahad S, Danish S et al. (2021) Mineral fertilizers improve the quality of turmeric and soil. *Sustainability 13*:9437. https://doi.org/10.3390/su13169437

Jan M, Muhammad Anwar-ul-Haq, Adnan NS, Muhammad Y, Javaid I, Xiuling L, Depeng W, Fahad S (2019) Modulation in growth, gas exchange, and antioxidant activities of salt-stressed rice (Oryza sativa L.) genotypes by zinc fertilization. *Arabian J Geosci* 12:775. https://doi.org/10.1007/s12517-019-4939-2

Kamaran M, Wenwen C, Irshad A, Xiangping M, Xudong Z, Wennan S, Junzhi C, Shakeel A, Fahad S, Qingfang H, Tiening L (2017) Effect of paclobutrazol, a potential growth regulator on stalk mechanical strength, lignin accumulation and its relation with lodging resistance of maize. *Plant Growth Regul* 84:317–332. https://doi.org/10.1007/ s10725-017-0342-8

Khadim D, Fahad S, Jahangir MMR, Iqbal M, Syed SA, Shah AK, Ishaq AM, Rahul D et al. (2021a) Biochar and urease inhibitor mitigate NH_3 and N_2O emissions and improve wheat yield in a urea fertilized alkaline soil. *Sci Rep* 11:17413

Khadim D, Saif-ur-R, Fahad S, Syed SA, Shah AK et al. (2021b) Influence of variable biochar concentration on yield-scaled nitrous oxide emissions, wheat yield and nitrogen use efficiency. *Sci Rep* 11:16774

Khan MMH, Niaz A, Umber G, Muqarrab A, Muhammad AA, Muhammad I, Shabir H, Shah F, Vibhor A, Shams HA-H, Reham A, Syed MBA, Nadiyah MA, Ali TKZ, Subhan D, Rahul D (2021) Synchronization of boron application methods and rates is environmentally friendly approach to improve quality attributes of *Mangifera indica* L. on sustainable basis. *Saudi J Bio Sci* 29(3):1869–1880 https://doi.org/10.1016/j.sjbs.2021.10.036

Khan TA, Mazid M, Mohammad F (2011a) Role of ascorbic acid against pathogenesis in plants. *J Stress Physiol Biochem* 7: 222–234

Khatun M, Sarkar S, Era FM, Islam AKMM, Anwar MP, Fahad S, Datta R, Islam AKMA (2021) Drought stress in grain legumes: Effects, tolerance mechanisms and management. *Agronomy* 11:2374. https://doi.org/10.3390/agronomy11122374

Mahar A, Amjad A, Altaf HL, Fazli W, Ronghua L, Muhammad A, Fahad S, Muhammad A, Rafiullah, Imtiaz AK, Zengqiang Z (2020) Promising technologies for Cd-contaminated soils: Drawbacks and possibilities. In: Fahad S, Hasanuzzaman M, Alam M, Ullah H,Saeed M, Khan AK, Adnan M (Eds.) *Environment, Climate, Plant and Vegetation Growth*. Springer Nature Switzerland, pp. 63–92. https://doi.org/10.1007/978-3-030-49732-3

Mahmood Ul H, Tassaduq R, Chandni I, Adnan A, Muhammad A, Muhammad MA, Muhammad H-ur-R, Mehmood AN, Alam S, Fahad S (2021) Linking plants functioning to adaptive responses under heat stress conditions: a mechanistic review. *J Plant Growth Regul* 41:2596–2613. https://doi.org/10.1007/s00344-021-10493-1

Malusà E, Pinzari F, Canfora L (2016) Efficacy of bio-fertilizers: challenges to improve crop production. In: Singh DP, et al. (Eds.) *Microbial Inoculants in Sustainable Agricultural Productivity*. Springer, India

Malusà E, Vassilev N (2014) A contribution to set a legal framework for biofertilisers. *Appl Microbiol Biotechnol* 98, 6599–6607

Manzer HS, Saud A, Soumya M, Abdullah A. Al-A, Qasi DA, Bander MA. Al-M, Hayssam MA, Hazem MK, Fahad S, Vishnu DR, Om PN (2021) Molybdenum and hydrogen sulfide synergistically mitigate arsenic toxicity by modulating defense system, nitrogen and cysteine assimilation in faba bean (*Vicia faba* L.) seedlings. *Environ Pollut* 290:117953

Marek-Kozaczuk M, Wielbo J, Pawlik A, Skorupska A (2014) Nodulation competitiveness of Ensifer meliloti alfalfa nodule isolates and their potential for application as inoculants. *Pol J Microbiol* 63: 375–386

Mazid M, Khan TA, Mohammad F (2011a) Potential of NO and H_2O_2 as signalling molecules in tolerance to abiotic stress in plants. *J Ind Res Technol* 1: 56–68

Mazid M, Khan TA, Mohammad F (2011b) Cytokinins, a classical multifaceted hormone in plant system. J Stress Physiol *Biochemistry* 7(4):347–368

Md Jakir H, Allah B (2020) Development and applications of transplastomic plants; A way towards eco-friendly agriculture. In: Fahad S, Hasanuzzaman M, Alam M, Ullah H, Saeed M, Khan AK, Adnan M (Eds.) *Environment, Climate, Plant and Vegetation Growth.* Springer Nature Switzerland, pp. 285–322. https://doi.org/10.1007/978-3-030-49732-3

Mehmood K, Bao Y, Saifullah, Bibi S, Dahlawi S, Yaseen M, Abrar MM, Srivastava P, Fahad S, Faraj TK (2022) Contributions of open biomass burning and crop straw burning to air quality: Current research paradigm and future outlooks. *Front Environ Sci* 10:852492. doi: 10.3389/fenvs.2022.852492

Mohammad I. Al-Wabel, Abdelazeem S, Munir A, Khalid E, Adel RAU (2020b) Extent of climate change in Saudi Arabia and its impacts on agriculture: A case study from Qassim Region. In: Fahad S, Hasanuzzaman M, Alam M, Ullah H, Saeed M, Khan AK, Adnan M (Eds.) *Environment, Climate, Plant and Vegetation Growth.* Springer Nature Switzerland, pp. 635–658. https://doi.org/10.1007/978-3-030-49732-3

Mohammad I. Al-Wabel, Munir Ahmad, Adel R. A. Usman, Mutair Akanji, and Muhammad Imran Rafique (2020a) Advances in pyrolytic technologies with improved carbon capture and storage to combat climate change. In: Fahad S, Hasanuzzaman M, Alam M, Ullah H, Saeed M, Khan AK, Adnan M (Eds.) *Environment, Climate, Plant and Vegetation Growth.* Springer Nature Switzerland, pp. 535–576. https://doi.org/10.1007/978-3-030-49732-3

Mohammadi K, Sohrabi Y (2012) Bacterial bio-fertilizers for sustainable crop production: a review. *J Agric Biol Sci* 7: 307–316

Mordor Intelligence LLP (2019) *Global Biofertilizer Market (2019–2024).* Mordor Intelligence LLP, India.

Mubeen M, Ashfaq A, Hafiz MH, Muhammad A, Hafiz UF, Mazhar S, Muhammad Sami ul Din, Asad A, Amjed A, Fahad S, Wajid N (2020) Evaluating the climate change impact on water use efficiency of cotton-wheat in semi-arid conditions using DSSAT model. *J Water Clim Change* 11(4):1661–1675. doi/10.2166/wcc.2019.179/622035/jwc2019179.pdf

Muhammad I, Khadim D, Fahad S, Imran M, Saud A, Manzer HS, Shah S, Jabar ZKK, Shamsher A, Shah H, Taufiq N, Hafiz MH, Jan B, Wajid N (2022) Exploring the potential effect of *Achnatherum splendens* L.-derived biochar treated with phosphoric acid on bioavailability of cadmium and wheat growth in contaminated soil. *Environ Sci Pollut Res* 29(25):37676–37684. https://doi.org/10.1007/s11356-021-17950-0.

Muhammad N, Muqarrab A, Khurram S, Fiaz A, Fahim N, Muhammad A, Shazia A, Omaima N, Sulaiman AA, Fahad S, Subhan D, Rahul D (2021) Kaolin and Jasmonic acid improved cotton productivity under water stress conditions. *J Saudi Soc Agr Sci* 28 (2021) 6606–6614. https://doi.org/10.1016/j.sjbs.2021.07.043

Muhammad Tahir ul Qamar, Amna F, Amna B, Barira Z, Xitong Z, Ling-Ling C (2020) Effectiveness of conventional crop improvement strategies vs. omics. In: Fahad S, Hasanuzzaman M, Alam M, Ullah H, Saeed M, Khan AK, Adnan M (Eds.), *Environment, Climate, Plant and Vegetation Growth.* Springer Nature Switzerland, pp. 253–284. https://doi.org/10.1007/978-3-030-49732-3

Muhammad Z, Abdul MK, Abdul MS, Kenneth BM, Muhammad S, Shahen S, Ibadullah J, Fahad S (2019) Performance of Aeluropus lagopoides (mangrove grass) ecotypes, a potential turfgrass, under high saline conditions. *Environ Sci Pollut Res* 26(13):13410–13412. https://doi.org/10.1007/s11356-019-04838-3

Mushtaq S, Ahmed N, Khan IA, Shah B, Khan A, Ali M, Rashid MT, Adnan M, Junaid K, Zaki AB, Ahmed S (2015) Study on the efficacy of ladybird beetle as a biological control agent against aphids (*Chaitophorus spp.). J Ento Zool Stud* 3(6):117–119

Muzammal R, Fahad S, Guanghui D, Xia C, Yang Y, Kailei T, Lijun L, Fei-Hu L, Gang D (2021) Evaluation of hemp (Cannabis sativa L.) as an industrial crop: a review. *Environ Sci Pollution Res* 28(38):52832–52843. https://doi.org/10.1007/s11356-021-16264-5

Niaz A, Abdullah E, Subhan D, Muhammad A, Fahad S, Khadim D, Suleyman T, Hanife A, Anis AS, Mohammad JA, Emre B, ¨Omer SU, Rahul D, Bernard RG (2022) Mitigation of lead (Pb) toxicity in rice cultivated with either ground water or wastewater by application of acidified carbon. *J Environ Manage* 307:114521

Noor M, Naveed ur R, Ajmal J, Fahad S, Muhammad A, Fazli W, Saud S, Hassan S (2020) Climate change and coastal plant lives. In: Fahad S, Hasanuzzaman M, Alam M, Ullah H,Saeed M, Khan AK, Adnan M (Eds.) *Environment, Climate, Plant and Vegetation Growth.* Springer Nature Switzerland, pp. 93–108. https://doi.org/10.1007/978-3-030-49732-3

Park J, Bolan N, Megharaj M, Naidu R (2010) Isolation of phosphate-solubilizing bacteria and characterization of their effects on lead immobilization. *Pedologist* 53: 67–75

Puente ML, García JE, Alejandro P (2009) Effect of the bacterial concentration of Azospirillum brasilense in the inoculum and its plant growth regulator compounds on crop yield of corn (Zea mays L.) in the field. *World J Agric Sci* 5(5):604–608

Qadeem W, Sattar S, Adnan M, Zaman M, Ali I, Shah SRA, Junaid K, Said F, Rahman IU, Saleem N (2015) Efficacy of botanical and microbial extracts against Angoumois grain moth *Sitotrogacerealella* (Olivier) (Lepidoptera: Gelechiidae) under laboratory conditions. *J Ento Zool Stud* 3(5):451–454

Qamar-uz Z, Zubair A, Muhammad Y, Muhammad ZI, Abdul K, Fahad S, Safder B, Ramzani PMA, Muhammad N (2017) Zinc biofortification in rice: leveraging agriculture to moderate hidden hunger in developing countries. *Arch Agron Soil Sci* 64:147–161. https://doi.org/10.1080/03650340.2017.1338343

Qin ZH, Nasib ur Rahman, Ahmad A, Yu-pei Wang, Sakhawat S, Ehmet N, Wen-juan Shao, Muhammad I, Kun S, Rui L, Fazal S, Fahad S (2022) Range expansion decreases the reproductive fitness of *Gentiana officinalis* (*Gentianaceae*). *Sci Rep* 12:2461. https://doi.org/10.1038/s41598-022-06406-1

Rai MK (2006) *Handbook of Microbial Bio-fertilizers.* Food Products Press, Haworth Press, New York

Raja N (2013) Biopesticides and bio-fertilizers: ecofriendly sources for sustainable agriculture. *J Biofertil Biopestici* 4: e112

Rajesh KS, Fahad S, Pawan K, Prince C, Talha J, Dinesh J, Prabha S, Debanjana S, Prathibha MD, Bandana B, Akash H, Gupta NK, Rekha S, Devanshu D, Dalpat LS, Ke L, Matthew TH, Saud S, Adnan NS, Taufiq N (2022) Beneficial elements: New players in improving nutrient use efficiency and abiotic stress tolerance. *Plant Growth Regulation* 100:237–265. https://doi.org/10.1007/s10725-022-00843-8

Rasheed MT, Inayatullah M, Shah B, Ahmed N, Khan A, Ali M, Ahmed S, Junaid K, Adnan M, Huma Z (2015) Relative abundance of insect pollinators on two cultivars of sunflower in Islamabad. *J Ento Zool Stud* 3(6):164–165

Rashid M, Qaiser H, Khalid SK, Mohammad I. Al-Wabel, Zhang A, Muhammad A, Shahzada SI, Rukhsanda A, Ghulam AS, Shahzada MM, Sarosh A, Muhammad FQ (2020) Prospects of biochar in alkaline soils to mitigate climate change. In: Fahad S, Hasanuzzaman M, Alam M, Ullah H,Saeed M, Khan AK, Adnan M (Eds.) *Environment, Climate, Plant and Vegetation Growth.* Springer Nature Switzerland, pp. 133–150. https://doi.org/10.1007/978-3-030-49732-3

Rehana S, Asma Z, Shakil A, Anis AS, Rana KI, Shabir H, Subhan D, Umber G, Fahad S, Jiri K, Sami Al Obaid, Mohammad JA, Rahul D (2021) Proteomic changes in various plant tissues associated with chromium stress in sunflower. *Saudi J Bio Sci* 29(4):2604–2612. https://doi.org/10.1016/j.sjbs.2021.12.042

Rehman M, Fahad S, Saleem MH, Hafeez M, Muhammad Habib ur Rahman, Liu F, Deng G (2020) Red light optimized physiological traits and enhanced the growth of ramie (Boehmeria nivea L.). *Photosynthetica* 58(4):922–931

Sadam M, Muhammad Tahir ul Qamar, Ghulam M, Muhammad SK, Faiz AJ (2020) Role of biotechnology in climate resilient agriculture. In: Fahad S, Hasanuzzaman M, Alam M, Ullah H, Saeed M, Khan AK, Adnan M (Eds.) *Environment, Climate, Plant and Vegetation Growth.* Springer Nature Switzerland, pp. 339–366. https://doi.org/10.1007/978-3-030-49732-3

Safi UK, Ullah F, Mehmood S, Fahad S, Ahmad Rahi A, Althobaiti F, et al. (2021) Antimicrobial, antioxidant and cytotoxic properties of Chenopodium glaucum L. *PLoS One* 16(10): e0255502. https://doi.org/10.1371/journal. pone.0255502

Sahrish N, Shakeel A, Ghulam A, Zartash F, Sajjad H, Mukhtar A, Muhammad AK, Ahmad K, Fahad S, Wajid N, Sezai E, Carol Jo W, Gerrit H (2022) Modeling the impact of climate warming on potato phenology. *European J AgroN* 132:126404

Saikia SP, Bora D, Goswami A, Mudoi KD, Gogoi A (2013) A review on the role of Azospirillum in the yield improvement of non-leguminous crops. *Afr J Microbiol Res* 6: 1085–1102

Saikia SP, Jain V (2007) Biological nitrogen fixation with non-legumes: an achievable target or a dogma. *Curr Sci* 92: 317–322

Sajid H, Jie H, Jing H, Shakeel A, Satyabrata N, Sumera A, Awais S, Chunquan Z, Lianfeng Z, Xiaochuang C, Qianyu J, Junhua Z (2020) Rice production under climate change: Adaptations and mitigating strategies. In: Fahad S, Hasanuzzaman M, Alam M, Ullah H, Saeed M, Khan AK, Adnan M (Eds.) *Environment, Climate, Plant and Vegetation Growth.* Springer Nature Switzerland, pp. 659–686. https://doi.org/10.1007/978-3-030-49732-3

Sajjad H, Muhammad M, Ashfaq A, Fahad S, Wajid N, Hafiz MH, Ghulam MS, Behzad M, Muhammad T, Saima P (2021a) Using space–time scan statistic for studying the effects of COVID-19 in Punjab, Pakistan: a guideline for policy measures in regional Agriculture. *Environ Sci Pollut Res* 30: 42495–42508. https://doi.org/10.1007/s11356-021-17433-2

Sajjad H, Muhammad M, Ashfaq A, Nasir M, Hafiz MH, Muhammad A, Muhammad I, Muhammad U, Hafiz UF, Fahad S, Wajid N, Hafiz MRJ, Mazhar A, Saeed AQ, Amjad F, Muhammad SK, Mirza W (2021b) Satellite-based evaluation of temporal change in cultivated land in Southern Punjab (Multan region) through dynamics of vegetation and land surface temperature. *Open Geo Sci* 13:1561–1577

Sajjad H, Muhammad M, Ashfaq A, Waseem A, Hafiz MH, Mazhar A, Nasir M, Asad A, Hafiz UF, Syeda RS, Fahad S, Depeng W, Wajid N (2019) Using GIS tools to detect the land use/land cover changes during forty years in Lodhran district of Pakistan. *Environ Sci Pollut Res* 27:39676–39692. https://doi.org/10.1007/s11356-019-06072-3

Saleem MH, Fahad S, Adnan M, Mohsin A, Muhammad SR, Muhammad K, Qurban A, Inas AH, Parashuram B, Mubassir A, Reem MH (2020a) Foliar application of gibberellic acid endorsed phytoextraction of copper and alleviates oxidative stress in jute (Corchorus capsularis L.) plant grown in highly copper-contaminated soil of China. *Environ Sci Pollut Res* 27:37121–37133. https://doi.org/10.1007/s11 356-020-09764-3

Saleem MH, Fahad S, Shahid UK, Mairaj D, Abid U, Ayman ELS, Akbar H, Analía L, Lijun L (2020c) Copper-induced oxidative stress, initiation of antioxidants and phytoremediation potential of flax (Linum usitatissimum L.) seedlings grown under the mixing of two different soils of China. *Environ Sci Poll Res* 27:5211–5221. https://doi.org/10.1007/s11356-019-07264-7

Saleem MH, Rehman M, Fahad S, Tung SA, Iqbal N, Hassan A, Ayub A, Wahid MA, Shaukat S, Liu L, Deng G (2020b) Leaf gas exchange, oxidative stress, and physiological attributes of rapeseed (Brassica napus L.) grown under different light-emitting diodes. *Photosynthetica* 58(3): 836–845

Saman S, Amna B, Bani A, Muhammad Tahir ul Qamar, Rana MA, Muhammad SK (2020) QTL mapping for abiotic stresses in cereals. In: Fahad S, Hasanuzzaman M, Alam M, Ullah H, Saeed M, Khan AK, Adnan M (Eds.) *Environment, Climate, Plant and Vegetation Growth.* Springer Nature Switzerland, pp. 229–252. https://doi.org/10.1007/978-3-030-49732-3

Sana A, Ahmed N, Khan IA, Shah B, Khan A, Adnan M, Junaid K, Rasheed MT, Zaki AB, Huma Z, Ahmed S (2015) Evaluating larvicidal action of *Coriandrumsativum* (Dhania) and Mentha (mint) plant extracts against *Aedesaegypti*and *Aedesalbopictus*larvae under laboratory conditions. *J Ento Zool Stud* 3(6):156–159

Sana U, Shahid A, Yasir A, Farman UD, Syed IA, Mirza MFAB, Fahad S, F. Al-Misned, Usman A, Xinle G, Ghulam N, Kunyuan W (2022) Bifenthrin induced toxicity in Ctenopharyngodon idella at an acute concentration: A multi-biomarkers based study. *J King Saud Uni Sci* 34 (2022):101752

Saud S, Fahad S, Hassan S (2022a) Developments in the investigation of nitrogen and oxygen stable isotopes in atmospheric nitrate. *Sustainable Chem Clim Action* 1:100003

Saud S, Li X, Jiang Z, Fahad S, Hassan S (2022b) Exploration of the phytohormone regulation of energy storage compound accumulation in microalgae. *Food Energy Secur* 2022;00:e418

Saud S, Chen Y, Fahad S, Hussain S, Na L, Xin L, Alhussien SA (2016) Silicate application increases the photosynthesis and its associated metabolic activities in Kentucky bluegrass under drought stress and post-drought recovery. *Environ Sci Pollut Res* 23(17):17647–17655. https://doi.org/10.1007/s11356-016-6957-x

Saud S, Chen Y, Long B, Fahad S, Sadiq A (2013) The different impact on the growth of cool season turf grass under the various conditions on salinity and drought stress. *Int J Agric Sci Res* 3:77–84

Saud S, Fahad S, Cui G, Chen Y, Anwar S (2020) Determining nitrogen isotopes discrimination under drought stress on enzymatic activities, nitrogen isotope abundance and water contents of Kentucky bluegrass. *Sci Rep* 10:6415 | https://doi.org/10.1038/s41598-020-63548-w

Saud S, Fahad S, Yajun C, Ihsan MZ, Hammad HM, Nasim W, Amanullah Jr, Arif M and Alharby H (2017) Effects of nitrogen supply on water stress and recovery mechanisms in Kentucky bluegrass plants. Front *Plant Sci* 8:983. doi: 10.3389/fpls.2017.00983

Saud S, Li X, Chen Y, Zhang L, Fahad S, Hussain S, Sadiq A, Chen Y (2014) Silicon application increases drought tolerance of Kentucky bluegrass by improving plant water relations and morph physiological functions. *Sci World J* 2014:1–10. https://doi.org/10.1155/2014/ 368694

Senol C (2020) The effects of climate change on human behaviors. In: Fahad S, Hasanuzzaman M, Alam M, Ullah H, Saeed M, Khan AK, Adnan M (Eds.) *Environment, Climate, Plant and Vegetation Growth.* Springer Nature Switzerland, pp. 577–590. https://doi.org/10.1007/978-3-030-49732-3

Shafi MI, Adnan M, Fahad S, Fazli W, Ahsan K, Zhen Y, Subhan D, Zafar-ul-Hye M, Martin B, Rahul D (2020) Application of single superphosphate with humic acid improves the growth, yield and phosphorus uptake of wheat (Triticum aestivum L.) in calcareous soil. *Agronomy* 10:1224. doi:10.3390/agronomy10091224

Shah F, Lixiao N, Kehui C, Tariq S, Wei W, Chang C, Liyang Z, Farhan A, Fahad S, Huang J (2013) Rice grain yield and component responses to near 2°C of warming. *Field Crop Res* 157:98–110

Shah S, Shah H, Liangbing X, Xiaoyang S, Shahla A, Fahad S (2022) The physiological function and molecular mechanism of hydrogen sulfide resisting abiotic stress in plants. *Braz J Bot* 45:563–572. https://doi.org/10.1007/s40415-022-00785-5

Sheraz SM, Hassan GI, Samoon SA, Rather HA, Showkat AD, Zehra B (2010) Bio-fertilizers in organic agriculture. *J Phytol* 2(10). Retrieved from https://updatepublishing.com/journal/index.php/jp/article/view/2180

Sidra K, Javed I, Subhan D, Allah B, Syed IUSB, Fatma B, Khaled DA, Fahad S, Omaima N, Ali TKZ, Rahul D (2021) Physio-chemical characterization of indigenous agricultural waste materials for the development of potting media. *J Saudi Soc Agr Sci* 28(12):7491–7498. https://doi.org/10.1016/j.sjbs.2021.08.058

Simarmata T, Hersanti T, Turmuktini BN, Fitriatin MR, Setiawati P (2016) Application of bioameliorant and bio-fertilizers to increase the soil health and rice productivity. *Hayati J Biosci* 23: 181–184

Singh JS, Pandey VC, Singh DP (2011) Efficient soil microorganisms: a new dimension for sustainable agriculture and environmental development. *Agric Ecosyst Environ* 140: 339–353

Smith S, Lakobsen I, Gronlund M, Smith FA (2011) Roles of arbuscular mycorrhizas in plant phosphorus nutrition: interactions between pathways of phosphorus uptake in arbuscular mycorrhizal roots have important implications for understanding and manipulating plant phosphorus acquisition. *Plant Physiol* 156: 1050–1057

Subhan D, Zafar-ul-Hye M, Fahad S, Saud S, Martin B, Tereza H, Rahul D (2020) Drought stress alleviation by ACC deaminase producing *Achromobacter xylosoxidans* and *Enterobacter cloacae*, with and without timber waste biochar in maize. *Sustainability* 12(15):6286. doi:10.3390/su12156286

Suhag M (2016) Potential of bio-fertilizers to replace chemical fertilizers. *IARJSET* 3(5):163–167

Sumita P, Singh HB, Farooqui A, Rakshit A (2015) Fungal bio-fertilizers in Indian agriculture: perception, demand and promotion. *J Eco-friendly Agri* 10: 101–113

Tariq M, Ahmad S, Fahad S, Abbas G, Hussain S, Fatima Z, Nasim W, Mubeen M, ur Rehman MH, Khan MA, Adnan M (2018) The impact of climate warming and crop management on phenology of sunflower-based cropping systems in Punjab, Pakistan. *Agri For Met* 15;256:270–282

Unsar Naeem-U, Muhammad R, Syed HMB, Asad S, Mirza AQ, Naeem I, Muhammad H ur R, Fahad S, Shafqat S (2020) Insect pests of cotton crop and management under climate change scenarios. In: Fahad S, Hasanuzzaman M, Alam M, Ullah H, Saeed M, Khan AK, Adnan M (Eds.) *Environment, Climate, Plant and Vegetation Growth.* Springer Nature Switzerland, pp. 367–396. https://doi.org/10.1007/978-3-030-49732-3

Wahid F, Fahad S, Subhan D, Adnan M, Zhen Y, Saud S, Manzer HS, Martin B, Tereza H, Rahul D (2020) Sustainable management with mycorrhizae and phosphate solubilizing bacteria for enhanced phosphorus uptake in calcareous soils. *Agriculture* 10(8):334. doi:10.3390/agriculture10080334

Wajid N, Ashfaq A, Asad A, Muhammad T, Muhammad A, Muhammad S, Khawar J, Ghulam MS, Syeda RS, Hafiz MH, Muhammad IAR, Muhammad ZH, Muhammad Habib ur R, Veysel T, Fahad S, Suad S, Aziz K, Shahzad A (2017) Radiation efficiency and nitrogen fertilizer impacts on sunflower crop in contrasting environments of Punjab. *Pak Environ Sci Pollut Res* 25:1822–1836. https://doi.org/10.1007/s11356-017-0592-z

Wiqar A, Arbaz K, Muhammad Z, Ijaz A, Muhammad A, Fahad S, (2022) Relative efficiency of biochar particles of different sizes for immobilising heavy metals and improving soil properties. *Crop Pasture Sci* 42(2):112–120. https://doi.org/10.1071/CP20453.

Wu C, Kehui C, She T, Ganghua L, Shaohua W, Fahad S, Lixiao N, Jianliang H, Shaobing P, Yanfeng D (2020) Intensified pollination and fertilization ameliorate heat injury in rice (Oryza sativa L.) during the flowering stage. *Field Crops Res* 252:107795

Wu C, Tang S, Li G, Wang S, Fahad S, Ding Y (2019) Roles of phytohormone changes in the grain yield of rice plants exposed to heat: a review. *Peer J* 7:e7792. DOI 10.7717/peerj.7792

Xue B, Huang L, Li X, Lu J, Gao R, Kamran M, Fahad S (2022) Effect of clay mineralogy and soil organic carbon in aggregates under straw incorporation. *Agronomy* 12:534. https://doi.org/10.3390/ agronomy12020534

Yang R, Dai P, Wang B, Jin T, Liu K, Fahad S, Harrison MT, Man J, Shang J, Meinke H, Deli L, Xiaoyan W, Yunbo Z, Meixue Z, Yingbing T, Haoliang Y(2022) Over-optimistic projected future wheat yield potential in the North China Plain: The role of future climate extremes. *Agronomy* 12:145. https://doi.org/10.3390/ agronomy12010145

Yang Z, Zhang Z, Zhang T, Fahad S, Cui K, Nie L, Peng S, Huang J (2017) The effect of season-long temperature increases on rice cultivars grown in the central and southern regions of China. *Front Plant Sci* 8:1908. https://doi.org/10.3389/fpls.2017.01908

Youssef MMA, Eissa MFM (2014) Bio-fertilizers and their role in management of plant parasitic nematodes. A review. *E3 J Biotechnol Pharm Res* 5: 1–6

Zafar-ul-Hye M, Muhammad T ahzeeb-ul-Hassan, Muhammad A, Fahad S,, Martin B, Tereza D, Rahul D, Subhan D (2020b) Potential role of compost mixed biochar with rhizobacteria in mitigating lead toxicity in spinach. *Scientific Rep* 10:12159. https://doi.org/10.1038/s41598-020-69183-9.

Zafar-ul-Hye M, Muhammad N, Subhan D, Fahad S, Rahul D, Mazhar A, Ashfaq AR, Martin B, Jiˇrí H, Zahid HT, Muhammad N (2020a) Alleviation of cadmium adverse effects by improving nutrients uptake in bitter gourd through cadmium tolerant Rhizobacteria. *Environments* 7(8), 54. doi:10.3390/environments7080054

Zafar-ul-Hye M, Akbar MN, Iftikhar Y, Abbas M, Zahid A, Fahad S, Datta R, Ali M, Elgorban AM, Ansari MJ et al. (2021) Rhizobacteria inoculation and caffeic acid alleviated drought stress in lentil plants. *Sustainability* 13: 9603. https://doi.org/10.3390/su13179603

Zahida Z, Hafiz FB, Zulfiqar AS, Ghulam MS, Fahad S, Muhammad RA, Hafiz MH, Wajid N, Muhammad S (2017) Effect of water management and silicon on germination, growth, phosphorus and arsenic uptake in rice. *Ecotoxicol Environ Saf* 144:11–18

Zahir SM, Zheng-HG, Ala Ud D, Amjad A, Ata Ur R, Kashif J, Shah F, Saud S, Adnan M, Fazli W, Saud A, Manzer HS, Shamsher A, Wajid N, Hafiz MH, Fahad S (2021) Synthesis of silver nanoparticles using Plantago lanceolata extract and assessing their antibacterial and antioxidant activities. *Sci Rep* 11:20754

Zaman I, Ali M, Shahzad K, Tahir, MS, Matloob A, Ahmad W, Alamri S, Khurshid MR, Qureshi MM, Wasaya A, Khurram SB, Manzer HS, Fahad S, Rahul D (2021) Effect of plant spacings on growth, physiology, yield and fiber quality attributes of cotton genotypes under nitrogen fertilization. *Agronomy* 11: 2589. https://doi.org/ 10.3390/agronomy11122589

Zia-ur-Rehman M (2020) Environment, climate change and biodiversity. In: Fahad S, Hasanuzzaman M, Alam M, Ullah H, Saeed M, Khan AK, Adnan M (Eds.) *Environment, Climate, Plant and Vegetation Growth.* Springer Nature Switzerland, pp. 473–502. https://doi.org/10.1007/978-3-030-49732-3

15 Bio-fertilizer Effects on Plant-parasitic Nematodes

Taufiq Nawaz[*1], *Muhammad Junaid*[2], *Mehwish Kanwal*[3], *Saeed Ahmed*[4], *Nazeer Ahmed*[*5], *Rafi Ullah*[5], *Muhammad Adnan*[5], *Muhammad Saeed*[5], *Muhammad Romman*[6], *Shah Fahad*[7], *Maid Zaman*[8], *Muhammad Haroon*[5], *Muhammad Shahab*[5], *Shah Saud*[9], *and Shah Hassan*[10]

[1] Department of Food Science and Technology, The University of Agriculture, Peshawar, Pakistan
[2] Graduate School of Life and Environmental Sciences, University of Tsukuba, Tsukuba, Ibaraki, Japan
[3] Ministry of National Food Security & Research, Pakistan Tobacco Board, Peshawar, Khyber Pakhtunkhwa, Pakistan
[4] Agricultural Research Center, Londrina State University, Londrina, Brazil
[5] Department of Agriculture, University of Swabi, Swabi, Khyber Pakhtunkhwa, Pakistan
[6] Department of Botany, University of Chitral, Khyber Pakhtunkhwa, Pakistan
[7] Department of Agronomy, Abdul Wali Khan University Mardan, Khyber Pakhtunkhwa, Pakistan [8]Department of Entomology, The University of Haripur, Khyber Pakhtunkhwa, Pakistan
[9] College of Life Science, Linyi University, Linyi, Shandong, China
[10] Department of Agricultural Extension Education and Communication, The University of Agriculture, Peshawar, Pakistan
** Correspondence: taufiqnawaz85@gmail.com (T.N.) drnazeerento@gmail.com (N.A.)

CONTENTS

DOI: 10.1201/9781003286233-15

15.1 INTRODUCTION

In agriculture, increased yields can be obtained by the increased use of fertilizers. However, they are also expensive and depreciate the environment by causing adverse effects on living things and soils' physio-chemical properties. Due to fractional uptake of fertilizers by plants, the leftover fertilizers leach down with rainwater to the waterbodies, causing eutrophication and affecting living organisms, including microorganisms responsible for growth inhibition. In addition, these fertilizers adversely affect soil fertility causing a disparity of soil nutrients and reducing the water-holding capacity of soils (Youssef and Eissa 2014). Eco-friendly and cost-efficient fertilizers that do not disturb natural resources are therefore required to be produced (Deepali and Gangwar 2010). Hence, several fertilizers have been developed recently that work as natural stimulators for the development and growth of plants (Khan et al. 2009). The knowledge of microbial inocula or natural stimulators has an extensive history, with the use of small-scale compost production passed down through generations of farmers (Halim 2009). This particular type of fertilizer includes products based on microorganisms promoting plant growth called "microbial nutrients" or "bio-fertilizers" that contain live efficient strains of phosphate-solubilizing, nitrogen-fixing, or cellulolytic microorganisms. These microorganisms are applied in seed, soil, or composting areas to increase their numbers and hasten microbial activities to aid the availability of nutrients to plants in a readily available form (Khosro and Yousef, 2012). These types of bio-fertilizer are essential factors in the integrated nutrient management of soils. At the same time, they have an important role in soil sustainability and productivity and also play a vital part in resisting different parasites, pathogens, and pests in susceptible crops (Ashry et al. 2018; Abd-Elgawad 2020). Bio-fertilizers are replacing chemical fertilizers in modern agriculture because they are ecological, economical, and renewable. This chapter highlights the need for bio-fertilizers and their roles in managing plant-parasitic nematodes.

15.2 ARTIFICIAL FERTILIZERS VS. BIO-FERTILIZERS VS. ORGANIC FERTILIZERS

NPK is the most significant and indispensable of plant nutrients. Nitrogen (N) is used in solid form and the conversion of gaseous N to the solid state is through nitrogen fixation. However, N fixation in chemical fertilizers is known as artificial nitrogen fixation (Gaur 2010).

Microorganisms in bio-fertilizers colonize the rhizosphere of soil or plant surfaces or within plant surfaces after application, and improve plant growth by providing essential primary nutrients for host plants. The source of organic manure is plant and animal wastes after the complete breakdown of raw material. It comprises all essential nutrients in small amounts required in the maximum amount for crop growth and could provide a suitable medium for bio-fertilizers (Gaur 2010; Zaki et al. 2015).

15.3 IMPORTANCE OF BIO-FERTILIZERS

Bio-fertilizers are considered the safest substitute for chemical fertilizers to decrease environmental hazards. They are eco-friendly, cost-effective, and can be made at any time. They are the most economical source of microelements and micronutrients, stabilize chemical fertilizers' negative effects, and enhance the release of growth hormones. They can improve crop production by up to 40% and fix nitrogen by 50%. One of the advantages of these fertilizers is that after 34 years of continuous application, there is no further need for bio-fertilizer applications, as the parental inocula will be sufficient for the growth, multiplication, and improvement of soil texture, pH, and other soil properties (Gaur 2010).

15.4 PLANT-PARASITIC NEMATODES (PPNS)

Plant roots contain various beneficial and harmful organisms that interact with each other to affect plant growth. Crop production and plant growth are adversely affected by harmful organisms (Oerke 2006; Khan et al. 2015) and among these harmful organisms, plant-parasitic nematodes (PPNs) cause significant damage to crop plants worldwide (Kenney and Eleftherianos 2016). PPNs affect agricultural and horticultural crops, including major crops, adversely. Losses are predicted to be up to 15% in agriculture production, amounting to USD 125–157 billion (Watkins et al. 2012; Singh et al. 2015).

15.5 CLASSIFICATION OF PPNS

More than 4100 species of PPNs have been discovered to date, most of which are major plant pathogens. However, some are specific to a limited range of crops, which cause huge effects on economically important crops. PPN species classification was established based on the effects of nematodes on crops. A researcher survey enlisted the "top 10" PPN based on scientific and economic importance, and these are *Aphelenchoides besseyi* (foliar nematode), *Xiphinema index* (dagger nematode; a virus vector nematode), *Nacobbus aberrans* (false root-knot nematode), *Bursaphelenchus xylophilus* (pine wilt nematode), *Rotylenchulus reniformis* (reniform nematode), *Radopholus similis* (burrowing nematodes), *Ditylenchus dipsaci* (stem and bulb nematode), *Heterodera* and *Globodera* spp. (cyst nematodes), *Pratylenchus* spp. (root lesion nematodes), and *Meloidogyne* spp. (root-knot nematodes) (Jones et al. 2013).

15.6 ATTACK MECHANISM OF PPNS

The PPN infestation symptoms are sometimes mistaken for water stress, soil-related conditions, nutrient deficiencies, or bacterial or fungal infections (Mesa-Valle et al. 2020). PPNs inject enzymes while attacking their host with their saliva, and a feeding site starts to be developed in the plant tissues; consequently, physiological changes start. Cell walls and plasma membrane areas may increase, and these changes begin to transport plant nutrients to the parasites. Cysts, galls, or lesions are formed. These morphological and biochemical modifications result in water and nutrient uptake and efficacy changes. The development of the plant reduces, its productivity decreases, and the plant starts to wilt and becomes yellow. Moreover, potential alarming characteristics of PPN infections are that the chances of secondary infections by soil-borne pathogens also increase because of mechanical deformations. Resistance to infections also is reduced. Therefore, the damage by these soil-borne pathogens is also increased (Rodiuc et al. 2014).

15.7 CONTROL STRATEGIES FOR PPNS

The economic impact of parasitic nematodes has resulted in many nematode control strategies being developed in agriculture, including the use of chemical nematicides. The need for nematode resistance increased with the loss of pesticides due to EU regulations (EC No. 1107/2009), as they are unsafe for human health and contaminate the environment (Zhang et al. 2014, 2017). Strategies linked with biological control are believed to be a safe alternative and highly realistic for plant-parasitic nematodes management. Bio-control of nematodes is defined as the regulation of populations of nematodes and a reduction in nematode damage by the action of organisms antagonistic to them which occurs naturally or by the environmental manipulation or the introduction of antagonists (D'Addabbo et al. 2019; Hadi et al. 2014; Khan and Kim 2007). Among the different biological control strategies, the application of bio-fertilizers to combat and control various species of PPNs has been found to be very efficient, effective, and economical.

15.8 MICROORGANISMS AS BIO-FERTILIZERS

Fungi, actinomycetes, bacteria, and other microorganisms are host specific and can kill PPNs. Among these microorganisms, soil-borne bacteria and fungi can more effectively control nematodes. They are efficient in promoting plant growth, and they secrete plant growth promoters such as auxin, gibberellic acid, cytokinins, abscisic acid, and ethylene, and improve the germination of seeds and the growth of roots. They decay organic matter and enhance compost formation in soil. These microbes efficiently control infestations of nematodes in different crops. Fatty acids, volatile compounds, hormones, enzymes, hydrogen sulphide, phenolic compounds, and alcohol are among the bacterial products used to control PPNs (Blyuss et al. 2019; El-Eslamboly et al. 2019).

15.9 FUNGI AS BIO-FERTILIZERS AGAINST PPNS

Nematophagous fungi have a specific feature to compete against PPNs and particularly against the root-knot nematode, *Meloidogyne* spp. (Collange et al. 2011). Some nematophagous fungi can trap nematodes, and their various trapping systems like *Arthrobrotrys oligospora* and *A. superb* have network trapping systems. Fungi feed on nematodes such as *A. dactyloides*, *Dactylaria brochopaga*, and *A. anchonia*, forming constrictive rings. These fungi have a saprophytic nature but can trap nematodes in their first larval or adult stage in the soil (de Freitas Soares et al. 2018). Most fungi are saprophytic and can attack eggs. Some of these fungi belong to the genera *Paecilomyces*, *Paecilomyces lilacinus*, *Pochonia*, and *Verticillium*. *Pochonia chlamydosporia* is thought to be the most active egg parasite. *P. lilacinus* has been already reported to manage root-knot nematodes effectively, including *M. incognita* and *M. javanica* on brinjal, tomato, and several other vegetable crops (Van Damme et al. 2005; Goswami et al. 2006; Haseeb and Kumar 2006; Khan et al. 2013).

Root nematodes inoculate tomato roots, where the mycorrhizal fungus *Glomus fasciculatus* increases infestation and spore production. The size and number of the root-knots created by the root-knot nematode decreases with the inoculation of the fungus *Meloidogyne incognita* and the improvement of crop growth criteria (Bagyaraj et al. 1979). Suresh et al. (1885) reported that *M. incognita* causes a remarkably low number of giant cells, and also the formation of giant cells in mycorrhizal produced in minimal amounts. In addition, mycorrhizal plant root extract produced nearly 50% mortality in nematode larvae in 96 hours. Hajra et al. (2013) evaluated an increase in vegetative growth of crops while infested plants showed retardation in growth. Many carbohydrates are present in the roots of mycorrhizal plants, which indicates the transfer of photosynthesis to fungal partners. Less carbohydrates are recorded in nematode-infested roots, displaying a great carbon sink to the rhizosphere. Classification of nematophagous fungi was done based on their mode of inhibitory action against PPNs (Cayrol et al. 1992). Important fungal strains that act as bio-fertilizers for PPN control and management are summarized in Table 15.1.

15.10 BACTERIA AS BIO-FERTILIZERS AGAINST PPNS

Species of the genus *Pasteuria* are cosmopolitan and are used as biological control agents against PPNs (Table 15.2). *Pasteuria* is a mycelial, endospore-forming bacterial parasite that destroys and reduces the nematode population (Bekal et al. 2001) and causes considerable damage to crop plants globally (Bird et al. 2003). *P. penetrans* has a symbiotic relationship with PPNs, especially with the root-knot nematodes, significantly damaging crop plants (Charles et al. 2005). *Pasteuria* species effectively infect and kill nematodes where root lesion nematodes (*Pratylenchus* spp.) are

TABLE 15.1
List of Some Important Fungal Species used as Bio-fertilizers against Plant Parasitic Nematodes (PPNs) and Their Modes of Action

Fungal Strains	PPNs	Mode of Action	References
Paecilomyces lilacinus	*Heterodera zeae*, *Meloidogyne javanica*	Reduced cyst population in soil; improves growth and protection against nematodes	Ashraf and Khan, 2010; Baheti et al. 2017
Trichoderma harzianum	*M. incognita*	Decreased root galls, egg masses, and nematode population, including root-knot nematode	Feyisa et al. 2015; Devi et al. 2016
T. viride	*Meloidogyne* spp., *M. graminicola*, *M. incognita*	Increased plant growth and yield; decreased root galls and egg masses	Narasimhamurthy et al. 2017
Pochonia chlamydosporia	*Globodera* spp., *H. zeae*, *M. incognita*	Lowered cyst nematode population; reduced eggs and juveniles in soils	Baheti et al. 2017

infected and killed by *P. thornei*, *P. penetrans* successfully controls the population of root-knot nematodes (*Meloidogyne* spp.) and the cyst nematodes are killed by *P. nishizawae* (Atibalentja et al. 2000).

Three blue-green algae, *Nostoc calcicola*, *Anabena oryzae*, and *Spirulina* sp., were reported by Youssef and Ali (1998) to decrease the number of galls and egg masses created by the root-knot nematode, *Meloidogyne incognita* infesting cowpea and enhanced plant growth criteria. The penetration of *Pratylenchus penetrans* and *M. chitwoodi* was reduced by *B. megaterium* in potato roots by 50% (Al- Rehiayani et al. 1999). Padgham and Sikora (2007) mentioned that *B. megatherium* treatment resulted in a 40% decrease in rice nematode penetration and gall formation. Compared to chemical fertilizers, plant growth microbe-based bio-fertilizers, such as phosphate-solubilizing microbes, can effectively manage nematode-caused diseases (Khan et al. 2007).

15.10.1 Effects of Some Bacterial Bio-fertilizers on the Root-Knot Nematode, *Meloidogyne incognita* Infecting Some Vegetable Crops

Significant increases in growth, quality, and yield parameters were observed by Khan et al. (2012) when PPN-infested chilli (*Capsicum annum* L.) was inoculated with the nitrogen fixation biological agents *Azosperillum* and *Azotobacter*. *Azosperillum* performance was found to be better in comparison to *Azotobacter*. Repetitive inoculations with bio-fertilizers increased plant growth, quality, and yield. The nematocidal effects of various bio-fertilizers bacteria, including *Paenibacillus polymyxa*, and phosphate- and potassium-solubilizing bacteria *Bacillus megatherium*, were tested separately on tomatoes in sandy soils for infestation by the root-knot nematode *M. incognita* (El-Haddad et al. 2011). Relating with the un-inoculated nematode-infested control, the *B. megatherium* PSB2, *P. polymyxa* NFB7, and *B. circulans* KSB2 inoculations augmented the number of total bacterial spores in plants potted in soil from 1.2 to 2.6 times in about 60 days after inoculation. Likewise, *P. polymyxa* NFB7 inoculation resulted in enhanced plant growth, quality, and yield parameters, and within 60 days of inoculation a significant reduction in the nematode population was observed. The strains *B. megatherium* PSB2, *P. polymyxa* NFB7, and *B. circulans* KSB2 produced a better result in nematode population control. Hence, bacterial-based bio-fertilizers enhance the soil nutrient (NPK) mobility and provide biological control against parasitic nematodes, including *M. incognita*.

TABLE 15.2
List of Some Important Bacterial Species Used against Plant Parasitic Nematodes (PPNs) and Their Modes of Action

Bacterial Strains	PPNs	Mode of Action	References
Bacillus firmus	*Meloidogyne incognita*, *Ditylenchus dipasi*, *Radopholus similis*, *Heterodera*, and *Pratylenchus* spp.	Secondary metabolites and Sep 1 protease	Xiong et al. 2015; Geng et al. 2016
B. licheniformis	*Bursaphelenchulus xylophilus*, *M. incognita*	Protease and chitinase	Jeong et al. 2015; El-Nagdi et al. 2019
B. cereus	*Heterodera avenae*, *M. incognita*, *M. javanica*	Secondary metabolites, sphingosine, protease, chitinase, antibiotic production, and ISR	Gao et al. 2016; Ahmed 2019; Jiang et al.2 020
B. thuringiensis	*H. glycines*, *M. incognita*	Bt crystal protein (toxin protein) and Thuringiensin (β-exotoxin)	Wei et al. 2003; Mohammed et al. 2008
B. megaterium	*H. glycines*, *M. incognita*, *M. graminicola*	Secondary metabolites and protease	Padgham et al. 2007; Mostafa et al. 2018
B. subtilis	*Rotylenchulus reniformis*, *Helicotylenchus multicinctus*, *M. graminicola*, *M. incognita*, *M. javanica*,	Secondary metabolites, lipopeptide antibiotics, and hydrolytic enzymes	Mazzuchelli et al. 2020
B. coagulans	*M. incognita*	Hydrolytic enzymes	Xiang et al. 2018
B. pumilus & *B. pumilus L1*	*H. glycines*, *M. arenaria*	Protease, chitinase, ISR and SAR	Lastochkina et al. 2017; Forghani and Hajihassani et al. 2020
Rhizobium etli	*Meloidogyne* spp	ISR	Reitz et al. 2000
Serratia marcescens	*M. incognita*, *M. javanica*, *R. similis*	Volatile metabolites, prodigiosin	Rahul et al. 2014
Pseudomonas aeruginosa	*Caenorhabditis elegans*, *M. incognita*, *M. javanica*	Hydrogen cyanide (HCN), ISR, and SAR	Fatima and Anjum, 2017; Singh and Siddiqui, 2010
P. fluorescens CHA0	*M. incognita*, *M. javanica*	Pyoluteorin, Extracellular protease, HCN, 2,4-diacetylphloroglucinol (DAPG)	Siddiqui et al. 2005
P. stutzeri	*M. incognita*	HCN	Khan et al. 2016
P. fluorescens Wood1R	*M. incognita*	DAPG	Timper et al.2 009
Corynebacterium paurometabolum	*M. incognita*	Hydrogen sulphide, chitinase	Mena and Pimentel 2002

15.10.2 Effects of Some Bacterial Bio-fertilizers on the Citrus Nematode, *Tylenchulus semipenetrans* Infesting Citrus Trees

Bacterial strains *Pseudomonas fluorescens* strain 843 and *Azospirillum brasilense* strain W24 effectively improved soil quality and orange production (Shamseldin et al. 2010). The inoculation of bio-fertilizers with *P. fluorescens* strain 843 in Washington navel orange as a growth-inducing rhizobacterium significantly enhanced plant growth parameters, including fruit length, weight, and yield, total soluble solids, and juice volume. In contrast, inoculations with *A. brasilense* strain W24 did not significantly improve these parameters. Three types of nematodes mainly attack the roots of Washington navel orange: saprophytic nematodes, *Tylenchulus* spp., and *Pratylenchus* spp., Meanwhile, *Tylenchulus semipenetrans* is the most abundant species. The use of bio-fertilizers improves and maintains soil fertility and soil health. According to researchers, bio-fertilizers effectively control plant-parasitic nematodes and contribute to achieving sustainable agriculture globally.

15.10.3 Effects of Some Commercial Bio-fertilizers and Nutrients on the Root-Knot Nematode, *M. incognita* Infesting Some Field and Vegetable Crops

Ismail and Hasabo (2000) tested six commercial Egyptian bio-fertilizers and five commercial Egyptian plant nutrients to control *M. incognita* in sunflower cv. Giza 101 under greenhouse conditions. These products considerably decreased ($P \leq 0.05$ and 0.01) the numbers of egg masses, juveniles, females, nematode build-up rate, gall formation on roots, and therefore gall and egg mass indices. Nematode populations and galls were significantly suppressed by the seed coating of rizobacterin, followed by phosphorine and nitrobein as bio-fertilizers. Menawhile, nameless biocides and blue-green algae resulted in a minimal nematode population. Moreover, plant nutrients, kapronite, Kotangein, potassein F, and citrein controlled the nematode population earlier. In other research into the control of *M. incognita* in cowpea, with the use of phosphorine and nirobien as bio-fertilizers, no significant results were found to control nematode infestation or improve plant growth, however, nitrobien performed better than the phosphorine.

15.11 CONCLUSION

Plant-parasitic nematode biocontrol techniques are an effective alternative to hazardous chemical nematodes. The use of bio-fertilizers to control plant-parasitic nematodes is one of the most important biological control strategies that leads not only to soil enrichment but also is compatible with long-term sustainability. Furthermore, they are environmentally friendly and do not threaten the environment. Thus, they can be used in place of chemical fertilizers. In conclusion, bio-fertilizers are a promising long-term biocontrol method for plant-parasitic nematodes in agriculture, owing to the vast variety of modes of action that, in most cases, function in concert. However, further research is needed to answer certain fundamental problems.

ACKNOWLEDGMENT

The authors would like to thank to the staff of the Department of Agriculture, University of Swabi, KP, Pakistan.

REFERENCES

Abd-Elgawad MM (2020) Plant-parasitic nematodes and their biocontrol agents: current status and future vistas. In: *Management of Phytonematodes: Recent Advances and Future Challenges* (pp. 171–203). Springer, Singapore. https://doi.org/10.1007/978-981-15-4087-5_8

Ahmed S, Liu Q, Jian H (2019) Bacillus cereus a potential strain infested cereal cyst nematode (Heterodera avenae). *Pak J Nematol* 37:53–61

Al-Rehiayani S, Hafez SL, Thornton M, Sundararaj P (1999). Effects of Pratylenchus neglectus, Bacillus megaterium, and oil radish or rapeseed green manure on reproductive potential of Meloidogyne chitwoodi on potato. *Nematropica* 29(1):37–49

Ashraf MS, Khan TA (2010) Integrated approach for the management of *Meloidogyne javanica* on eggplant using oil cakes and biocontrol agents. Arch. *Phytopathol. Pflanzenschutz* 43(6):609–614

Ashry NA, Ghonaim MM, Mohamed HI, Mogazy AM (2018) Physiological and molecular genetic studies on two elicitors for improving the tolerance of six Egyptian soybean cultivars to cotton leaf worm. *Plant Physiol Biochem* 130: 224–234

Atibalentja N, Noel GR, Domier LL (2000) Phylogenetic position of the North American isolate of Pasteuria that parasitizes the soybean cyst nematode, Heterodera glycines, as inferred from 16S rDNA sequence analysis. *Int J Syst Evol* 50(2):605–613

Bagyaraj DJ, Manjunath A, Reddy DDR (1979) Interaction of vesicular arbuscular mycorrhiza with root knot nematodes in tomato. *Plant Soil* 51(3):397–403

Baheti BL, Dodwadiya M, Bhati SS (2017) Eco-friendly management of maize cyst nematode, Heterodera zeae on sweet corn (*Zea mays* L. saccharata). *J Entomol Zool Stud* 5:989–993

Bekal S, Borneman J, Springer S, Giblin-Davis RM, Becker JO (2001) Phenotypic and molecular analysis of a Pasteuria strain parasitic to the sting nematode. *J Nematology* 33(2–3):110

Bird DM, Opperman CH, Davies KG (2003) Interaction between bacteria and plant-parasitic nematodes: now and then. *Int J Parasitol* 33:1269–1276

Blyuss KB, Fatehi F, Tsygankova VA, Biliavska LO, Iutynska GO, Yemets AI, Blume YB (2019) RNAi-based biocontrol of wheat nematodes using natural poly-component biostimulants. *Front Plant Sci* 10: 483

CayrolJC,Djian-CaporalinoC,Panchaud-MatteiE(1992)Laluttebiologiquecontrelesne´matodesphytoparasites. *Courr Cellule Environ l'INRA* 17:31–44

Charles L, Carbone I, Davies KG, Bird D, Burke M, Kerry BR, Opperman CH (2005) Phylogenetic analysis of Pasteuria penetrans by use of multiple genetic loci. *J Bacteriol Res* 187(16):5700–5708

Collange B, Navarrete M, Peyre G, Mateille T, Tchamitchian M (2011) Root-knot nematode (Meloidogyne) management in vegetable crop production: The challenge of an agronomic system analysis. *Crop Prot* 30(10):1251–1262

D'Addabbo T, Laquale S, Perniola M, Candido V (2019) Biostimulants for plant growth promotion and sustainable management of phytoparasitic nematodes in vegetable crops. *Agronomy* 9(10):616

de Freitas Soares FE, Sufiate BL, de Queiroz JH (2018) Nematophagous fungi: Far beyond the endoparasite, predator and ovicidal groups. *Agric Nat Resour* 52(1):1–8

Deepali GK, Gangwar K (2010) Biofertilizers: An eco-friendly way to replace chemical fertilizers. Available online (accessed on 7 February 2021).

Devi TS, Mahanta B, Borah A (2016) Comparative efficacy of Glomus fasciculatum, Trichoderma harzianum, carbofuran and carbendazim in management of Meloidogyne incognita and Rhizoctonia solani disease complex on brinjal. *Indian J Nematol* 46(2):161–164

El-Eslamboly AASA, El-Wanis A, Mona M, Amin A W (2019) Algal application as a biological control method of root-knot nematode Meloidogyne incognita on cucumber under protected culture conditions and its impact on yield and fruit quality. *Egypt J Biol Pest Control* 29(1):1–9

El-Hadad ME, Mustafa MI, Selim SM, El-Tayeb TS, Mahgoob AEA, Aziz NHA (2011) The nematicidal effect of some bacterial biofertilizers on Meloidogyne incognita in sandy soil. *Braz J Microbiol* 42(1):105–113

El-Nagdi WMA, Abd-El-Khair H, Soliman GM, Ameen HH, El-Sayed GM (2019). Application of protoplast fusants of Bacillus licheniformis and Pseudomonas aeruginosa on Meloidogyne incognita in tomato and eggplant. *Middle East J Appl Sci* 9: 622–629

Fatima S, Anjum T (2017) Identification of a potential ISR determinant from Pseudomonas aeruginosa PM12 against Fusarium wilt in tomato. *Front Plant Sci* 8: 848

Feyisa B, Lencho A, Selvaraj T, Getaneh G (2015) Evaluation of some botanicals and Trichoderma harzianum for the management of tomato root-knot nematode (Meloidogyne incognita (Kofoid and White) Chit Wood). *Adv Crop Sci Technol* 2–10

Forghani F, Hajihassani A (2020) Recent advances in the development of environmentally benign treatments to control root-knot nematodes. *Front Plant Sci* 1125

Gao H, Qi G, Yin R, Zhang H, Li C, Zhao X (2016) Bacillus cereus strain S2 shows high nematicidal activity against Meloidogyne incognita by producing sphingosine. *Sci Rep* 6(1):1–11

Gaur V (2010) Biofertilizer – necessity for sustainability. *J Adv Dev* 1(7):8

Geng C, Nie X, Tang Z, Zhang Y, Lin J, Sun M, Eng D (2016) A novel serine protease, Sep1, from Bacillus firmus DS-1 has nematicidal activity and degrades multiple intestinal-associated nematode proteins. *Sci Rep* 6(1):1–12

Goswami BK, Pandey RK, Rathour KS, Bhattacharya C, Singh L (2006) Integrated application of some compatible biocontrol agents along with mustard oil seed cake and furadan on Meloidogyne incognita infecting tomato plants. *J Zhejiang Univ Sci B* 7(11):873–875

Hadi F, Rahman AU, Ibrar M, Dastagir G, Arif M, Naveed K, Adnan M (2014) Weed diversity with special reference to their ethnomedicinal uses in wheat and maize at Rech valley, Hindokush Range, Chitral, Pakistan. *Pak J Weed Sci Res* 20(3):335–346

Hajra N, Shahina F, Firoza K (2013) Biocontrol of root-knot nematode by arbuscular mycorrhizal fungi in Luffa cylindrica. *Pak J Nematol* 31(1)

Halim NA (2009) Effects of using enhanced biofertilizer containing N-fixer bacteria on patchouli growth (Doctoral dissertation, UMP).

Haseeb A, Kumar V (2006) Management of Meloidogyne incognita-Fusarium solani disease complex in brinjal by bio-control agents and organic additives. *Ann Plant Sci* 14(2):519–521

Ismail AE, Hasabo SA (2000) Evaluation of some new Egyptian commercial biofertilizers, plant nutrients and a biocide against Meloidogyne incognita root knot nematode infecting sunflower. *Pak J Nematol (Pakistan)* 18(1–2):39–49

Jeong MH, Yang SY, Lee YS, Ahn YS, Park YS, Han HR, Kim KY (2015) Selection and characterization of Bacillus licheniformis MH48 for the biocontrol of pine wood nematode (Bursaphelenchus xylophilus). *J Korean For Soc* 104(3):512–518

Jiang C, Fan Z, Li Z, Niu D, Li Y, Zheng M, ... Guo J (2020) Bacillus cereus AR156 triggers induced systemic resistance against Pseudomonas syringae pv. tomato DC3000 by suppressing miR472 and activating CNLs-mediated basal immunity in Arabidopsis. *Mol Plant Pathol* 21(6):854–870

Jones JT, Haegeman A, Danchin EG, Gaur HS, Helder J, Jones MG, ... Perry RN (2013) Top 10 plan-parasitic nematodes in molecular plant pathology. *Mol Plant Pathol* 14(9):946–961

Kenney E, Eleftherianos I (2016) Entomopathogenic and plant pathogenic nematodes as opposing forces in agriculture. *Int J Parasitol* 46(1):13–19

Khan A, Sharif M, Ali AS, Shah NM, Mian IA, Wahid F, Jan B, Adnan M, Nawaz S, Ali N (2013) Potential of AM fungi in phytoremediation of heavy metals and effect on yield of wheat crop. *Am J Plant Sci* 5:1578–1586

Khan MA, Basir A, Adnan M, Saleem N, Khan A, Shah SRA, Shah JA, Ali K (2015) Effect of tillage, organic and inorganic nitrogen on maize yield. *Am-Eurasian J Agric & Environ Sci* 15(12):2489–2494

Khan MR, Khan SM, Mohiddin FA, Askary TH (2007) Effect of certain phosphate-solubilizing bacteria on root-knot nematode disease of mungbean. *In: First International Meeting on Microbial Phosphate Solubilization,* pp. 341–346. Springer, Dordrecht

Khan MR, Mohidin FA, Khan U, Ahamad F (2016) Native Pseudomonas spp. suppressed the root-knot nematode in in vitro and in vivo, and promoted the nodulation and grain yield in the field grown mungbean. *Biol Control* 101: 159–168

Khan W, Rayirath UP, Subramanian S, Jithesh MN, Rayorath P, Hodges DM, ... Prithiviraj B (2009) Seaweed extracts as biostimulants of plant growth and development. *J Plant Growth Regul* 28(4):386–399

Khan Z, Kim YH (2007) A review on the role of predatory soil nematodes in the biological control of plant parasitic nematodes. *Appl Soil Ecol* 35(2):370–379

Khan Z, Tiyagi SA, Mahmood I, Rizvi R (2012) Effects of N fertilization, organic matter, and biofertilizers on the growth and yield of chilli in relation to management of plant-parasitic nematodes. *Turk J Bot* 36(1):73–81

Khosro M, Yousef S (2012) Bacterial bio-fertilizers for sustainable crop production: A review. *APRN J Agr Biol Sci* 7(5):237–308

Lastochkina O, Pusenkova L, Yuldashev R, Babaev M, Garipova S, Blagova DY, ... Aliniaeifard S (2017) Effects of Bacillus subtilis on some physiological and biochemical parameters of Triticum aestivum L.(wheat) under salinity. *Plant Physiol Biochem* 121: 80–88

Mazzuchelli RDCL, Mazzuchelli EHL, de Araujo FF (2020) Efficiency of Bacillus subtilis for root-knot and lesion nematodes management in sugarcane. *Biol Control* 143: 104185

Mena J, Pimentel E (2002) Mechanism of action of *Corynebacterium pauronetabolum strain* C-924 on nematodes. *Nematology* 4:287

Mesa-Valle CM, Garrido-Cardenas JA, Cebrian-Carmona J, Talavera M, Manzano-Agugliaro F (2020) Global research on plant nematodes. *Agronomy* 10(8):1148

Mohammed SH, El Saedy MA, Enan MR, Ibrahim NE, Ghareeb A, Moustafa SA (2008) Biocontrol efficiency of Bacillus thuringiensis toxins against root-knot nematode, Meloidogyne incognita. *J Cell Mol Biol* 7(1):57–66

Mostafa FA, Khalil AE, El-Deen AHN, Ibrahim DS (2018) The role of Bacillus megaterium and other bio-agents in controlling root-knot nematodes infecting sugar beet under field conditions. *Egypt J Biol Pest Control* 28(1):1–6

Narasimhamurthy HB, Ravindra H, Sehgal M (2017) Management of rice rootknot nematode, Meloidogyne graminicola. *Int J Pure Appl Biosci* 5:268–276

Oerke EC (2006) Crop losses to pests. *J Agr Sci* 144(1):31–43

Padgham JL, Sikora RA (2007) Biological control potential and modes of action of Bacillus megaterium against Meloidogyne graminicola on rice. *Crop Prot* 26(7):971–977

Rahul S, Chandrashekhar P, Hemant B, Chandrakant N, Laxmikant S, Satish P (2014) Nematicidal activity of microbial pigment from Serratia marcescens. *Nat Prod Res* 28(17):1399–1404

Reitz M, Rudolph K, Schroder I, Hoffmann-Hergarten S, Hallmann J, Sikora R (2000) Lipopolysaccharides of Rhizobium etli strain G12 act in potato roots as an inducing agent of systemic resistance to infection by the cyst nematode Globodera pallida. *Appl Environ Microbiol* 66(8):3515–3518

Rodiuc N, Vieira P, Banora MY, de Almeida Engler J (2014) On the track of transfer cell formation by specialized plant-parasitic nematodes. *Front Plant Sci* 5: 160

Shamseldin A, El-Sheikh MH, Hassan HSA, Kabeil SS (2010) Microbial biofertilization approaches to improve yield and quality of Washington navel orange and reducing the survival of nematode in the soil. *Am J Sci* 6(12):264–271

Siddiqui IA, Haas D, Heeb S (2005) Extracellular protease of Pseudomonas fluorescens CHA0, a biocontrol factor with activity against the root-knot nematode Meloidogyne incognita. *Appl Environ Microbiol* 71(9):5646–5649

Singh P, Siddiqui ZA (2010) Biocontrol of root-knot nematode Meloidogyne incognita by the isolates of Pseudomonas on tomatoArch. *Phytopathol Pflanzenschutz* 43: 1423–1434

Singh S, Singh B, Singh AP (2015) Nematodes: A threat to sustainability of agriculture. *Procedia Environ Sci* 29: 215–216

Suresh CK, Bagyaraj DJ, Reddy DDR (1985) Effect of vesicular-arbuscular mycorrhiza on survival, penetration and development of root-knot nematode in tomato. *Plant Soil* 87(2):305–308

Timper P, Kone D, Yin J, Ji P, Gardener BBM (2009) Evaluation of an antibiotic-producing strain of Pseudomonas fluorescens for suppression of plant-parasitic nematodes. *J Nematol* 41(3):234

Van Damme V, Hoedekie A, Viaene N (2005) Long-term efficacy of Pochonia chlamydosporia for management of Meloidogyne javanica in glasshouse crops. *Nematology* 7(5):727–736

Watkins PR, Huesing JE, Margam V, Murdock LL, Higgins TJV (2012) Insects, nematodes, and other pests. In: *Plant Biotechno Agri,* pp. 353–370. Academic Press

Wei JZ, Hale K, Carta, L, Platzer E, Wong C, Fang SC, Aroian RV (2003) Bacillus thuringiensis crystal proteins that target nematodes. *PNAS* 100(5):2760–2765

Xiang N, Lawrence KS, Donald PA (2018) Biological control potential of plant growth-promoting rhizobacteria suppression of Meloidogyne incognita on cotton and Heterodera glycines on soybean: A review. *Pak J Phytopathol* 166(7–8):449–458

Xiong J, Zhou Q, Luo H, Xia L, Li L, Sun M, Yu Z (2015) Systemic nematicidal activity and biocontrol efficacy of Bacillus firmus against the root-knot nematode Meloidogyne incognita. *World J Microbiol Biotechnol* 31(4):661–667

Youssef MMA, Ali MS (1998) Management of Meloidogyne incognita infecting cowpea by using some native blue green algae. *Anz Schädlingskd, Pflanz, Umweltschutz* 71(1):15–16

Youssef MMA, Eissa MFM (2014) Biofertilizers and their role in management of plant parasitic nematodes. A review. *J Biotech Pharm Res* 5(1):1–6

Zaki AB, Ahmad N, Khan IA, Shah B, Khan A, Rasheed MT, Adnan M, Junaid K, Huma Z, Ahmed S (2015) Adulticidal efficacy of *Azadirachtaindica* (neem tree), *Sesamumindicum*(til) and *Pinussabinaena*(pine tree) extracts against *Aedesaegypti*under laboratory conditions. *J Ento Zool Stud* 3(6):112–116

Zhang S, Gan Y, Ji W, Xu B, Hou B, Liu J (2017) Mechanisms and characterization of Trichoderma longibrachiatum T6 in suppressing nematodes (Heterodera avenae) in wheat. *Front. Plant Sci* 8: 1491

Zhang S, Gan Y, Xu B, Xue Y (2014) The parasitic and lethal effects of *Trichoderma longibrachiatum* against *Heterodera avenae*. *Biol Control* 72: 1–8

Index

T

V

Z